ACETYLENES AND ALLENES

Addition, Cyclization, and Polymerization Reactions

THOMAS F. RUTLEDGE
Chemical Research Department
Atlas Chemical Industries, Inc.
Wilmington, Delaware

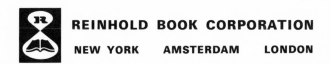

REINHOLD BOOK CORPORATION

NEW YORK   AMSTERDAM   LONDON

PREFACE

When the editors of the Reinhold Book Corporation first invited me to compose two books on the recent chemistry of acetylenic compounds, I agreed only that such books would be timely. The most modern book is R. A. Raphael's "Acetylenic Compounds in Organic Synthesis," published in 1955. The chemistry of acetylenic compounds has received so much attention during the last ten to fifteen years that I hesitated to tackle the task of assembling, organizing, and presenting even a part of the result. The timeliness argument finally prevailed. "Acetylenic Compounds" published in 1968 was the first. In the first book, the preparation, substitution reactions, and some uses of acetylenic compounds are discussed. In this book, preparation of allenic compounds and the addition reactions involving acetylenic and allenic bonds are covered.

Comprehensive coverage of the recent progress in acetylene chemistry is obviously impossible in two books of reasonable size. I have not attempted to assemble all the data and list all of the reactions and properties of acetylenes. My approach finally evolved into an effort to include enough details of the most important reactions of all kinds of acetylenic compounds to furnish interested chemists with a good background and with leads into the pertinent literature. In order to cover the maximum number of facts in the least number of words, I adopted a form of "informative writing" which is certainly not fascinating prose. I feel that this is a reasonable sacrifice.

Where possible, the subject matter is organized by kind of reaction. Mechanisms are emphasized because they furnish ideas for new reactions. Long chapters are divided into smaller parts, and literature references are listed at the end of each part in the order in which they appear in the text. This minimizes the annoyance of thumbing through page after page of references.

This book first describes the preparation and addition reactions of allenes and higher cumulenes because the chemistry of these compounds is intimately related to the chemistry of acetylenes. Addition to acetylenic bonds in different kinds of molecules is covered next, followed by discussion of the special additions usually referred to as vinylations. Metal complexes and oligomerization, cyclization, and polymerization reactions are reviewed in the last two chapters.

The information in the literature through 1967 and into mid 1968 is covered as completely as possible.

I gratefully acknowledge the cooperation of my employer, Atlas Chemical Industries, Inc., in permitting me to write this book as a personal project. Dr. J. W. LeMaistre read the text and made many valuable comments. I thank my wife, Betty, for patiently enduring the writing years and dedicate this book to her and to my children Tom, Mark, and Lee Anna.

January, 1969   THOMAS F. RUTLEDGE

INTRODUCTION

Some acetylenic and allenic compounds have utility *per se*. Thus, the acetylenic bond exerts a powerful influence on the physiological activity of some compounds (e.g., carbamates) and makes these compounds more effective as drugs. Frequently, however, acetylenic and allenic compounds are utilized as precursors to other materials. Addition of various reagents to the acetylenic or allenic bonds is the route chosen in such cases.

The main purpose of this book is to summarize the recent chemistry of the most important addition reactions to acetylenic and allenic carbon atoms. In addition, the preparation and isomerization reactions of allenes and the higher cumulenes will be covered.

There is no single recent summary of methods for preparing allenes and cumulenes. Chapter 1 attempts to rectify this situation by providing a summary and discussion of the preparation of allenes and higher cumulenes. In the summary and discussion, preparative methods are organized by kind of reaction leading to allenic compounds of many types. Elimination reactions and elimination-rearrangement reactions are most frequently used. The literature contains a somewhat bewildering array of preparative methods. The purpose of the summary and discussion is to simplify the task of choosing the best method for preparing a given allenic compound.

Isomerization of acetylenic compounds to allenic compounds, and vice versa, is also discussed. This section of Chapter 1 is organized according to reagents (metal hydroxide, metal amide, etc.). The isomerizations often complicate reactions designed to produce propargylic compounds, and thus are of vital interest to the preparative chemist. Preparation and addition reactions of the higher cumulenes such as the butatrienes and pentatetraenes are discussed (briefly) because of their similarity to the allenes (propadienes). The higher cumulenes are best prepared from the corresponding *t*-acetylenic glycols by reduction methods.

Chapter 2, "Addition Reactions of Acetylenic Compounds," is divided into nine parts. References are found after each part. This summary of additions to acetylenic bonds is organized partly by kind of reaction (hydrogenation, hydration, free radical addition and oxidation), and partly by kind of acetylenic compound ("nonactivated" acetylenes, enynes and polyynes, acetylenic ethers, and "activated" acetylenes). This arrangement is dictated by the large number of reagents which have been used as addenda to acetylenic compounds. Organization entirely by kind of reaction or by kind of compound would be less effective and more confusing. Because of the symmetrical electronic structure of the acetylenic bond, acetylenic carbon atoms are more susceptible to nucleophilic attack than ethylenic carbons are. Conversely, acetylenic carbon atoms are less susceptible to electrophilic attack. As expected, the literature contains many references to nucleophilic addition reactions, and fewer references to electrophilic additions. A few free radical addition reactions have also been studied in some detail.

Chapter 3, "Vinylation," covers addition reactions to acetylenic bonds, which are usually called "vinylations." Most of the literature on vinylations consists of patents. If one considers the industrial importance of vinylation reactions, it is surprising that so little work has been published on mechanisms. The investigation of mechanisms is emphasized in this text. Also included are discussions of the acrylate synthesis, even though this is not, by strict definition, a vinylation.

Since many addition reactions involve complexes of acetylenes with metal compounds, Chapter 4 discusses the various kinds of complexes which can be found. The application of the complexes to cyclotrimerization, cyclotetramerization and oligomerization reactions is the most important aspect of the discussion. In Chapter 5, the application of metal complexes of acetylenes and allenes to polymerizations is covered. Other polymerization systems are included. For example, 1,3-dipolar addition and Diels–Alder reactions have been used as polymer-forming reactions. The preparation of polymers which contain acetylenic bonds is also included in the text. Polyesters, polyurethanes, polyacrylates and polystyrenes containing acetylenic bonds have properties quite different from polymers containing olefinic or saturated bonds. The acetylenic bond in polymers is a convenient reactive site, and polymers can be modified by adding various reagents to the acetylenic bonds. Many of the polymers prepared from acetylenic compounds have properties which could allow practical applications.

Although numerous interesting and new addition reactions have been discovered and developed within the last ten or twenty years, few of them have been commercialized. In too many instances, cost factors are overwhelming. Progress in developing commercial systems almost certainly depends on finding new, efficient, simple catalytic systems for some important addition reactions. The present knowledge about complexes of acetylenic compounds with metal compounds must be extended if these new systems are to become realities.

CONTENTS

Chapter One
ALLENE AND THE HIGHER CUMULENES

Introduction

Many acetylenes and allenes are interconvertible. The acetylene-allene equilibrium in many compounds is well known. Reactions of one kind of derivative often give products of another: propargyl Grignard reagents sometimes react as if they were largely allenic Grignards; allene reacts with lithium amide in ammonia to give lithium propynylide; tertiary ethynylcarbinols and alkyldichlorophosphines give mixed derivatives, the propargyl allenylalkylphosphonates. Propargyl alcohols can be converted to allenic halides in one step. Grignard reagents of 1,3-dienes can react to give allenic products, but 1,3-dienes are usually more stable than acetylenes or allenes, and isomerizations to 1,3-dienes are ordinarily irreversible. Thus, the chemistry of allenes is intimately related to the chemistry of acetylenes.

In 1964, Fischer[1] published a detailed review of the synthesis and reactions of allenic compounds and cumulenes, and Petrov and Fedorova[2] reviewed the allenic hydrocarbons. The synthesis and chemistry of allene itself was reviewed by Griesbaum in 1966.[3] Three other recent reviews on allenes are also available.[4,5,5a]

This chapter is divided into four parts:
(1) Preparation of allenic compounds,
(2) Acetylene-allene isomerizations,
(3) Reactions of allenes,
(4) The higher cumulenes.

Pertinent references are listed at the end of each part.

Historical Background. Some unknown mixtures were assigned the allene structure a hundred years ago, but the chemists who actually prepared the first allenic compound thought it was an acetylene. In 1887, Burton and Pechmann[6] made glutinic acid, but believed it was the acetylenic acid $HO_2CC{\equiv}CCH_2CO_2H$, and not the allenic acid

$$\underset{\quad}{HO_2C\overset{\overset{\displaystyle H}{|}}{C}=C=\overset{\overset{\displaystyle H}{|}}{C}CO_2H.}$$

In 1954 Jones[7] confirmed the allenic structure. In 1888, several papers reported the synthesis of allene and some substituted allenes. Chemists generally thought that cumulenes with more than two double bonds would be unstable, so it was not until 1921 that the next higher homolog, a butatriene, was made.[8] In 1938, Kuhn[9] made a higher cumulene series, represented by tetraphenylhexapentaene. Compounds with more than five cumulative double bonds so far have been obtained only in solution.

Staudinger and Ruzicka[10] isolated the first naturally occurring allene, pyrethrolone, in 1924. Their structure has since been confirmed:

The allene structure was tentative for years, because allenes in nature were simply unheard of. In 1952, however, Celmer and Solomans[11] isolated the very unstable antibiotic mycomycin, $HC{\equiv}CC{\equiv}CCH=C=CH(CH=CH)_2CH_2CO_2H$, and proved that the allene group causes its optical activity. Many naturally occurring allenes are now known.[12] By 1966, chemists had isolated 18 allene-polyacetylenes from microorganisms. In 1965, Bohlmann[13] reported the first naturally occurring 3-cumulene, isolated from *Asterae* plants

Acetylene, Allene and Cumulene Bonds. Explanation of certain molecular properties by hyperconjugation has caused much disagreement, mainly on whether (a) electron delocalization through σ and π bonds or (b) hybridizational changes in the σ bond offer the best explanation of physical properties and chemical reactivity. Electron delocalization may be the more likely explanation in allenes and ketenes because of some features of their NMR spectra.[14] The structure of allenes and ketenes locks the hydrogens in the same symmetry plane as the vicinal π electron systems. The C=C bonds are shorter than in ethylene. This should favor hyperconjugative electron delocalization. Dimethylallene-3-d was prepared from 3-chloro-3-methyl-1-butyne-3-d and lithium aluminum hydride, and ketene-2-d was made by pyrolyzing deuterated acetic anhydride. The spin–spin coupling constant over five bonds in the allene was ± 3.1 cycles/sec, due primarily to the hyperconjugative term from electron delocalization.[14] The indicated structures are:

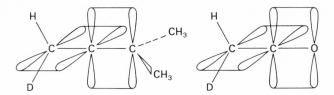

In cumulenes with an even number of double bonds, the four groups attached to the end carbons must be in perpendicular planes, as in allenes (Figure 1-1). In cumulenes with an odd number of double bonds, the end groups must be in the same plane (Figure 1-2). Even-numbered cumulenes should occur in optically active forms, and odd-numbered cumulenes should show *cis-trans* isomerism.[1]

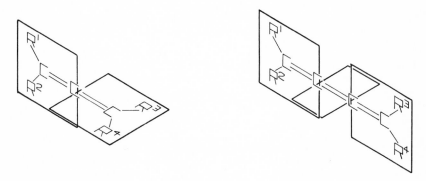

Figure 1-1   **Figure 1-2**

Extended Hückel calculations show that anions and cations of cumulenes, and the first excited state of cumulenes, prefer twisting of the terminal CH_2 group. The excited state of allene is bent. Polyacetylenes and polyenes show bond alteration, but cumulenes do not.[15]

Propargyl-Allenyl Cations. Pittman and Olah[16] prepared solutions of tertiary ethynyl-carbinols in antimony pentafluoride-fluorosulfonic acid-sulfur dioxide, and observed their NMR spectra. Secondary ethynyl alcohols gave poor spectra, because of extensive by-product formation. The tertiary ethynylcarbinols gave clean spectra, and the extent of downfield shifts compared to the parent carbinol indicated formation of propargyl cations. The spectra strongly support the idea that allenyl carbonium ions contribute significantly to the ion structure:

$$-C\equiv C-\overset{+}{C}\diagup \quad \longleftrightarrow \quad -\overset{+}{C}=C=C\diagup$$

The equilibrium delocalizes the charge further, causing the observed downfield shift.

Earlier in 1965, Richey[17] had reported the first ethynyl carbonium ions, prepared from tertiary acetylenic alcohols in concentrated sulfuric acid. Richey's NMR data also indicate that the allenyl cation is part of the resonance structure of the propargyl cation.

Propargyl-Allenyl Anions. The structure of the propargyl-allenyl anion is uncertain. It may be similar to the allylic anion.[18] The anion is nonplanar, since allene itself is nonplanar (in the manner shown in Figure 1-1).

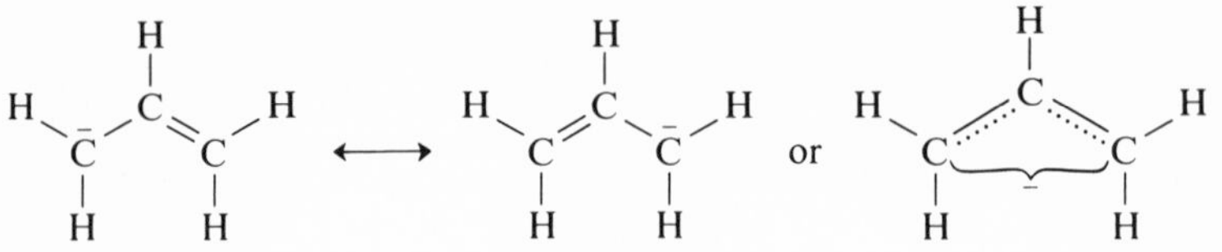

allylic anion

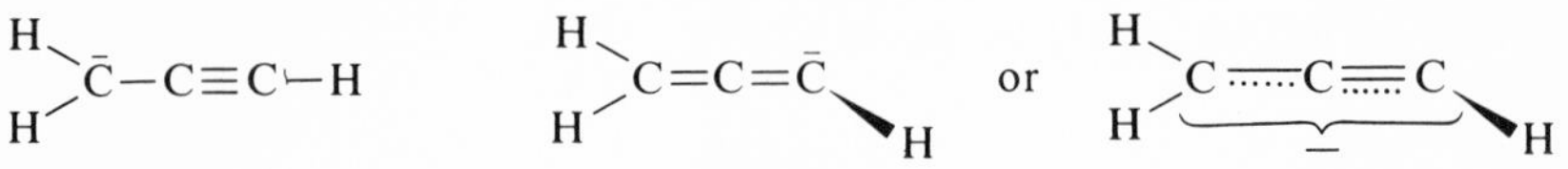

propargyl anion

allenyl anion

compromise structure

PART ONE

Preparation of Allenic Compounds

1. SUMMARY OF METHODS OF PREPARING ALLENES

The many possible ways of preparing allenic compounds are arranged in this summary by kind of reaction. A large number of substituted allenic compounds have been made by one or more of these methods, including hydrocarbons, carbinols, ketones, acids, ethers, thioethers, phosphonates, boronates and malonic ester derivatives. The list of products is complete enough to serve as a guide to methods for preparing new kinds of allenic compounds. A discussion of some reactions which form allenes follows the summary.

1.1. Elimination Reactions to Form Allenes

Allenic Compounds	Method	% Yield	Reference
General synthesis	$RR'C=CR''R''' + :CBr_2$, then Na or $RLi \longrightarrow RR'C=C=CR''R'''$ (acetylenes are minor by-products)	Acyclic— 60 Cyclic— 40–90	19, 20, 21 19, 20, 22, 23

Allene	2,3-Dibromopropene, $-2HBr(KOH)$, then Zn-BuOAc)	95	24
	Diketene, 550°	98	25
	i-Butene or propene, 1000°	50 (+ propyne)	26, 27
	2,3-Dichloropropene, Zn-methyl "Cellosolve"	50	28
Allene hydrocarbon	Propargyl halides, Zn-Cu couple	Good laboratory synthesis	29
	Wittig reaction: $RR'C{=}C{=}PPh_3 +$ carbonyl compound $\longrightarrow$ $RR'C{=}C{=}CR''R'''$	Good	30, 31
Tetraphenylallene	Dry distillation of $Ba(Ph_2CHCO_2)_2$	Good	32
Tetraarylallenes	$ArAr'C{=}CH_2 + ClCHAr''Ar'''$, then $-HCl$	Good	33
Phenoxyallene	$PhOCH{=}CBrCH_3$, KOH	Good, less acetylenic if H and Br are *cis*	34
Tetrachloroallene	$Cl_2C{=}CClCHCl_2$, KOH	Good	35
1,1-Difluoroallene	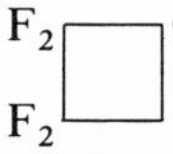 $F_2\text{—}\boxed{}\text{—}CH_2OAc$, 900° F_2	35	36
Tetrafluoroallene	$CF_2Br_2 + CH_2{=}CF_2 \longrightarrow$ $CF_2BrCH_2CF_2Br, -2HBr(KOH)$	—	37
α-Allenic acids and esters	$Ph_3P{=}CR'' + 2RR'CHCOCl \longrightarrow$ $RR'C{=}C{=}CR''CO_2H$ (or R''')	50–80	38
Phenylsulfonyl-allenes	$PhSO_2CH_2CCl{=}CH_2 + MeO^- \longrightarrow$ $PhSO_2CH{=}C{=}CH_2 +$ $PhSO_2CH_2C{\equiv}CH$	Allene mostly	39

1.2. Elimination-Rearrangement

α-Allenic aldehydes	Pyrolysis of propargyl vinyl ethers	10–90	40
	Pyrolysis of dipropargyl acetals	—	41
α,α′-Ketoallenyl-alcohols	$R'R''C(OH)C{\equiv}CCH{=}C(R)OEt$, $HClO_4 \longrightarrow$ $\overset{\displaystyle O}{\overset{\|}{R'R''C(OH)CH{=}C{=}CHCR}}$	20–80	42
α-Allenic nitriles	$\overset{\displaystyle O}{\overset{\|}{PrC}}{-}NHC{\equiv}CH \xrightarrow[\text{(2) Et}_3\text{N}]{\text{(1) phosgene}}$ $CH_2{=}C{=}CHC(Et)(H)CN$	30–77	43

1.3. Addition

α-Allenylacetyl-acetone	Allene carbene + acetylacetone anion	—	46
Hexapentaenes	Allene carbene + propargylic cation	—	44
α-Allenyl-malonates	Allene carbene + malonic ester anion	—	45

1.4. Addition-Elimination

$RC{=}C{=}CH_2$ 1-Alkyne $\xrightarrow[\text{(2) NaOH-H}_2]{\text{(1) PCl}_5\text{-benzene}}$ 60   47

                  1,2-alkadienes (1-octyne $\longrightarrow$
                  1,2-octadiene)

1.5. Addition-Rearrangement

1-Allenyl-α-ethynyl-carbinols	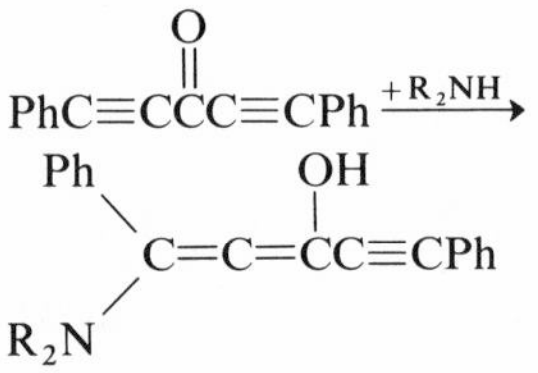 $\overset{\displaystyle O}{\overset{\displaystyle \|}{PhC{\equiv}CCC{\equiv}CPh}} \xrightarrow{+R_2NH}$ $\underset{R_2N}{\overset{Ph}{\diagdown}}C{=}C{=}\overset{OH}{\overset{\|}{C}}C{\equiv}CPh$	48
1,3-Disubstituted allenes	1,4-Addition to enynes: 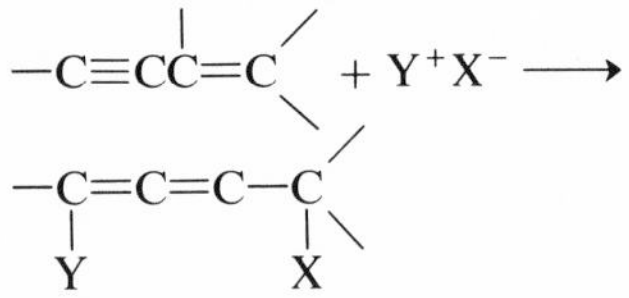 $-C{\equiv}CC{=}C{\diagup}^{\diagup} + Y^+X^- \longrightarrow$ $-C{=}C{=}C{-}\underset{X}{\overset{}{C}}{\diagup}$ $\quad\ \underset{Y}{\overset{}{\|}}\quad\ \|$	1, 49, 50, 51, 52, 53, 54 55

Examples:

Allenic hydro-carbons	$RC{\equiv}CCH{=}CH_2 + R'Li \longrightarrow$ $RCH{=}C{=}CHCH_2R'$ Diacetylene + R'Li $\longrightarrow$ conjugated vinylallenes	56
Allenyl-cyclo-propanes	Allene carbene + olefin	56
β-Allenic phosphines	$EtC{\equiv}CCH{=}CH_2 + Et_2PLi \longrightarrow$ $Et_2PCH_2CH{=}C{=}CH_2$	58
Tetramethoxy-allene	$(MeO)_2C{=}C(OMe)_2 \xrightarrow[\text{2. } n\text{-BuLi}]{\text{1. HCX}_3\text{-KO-}t\text{-Bu}}$ $72\%(MeO)_2C{=}C{=}C(OMe)_2$	58a

1.6. Rearrangement

Hydrocarbons	All possible allenes and acetylenes by KO-t-Bu isomerization of any allene or acetylene	Equilibrium mixtures	59, 60, 61
Conjugated vinyl-allenes	Alkali rearrangement of 1,4-enynes	General method	62
	OH $\|$ $RR'CC\equiv CCH_2CH=CHR''$, 5% NaOH, reflux $\longrightarrow$ $RR'C(OH)CH=C=CHCH=CHR''$	60–80	63
1,3-Diarylallenes	$ArC\equiv CCH_2Ar' \xrightarrow{alumina}$ $ArCH=C=CHAr'$	General method	64
Triarylallenes	$PhC\equiv CCHPh_2 \xrightarrow[20°]{alumina} PhCH=C=CPh_2$	84 General method	65, 66
Negatively substituted cyclobutyl-allenes		55	66a

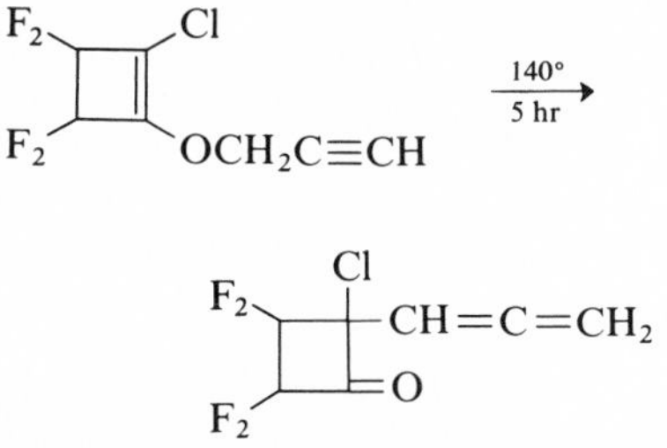

$$F_2 \diagdown \diagup Cl$$
$$\square\xrightarrow[5\ hr]{140°}$$
$$F_2 \diagup \diagdown OCH_2C\equiv CH$$

$$F_2 \diagdown \overset{Cl}{\underset{}{\diagup}}\!\!-CH=C=CH_2$$
$$\square$$
$$F_2 \diagup =O$$

α-Allenic acids	$HO_2CC\equiv CCH_2Am \xrightarrow[NH_3]{NaNH_2}$ $HO_2CCH=C=CHAm$ $(Am = n\text{-}C_5H_{11})$	100 General method	67
1-Allenyl ethers	$R_2CHC\equiv CCH_2OEt \xrightarrow[NH_3]{NaNH_2}$ $R_2CHCH=C=CHOEt$	80–95	68
1-Allenyl thio-ethers	$RCH_2C\equiv CCH_2SEt \xrightarrow[NH_3]{NaNH_2}$ $RCH=C=CHSEt$ (Et can be R′)	90	69
1-Allenylphos-phine oxides	$RCH-C\equiv CR'' \xrightarrow{R_3''N}$ $\|$ $P(OR')_2$	40–90	69a

$$R=C=C=C \diagup^{R''}_{\diagdown P(OR')_2}$$

1.7. Substitution-Rearrangement

1-Substituted allenes	$RR'C(X)C{\equiv}CR'' \xrightarrow[(2)\ Z^+]{(1)\ Mg}$ $RR'C{=}C{=}C(Z)R''$ (Z = H, CO_2H, etc.)	General method	1
Allene hydro-carbons	Propargylic bromides, + $LiAlH_4$	General, easy for laboratory	70, 71
	$HC{\equiv}CCH_2Br + RMgBr \longrightarrow$ $RCH{=}C{=}CH_2$	—	72, 72a
	(R = β-methallyl)	58	73
	2-Butynyl ethers + $RMgBr \longrightarrow$ $MeC(R){=}C{=}CH_2$	40–50	74
α-Allenyl-carbinols	$ClCH_2C{\equiv}CCH_2OH + LiAlH_4 \longrightarrow$ $CH_2{=}C{=}CHCH_2OH$	Good	75
α-Acetylenic allenes	$RC{\equiv}CCu + XC(R'R'')C{\equiv}CR'''$, aqueous base $\longrightarrow$ diyne + $RC{\equiv}CC(R'''){=}C{=}CR'R''$	20–75	76
α-Allenyl ketones	$RCH(OH)CH_2C{\equiv}CH$, $CrO_3 \longrightarrow$ $RC(O)CH{=}C{=}CH_2$	General	1
β-Allenyl carbinols	1,3-Diene-2-Grignard + carbonyl compound $\longrightarrow$ $\overset{\diagdown}{\underset{\diagup}{}}C{=}C{=}\overset{\mid}{\underset{\mid\ \mid}{C}}CH(OH)R$	90	77
1-Allenyl-boronates	$HC{\equiv}CCH_2Br \xrightarrow[(2)\ B(OMe)_3]{(1)\ Al\ or\ Mg}$ $CH_2{=}C{=}CHB(OMe)_2$	35	78
α-Allenic α'-acetylenic carbinols	$HC{\equiv}CCH_2Br \xrightarrow[(2)\ RC{\equiv}CCHO]{(1)\ Al\ or\ Zn}$ $RC{\equiv}CCH(OH)CH{=}C{=}CH_2$	50–70	79
α-Allenyl amides	Pyrrolidone + $BrCH_2C{\equiv}CH \xrightarrow{NaH}$	—	79a
1-Allenic nitriles	$RR'C(OH)C{\equiv}CH \xrightarrow[NaCN]{CuCN}$ $RR'C{=}C{=}CHCN$	20–90	1, 80, 81, 82
1-Haloallenes	t-Ethynyl alcohol, $SOCl_2$ or $HCl \longrightarrow$ $\overset{\diagdown}{\underset{\diagup}{}}C{=}C{=}CHCl$	General 60	83, 84, 85
1-Aminoallenes	t-Ethynyl chloride, $R_3N \longrightarrow$ $RR'C{=}C{=}CHNR_3Cl$	General if R, R' > Me	86
Propargyl allenyl-phosphonates	Ethynylcarbinol + PCl_3, $PhPCl_2$, or $(RO)_2PCl$	Good	87, 88

(pyrrolidone structure with CH=C=CH2 on nitrogen)

Tetraiodoallene $HC\equiv CCH_2Br + KOH-I_2$ 89

Fluorinated $\underset{\underset{OH}{|}}{\overset{\overset{CF_2X'}{|}}{XF_2C-C}}-C\equiv CH \xrightarrow{\ SF_4\ }$ 30–84   91a
allenes

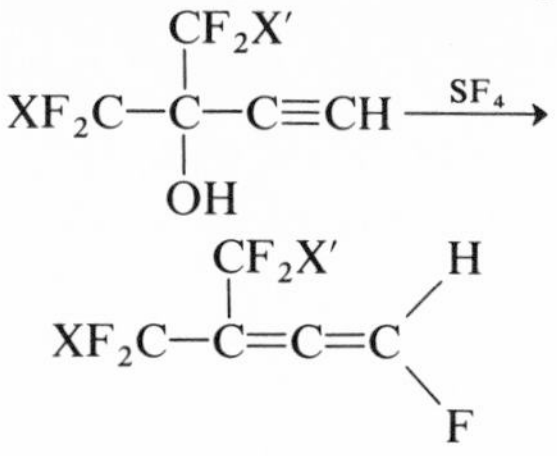

$$XF_2C-\underset{\underset{}{|}}{\overset{\overset{CF_2X'}{|}}{C}}=C=C\overset{\diagup H}{\underset{\diagdown F}{}}$$

(X, X' can be F or Cl)

1.8. Substitution (Allene ⟶ Allene)

α-Allenyl $CH_2=C=C(R)CH_2Br \xrightarrow[\text{(2) NaOH-MeOH-H}_2\text{O}]{\text{(1) NaOAc-HOAc}}$ 50   90
alcohols $CH_2=C=C(R)CH_2OH$

Conjugated $RR'C=C=CHX + HC\equiv C(CH_2)_nR''$ 18–82   91
allenynes $\xrightarrow[\text{base}]{Cu^+} RR'C=C=CHC\equiv C(CH_2)_nR''$ general

2. DISCUSSION OF METHODS OF FORMING ALLENES

2.1. Elimination to Allenes

2.1.1. ALLENE

Slobodin[92] developed one of the best methods for making allene. When he used butyl or isoamyl acetate as solvent during dehydrobromination of 2,3-dibromo-1-propene by KOH and then by zinc, he obtained 95–98% yield of allene. The allene contained no 2-bromopropene which was present in allene from other systems and which could not be removed by distillation. Dehydrohalogenation of 2-halopropenes is another good method.[1] Pyrolysis of diketene by passing it through a copper tube at 550° gives allene in 98% yield. The allene contains no propyne. This is an easy laboratory reaction.[25] Allene is potentially a cheap chemical commodity. Allene has been prepared by cracking of various hydrocarbons.[3,26–28]

2.1.2. TETRAARYL- AND TRIARYL-t-BUTYLALLENES

Tetrasubstituted allyl alcohols are dehydrated by perchloric acid in pyridine to form allenes.[93,94]

$$RR'C=CHC(OH)R''R''' \xrightarrow[-H_2O]{HClO_4} [RR'C=CH\overset{+}{C}R''R'''\ ClO_4^-] \xrightarrow{\text{pyridine}}$$

$$RR'C=C=CR''R'''$$

No R can be H, because the product is then a mixture of dimers of unknown structure. If the R's are alkyl, the products are conjugated dienes, but one R can be t-butyl if the rest are aryl. If the R's are p-amino or p-methoxyphenyl, the intermediate carbonium salt can be isolated. Tetraarylallenes should not come in contact with acid because they cyclize easily to form triarylindenes.[1]

2.1.3. TETRAPHENYLALLENE AND α-ALLENIC ACIDS BY THE WITTIG REACTION

Gilman[30] made tetraphenylallene by a Wittig reaction:

$$Ph_2C{=}CHP^+Ph_3, Br^- + PhLi \longrightarrow Ph_2C{=}\overset{-}{C}{-}\overset{+}{P}Ph_3 \xrightarrow{+\text{benzophenone}}$$

$$Ph_2C{=}C{=}CPh_2 + Ph_3P{=}O$$
$$(54\%)$$

He felt that this might be developed into a general method for allenes with any degree of substitution by proper choice of ylide and ketone or aldehyde:

$$R^1R^2C{=}C{=}PPh_3 + R^3R^4C{=}O \longrightarrow R^1R^2C{=}C{=}CR^3R^4 + Ph_3P{=}O$$

A similar reaction which is a new synthesis of α-allenic acids and esters was reported later.[38] Triphenylphosphinealkylenes which have no hydrogen on the carbon attached to phosphorus will react with acyl chlorides to give betaines, which lose triphenylphosphine oxide by γ-elimination and form α-allenic acid esters:

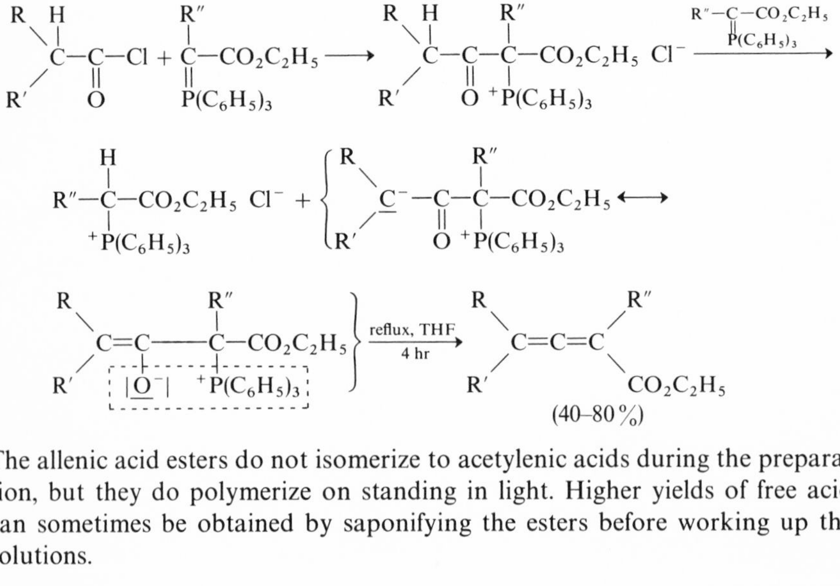

The allenic acid esters do not isomerize to acetylenic acids during the preparation, but they do polymerize on standing in light. Higher yields of free acid can sometimes be obtained by saponifying the esters before working up the solutions.

Ylides which have a methylene group beta to phosphorus also react with benzylidene aniline to give allenes:[31]

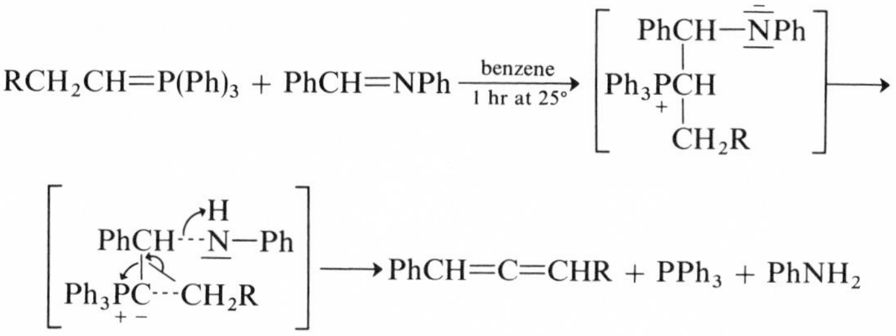

$$RCH_2CH{=}P(Ph)_3 + PhCH{=}NPh \xrightarrow[\text{1 hr at 25°}]{\text{benzene}} \begin{bmatrix} PhCH{-}\overset{-}{N}Ph \\ | \\ Ph_3\overset{+}{P}CH \\ | \\ CH_2R \end{bmatrix} \longrightarrow$$

$$\begin{bmatrix} PhCH\cdots\overset{H}{N}{-}Ph \\ Ph_3\overset{+}{P}C\cdots CH_2R \end{bmatrix} \longrightarrow PhCH{=}C{=}CHR + PPh_3 + PhNH_2$$

2.1.4. 1,1-DIFLUOROALLENE

Allyl acetate reacts with tetrafluoroethylene to form 1-acetoxymethyl-2,2,3,3-tetrafluorocyclobutane (65–75% yield). Pyrolysis at 800–950° gives methylene tetrafluorocyclobutane, which also pyrolyzes to 1,1-difluroallene.[36]

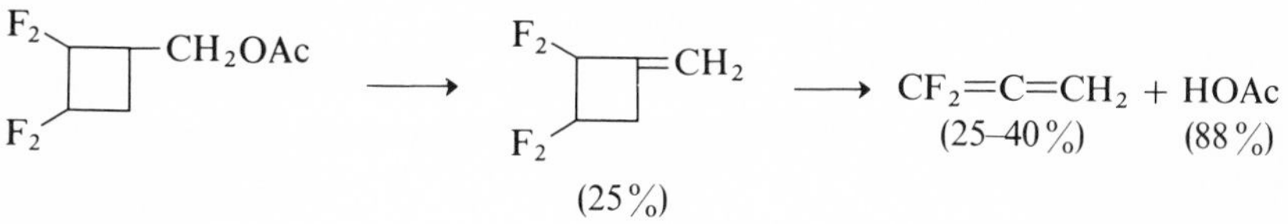

$$CF_2{=}C{=}CH_2 + HOAc$$
$$(25{-}40\%) \quad (88\%)$$
$$(25\%)$$

Difluoroallene can be dangerous. In detonation tests, it gave pressure rises comparable to acetylene, but the rate was faster. Properly barricaded pressure vessels must be used.

2.1.5. TETRAFLUOROALLENE

Banks[95] improved earlier syntheses[96] and obtained better yields of tetrafluoroallene:

$$CF_2Br_2 + CH_2{=}CF_2 \xrightarrow[\text{peroxide}]{\text{benzoyl}} CF_2BrCH_2CF_2Br \xrightarrow{\text{KOH}}$$
$$(51\%)$$

$$CF_2BrCH{=}CF_2 \xrightarrow{\text{KOH}} CF_2{=}C{=}CF_2$$
$$(89\%) \quad (33\%)$$

Tetrafluoroallene boils at $-37.6°$ (allene boils at $-34.3°$), is colorless, and has a characteristic fluorocarbon odor rather than the garlic odor of most allenes. It polymerizes easily, so it is stored at $-196°$ under vacuum.

2.1.6. ALLENIC KETO ALCOHOLS

Bertrand[42] synthesized many allenic keto alcohols by the reaction:

$$
\underset{R'R''C-C\equiv CCH=COEt}{\overset{\overset{OH}{|}\quad\quad\quad\overset{R}{|}}{}} \xrightarrow[\text{HClO}_4]{\text{cold } 0.1M} \underset{R'R''C-CH=C=CH-CR}{\overset{\overset{OH}{|}\quad\quad\quad\overset{O}{\|}}{}} + \text{EtOH}
$$
$$(25\text{--}85\%)$$

The R groups were hydrogen and/or lower alkyl in these reactions.

2.1.7. α-ALLENIC NITRILES

α-Allenic nitriles are made by treating N-propargyl amides with phosgene followed by triethylamine:

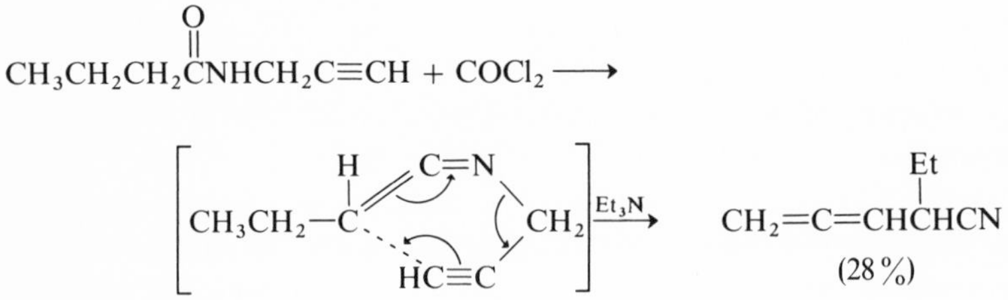

$$
\underset{CH_3CH_2CH_2\overset{\overset{O}{\|}}{C}NHCH_2C\equiv CH} + COCl_2 \longrightarrow
$$

$$
\left[\underset{CH_3CH_2-C}{\overset{H\quad C=N}{\underset{HC\equiv C}{}}}\ CH_2 \right] \xrightarrow{Et_3N} \quad \underset{CH_2=C=CHCHCN}{\overset{Et}{|}}
$$
$$(28\%)$$

N-Allylbutyramide gives 77% yield of 2-ethyl-4-pentenenitrile.[43]

2.1.8. PHENYLSULFONYLALLENES

Nucleophilic displacements in 2-chloro-3-phenylsulfonylpropene give products from three reaction paths:[39]

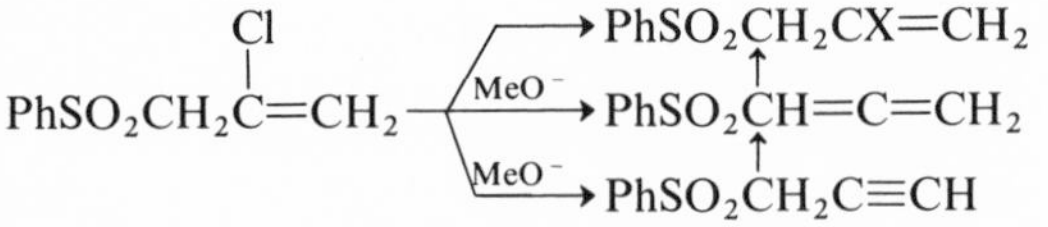

$$
\underset{PhSO_2CH_2C=CH_2}{\overset{\overset{Cl}{|}}{}}
\begin{cases}
\longrightarrow PhSO_2CH_2CX=CH_2 & (1) \\
\xrightarrow{MeO^-} PhSO_2CH=C=CH_2 & (2) \\
\xrightarrow{MeO^-} PhSO_2CH_2C\equiv CH & (3)
\end{cases}
$$

Reaction (2) appears to be the major path for methoxide (65% allenic product), but the allene might also form via reaction (3). The reaction half-life at 20° is only 30 seconds.

2.1.9. BIS(BIPHENYLENE)ALLENE AND DIPHENYLBIPHENYLENEALLENE

Fischer prepared these products by standard dehydrohalogenation reactions:

BiphC=CHCHBiph + Cl₂, then Et₃N ⟶

$$BiphC{=}CHCHBiph + Cl_2, \text{ then } Et_3N \longrightarrow$$

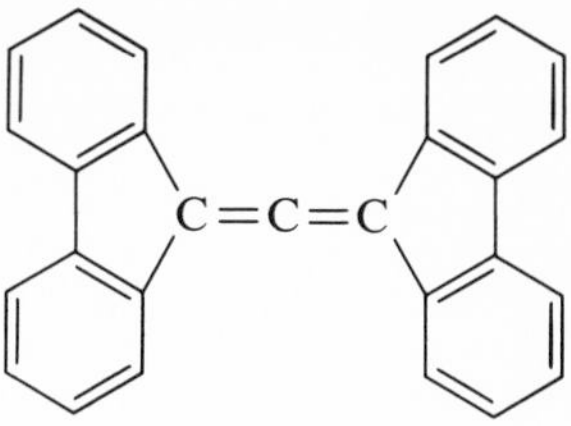

Ph₂CClCHClCHBiph $\xrightarrow[\text{Et}_3\text{N}]{\text{HOAc,}}$ Ph₂C=C=C

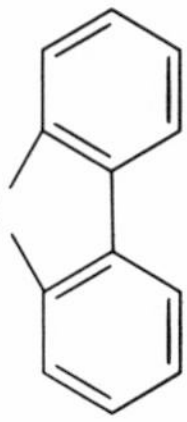

According to the Hückel theory, tetraphenylallenes should contain two independently acting olefinic groups. In addition reactions, Fischer's allenes added HBr, PhLi, and dimerized, only at one double bond. In the ultraviolet, the products absorb like bis(biphenylene)propene. Apparently the two double bonds do act independently.

2.1.10. ALLENES FROM OLEFINS VIA ADDITION OF CARBENES AND ELIMINATION

Doering[21] found that dibromocyclopropanes can be converted to allenes very easily on treatment with high-surface sodium (dispersed on alumina). He made the dibromocyclopropanes from olefins and bromoform. Under vacuum, neat 1,1-dibromo-2-n-propylcyclopropane over sodium-alumina gave 96% yield of products. The major product by GLC (gas-liquid chromatography) analysis was 1,2-hexadiene (64% of the products). The two minor components were acetylenes. Although a number of reaction paths can be written, the most likely one is simultaneous loss of Br⁻ and ring opening to the allene:[21]

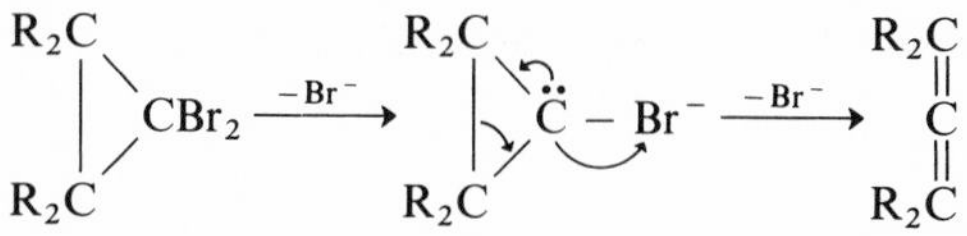

$$R_2C{\diagdown \atop \diagup}CBr_2 \xrightarrow{-Br^-} R_2C{\diagdown \atop \diagup}\ddot{C}-Br^- \xrightarrow{-Br^-} R_2C{\parallel \atop \parallel}C \atop R_2C$$

Later, Untch[22] reported that methyllithium causes a similar reaction in ether at $-65°$. 1,5-Cyclooctadiene gave 1,2,6-cyclononatriene in 37–71% yield. cis-Cyclooctene gave 1,2-cyclononadiene in 74% yield. Addition of dichlorocarbene followed by rearrangement by butyllithium has been used to enlarge the rings of cyclic allenes:[98]

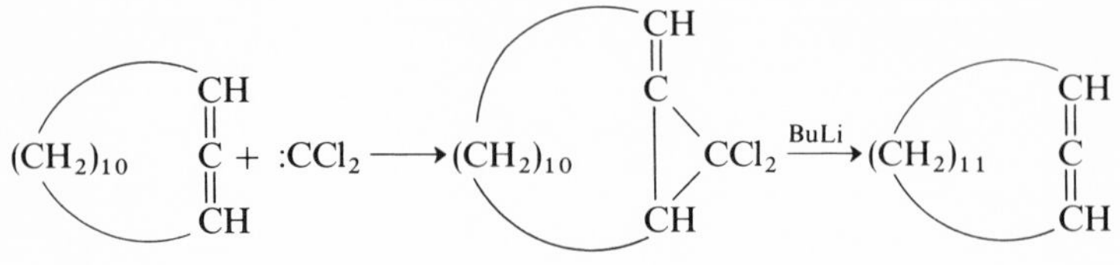

Hydration of the cyclic allenes with aqueous acid gives the cyclanone.

The addition of carboxyethyl carbene to olefins, followed by reactions leading to the ethyl N-nitroso-N-(*trans*-2,3-disubstituted cyclopropyl)carbamates, and finally treatment with sodium methoxide, is a new general synthesis of optically active 1,3-disubstituted allenic hydrocarbons.[98a] Yields are generally high:

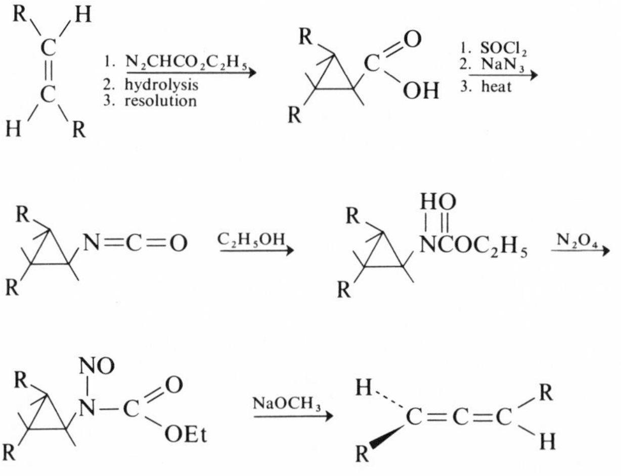

$$R = C_6H_5, (CH_2)_3CH_3, C_2H_5, \text{ or } CH_3$$

2.2. Substitution-Rearrangement Reactions

2.2.1. 1-BROMOALLENES

Jacobs[83] reacted 3-methyl-1-butyn-3-ol with PBr_3 in pyridine and obtained the propargylic bromide in 30% yield, along with isopropenylacetylene and bromobutadiene. Thionyl bromide gave the same propargylic bromide, 39% yield. Cuprous bromide isomerized the propargylic bromide to the 1-bromoallene. The bromoallene isomerized to bromobutadiene at 70°, and dimerized in the presence of peroxide to form the cyclobutane:

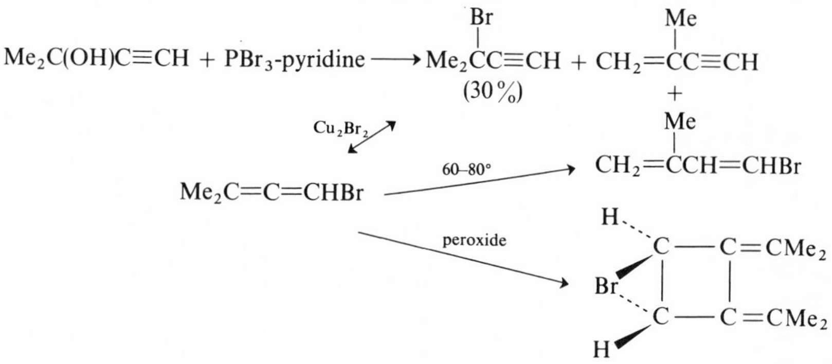

$$Me_2C(OH)C{\equiv}CH + PBr_3\text{-pyridine} \longrightarrow Me_2\overset{\overset{\displaystyle Br}{|}}{C}C{\equiv}CH + CH_2{=}\overset{\overset{\displaystyle Me}{|}}{C}C{\equiv}CH$$

$$(30\%) \qquad +$$

$$Me_2C{=}C{=}CHBr \underset{\underset{\displaystyle peroxide}{\searrow}}{\overset{\overset{\displaystyle Cu_2Br_2 \nearrow}{\xrightarrow{60\text{–}80^\circ}}}{\quad}} CH_2{=}\overset{\overset{\displaystyle Me}{|}}{C}CH{=}CHBr$$

t-Acetylenic alcohols which have large substituents, such as phenyl, t-butyl, mesityl, and α-naphthyl, give mostly the allenic bromide, which reacts with nucleophiles to form other allenic compounds.[83a]

The reaction of acetylenic alcohols with HBr to form 1-bromoallenes is accelerated by equivalent amounts of cuprous bromide and ammonium bromide:[99]

$$R_2C(OH)C{\equiv}CH \longrightarrow R_2C{=}C{=}CHBr$$
$$(45\text{–}90\%)$$

Low molecular weight t-ethynylcarbinols react in 1–2 hours at 35–40°. Higher molecular weight or hindered alcohols require 2–20 hours. Secondary alcohols react best at room temperature for 20–30 hours. The tertiary alcohols give pure 1-bromoallenes, but secondary alcohols give products containing up to 5% of the 3-bromoacetylene.

2.2.2. 1-CHLOROALLENES

Ethynylcyclohexanols with no hydrogen on the carbons alpha to the carbon bearing the hydroxyl and ethynyl groups cannot dehydrate, so reaction with thionyl chloride gives allenic chloride exclusively:[85]

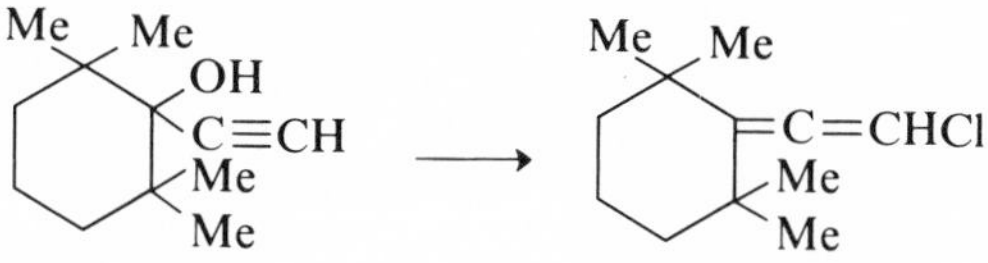

$$(83\%)$$

As the number of α-hydrogen atoms increases, and steric hindrance of the α-hydrogens decreases, the amount of dehydration to enyne increases. The reaction probably goes via a chlorosulfite ester, and pyridine increases the amount of dehydration to enyne. Enyne probably forms via E_2 elimination.

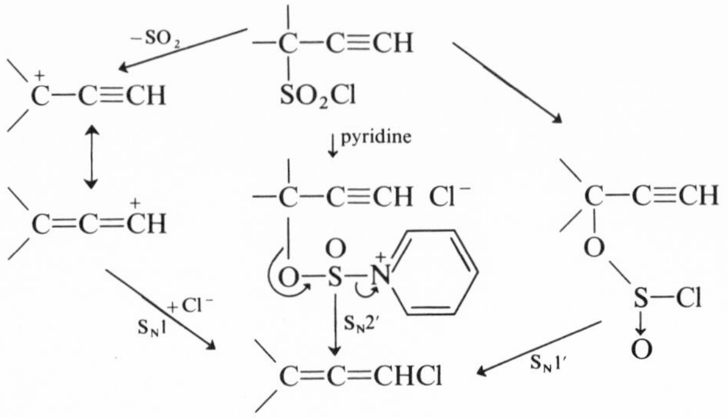

Acyclic tertiary acetylenic carbinols give similar results.[84] For example:

Dry HCl, PCl_3 or thionyl chloride gives the same product from
$Ph_2C(OH)C{\equiv}CPh$, namely, $Ph_2C{=}C{=}\overset{\displaystyle Cl}{\overset{|}{C}}Ph$. No propargylic chloride forms.

Under some conditions, the chloroallenes dimerize and cyclize. Landor[100,101] reacted thionyl chloride with 1,1,3-triphenylprop-2-yn-1-ol and obtained 5-chloro-5,12-dihydro-5,6,11,12-tetraphenyl-naphthacene and 1,3,3-triphenylprop-2-en-1-one as major products.

2.2.3. IODALLENES AND PROPARGYLIC IODIDES

Addition of secondary acetylenic alcohols to crystalline triphenyl phosphite methiodide followed by distillation gives iodoallenes of variable yield and purity. Low-boiling iodides such as iodopropadiene and iodobutadiene can be distilled away from the by-product phenol.[102] Iodoallenes are best made in dimethylformamide at 50–100°, followed by distillation. The iodoallene-dimethylformamide mixture distills over first, and the DMF is removed by water washing. Propargylic iodides are best made at low temperature and in methylene chloride. Prop-2-ynyl iodide can be prepared 90% pure and 3-iodobutyne 43% pure. Pent-4-en-1-yn-3-ol gave a mixture of unstable *cis*- and *trans*-1-iodopent-2-en-4-ynes.

The mechanism of formation of the allenic and propargylic iodides can be: (a) a two-step reaction in which the first step is acetylenic carbinol to iodide, and rearrangement of the iodide to the equilibrium mixture of acetylenic and allenic iodide, or (b) independent formation by different mechanisms, affected to different degrees by conditions and solvent. The second possibility is preferred,[102] but no definite choice is possible.

Propargylic and allenic iodides are fairly stable when pure and can be stored for several weeks. At room temperature in DMF and with triphenyl phosphite methiodide present, 3-iodobutyne rearranges to 1-iodobuta-1,2-diene.

2.2.4. TRIIODOALLENE AND TETRAIODOALLENE

These reactions give the polyiodoallenes:[89]

$$HC \equiv CCH_2Br + I_2 + KOH \longrightarrow IC \equiv CCH_2Br \xrightarrow{+KI} IC \overset{\cdot}{=} CCH_2I \xrightarrow{KOH}$$
$$+$$
$$KI$$

$$I_2C = C = CHI \xrightarrow{I_2 + KOH} I_2C = C = CI_2 \xrightarrow[ROH]{air, RT} I_2C = C(I)CO_2R$$

The oxidation of tetraiodoallene in air is very easy. Arylthiotriiodoallenes are made easily from propargyl thioethers:

$$R - Ph - S - CH_2C \equiv CH \xrightarrow[KOH]{I_2} R - Ph - S - CI = C = CI_2$$
$$(75\% \text{ yield})$$

1-iodoallenes react with phenylthio propargyl ethers to give the same products. The iodoacetylene is converted to the free acetylene.

2.2.5. ALLENIC COMPOUNDS FROM PROPARGYLIC HALIDES

The conversion of propargylic halides to allenic derivatives during reaction with a nucleophilic reagent is a general preparative method. The amount of propargylic and allenic substitution products depends on the reagent and on the size of the groups attached to the propargylic carbon.[83a] Shiner and Humphrey[103] studied these reactions in detail, and arrived at a reasonable explanation.

The reaction of 3-bromo-3-methyl-1-butyne and 1-bromo-3-methyl-1,2-butadiene with sodium thiophenoxide in aqueous ethanol without base gave these results:

	SPh | $Me_2CC \equiv CH$	$Me_2C = C = CHSPh$	Half-life (min)
$Me_2CBrC \equiv CH$	0	90	2.5
$Me_2C = C = CHBr$	45	55	180

With base added, the propargyl:allenyl ratio in the product was 50:30, and the half-lives were 0.3 and 9.5 seconds, respectively. Thus, base accelerated the reaction and promoted the formation of the propargylic substitution product. Sodium azide in the presence of sodium hydroxide gives the same products from either bromide, but the allenic bromide reacts more slowly. Other workers have shown that hydroxide and ethoxide give the propargyl product only. Azide favors the propargyl product by a factor of 8, and thiophenoxide favors propargyl by a factor of 1.6. Bromide attacks the propargylic and allenic positions equally.

The explanation is that some of the attacking anions stabilize the allenic carbanion so that it becomes more stable than the propargylic anion, which is normally the more stable of the two.

Bramwell[46] and Hartzler[57] refer to the different attacks as "α-attack" and "γ-attack":

$$
\begin{array}{c}
Me_2C(Cl)C\equiv CH \\
or \\
Me_2C=C=CHCl
\end{array}
\longrightarrow
\left[
\underset{\text{α-attack}}{Me_2\overset{\downarrow}{\overset{+}{C}}C\equiv C^-}
\rightleftharpoons
\underset{\text{γ-attack}}{Me_2C=C=\overset{\downarrow}{\overset{+}{C}}:}
\right]
$$

The acetylene-allene equilibrium mixture can be made from either halide. Diethyl sodiomethylmalonate reacts with either to form the same products in almost the same ratio. Acetoacetate anion gives mostly α-attack. Acetylacetone anion gives mainly γ-attack.

Weakly basic methyllithium does not remove halogen from propargyl halides at low temperature. $R_2C(X)C\equiv C-Li$ forms instead and reacts normally with carbonyl compounds, carbon dioxide and esters.[104]

2.2.5.1. *1-Cyanoallenes.* Propargylic chlorides react with NaCN in the presence of cuprous ion to form propargylic nitriles, which undergo prototropic rearrangement to form the 1-cyanoallenes:[1]

$$CH_3C\equiv CCH_2Cl + NaCl + Cu^+ \longrightarrow CH_3C\equiv CCH_2CN \xrightarrow{\text{base}} CH_3CH=C=CHCN^{81}$$

Greaves[82] in 1965 summarized the new methods for preparing cyanoallenes. Most tertiary acetylenic carbinols form cyanoallenes when treated with 1.5 equivalents of cuprous cyanide, a trace of copper, 1 equivalent of KCN, and 2.5 equivalents of 48% HBr for three days. HCl instead of HBr gives lower yields of cyanoallenes plus equal amounts of chloroallenes. Other strong acids cannot be used. These results are interpreted to mean that acetylenic chloride or bromide forms slowly and changes rapidly into allenic cyanide via a cuprous cyanide π complex. Sterically hindered or polysubstituted cyanoallenes cannot be made from alcohols, but they can be made from bromoallenes by reaction with dry CuCN at 110°. If cyanoallenes have a suitable proton, they eliminate HCN to form enynes. All of the methods which start with α-haloacetylenes give poor yields of mixtures of acetylenes and allenes.[80] Products and yields are listed in Table 1-1.

2.2.5.2. *1-Aminoallenes.* Reaction of ammonia and amines with *t*-propargylic halides usually gives propargylic amines, some of which are characterized by a remarkably high steric crowding around the nitrogen.[105-108] Most attempts to find halide-amine combinations in which steric effects would favor allene amines have failed. Hennion[86] found some such combinations. While studying reactions of *t*-propargylic halides RR′C(Cl)C≡CH with trimethylamine, he observed that if R and/or R′ was small, or if RR′ was cyclohexyl, propargylic

TABLE 1-1
1-Cyanoallenes from Acetylenic Alcohols[82]

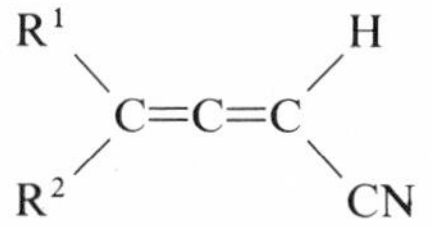

$$\begin{array}{c} R^1 \qquad\qquad H \\ \diagdown \qquad\quad \diagup \\ C=C=C \\ \diagup \qquad\quad \diagdown \\ R^2 \qquad\qquad CN \end{array}$$

R^1	R^2	% Yield
Et	H	60
Pr	H	60
Me	Me	30
		22
Me	Et	51
		31
Et	Et	75
		18
Me	i-Bu	40
Me	t-Bu	60
i-Pr	i-Pr	60
i-Bu	i-Bu	60
t-Bu	t-Bu	90

amines formed. If R and/or R′ was larger, the product was the 1-aminoallene. Less basic amines gave more total substitution than strongly basic amines, which tended to give enyne plus quaternary ammonium salts.[109] The effect of the size of R and R′ on the amine products is shown in Table 1-2. Solvation effects were important. Acetone plus a little isopropanol or water was better than pure dry acetone.

Attempts to hydrogenate the sterically crowded propargylamines caused hydrogenolysis to trimethylamine·HCl, as usual with crowded propargylic compounds. Hydrogenation of the aminoallenes gave allylic amines. This is one of the few examples of semihydrogenation of an allenic bond.

2.2.5.3. Propargyl and 1-Allenyl Derivatives of Group IVB Elements. Propargyl- and 1-allenylstannanes can be prepared by (1) reacting propyne Grignard reagent with triphenyltin iodide or (2) reacting propargyl bromide with triphenyltin lithium:[110]

$$CH_3C{\equiv}CMgBr + Ph_3SnI \xrightarrow[20°, 2\ hr]{ether} HC{\equiv}CCH_2SnPh_3 \tag{1}$$
$$(75\%)$$
$$+$$
$$CH_2{=}C{=}CHSnPh_3$$
$$(10\%)$$

$$HC{\equiv}CCH_2Br + Ph_3SnLi \longrightarrow CH_2{=}C{=}CHSnPh_3 \tag{2}$$
$$(30\%)$$

TABLE 1-2
Reaction of Tertiary Acetylenic Chlorides
with Trimethylamine[209]

$$RR'C(Cl)C\equiv CH + Me_3N \longrightarrow RR'C(\overset{+}{N}Me_3)-C\equiv CH \quad (Cl^-)$$

or

$$RR'C=C=CH-\overset{+}{N}Me_3 \quad (Cl^-)$$

R	R'	% Yield
Propargylic Amines		
CH_3	CH_3	83
CH_3	C_2H_5	55
CH_3	$i\text{-}C_4H_9$	50
C_2H_5	C_2H_5	52
$-CH_2(CH_2)_3CH_2-$		100
1-Aminoallenes		
C_2H_5	C_2H_5	88
C_2H_5	C_5H_{11}	60

Method (2) gives the allene derivative exclusively. The reaction of methyl-substituted propargyl Grignards yields two substitutions:

$$Ph_3SnI + BrMg\overset{\overset{\textstyle Me}{|}}{C}HC\equiv CH \longrightarrow Ph_3Sn\overset{\overset{\textstyle Me}{|}}{C}HC\equiv CH + Ph_3SnCH=C=CHMe$$
$$\text{"normal"} \qquad\qquad\qquad \text{"reverse"}$$

The products obtained by the Grignard method are given in Table 1-3.

These triphenyltin derivatives of allenes are more soluble in organic solvents than the propargylic derivatives. All of the products are relatively stable to

TABLE 1-3

Reaction of Propargylic and Propynyl
Grignard Reagents with R_3SnI[110]

Product	% Yield	Ratio
$Ph_3SnCH_2C\equiv CH$	80	90
$Ph_3SnCH=C=CH_2$		10
$Ph_3SnC(CH_3)HC\equiv CH$	70–75	80
$Ph_3SnCH=C=CHCH_3$		20
$Ph_3SnCH_2C\equiv CCH_3$	60	—
$Me_3SnCH_2C\equiv CH$	30–40	30
$Me_3SnCH=C=CH_2$		70

moisture and air, and can be stored at 0°. Acid hydrolyzes the C—Sn bond. The propargyl derivatives easily isomerize to the allenes on boiling for a few minutes in alcohol solvents.

The reaction of propargylmagnesium bromide with other R_3MX gives a mixture of propargyl and reversed allenyl derivatives:[110a]

$$R_3MX + \underset{|}{\overset{R'}{BrCH}}-C\equiv CR'' + Mg \xrightarrow[\text{exothermic reaction}]{\text{ether, 25°, 2 hr}} R_3M\underset{|}{\overset{R'}{C}}H-C\equiv CR'' \qquad (1)$$

$$+$$

$$R_3M C=C=CH\underset{|}{\overset{R''}{}} R' \qquad (2)$$

$$\text{(reversed allenic product)}$$

For the reaction with propargyl bromide itself, the amount of allenic product varies with the Group IVB element and with R:

R	M =	C	Si	Ge	Sn	Pb
			% Allenic Product (2)			
Me, Et		—	10	20	70	95
Ph		0	10	20	10	92

For the reaction of other propargylic bromides with R_3MX, the amount of allenic product varies with R' and R'':

$$\underset{|}{\overset{R'}{BrCH}}-C\equiv CR''$$

R'	R''	Ph₃CCl	Me₃SiCl	Ph₃GeBr	Ph₃SnI	Ph₃PbI
			% Allenic product (2) from Reaction with:			
Me	H	5	20	20	50	80
H	Me	3	10	10	10	50
H	Ph					40

The largest proportion of allenic product is formed when R in R_3MX is phenyl, and when M is lead.

2.2.5.4. *Allenic Ethers, Thioethers, Selenoethers and Telluroethers.* The allenic ethers have received little attention. Propargylic halides react with R—M⁻

(M = O, S, Se or Te) to form $RM\underset{|}{\overset{R'}{C}}HC\equiv CR'$ and $RM\underset{|}{\overset{R''}{C}}=C=CHR''$.[111] Bases isomerize the propargylic ethers to allenyl and 1-propynyl ethers. In the series O, S, Se and Te, the ratio of allenic to propargylic ether increases from 0–100% in the reaction of PhM⁻ with propargylic halides (Table 1-4). PhSCH₂C≡CH isomerizes faster to allenic ether than PhSeCH₂C≡CH does. After 7 minutes at 60° with sodium ethoxide, about 90 % yield of the allenic thioether is formed, and only 30 % yield of the allenic selenoether. Some properties of the allenic ethers are given in Table 1-5. The propargylic, allenic and 1-propynyl thioethers

TABLE 1-4[111]

$$PhM^- + R''CHC{\equiv}CR' \longrightarrow Ph\overset{\overset{\displaystyle X}{|}}{M}CHC{\equiv}CR' + Ph\overset{\overset{\displaystyle R'}{|}}{M}C{=}C{=}CHR''$$

			Ratio	
R'	R''	M	Propargylic	Allenyl
H	H	O	100	0
H	CH_3	O	99.5–100	Traces
CH_3	H	O	100	0
H	H	S	100	0
CH_3	H	S	100	0
H	CH_3	S	98.5	1.5
H	H	Se	96–98	2–4
H	CH_3	Se	90–95	5–10
CH_3	H	Se	100	0
H	H	Te	5–25	75–95
H	CH_3	Te	0–2	98–100

TABLE 1-5
Allenic Ethers, Thioethers, Selenoethers and Telluroethers[111]

Allenic Ether	b.p. (°C, @ mm Hg)	% Yield	Preparative Method[a]
$C_6H_5OCH{=}C{=}CH_2$	98, @ 35	67.5	a
$C_6H_5OC(CH_3){=}C{=}CH_2$	52, @ 2	78	a
$C_6H_5SCH{=}C{=}CH_2$	68, @ 0.1	46	b
$C_2H_5SCH{=}C{=}CH_2$	54, @ 55	43	b
$C_6H_5SC(CH_3){=}C{=}CH_2$	59, @ .01	67	a
$C_6H_5SeCH{=}C{=}CH_2$	70, @ 0.5	58.5	b
$C_6H_5SeC(CH_3){=}C{=}CH_2$	84, @ 2	72	a
$C_6H_5TeCH{=}C{=}CH_2$ (5–6 % propargylic isomer)	50–55, @ 10^{-3}	27	a
$C_6H_5SC{\equiv}CCH_3$	82, @ 1	90	a
$C_2H_5SC{\equiv}CCH_3$	75, @ 105	66	a
$(CH_3)_3CSC{\equiv}CCH_3$	69, @ 35	67	a
$CH_3(CH_2)_{11}SC{\equiv}CCH_3$	134, @ 0.8	72	a
$C_6H_5SeC{\equiv}CCH_3$	77, @ 0.5	45	c

[a] Preparative methods: (a) direct synthesis, (b) separated via a derivative, (c) separated by GLC.

are easily oxidized to sulfones. The acetylenic hydrogens in propargylic ethers undergo the normal Chodkiewicz–Cadiot coupling, KOH and Grignard alkynylation reactions, and hydration reactions.

2.2.5.5. *α-Acetylenic Allenes.* Propargyl halides or tosylates react with cuprous salts of terminal acetylenes to give acetylenic and allenic products:[76]

$$RC{\equiv}CCu + XC(R'R'')C{\equiv}CR''' \xrightarrow[\text{NH}_2\text{OH·HCl}]{\text{aqueous base}} RC{\equiv}CCR'''{=}C{=}CR''R' + RC{\equiv}C\overset{\overset{\displaystyle R'}{|}}{\underset{\underset{\displaystyle R''}{|}}{C}}C{\equiv}CR'''$$

Example:

$$Me_2\overset{\overset{\displaystyle OH}{|}}{C}C{\equiv}CCu + BrCH_2C{\equiv}CH \longrightarrow Me_2\overset{\overset{\displaystyle OH}{|}}{C}C{\equiv}CCH{=}C{=}CH_2$$
$$(75\%)$$

2.2.5.6. *Allenic Hydrocarbons.* Dehalogenation of *t*-acetylenic carbinyl chlorides by lithium aluminum hydride is S_N2 attack by hydride ion of AlH_4^- on the terminal acetylenic carbon, with shift of bonds to an allenic system and elimination of halogen. The same mechanism should apply to the acetylenic bromides:[70]

$$Li^+H_3AlH^- \rightarrow HC{\equiv}C{-}\overset{\overset{\displaystyle Me_2}{|}}{C}{-}Br \longrightarrow H_2C{=}C{=}CMe_2 + Br^-$$

The reaction of alkylmagnesium bromides with propargyl bromide gives allenic hydrocarbons. The most likely mechanism is direct formation of the allenic hydrocarbon by an S_N2' reaction:[72a]

$$\underset{\underset{\displaystyle Br}{|}}{\overset{\displaystyle H{-}C{\equiv}C{-}CH_2}{R{-}Mg}}\ \ Br \longrightarrow RCH{=}C{=}CH_2 + MgBr_2$$

Trisubstituted propargyl halides react with organolithium compounds to form tetrasubstituted allenes in which some of the substituents are alkyl. Different groups can be present at the ends of the allenic system. This is probably the best method for preparing these compounds. For example:[72b]

$$Me_2CClC{\equiv}CEt + MeLi \xrightarrow{\text{ether}} 88\% \text{ yield } Me_2C{=}C{=}\overset{\nearrow Me}{\underset{\searrow Et}{C}} + Me_3CC{\equiv}CEt$$
$$(89\%)\qquad\qquad (11\%)$$

Reactions of the propargyl halides with Grignard reagents are less selective and form higher ratios of acetylenic products.

2.2.5.7. *α- and β-Allenic Alcohols.* Reduction of propargyl halides by the Zn-Cu couple was used to prepare 3,4-hexadien-1-ol and 2,3-pentadien-1-ol.[126]

Preparation of the hexadienol illustrates the reactions involved (R = 2-tetra-hydropyranyl):

$$CH_3\overset{\underset{\displaystyle OR}{|}}{CH}C\equiv CLi + CH_2\overset{\displaystyle O}{-}CH_2 \xrightarrow{NH_3} CH_3\overset{\underset{\displaystyle OR}{|}}{CH}C\equiv CCH_2CH_2OH \xrightarrow[\text{pyridine}]{Ac_2O}$$

$$\text{acetate} \xrightarrow{\text{hydrolysis}} ROH + CH_3\overset{\underset{\displaystyle OH}{|}}{CH}C\equiv CCH_2CH_2OAc \xrightarrow{SOCl_2}$$

$$CH_3\overset{\underset{\displaystyle Cl}{|}}{CH}C\equiv CCH_2CH_2OAc \xrightarrow{Zn\text{-}Cu} CH_3CH=C=CHCH_2CH_2OH$$
$$(46\%)$$

2.2.6. 1-ALLENYLPHOSPHONATES

PCl_3, $PhPCl_2$ or $(EtO)_2PCl$ reacts with ethynylcarbinols in ether, in the presence of triethylamine. Products from $RPCl_2$ were the propargylic esters of 1-allenylphosphonates.[87]

$$R'R''C=C=CH-\overset{\displaystyle O}{\underset{\displaystyle R}{\overset{\uparrow}{P}}}-O-CR'R''C\equiv CH, \text{ monoester, type (1)}$$

$$R'R''C=C=CH-\overset{\displaystyle O}{\overset{\uparrow}{P}}(OCR'R''C\equiv CH)_2, \text{ diester, type (2)}$$

The products obtained in some reactions are listed in Table 1-6. Mild alkaline hydrolysis at 100° of type (2) diesters gave these salts:

$$CH_2=C=CH-\overset{\displaystyle O}{\underset{\displaystyle OBa}{\overset{\uparrow}{P}}}-OCH_2C\equiv CH \quad 53\% \text{ yield}$$

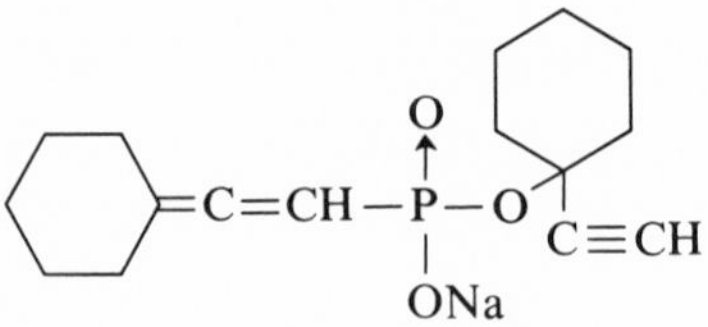

$$\text{cyclohexylidene}=C=CH-\overset{\displaystyle O}{\underset{\displaystyle ONa}{\overset{\uparrow}{P}}}-O\overset{\displaystyle cyclohexyl}{\underset{\displaystyle C\equiv CH}{}} \quad 64\% \text{ yield}$$

$$PhCH=C=CH-\overset{\displaystyle O}{\underset{\displaystyle OBa}{\overset{\uparrow}{P}}}-OEt \quad 57\% \text{ yield}$$

$$PhCH=C=CH-\overset{\displaystyle O}{\underset{\displaystyle OBa}{\overset{\uparrow}{P}}}-OEt \quad 47\% \text{ yield}$$

TABLE 1-6
Phosphonate Esters of Acetylenic Carbinols[37]

Phosphorus Halide	Acetylenic Carbinol	Product Type	% Yield	m.p. (°C) or b.p. (°C, @ mm Hg)
PCl_3	Propargyl alcohol	2	92	115, @ 0.25
PCl_3	1-Ethynyl-1-cyclohexanol	2	47	78–79
PCl_3	3-Methylbut-1-yn-3-ol	2	90	
$PhPCl_2$	Propargyl alcohol	1	23	130, @ 0.3
$PhPCl_2$	1-Ethynyl-1-cyclohexanol	1	34	
$PhPCl_2$	Methylbutynol	1	30	
$(EtO)_2PCl$	Hex-1-yn-3-ol	2	90	
$(EtO)_2PCl$	3-Phenylprop-1-yn-3-ol	2	90	

Acid hydrolysis of type (2) esters gave ethynyl- and allenyl-phosphonates:

$$CH_3C{\equiv}C{-}PO_3Ba \qquad 40\% \text{ yield}$$

$$CH_3CH{=}C{=}CHPO_3H_2 \quad 71\% \text{ yield}$$

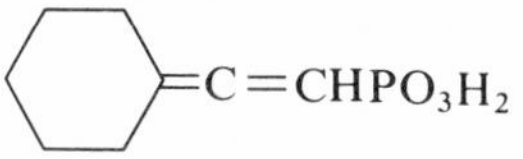

$=C{=}CHPO_3H_2 \quad 34\% \text{ yield}$

Acid hydrolysis of type (1) esters gave allenylphosphonates:

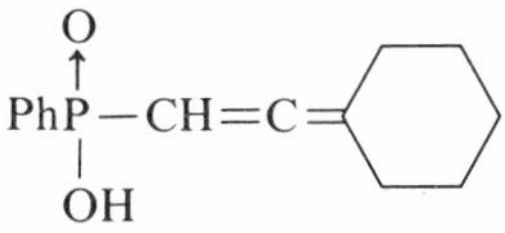

$10\% \text{ yield}$

$9\% \text{ yield}$

$33\% \text{ yield}$

1-Hexyn-3-yl(3-propylallenyl)phenylphosphinate is much more stable to acid hydrolysis than any of the carbinols listed here.[112]

If the ester group has a hydroxyl on the phosphorus atom, the product does not rearrange to allene, presumably because the phosphorus atom no longer has a lone pair of electrons. Thus, the monoester of 1-hexyn-3-ol did not rearrange when heated with a mole of phosphoric acid for 17 hours at 80° and 15 mm.

When 0.3 mole PCl_3 in ether was reacted with 0.9 mole propargyl alcohol at low temperature, two fractions were isolated: $(HC\equiv CCH_2O)_2P(O)H$, b.p. 80–82° at 0.3 mm of Hg, and 40% di(2-prop-yn-1-yl)phosphonate, b.p. 115–116° at 0.25 mm, which slowly rearranged to $MeC\equiv CP(O)(OCH_2C\equiv CH)_2$. Propargyl alcohol reacted with $SOCl_2$ under similar conditions to give 35% yield of $(CH\equiv CCH_2O)_2SO$, b.p. 55–56° at 0.3 mm. $AsCl_3$ gave 50% yield of $As(OCH_2C\equiv CH)_3$, b.p. 94–95° at 0.7 mm. Boric acid gave 28% yield of propargyl borate, b.p. 72–73° at 1 mm. None of the last three products rearranged.

Products from the reaction:

$$Me_2C(OH)C\equiv CH + Ph_2PCl \xrightarrow[\text{benzene, 8°, 5 hr}]{\text{pyridine}} Ph_2\overset{\overset{O}{\uparrow}}{P}CH=C=CMe_2$$

$$+ HC\equiv C\overset{\overset{Me_2}{|}}{C}-O\overset{\overset{O}{\uparrow}}{\underset{\underset{Ph}{|}}{P}}-CH=C=CMe_2$$

were claimed to be weed killers, insecticides, antioxidants and corrosion inhibitors.[88]

2.2.7. 1-ALLENYLDICHLOROPHOSPHITES

Propargylic chlorophosphites isomerize easily to 1-allenyldichlorophosphites.[113] This special kind of Arbuzov rearrangement goes through a cyclic intermediate:

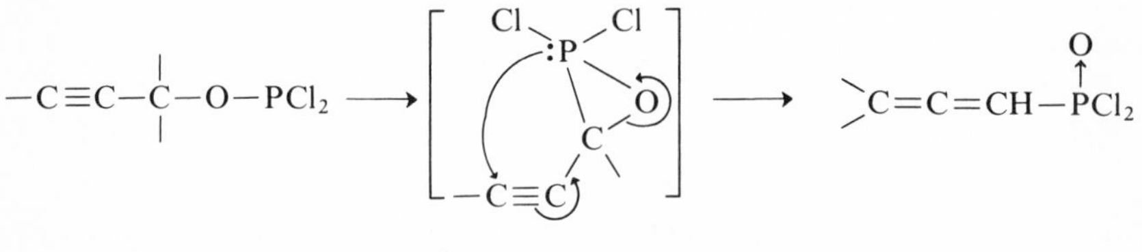

$$-C\equiv C-\overset{|}{\underset{|}{C}}-O-PCl_2 \longrightarrow \left[\begin{array}{c} Cl \diagdown \underset{}{:P} \diagup Cl \\ \diagup \diagdown O \\ C \\ -C\equiv C \end{array} \right] \longrightarrow \; {\diagup}C=C=CH-\overset{\overset{O}{\uparrow}}{P}Cl_2$$

2.2.8. ALLENIC KETONES AND ALLENIC ALCOHOLS

β-Acetylenic secondary alcohols are oxidized by CrO_3 to form α-allenic ketones.[114]

$$RCH(OH)CH_2C\equiv CH \longrightarrow R\overset{\overset{O}{\|}}{C}CH=C=CH_2$$
$$(40\text{--}45\%)$$

Lithium aluminum hydride reduces the allenic ketones to allenic alcohols in 50–85% yields. The allenic bond is not reduced.

α,β-Diacetylenic alcohols are oxidized by chromic acid-sulfuric acid to form mixtures of α,β-diacetylenic ketones and α-acetylenic α'-allenic ketones:[114a]

$$C_5H_{11}C{\equiv}CCH(OH)CH_2C{\equiv}CH \xrightarrow[\text{H}_2\text{SO}_4,\ 0°]{\text{CrO}_3} C_5H_{11}C{\equiv}C\overset{\displaystyle O}{\overset{\|}{C}}CH_2C{\equiv}CH$$
$$(18.7\%)$$

$$+ C_5H_{11}C{\equiv}C\overset{\displaystyle O}{\overset{\|}{C}}CH{=}C{=}CH_2$$
$$(36\%)$$
$$\text{(b.p. }67°\text{ at 0.1 mm)}$$

2.2.9. β-ALLENIC ALDEHYDES

Acetylenic carbinols react with aldehydes in the presence of an acid catalyst to give β-allenic aldehydes:[115]

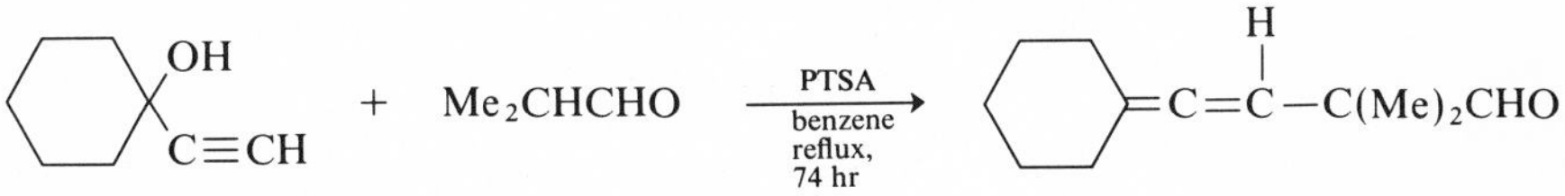

2.3. Addition Reactions to Form Allenes

2.3.1. 1,4-ADDITION TO ENYNES

2.3.1.1. *Allene Hydrocarbons.* 1,4-Addition to enynes is a convenient synthesis of allenes, but some acetylenic product frequently forms by addition across the double bond. Vinylalkylacetylenes react with equimolar RLi in ether under nitrogen to form good yields of allenes.[52] Infrared study of the products from BuLi and $EtC{\equiv}CCH{=}CH_2$ showed the intermediate was $EtC{=}C{=}CHCH_2Bu$ with a Li substituent.[116] Hydrolysis gave the allene hydrocarbon; carbonation gave the 1-allenic acid $EtC{=}C{=}CHCH_2Bu$ with a COOH substituent. With vinylisopropenylacetylenes, R^-Li^+ adds R to the vinyl group and not to the isopropenyl group. Products are isopropenylallenes $CH_2{=}C{-}CH{=}C{=}CHR$ with a Me substituent.[50] When triphenylchloromethane, enynes, excess mercury, and benzene were mixed and allowed to stand under CO_2 for 5–6 days, allene products formed in 50–65% yield.[117,118]

$$RC{\equiv}CCH{=}CHR' + Ph_3C{\cdot} \longrightarrow Ph_3C\overset{R}{C}{=}C{=}CH\overset{R'}{C}HCPh_3$$

When triphenylmethyl radical adds to divinylacetylene in benzene, 60% yield of 7,7,7-triphenyl-3-triphenylmethyl-1,3,4-heptatriene is obtained:[119]

$$\underset{\displaystyle |}{CPh_3}$$

$CH_2=CHC=C=CHCH_2CPh_3$. Other enynes give similar results. In substituted vinylacetylenes, the triphenylmethyl radical adds preferably to the unsubstituted side of the double bond.

2.3.1.2. *1-Allenic Silanes.* Allenic products form when RLi adds to trialkylvinylethynylsilanes:[49]

$$Me_3SiC{\equiv}CCH{=}CHR' + RLi \longrightarrow Me_3SiCH{=}C{=}CHCH_2R$$

$$+ \; Me_3SiC{\equiv}CCH_2CH_2R$$

Methyllithium gives 90% of the acetylenic product. The amount of acetylenic product decreases as the size of the R group in RLi increases. Butyllithium gives only 55% acetylenic product. Chloromethyl ethers add to vinylethynylsilanes to give mostly allenic products:[120]

$$\underset{\displaystyle Me_3SiC{=}C{=}CHCH_2Cl}{\overset{\displaystyle CH_2OR}{\overset{\displaystyle |}{}}}$$

2.3.1.3. *β-Allenic Carbinols.* Enynes react with excess EtLi, and then with epoxide, aldehyde or ketone, to form β-allenic carbinols.[51] For example

$$CH_2{=}CHC{\equiv}CEt + EtLi \xrightarrow{\;+RR'C{=}O\;} \underset{\displaystyle }{RR'C}{\overset{\displaystyle OH}{\overset{\displaystyle |}{}}}{-}\underset{\displaystyle }{C}{\overset{\displaystyle Et}{\overset{\displaystyle |}{}}}{=}C{=}CHPr$$

2.3.1.4. *1-Allenic Phosphines.* Lithium diethylphosphide adds to hex-1-en-3-yne in cold ether.[58] The product is 1-diethylphosphino-2,3-hexadiene:

$$Et_2PLi + EtC{\equiv}CCH{=}CH_2 \longrightarrow Et_2PCH_2CH{=}C{=}CHEt$$

2.3.2. ALLENE CARBENES AND THEIR ADDITION REACTIONS—ALLENE CARBENES FROM *t*-ACETYLENIC COMPOUNDS

Hartzler[57,121] proved that carbenes are intermediates in the base-catalyzed reactions of *t*-acetylenic chlorides. He added 3-chloro-3-methyl-1-butyne to an alcohol-free slurry of potassium *t*-butoxide in styrene, and obtained 1-(2-methylpropenylidene)-2-phenylcyclopropane:

$$Me_2C(Cl)C{\equiv}CH + PhCH{=}CH_2 + KO{-}t{-}Bu \longrightarrow PhCH{-}\underset{\displaystyle (48\%)}{C}{=}C{=}CMe_2$$

with cyclopropane ring CH_2 bridge shown above the C.

Reaction with cyclohexene gave

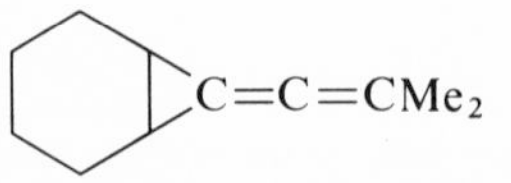

Pyrolysis of di-*t*-butyl-propargyl acetate gave di-*t*-butylvinylidenecarbene, 10–52% yield.[90] This carbene also added to olefins to give substituted allenes, and the additions were stereospecific. Most of the alkenylidenecyclopropanes are colorless liquids which oxidize rapidly in air. The product from tetramethylethylene is a low-melting air-stable solid with very simple IR and NMR spectra. All of the products have the characteristic allene IR absorption at $2020 \pm 20 \, \text{cm}^{-1}$.

Cadiot[122] noted in 1956 that acetic anhydride added to a slurry of KOH in an ether solution of diarylethynylcarbinols gave tetraarylhexapentaenes:

$$Ar_2C(OH)C\equiv CH + KOH + Ac_2O \xrightarrow{\text{ether}} Ar_2C=C=C=C=C=CAr_2$$

Hartzler[57] believes Cadiot's mechanism is not as good as an allene carbene mechanism. When Hartzler did the reaction by adding a solution of diphenylethynylcarbinyl acetate in styrene to a slurry of KO–*t*-Bu in styrene at $-5°$, he got 1-diphenylvinylidene-2-phenylcyclopropane (22% yield), and 1,1,4,4-tetraphenyl-2-butyne-1,4-diol (44% yield). This indicates that the carbene was present in Cadiot's system. The hexapentaene is probably formed by reaction of the carbene with the anion of the acetylenic acetate. Later, Hartzler[90] obtained additional evidence. Di-*t*-butylvinylidenecarbene gave 20% yield of hexapentaene if no olefin was present.

$$t\text{-Bu}_2C(OAc)C\equiv CH \xrightarrow{\text{KO}-t\text{-Bu}} [t\text{-Bu}_2C=C=C: \longleftrightarrow t\text{-Bu}_2\overset{+}{C}-C\equiv \overset{-}{C}]$$

$t\text{-Bu}_2C(OAc)C\equiv C^-$ (left branch) \qquad $^-O-t\text{-Bu}$ (right branch)

$t\text{-Bu}_2C=C=C=C=C=C-t\text{-Bu}_2$ \qquad\qquad $t\text{-Bu}_2C=C=CHO-t\text{-Bu}$
(20%) \qquad\qquad\qquad\qquad\qquad\qquad (19%)
aliphatic hexapentaene \qquad\qquad\qquad\qquad 1-allenic ether

Tertiary acetylenic halides or esters are more likely to form allenylcarbenes than primary halides are because competing bimolecular nucleophilic displacements are sterically hindered.[57] As substituents change from phenyl to alkyl to hydrogen, the elimination of X^- from $R_2C(X)C\equiv C^-$ should decrease. The differences in rate of elimination of X^- will not be as large as in neutral halides, since the transition state in the acetylenic anion series has less positive charge formation on the carbon bearing the X^-. The reaction of propargyl bromide with methoxide ion shows no deuterium isotope effect; thus γ-elimination is probably not involved.

Hartzler[123] also studied rates of addition of allene carbenes to olefins. He could generate the carbene either from substituted allenyl chloride or propargyl chloride, and get the same products and same rates of addition. This confirms

Hennion's mechanism of the solvolysis of *t*-carbinylacetylenic chlorides:[124]

$$(CH_3)_2C=C=CHCl \underset{B:H^+}{\overset{B:}{\rightleftarrows}} (CH_3)_2C=C=\overset{-}{C}Cl$$

$$(CH_3)_2\underset{\underset{Cl}{|}}{C}-C\equiv C^- \underset{-Cl^-}{\overset{+Cl^-}{\rightleftarrows}} \qquad +Cl^-\,\Big|\!\Big|\,-Cl^-$$

$$B:\,\Big|\!\Big|\,B:H^+ \quad [(CH_3)_2\overset{+}{C}-C\equiv C^- \longleftrightarrow (CH_3)_2C=C=C:]$$

$$(CH_3)_2\underset{\underset{Cl}{|}}{C}-C\equiv CH \qquad \Big|\, C_6H_5CH=CH_2$$

$$\begin{array}{c} CH_2 \\ \diagup \quad \diagdown \\ C_6H_5C\text{------}C=C=C(CH_3)_2 \\ H \end{array}$$

In 1965, Landor[45] reported the formation of allene carbenes from 1-bromo-allenes in the presence of strong bases. He used the same trapping reactions Hartzler used. Landor explained the condensation of malonates with 1-halo-allenes by an elimination-addition mechanism. 1-Haloallenes and 3-halo-acetylenes give products with the same acetylenic-to-allenic ratio, so it is reasonable to assume that there is a common intermediate.

Serratosa[72] reacted propargyl bromide with ethynylmagnesium bromide in ether (cuprous chloride catalyst) to make 1,4-pentadiyne. However, when he reacted alkylmagnesium bromides with propargyl bromide, he got not only the expected acetylene but also the corresponding allene:

$$RMgBr + BrCH_2C\equiv CH \longrightarrow RCH_2C\equiv CH + RCH=C=CH_2$$

$$\downarrow$$

$$BrCH_2C\equiv CMgBr \xrightarrow{-MgBr_2} H_2C=C=C: \text{------}\!\!\uparrow^{+R^+}$$

The allene carbene is probably the intermediate to the substituted allene.

2.4. Elimination-Rearrangement

Pyrolysis of propargyl vinyl ethers[125] and of dipropargyl acetals[41] gives allenic aldehydes. Propargyl vinyl ethers rearrange at only 120–150° to form allenic aldehydes.[40] Increasing substitution of the propargyl vinyl ether increases the yield (Table 1-7). Since increased substitution causes higher yield at lower temperature, steric hindrance in the transition state cannot be important. The rearrangement is a cyclic process in which a C—C bond forms as the C—O bond breaks. This is assisted by substitution at the carbon atoms in positions 2 and 5 because substitution favors partial radical formation.

TABLE 1-7
Allenic Aldehydes from Pyrolysis of
Propargyl Vinyl Ethers[40]

Aldehyde	b.p. (°C)	% Yield
$CH_2=C=CHCH_2CHO$	—	20–30
$MeCH=C=CHCH_2CHO$	—	10–20
$Me_2C=C=CHCH_2CHO$	—	10
$CH_2=C=CHCHMe_2CHO$	126–127	70
$MeCH=C=CHCMe_2CHO$	139–140	60
$Me_2C=C=CHCMe_2CHO$	156–158	76
$PrCH=C=CHCMe_2CHO$	178	93

The Cope rearrangement of diethyl isobutenylpropargylmalonate at 270° gives both rearrangement and fission, also by cyclic processes. Rearrangement:

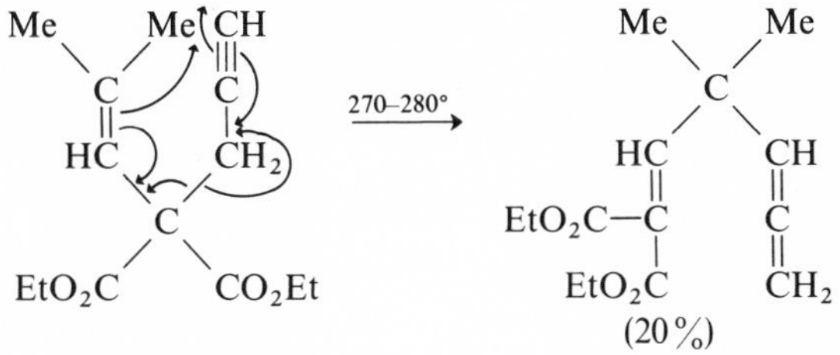

Fission:

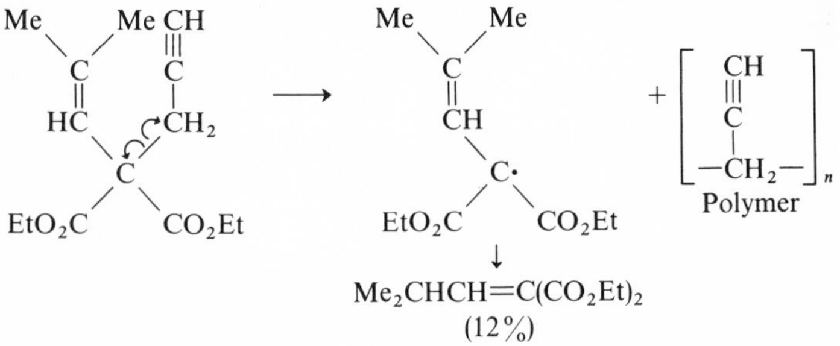

References

1. Fischer, H., in (Patai, S., editor), "The Chemistry of Alkenes," p. 1025, New York, Interscience Publishers, 1964.
2. Petrov, A. A., and Fedorova, A. V., *Usp. Khim.*, **33**, 3 (1964); *Russ. Chem. Rev.*, *English Transl.*, **33**, 1 (1964).

3. Griesbaum, K., *Angew. Chem., Intern. Ed.*, **5**, 933 (1966).
4. Cadiot, P., Chodkiewicz, W., and Rauss-Godineau, J., *Bull. Soc. Chim. France*, **28**, 2176 (1961).
5. Levina, R. Ya., and Viktorova, E. A., *Usp. Khim.*, **27**, 162 (1958).
5a. Taylor, D. R., *Chem. Rev.*, **67**, 317 (1967).
5b. Mavrov, M. V., and Kucherov, V. F., *Russ. Chem. Rev. (English Transl.)*, **36**, 233 (1967).
6. Burton, B. S., and Pechmann, H. V., *Chem. Ber.*, **20**, 145 (1887).
7. Jones, E. R. H., Mansfield, G. H., and Whiting, M. C., *J. Chem. Soc.*, 3208 (1954).
8. Brand, K., *Chem. Ber.*, **54**, 1987 (1921).
9. Kuhn, R., and Wallenfels, K., *Chem. Ber.*, **71**, 783 (1938).
10. Staudinger, H., and Ruzicka, L., *Helv. Chim. Acta*, **7**, 177 (1924).
11. Celmer, W. D., and Solomans, I. A., *J. Am. Chem. Soc.*, **74**, 1870, 2245, 3838 (1952); **75**, 1372, 3430 (1953).
12. Baker, C. S. L., Landor, P. D., and Landor, S. D., *J. Chem. Soc.*, 4659 (1965).
13. Bohlmann, F., Bornowski, H., and Arndt, C., *Chem. Ber.*, **98**, 2236 (1965).
14. Allred, E. L., Grant, D. M., and Goodlett, W., *J. Am. Chem. Soc.*, **87**, 673 (1965).
15. Hoffmann, R., *Tetrahedron*, **22**, 521 (1966).
16. Pittman, C. U., Jr., and Olah, G. A., *J. Am. Chem. Soc.*, **87**, 5632 (1965).
17. Richey, H. G., Phillips, J. C., and Rennick, L. E., *J. Am. Chem. Soc.*, **87**, 1381 (1965).
18. Cram, D. J., "Fundamentals of Carbanion Chemistry," pp. 19, 48, 54, 59, New York, Academic Press, 1965.
19. Moore, W. R., and Ward, H. R., *J. Org. Chem.*, **25**, 2037 (1960).
20. *Ibid.*, **27**, 4179 (1962).
21. Doering, W. von E., and LaFlamme, P. M., *Tetrahedron*, **2**, 75 (1959).
22. Untch, K. G., Martin, D. J., and Castellucci, N. T., *J. Org. Chem.*, **30**, 3572 (1965).
23. Moore, W. R., and Bertelson, R. C., *J. Org. Chem.*, **27**, 4182 (1962).
24. Speziale, A. J., *et al.*, *J. Am. Chem. Soc.*, **82**, 903 (1960); **84**, 1868 (1962).
25. Conley, R. T., and Rutledge, T. F. (to Air Reduction Co.), U.S. Patent 2,818,456 (Dec. 31, 1957); *Chem. Abstr.*, **52**, 6391 (1958).
26. Happel, J., and Marsel, C. J. (to National Lead Co.), U.S. Patent 3,198,848 (Aug. 3, 1965); *Chem. Abstr.*, **63**, 13073 (1965).
27. Happel, J., and Marsel, C. J. (to National Lead Co.), French Patent 1,389,102 (Feb. 12, 1965); *Chem. Abstr.*, **62**, 13044 (1965).
28. N.V. deBataafsche Petrol. Maatschappij, British Patent 828,989 (Feb. 24, 1960); *Chem. Abstr.*, **54**, 13733 (1960).
29. Ginzburg, Ya. I., *Zh. Obshch. Khim.*, **10**, 513 (1940); **15**, 442 (1945).
30. Gilman, H., and Tomasi, R. A., *J. Org. Chem.*, **27**, 3647 (1962).
31. Bestmann, H. J., and Seng, F., *Angew. Chem.*, **75**, 475 (1963).
32. Vorländer, D., and Siebert, C., *Chem. Ber.*, **39**, 1024 (1906).
33. Tadros, W., Sakla, A. B., and Helmy, A. A. A., *J. Chem. Soc.*, 2687 (1961).
34. Hatch, L. F., and Weiss, H. D., *J. Am. Chem. Soc.*, **77**, 1798 (1955).
35. Pilgram, K., and Korte, F., *Tetrahedron Letters*, 883 (1962).
36. Knoth, W. H., and Coffman, D. D., *J. Am. Chem. Soc.*, **82**, 3873 (1960).
37. Borisov, A. E., and Nesmeyanov, A. N., *Akad. Nauk SSSR Inst. Organ. Khim. Sintezy Organ. Soedin. Sb.*, **1**, 150 (1950); *Chem. Abstr.*, **47**, 8004 (1953).
38. Bestmann, H. J., and Hartung, H., *Chem. Ber.*, **99**, 1198 (1966).

39. Stirling, C. J. M., *J. Chem. Soc. Suppl.*, No. 1, 5875 (1964).
40. Black, D. K., and Landor, S. R., *J. Chem. Soc.*, 6784 (1965).
41. Jones, E. R. H., Loder, J. D., and Whiting, M. C., *Proc. Chem. Soc.*, 180 (1960).
42. Bertrand, M., and Rouvier, C., *Compt. Rend.*, **260**, 209 (1965).
43. Brannock, K. C., and Burpitt, R. D., *J. Org. Chem.*, **30**, 2564 (1965).
44. Hartzler, H. D., *J. Am. Chem. Soc.*, **83**, 4990, 4997 (1961).
45. Landor, P. D., and Landor, S. R., *J. Chem. Soc.*, 1015 (1956).
46. Bramwell, A. F., Crombie, L., and Knight, M. H., *Chem. Ind. (London)*, 1265 (1965).
47. Meisters, A., and Swan, J. M., *Australian J. Chem.*, **18**, 155 (1965); *Chem. Abstr.*, **62**, 13173 (1965).
48. Chauvelier, J., *Ann. Chim. (Paris)*, **3**, 393 (1948),
49. Petrov, A. A., and Stadnichuk, M. D., *Zh. Obshch. Khim.*, **30**, 2243 (1960); *Chem. Abstr.*, **55**, 13301 (1961).
50. Petrov, A. A., and Yakovleva, T. V., *Zh. Obshch. Khim.*, **30**, 2238 (1960); *Chem. Abstr.*, **55**, 13301 (1961).
51. Perepelkin, O. V., Kormer, V. A., and Bal'yan, Kh. V., *Zh. Obshch. Khim.*, **35**, 957 (1965); *Chem. Abstr.*, **63**, 9797 (1965).
52. Kormer, V. A., and Petrov, A. A., *Zh. Obshch. Khim.*, **30**, 216 (1960); *Chem. Abstr.*, **54**, 22313 (1960).
53. Carothers, W. H., and Berchet, G. J., U.S. Patent 2,073,363 (1937); *Chem. Abstr.*, **31**, 3503 (1937).
54. Carothers, W. H., Berchet, G. J., and Collins, A. M., *J. Am. Chem. Soc.*, **54**, 4066 (1932).
55. Vartanyan, S. A., Badanyan, Sh. O., and Agababyan, R. G., *Armyansk. Khim. Zh.*, **19**, 66 (1966); *Chem. Abstr.*, **65**, 2122 (1966).
56. Petrov, A. A., and Kormer, V. A., *Zh. Obshch. Khim.*, **30**, 216 (1960).
57. Hartzler, H. D., *J. Am. Chem. Soc.*, **83**, 4990, 4997 (1961).
58. Petrov, A. A., and Kormer, V. A., *Zh. Obshch. Khim.*, **30**, 1056 (1960); *Chem. Abstr.*, **55**, 358 (1961).
58a. Hoffmann, R. W., and Bressel, U., *Angew Chem. Internat. Ed.*, **6**, 808 (1967).
59. Prevost, C., and Smadja, W., *Compt. Rend.*, **255**, 948 (1962).
60. Smadja, W., *Ann. Chim. (Paris)*, **10**, 105 (1965); *Chem. Abstr.*, **63**, 6834 (1965).
61. Smadja, W., *Compt. Rend.*, **256**, 2426 (1963).
62. Mikalajczak, K. L., *et al.*, *J. Org. Chem.*, **29**, 318 (1964).
63. Bertrand, M., *Compt. Rend.*, **247**, 824 (1958).
64. Jacobs, T. L., *et al.*, *J. Org. Chem.*, **17**, 475 (1952); **22**, 1424 (1957).
65. Jacobs, T. L., Dankner, D., and Singer, S., *Tetrahedron*, **20**, 2177 (1964).
66. Cram, D. J., "Fundamentals of Carbanion Chemistry," p. 189, New York, Academic Press, 1965.
66a. Krespan, C. G., *Tetrahedron*, **23**, 4243 (1967).
67. Cymerman-Craig, J., and Moyle, M., *Proc. Chem. Soc.*, 283 (1962).
68. van Boom, J. H., *et al.*, *Rec. Trav. Chim.*, **84**, 31 (1965).
69. Brandsma, L., Jonker, C., and Berg, M. H., *Rec. Trav. Chim.*, **84**, 560 (1965).
69a. Sevin, A., and Chodkiewicz, W., *Tetrahedron Letters*, 2975 (1967).
70. Jacobs, T. L., Teach, E. G., and Weiss, D., *J. Am. Chem. Soc.*, **77**, 6254 (1955).
71. Bailey, W. J., and Pfeifer, C. R., *J. Org. Chem.*, **20**, 95 (1955).
72. Serratosa, F., *Tetrahedron Letters*, 895 (1964).

72a. Brandsma, L., and Arens, J. F., *Rec. Trav. Chim.*, **86**, 734 (1967).

72b. Jacobs, T. L., and Prempree, P., *J. Am. Chem. Soc.*, **89**, 6177 (1967).

73. Huntsman, W. D., DeBoer, J. A., and Woosley, M. H., *J. Am. Chem. Soc.*, **88**, 5846 (1966).

74. Mkryan, G. M., Mndzhoyan, Sh. L., and Gasparyan, S. M., *Armyansk. Khim. Zn.*, **19**, 37 (1966); *Chem. Abstr.*, **65**, 2112 (1966).

75. Bailey, W. J., and Pfeifer, C. R., *J. Org. Chem.*, **20**, 1337 (1955).

76. Sevin, A., Chodkiewicz, W., and Cadiot, P., *Tetrahedron Letters*, 1953 (1965).

77. Pasternak, Y., and Traynard, J.-C., *Bull. Soc. Chim. France*, 356 (1966).

78. Favre, E., and Gaudemar, M., *Compt. Rend.*, **262C**, 1332 (1966).

79. Gaudemar, M., and Travers, S., *Compt. Rend.*, **262C**, 139 (1966).

79a. Dickinson, W. B., and Lang, P. C., *Tetrahedron Letters*, 3035 (1967).

80. Kurtz, P., Gold, H., and Disselnkötter, H., *Ann. Chem.*, **624**, 1 (1959).

81. Smith, L. I., and Swenson, J. S., *J. Am. Chem. Soc.*, **79**, 2962 (1957).

82. Greaves, P. M., Landor, S. R., and Laws, D. R. J., *Chem. Commun.*, 321 (1965).

83. Jacobs, T. L., and Petty, W. L., *J. Org. Chem.*, 28, **1360** (1963).

83a. Jacobs, T. L., *et al.*, *J. Org. Chem.*, **32**, 2283 (1967).

84. Jacobs, T. L., and Fenton, D. M., *J. Org. Chem.*, **30**, 1808 (1965).

85. Bhatia, Y. R., Landor, P. D., and Landor, S. R., *J. Chem. Soc.*, 24 (1959).

86. Hennion, G. F., and DiGiovanna, C. V., *J. Org. Chem.*, **30**, 3696 (1965).

87. Cherbuliez, E., *et al.*, *Helv. Chim. Acta*, **48**, 632 (1965).

88. Boiselle, A. (to Lubrizol Corp.), U.S. Patent 3,189,636 (June 15, 1965); *Chem. Abstr.*, **63**, 18155 (1965).

89. Kai, F., and Seki, S., *Chem. Pharm. Bull.* (*Tokyo*), **13**, 1374 (1965); *Chem. Abstr.*, **64**, 6475 (1966); *Tetrahedron*, **24**, 415 (1968).

90. Michel, E., and Troyanowsky, C., *Compt. Rend.*, **262C**, 1705 (1966).

91. Baker, C. S. L., Landor, P. D., and Landor, S. R., *J. Chem. Soc.*, 4659 (1965).

91a. Dear, R. E. A., and Gilbert, E. F., *J. Org. Chem.*, **33**, 819 (1968).

92. Slobodin, Ya. M., and Khitrov, A. P., *Zh. Obshch. Khim.*, **31**, 3945 (1961).

93. Ziegler, K., *Ann. Chem.*, **434**, 34 (1923); *Chem. Ber.*, **54**, 3003 (1921).

94. Ziegler, K., and Ochs, C., *Chem. Ber.*, **55**, 2257 (1922).

95. Banks, R. E., Haszeldine, R. N., and Taylor, D. R., *J. Chem. Soc.*, 978 (1965).

96. Jacobs, T. L., and Bauer, R. S., *J. Am. Chem. Soc.*, **81**, 606 (1959).

97. Fischer, H., and Fischer, H., *Chem. Ber.*, **97**, 2975 (1964).

98. Muehlstaedt, M., and Graefe, J., *Z. Chem.*, **6**, 69 (1966); *Chem. Abstr.*, **64**, 17440 (1966).

98a. Walbrick, J. M., Wilson, J. W., Jr., and Jones, W. M., *J. Am. Chem. Soc.*, **90**, 2895 (1968).

99. Landor, S. R., *et al.*, *J. Chem. Soc.* (*C*), 1223 (1966).

100. Landor, P. D., and Landor, S. R., *Proc. Chem. Soc.*, 77 (1962).

101. Landor, P. D., and Landor, S. R., *J. Chem. Soc.*, 2707 (1963).

102. Baker, C. S. L., *et al.*, *J. Chem. Soc.*, 4348 (1965).

103. Shiner, R. V., Jr., and Humphrey, J. S., Jr., *J. Am. Chem. Soc.*, **89**, 622 (1967).

104. Battioni, J.-P., and Chockiewicz, W., *Compt. Rend.*, **263C**, 761 (1966).

105. Hennion, G. F., and Nelson, K. W., *J. Am. Chem. Soc.*, **79**, 2142 (1957).

106. Hennion, G. F., and Hanzel, R. S., *J. Am. Chem. Soc.*, **82**, 4908 (1960).

107. Hennion, G. F., and Teach, E. G., *J. Am. Chem. Soc.*, **75**, 1653 (1953).

108. Hennion, G. F., and DiGiovanna, C. V., *J. Org. Chem.*, **30**, 2645 (1965).
109. *Ibid.*, **31**, 1977 (1966).
110. LeQuan, M., and Cadiot, P., *Bull. Soc. Chim. France*, 45 (1965).
110a. Masson, J.-C., LeQuan, M., and Cadiot, P., *Bull. Soc. Chim. France*, 777 (1967).
111. Pourcelot, G., and Cadiot, P., *Bull. Soc. Chim. France*, 3016, 3024 (1966).
112. Cherbuliez, E., *et al.*, *Helv. Chim. Acta*, **49**, 2395 (1966).
113. Ignat'ev, V. M., Ionin, B. I., and Petrov, A. A., *Zh. Obshch. Khim.*, **36**, 1505 (1966); *Chem. Abstr.*, **66**, 1076 (1967).
114. Bertrand, M., *Compt. Rend.*, **244**, 1790 (1957).
114a. Queroix-Travers, S., and Gaudemar, M., *Bull. Soc. Chim. France*, 355 (1967).
115. Thompson, B. (to Eastman Kodak Co.), U.S. Patent 3,236,869 (Feb. 22, 1966); *Chem. Abstr.*, **64**, 17428 (1966).
116. Petrov, A. A., Kormer, V. A., and Savich, I. G., *Zh. Obshch. Khim.*, **30**, 3845 (1960); *Chem. Abstr.*, **55**, 20923 (1961).
117. Kheruze, Yu. I., and Petrov, A. A., *Zh. Obshch. Khim.*, **31**, 2559 (1961); *Chem. Abstr.*, **56**, 14121 (1962).
118. Petrov, A. A., Stadnichuk, M. D., and Kheruze, Yu. I., *Dokl. Akad. Nauk SSSR*, **139**, 1124 (1961); *Chem. Abstr.*, **56**, 494 (1962).
119. Kheruze, Yu. I., and Petrov, A. A., *Izv. Vysshikh Uchebn. Zavedenii Khim. i Khim. Tekhnol.*, **6**, 170 (1963).
120. vanMeeteren, H. W., and van der Plas, H. C., *Tetrahedron Letters*, 4517 (1966).
121. Hartzler, H. D., *J. Am. Chem. Soc.*, **81**, 2024 (1959).
122. Cadiot, P., *Ann. Chim. (Paris)*, **1**, 214 (1956).
123. Hartzler, H. D., *J. Org. Chem.*, **29**, 1311 (1964).
124. Hennion, G. F., and Maloney, D. E., *J. Am. Chem. Soc.*, **73**, 4735 (1951).
125. Black, D. K., and Landor, S. R., *J. Chem. Soc.*, 5225 (1965).
126. Landor, P. D., Landor, S. R., and Pepper, E. S., *J. Chem. Soc.* (*C*), 185 (1967).
127. Landor, S. R., Pepper, E. S., and Regan, J. P., *J. Chem. Soc.* (*C*), 189 (1967).

PART TWO

Acetylene-Allene Isomerations

1. ISOMERIZATIONS USING ALKALI METAL AMIDES

1.1. Alkynes

Potassium alcoholates isomerize 1-alkynes to 2-alkynes, and metals such as sodium isomerize 2-alkynes to 1-alkynes above 100°. The metal reacts with the 1-alkyne to form sodioalkyne and hydrogen. Sodamide isomerizes 2-alkynes to 1-alkynes at 150°, but cannot reduce the 1-alkyne. Sodamide isomerizes 1,2-alkadienes to 1-alkynes, and isomerizes 2-alkynes to 1-alkynes, probably via the 1,2-alkadiene.

TABLE 1-8
Isomerization of 2-Octyne with Sodamide at 125 and 175°[1]

Temperature (°C)	Time (hr)	Composition (%) 2-Octyne	1,2-Octadiene	1-Octyne
125	3	97.5	1.4	1.1
	4	96.8	1.6	1.6
	10	89.5	4	7
	24	61.2	14.6	24.2
	35	42.5	18	39.5
	48	23.5	19	57.5
	60	9	17.5	73.5
175	3	69.5	8	22.5
	4	62.5	10	27.5
	10	35.5	19.5	45
	16	19.5	24	56.5
	30	7	19	74
	48	1.5	4.5	94
	60	0	~1	99

Bainvel[1] isomerized 2-octyne, 3-octyne, and 1,2- and 2,3-octadiene with molar quantities of sodamide above 100°. The results in Table 1-8 were obtained with 2-octyne, and are typical. 1,2-Octadiene isomerized even at 50°, forming 99% 2-octyne after 48 hours, which slowly rearranged to 1-octyne. At 200°, the product was 97% 1-octyne after only 6 hours. 2,3-Octadiene isomerized slowly at 100° to give mostly 2-octyne, which decreased with time as 1-octyne increased. Thus, allenes are intermediates in the isomerization of acetylenes by sodamide. The allenes isomerize rapidly to an acetylene with the triple bond closer to the end of the chain:

$$2,3\text{-octadiene} \longrightarrow 2\text{-octyne} \longrightarrow 1,2\text{-octadiene} \longrightarrow 1\text{-octyne}$$

The equilibrium is disturbed to favor 1-alkyne because the 1-alkyne reacts with sodamide to form sodium alkynylide. These results are consistent with Jacob's isomerization mechanism:[2]

$$R-CH=C=CH-CH_3$$
$$\updownarrow$$
$$R-CH_2-C\equiv C-CH_3 \rightleftharpoons (R-\overset{..}{C}H-C\equiv C-CH_3 \longleftrightarrow R-CH=C=\overset{..}{C}-CH_3)$$
$$R-CH_2-C\equiv C-CH_3 \rightleftharpoons (R-CH_2-C\equiv C-\overset{..}{C}H_2 \longleftrightarrow R-CH_2-\overset{..}{C}=C=CH_2)$$
$$\updownarrow$$
$$R-CH_2-CH=C=CH_2$$
$$\updownarrow$$
$$R-CH_2-CH_2-C\equiv CH \rightleftharpoons (R-CH_2-\overset{..}{C}H-C\equiv CH \longleftrightarrow$$
$$R-CH_2-CH=C=\overset{..}{C}H)$$

With catalytic quantities of sodamide in ammonia at 25°, terminal acetylenes isomerize to nonterminal acetylenes.[3] Here the 1-alkyne is the excess reagent, so it cannot be converted completely to sodium alkynylide. 1-Hexyne with 0.2 mole of sodamide after 94 hours was converted to a mixture containing 0.5% 1,2-hexadiene, 71% 2-hexyne, and 22% unreacted 1-hexyne.

1.2. Conjugated Enynes

Alkali amide isomerization of conjugated enynes has been studied little. Petrov[4] reported the isomerization of 1,3-pentenyne to sodium pentenylide:

$$MeC \equiv CCH = CH_2 \xrightarrow{NaNH_2, NH_3} CH_2^- C \equiv C - CH = CH_2 \longleftrightarrow CH_2 = C = C = CHCH_2^-$$
$$\longrightarrow \bar{C} \equiv CCH = CHCH_3 \xrightarrow{+RI} RC \equiv CCH = CHCH_3$$

Note that a 3-cumulene is shown as one intermediate.

Later, van Boom[5] reported extensive studies on the isomerization of conjugated enynes. The reaction steps in one example are:

$$HC \equiv CCH = CH - n\text{-amyl} \xrightarrow[\text{(2) MeI in NH}_3]{\text{(1) LiNH}_2} CH_3C \equiv CCH = CH - n\text{-amyl} \xrightarrow[\text{(2) H}_2\text{O}]{\text{(1) KNH}_2}$$
$$HC \equiv CCH = CHCH_2 - n\text{-amyl}$$
$$(81\%)$$

In these reactions, the methylated enyne was added to excess KNH_2 in ammonia. After the reaction, the ammonia was evaporated, and ether or pentane was added. The mixture was then hydrolyzed. If all the ammonia was not removed, hydrolysis gave an impure product which contained conjugated ene-allene groups. No evidence for cumulenes was obtained. Cumulenes cannot be excluded however, because one mesomeric form of the cumulene anion could react with water to form the conjugated ene-allene system detected by infrared. Generally, the *cis/trans* ratio in the starting material and the end product was quite different. Thus, the *trans* form of ⬡—$CH_2CH = CHC \equiv CH$ methylated and isomerized to a product which was at least 90% *cis*.

Weakly basic phenyllithium at 0° also isomerizes conjugated enynes:[7] $PhCH_2CH = CHC \equiv CMe \longrightarrow cis$- and $trans$-$PhCH_2CH_2CH = CHC \equiv CH$. Diynes isomerize in an analogous way. Excess sodamide in ammonia isomerizes 2,4-hexadiyne to 1,3-hexadiyne. The product 1,3-hexadiyne is easily polymerized: when boiled at atmospheric pressure, it polymerizes explosively with "a phenomenon of light."[8]

1.3. Enynols

The course of the lithium amide isomerization of 6-methyl-l-octen-4-yn-6-ol was followed by GLC, NMR and infrared:[9]

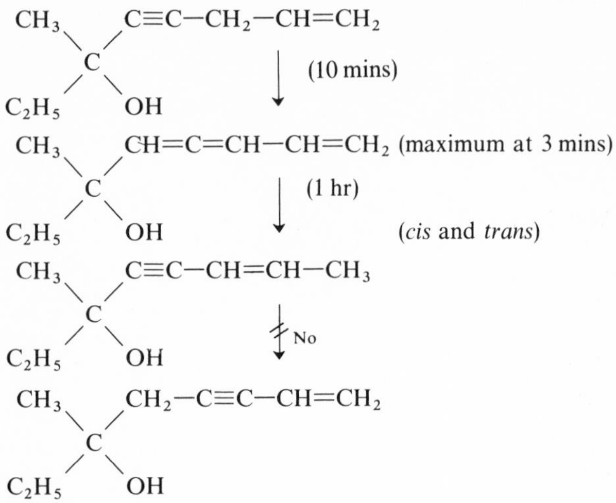

$$\underset{C_2H_5}{\overset{CH_3}{>}}\!\!C\!\!<\underset{OH}{\overset{C\equiv C-CH_2-CH=CH_2}{}}$$

↓ (10 mins)

$$\underset{C_2H_5}{\overset{CH_3}{>}}\!\!C\!\!<\underset{OH}{\overset{CH=C=CH-CH=CH_2}{}} \text{(maximum at 3 mins)}$$

↓ (1 hr)

(*cis* and *trans*)

$$\underset{C_2H_5}{\overset{CH_3}{>}}\!\!C\!\!<\underset{OH}{\overset{C\equiv C-CH=CH-CH_3}{}}$$

↓ No

$$\underset{C_2H_5}{\overset{CH_3}{>}}\!\!C\!\!<\underset{OH}{\overset{CH_2-C\equiv C-CH=CH_2}{}}$$

A derivative in which the allylic-propargylic hydrogens were replaced by butyl groups did not isomerize:

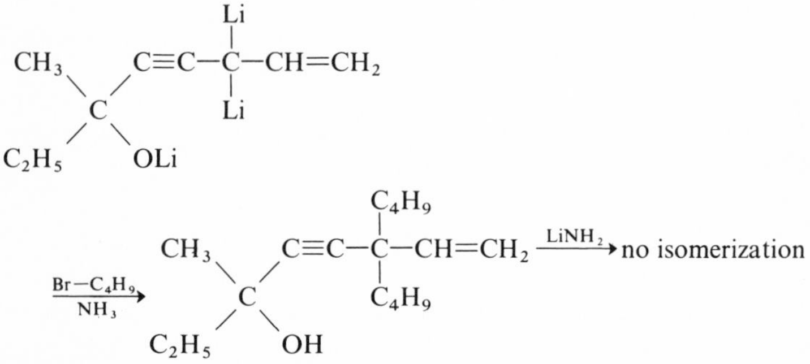

$$\underset{C_2H_5}{\overset{CH_3}{>}}\!\!C\!\!<\underset{OLi}{\overset{C\equiv C-\underset{Li}{\overset{Li}{C}}-CH=CH_2}{}}$$

$$\xrightarrow[NH_3]{Br-C_4H_9}\quad \underset{C_2H_5}{\overset{CH_3}{>}}\!\!C\!\!<\underset{OH}{\overset{C\equiv C-\underset{C_4H_9}{\overset{C_4H_9}{C}}-CH=CH_2}{}}\xrightarrow{LiNH_2}\text{no isomerization}$$

The isomerization requires at least one propargylic hydrogen, which is abstracted by lithium amide to form an intermediate carbanion:

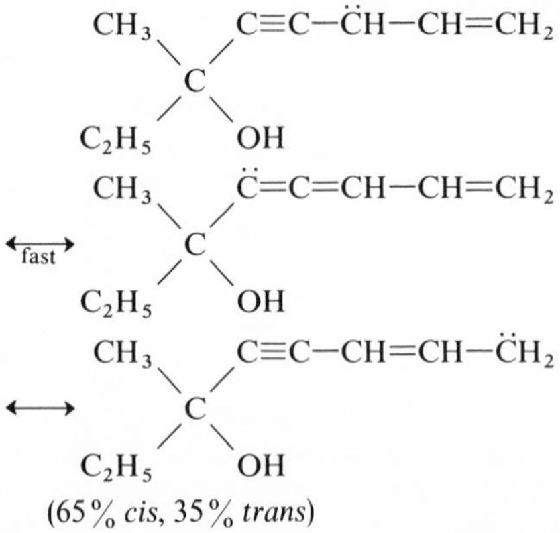

$$\underset{C_2H_5}{\overset{CH_3}{>}}\!\!C\!\!<\underset{OH}{\overset{C\equiv C-\overset{..}{C}H-CH=CH_2}{}}$$

⇄ fast

$$\underset{C_2H_5}{\overset{CH_3}{>}}\!\!C\!\!<\underset{OH}{\overset{\overset{..}{C}=C=CH-CH=CH_2}{}}$$

⟷

$$\underset{C_2H_5}{\overset{CH_3}{>}}\!\!C\!\!<\underset{OH}{\overset{C\equiv C-CH=CH-\overset{..}{C}H_2}{}}$$

(65% *cis*, 35% *trans*)

1.4. α-Acetylenic Acids

α-Acetylenic acids which have a γ-hydrogen atom are isomerized by sodamide in ammonia to the corresponding allenic acid in quantitative yield :[10]

$$HO_2CC\equiv CCH_2-amyl \longrightarrow HO_2CCH=C=CH-amyl$$

α-Allenic acids are often difficult to make by other methods.

1.5. Propynyl Ethers to Propargyl Ethers

Propynyl ethers ($CH_3C\equiv COR'$) isomerize easily to propargyl ethers ($HC\equiv CCH_2OR'$) in the presence of 1 or 2 equivalents of sodamide in ammonia.[11] Homologous 1-alkynyl ethers, $RCH_2CH_2C\equiv COR'$, and 2-alkynyl ethers, $RCH_2C\equiv CCH_2OR'$, treated with 2 moles of sodamide in ammonia give an enyne plus R'OH. If only 1 mole of sodamide is used, half of the starting ether remains unchanged.[12] The reaction may involve intermediate cumulenes: Brandsma noted a "lively play of intense colors"[12] in the reaction mixtures. The product enynes are *cis* and *trans* mixtures.

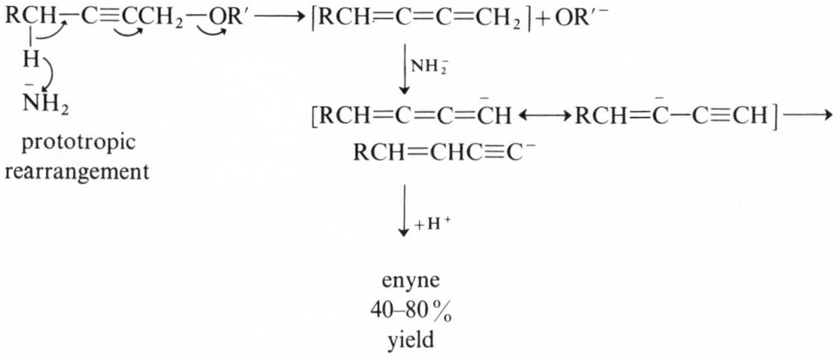

$$RCH_2-C\equiv CCH_2-OR' \longrightarrow [RCH=C=C=CH_2]+OR'^-$$

H

NH_2

prototropic
rearrangement

$$\downarrow NH_2^-$$

$$[RCH=C=C=\overset{-}{C}H \longleftrightarrow RCH=\overset{-}{C}-C\equiv CH] \longrightarrow$$
$$RCH=CHC\equiv C^-$$

$$\downarrow +H^+$$

enyne
40–80%
yield

In 2-alkynyl ethers with branching on the carbon atom in position 4, the elimination of alcohol and formation of enyne is slow, and products containing 85% or more of allenic ether are formed in 75–90% yields.[13] 2-Alkynyl ethers with no hydrogen on the carbon atom in position 4 give nearly quantitative yield of allenyl ethers:

$$RR'R''CC\equiv CCH_2OEt \xrightarrow{\text{NaNH}_2,\ \text{NH}_3} RR'R''CCH=C=CHOEt$$
$$(100\%)$$

Allenyl ethers are valuable reagents. For example, with dilute sulfuric acid they hydrolyze to *trans*-α,β-unsaturated aldehydes:

$$t\text{-BuCH}=C=CHOEt \longrightarrow t\text{-BuCH}=CH-CHO \text{ (good yields)}$$

1.6. Alkynyl Thioethers

Allenyl thioethers were first described in 1961 by Pourcelot[14] who prepared the products $CH_2{=}C{=}\overset{\overset{\textstyle R'}{|}}{C}{-}SR$, where R' = H or alkyl. These allenyl thioethers are intermediates or end products from isomerization of propargyl thioethers, $HC{\equiv}C\overset{\overset{\textstyle R'}{|}}{C}H{-}SR$, by sodium ethoxide or hydroxide in ethanol at 20–80°. The solutions of allenyl thioethers slowly isomerize to the propargyl thioethers. The progress of the reaction can be followed by infrared. The allene absorbs at 1940 cm^{-1}, and the propargyl form at 3300 cm^{-1}.

1-Alkynyl thioethers are isomerized readily into the allenyl thioethers by lithium amide or sodium amide in ammonia.[15] Alkylthio-1-propynes isomerize very rapidly to allenyl thioethers:

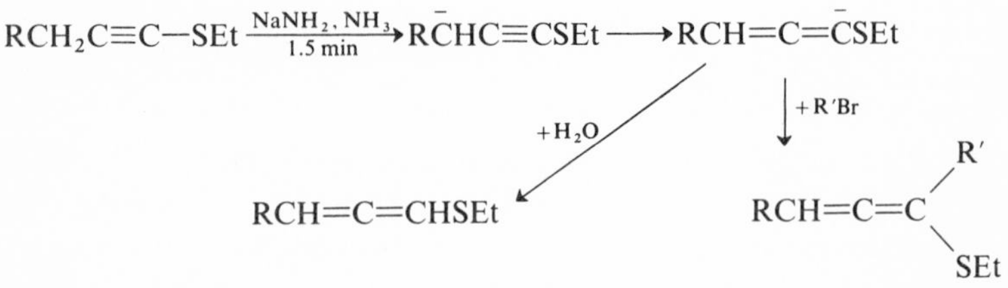

$$RCH_2C{\equiv}C{-}SEt \xrightarrow[\text{1.5 min}]{\text{NaNH}_2, \text{NH}_3} RCHC{\equiv}\overset{-}{C}SEt \longrightarrow RCH{=}C{=}\overset{-}{C}SEt$$

with branches: $+H_2O$ giving $RCH{=}C{=}CHSEt$, and $+R'Br$ giving $RCH{=}C{=}C\underset{\textstyle SEt}{\overset{\textstyle R'}{\diagup}}$

Allenyl thioethers are colorless liquids with a peculiar odor. They react rapidly with traces of oxygen from air. They form viscous liquids when shaken with oxygen.

1.7. N-2-Propynyl Heterocycles to N-Substituted Allenes

Potassium amide on alumina causes rapid isomerization of N-2-propynyl heterocycles to N-substituted allenes. Very little N-1-propynyl by-product is obtained[15a]:

$$\underset{1}{RCH_2{\cdot}C{\equiv}CH} \rightleftharpoons \underset{2}{RCH{=}C{=}CH_2} \rightleftharpoons \underset{3}{RC{\equiv}C{\cdot}CH_3}$$

R	% Yield	
	2	3
Pyrrol-1-yl	20	0
Imidazol-1-yl	40	0
Pyrazol-1-yl	20	0
Indol-1-yl	14	6
Carbazol-9-yl	35	15

2. ISOMERIZATIONS USING ALKALI METAL HYDROXIDES AND ALKOXIDES

2.1. Acetylene-Allene-Conjugated Diene Interconversions

Prevost and Smadja[16] equilibrated 1-heptyne by heating with 0.47N sodium t-butoxide in t-butanol at 200° for 7 hours. The products were 2-heptyne, 2,3-heptadiene, 3-heptyne, and 3,4-heptadiene. The amount of 3-heptyne increased as temperature increased from 115–250°. Smadja[17,18] was able to prepare all of the possible allenes and acetylenes from a given starting allene or acetylene by simple base isomerization. 1-Decyne isomerized completely at 160° to give mostly 2-decyne. At 240° the product was mostly 5-decyne. Smadja worked extensively on the twelve position isomers of heptyne and heptadiene. Some conjugated dienes formed irreversibly. Equilibration at 196° for 20 hours gave 1-heptyne, 1,2-heptadiene, 40% 2-heptyne, 9.6% 2,3-heptadiene, 44.5% 3-heptyne, and 5.7% 3,4-heptadiene.

Potassium t-butoxide in dimethyl sulfoxide isomerizes 1-hexyne to a mixture of conjugated hexadienes.[19] After 92 hours at 72°, the product contained 2.3% 1,3-hexadiene, 34.1% trans, trans-2,4-hexadiene, 52% cis, trans-2,4-hexadiene, and 10.2% 3-hexyne. At 25°, the same system isomerized 2-dialkylaminopropynes to dialkylaminobutadienes. The starting propargylamines are available from the cuprous-catalyzed reaction of acetylene with secondary amines.

2.2. Arylpropynes

The first attempt in 1922 to prepare triarylallenes by usual dehydrohalogenation methods probably gave the correct products, but they dimerized during molecular weight determinations.[20] Jacobs[21] succeeded in preparing 1,2-diarylallenes by base isomerization of 1,3-diarylpropynes, using alumina, sodium hydroxide or potassium hydroxide. In 1964, Jacobs[22] reported the first preparation of a triarylallene:

$$PhC\equiv CCHPh_2 \xrightarrow[RT]{Al_2O_3} PhCH=C=CPh_2.$$
$$(89\%)$$

3-(p-Biphenyl)-1,3-diphenylpropyne did not isomerize over alumina. Over alumina impregnated with 20% NaOH, it isomerized to form 62% yield of moderately pure 1-(p-biphenyl)-1,2-diphenylallene. A few drops of strong acid added to a glacial acetic acid solution of triphenylallene gave the dimer reported in 1922 (see above).[20]

1-Phenylallene, 1-phenyl-1-propyne, and 1-phenyl-2-propyne form when propargyl bromide reacts with phenylmagnesium bromide in ether. Refluxing the mixture for 10 minutes with KOH in tetrahydrofuran gives 82% yield of 1-phenyl-1-propyne, 95% pure.[23,24]

2.3. Intramolecular Proton Transfer in Isomerizations

Cram[25] studied the degree of intramolecular versus intermolecular proton transfer in allylic rearrangements. Intramolecular proton transfer was determined by rearranging the allylic compounds in deuterated solvents in the presence of a base, such as alumina, potassium t-butoxide or amines. Cram used this technique to study the propargyl-allene system; he concluded that the rearrangement is like the allylic rearrangement except that the three carbons involved in proton transfers are linear and are farther apart than in allylic rearrangements.

Cram has proposed a "conducted tour" mechanism:

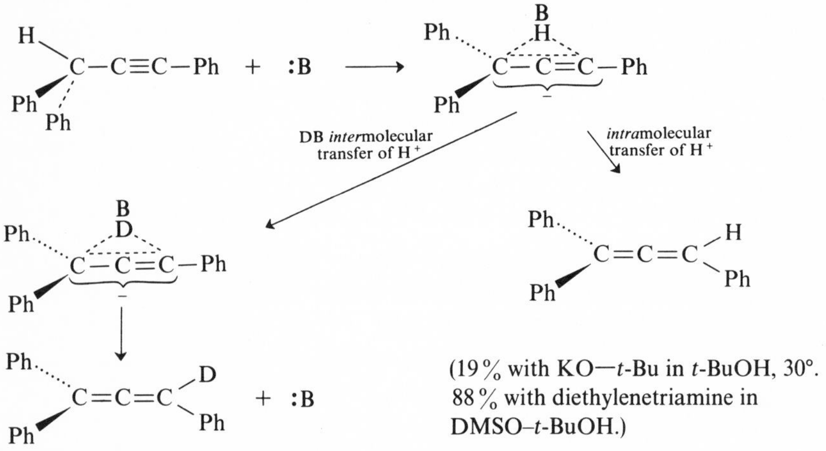

(19% with KO—t-Bu in t-BuOH, 30°.
88% with diethylenetriamine in
DMSO–t-BuOH.)

2.4. 1,4-Enynes to Conjugated Ene-Allenes

Rearrangement of *trans*-1,4-enynes by alkali usually gives conjugated ene-allenes instead of conjugated trienes. Mikolajczak[26] isolated a new poly-unsaturated fatty acid, crepenynic acid, from the seed oil of *Crespis foetida*, and showed it to be *cis*-9-octadecen-12-ynoic acid. Isomerization by KOH in ethylene glycol gave 70% yield of an 8,10,12-octadecatrienoic acid,[27] and 30% yield of isomeric conjugated trienoic acids. Potassium t-butoxide rearranged crepenynic acid to the ene-allene, which isomerized to a conjugated trienoic acid when heated:

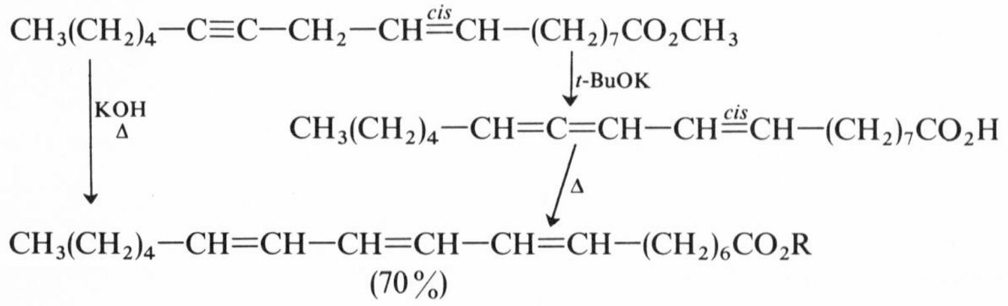

The proposed mechanism is:

$$-C \overset{12}{\equiv} C - CH_2 - CH = \overset{9}{C}H - CH_2 - \xrightarrow{-H^+} \left[\overset{\delta^-}{C} \equiv C = \overset{\delta^-}{C}H = CH = \overset{\delta^-}{C}H - CH_2 \right] \xrightarrow{+H^+}$$

$$-CH = \overset{12}{C} = CH - CH = \overset{9}{C}H - CH_2 - \xrightarrow{\Delta}
\left[
\begin{array}{c}
CH \\
\parallel \\
{}_{12}C \diagdown {}^{H} \diagdown CH \\
\vdots \qquad \vdots \\
HC \diagup {}_{C} \diagup CH{}_{9} \\
\quad H
\end{array}
\right] \xrightarrow{\Delta}$$

$$-CH = CH - CH = CH - CH = CH -$$

trans, cis, trans and *trans, cis, cis*

The enyne loses a proton to form a hybrid carbanion which adds a proton to form the conjugated ene-allene. The enyne to ene-allene step is base catalyzed, since rearrangement does not occur with heat alone. The ene-allene to conjugated triene step is thermal, and the newly formed Δ^{10} double bond is *cis*. A similar thermal rearrangement at 115° of a conjugated ene-allene to a conjugated triene with a *cis* center double bond was noted by Crowley.[28] Isomerization of allylacetylene in MeOH-NaOH involves a similar hybrid anion, $HC \equiv C - \overline{CHCH = CH_2}$ in which the negative charge resonates between carbons 1 and 3. If methanol attacks at carbon 1, the product is vinylallene. If methanol attacks at carbon 3, allylacetylene is regenerated.[29]

2.5. 1,5-Enynes to Allenes and Conjugated Olefins

1,5-Enynecarboxylic acids rearrange to conjugated trienoic acids.[30] Sondheimer[31] studied the base-catalyzed rearrangement of linear 1,5-enynes to conjugated olefins, in order to develop a method applicable to the same grouping in macrocyclic rings. 1,5-Enyne and 1,5-diyne hydrocarbons can be prepared by heating halogen compounds with alcoholic KOH around 100°, but heating to 125–175° gives equilibrium mixtures of 1-alkynes, 1,2-allenes and 2-alkynes. Linear 1,5-enynes and 1,5-diynes isomerize best in the presence of potassium *t*-butoxide in *t*-butanol. These prototropic rearrangements probably go through the allenes. The preparation and rearrangement of 1-penten-5-yne is typical of Sondheimer's reactions:

$$CH_2 = CHCH_2CH_2CH = CH_2 \xrightarrow{1Br_2} BrCH_2CHBr - CH_2CH_2CH = CH_2 \xrightarrow[NH_3]{NaNH_2}$$

$$HC \equiv CCH_2CH_2CH = CH_2 + HC \equiv CCH = CHCH_2CH_3 \text{ (8\% impurity)}$$

$$\downarrow \begin{array}{l} + KO - t\text{-}Bu \\ t\text{-}BuOH, 65° \end{array}$$

$$H(CH = CH)_3H$$
$$(40\%)$$

Isomerization of *trans*-5-decene-1,9-diyne and the corresponding ene-allenyne gave the same product in the same yield. This is further evidence that the allene is intermediate in the acetylene rearrangement:

$$\left.\begin{array}{l} HC{\equiv}CCH_2CH_2CH{=}CHCH_2CH_2C{\equiv}CH \\ CH_2{=}C{=}CHCH_2CH{=}CHCH_2CH_2C{\equiv}CH \end{array}\right] \longrightarrow H(CH{=}CH)_5H$$
$$(16\%)$$

Coupling of 1-hexen-5-yne to dodeca-1,11-diene-5,7-diyne and isomerization gave 10% yield of fully conjugated dodecahexaene. A similar sequence starting with 1-hexen-5-yne coupled with 5-decene-1,9-diyne and followed by isomerization of the triyne from co-coupling gave 2.5% yield of fully conjugated hexadecaoctaene, an unstable yellow-orange solid.

The symmetrical coupling product from 5-decene-1,9-diyne rearranged to the fully conjugated C_{20} polyolefin. With this work, all of the conjugated polyenes $H(CH{=}CH)_nH$ up to $n = 10$ had been made, except the one where $n = 9$.

2.6. Non-Conjugated Diynes

2.6.1. 1,4-DIYNES

1,4-Nonadiyne treated with alkali at room temperature rearranged to 2,5-nonadiyne, via the conjugated ethynylallene,[32] which can be isolated. A detailed study of the rates of isomerization of 1-phenyl-1,4-pentadiyne, 1,5-diphenyl-1,4-pentadiyne, and 1-phenyl-1,4-nonadiyne catalyzed by ethoxide anion in ethanol showed that conjugated allenynes are intermediates.[33] The allenynes were not isolated, but were detected spectrally. The skipped diyne (1,4-diyne) and the allenyne disappeared completely during the isomerization. The reaction probably involves carbanions, and exchange reactions showed Cram's[25] "conducted tour" mechanism does not apply in this case:

Delocalization
Energies

$$C_2H_5O^- + C_2H_5OH$$
$$+$$

0

$$RC{\equiv}C{-}CH_2{-}C{\equiv}CR'$$
$$\downarrow\uparrow$$

1.46β

$$(R\overset{\cdots\cdots\cdots\cdots\cdots}{C{=}C{-}CH{-}C{=}CR'})^-$$
$$\downarrow\uparrow$$

0.47β

$$RCH{=}C{=}CH{-}C{\equiv}CR'$$
$$\downarrow\uparrow$$

1.94β

$$(RCH{-}C{=}C{-}C{=}CR)^-$$

0.94β $RCH{=}C{=}C{=}C{=}CHR$ (?) $RCH_2{-}C{\equiv}C{-}C{\equiv}CR'$

1.94β

$$(RCH{=}\overset{\cdots\cdots\cdots\cdots\cdots}{C{-}C}{=}C{-}CH{-}CHR'')^-\ (R' = R''CH_2)\ \left.\begin{array}{l}\text{further}\\ \text{possibili-}\end{array}\right.$$

0.99β $RCH{=}CH{-}C{\equiv}C{-}CH{=}CHR''$ $RCH{=}C{=}C{=}CH{-}CH{=}CHR''$ ties

2.6.2. 1,5-DIYNES

1,5-Diynes rearrange to polyenynes. For example, dipropargyl (1,5-hexadiyne) rearranged at 70° in the presence of potassium t-butoxide in t-butanol to form 1,3-hexadien-5-yne.[34] The isolated yield of this low-boiling, unstable product was 33%, while the yield by spectroscopic analysis was 65%. 1,5-Hexadiyne isomerized faster than 1-hexen-5-yne, and required less potassium t-butoxide.[35] This is an important property of 1,5-hexadiynes which proved very useful in rearrangements in macrocyclic compounds.

Refluxing alcoholic KOH isomerized 1,5-hexadiyne to 2,4-hexadiyne, in the classical rearrangement of terminal to nonterminal acetylenes. 1,5-Hexadiyne in refluxing potassium t-butoxide in 2,2′-dimethoxydiethyl ether formed a vinyl ether.[36] Under these conditions, other diynes cyclized to substituted benzenes.

2.7. Triynes and Tetraynes

Sondheimer[34] also isomerized a triyne, 1,5,9-decatriyne, using potassium t-butoxide in t-butanol.

$$HC\equiv CCH_2CH_2C\equiv CCH_2CH_2C\equiv CH \longrightarrow$$

$$27\%\begin{cases} CH_2{=}CHCH{=}CHC\equiv CCH{=}CHCH{=}CH_2 \\ \text{and/or} \\ CH_2{=}CHCH{=}CHCH{=}CHC\equiv CCH{=}CH_2 \end{cases}$$

$$+$$

$$20\%\ CH_2{=}CHCH{=}CHCH{=}CHCH{=}CHC\equiv CH$$

1,5,7,11-Dodecatetrayne isomerized to products which were more complex. Coupling of 1,5-hexadiyne gave the starting material in 40% yield. Isomerization by potassium t-butoxide in t-butanol gave at least 10% of $CH_2{=}CHCH{=}CHC\equiv CC\equiv CCH{=}CHCH{=}CH_2$.

2.8. Diacetylenes and Enynes to Aromatics

In 1954, Eglinton[37] tried to duplicate the rearrangement of hepta-1,6-diyne-4-carboxylic acid to m-toluic acid as reported in 1907 by Perkin. He obtained 100% yield of m-toluic acid when he boiled the acid in KOH solution.

Diyne hydrocarbons aromatized best in the presence of potassium t-butoxide in diglyme. 1,6-Heptadiyne formed toluene rapidly. With a series of long-chain diacetylenes $HC\equiv C(CH_2)_nC\equiv CH$, $n = 4, 5, 6, 10$ and 17, the products were mainly isomeric o-dialkylbenzenes, with total conversions around 65%. The acetylene groups do not have to be terminal: 2,7- and 1,8-nonadiyne gave the same proportions of n-propylbenzene and o-ethyltoluene. The acetylene groups can be conjugated: 3,5- and 1,7-octadiyne gave similar amounts of ethylbenzene and o-xylene. The decadiynes gave relatively complicated products, so Eglinton studied the simpler octadiynes in more detail.

1,7-Octadiyne gave eight products. Major products were ethylbenzene and o-xylene, along with m-xylene and p-xylene. Hydrogenation showed the presence of compounds with seven and eight-membered rings. Under milder isomerization conditions, two acyclic products were tentatively identified: 1,3,5,6-octatetraene and 2,6-octadiyne. If the diglyme-butoxide solutions were filtered to remove solid potassium t-butoxide, the products were much simpler. 1,7-Octadiyne gave a 9:5 mixture of o-xylene and ethylbenzene, and less than 1% of other products.

The simplest explanation for the reaction is a multiple prototropic rearrangement to form a cis-allene-diene

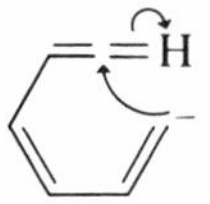

followed by either intramolecular Diels-Alder reaction or internal attack on the allene by a terminal carbanion, as shown above. This mechanism requires the diyne have at least seven carbons. 1,5-Hexadiyne under aromatization conditions did not form benzene.

Similar allene-olefins are intermediates in the thermal cyclization of alk-1-en-5-ynes.[38] 1-Hexen-5-yne at 340° gives three products:

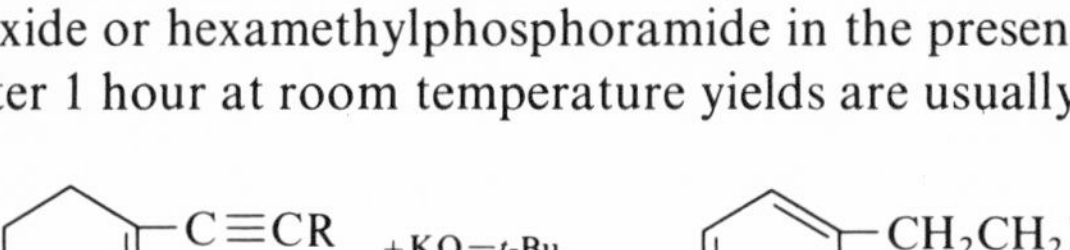

(8–67%)   (3–38%)   (2–30%)

1-Acetylenic-1-cyclohexenes isomerize easily to substituted benzenes in dimethyl sulfoxide or hexamethylphosphoramide in the presence of potassium t-butoxide. After 1 hour at room temperature yields are usually 70–90%:[39a]

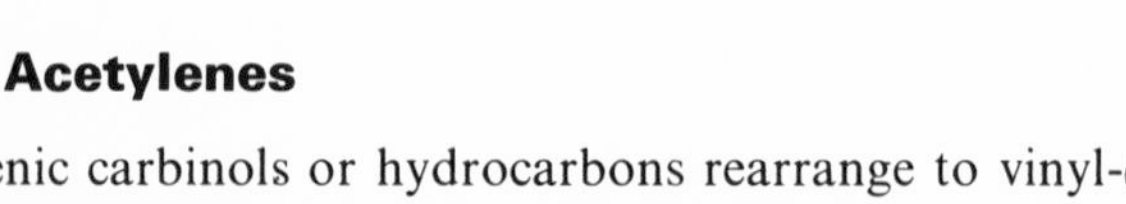

2.9. Allylic Acetylenes

Allylic acetylenic carbinols or hydrocarbons rearrange to vinyl-α-allenic products when refluxed in 5% NaOH in methanol.[39]

$$RC{\equiv}CCH_2CH{=}CHR' \longrightarrow RCH{=}C{=}CHCH{=}CHR'$$

The reaction applied to tertiary alcohols and to hydrocarbons gives 60–88% yields of vinylallenes.

2.10. Acetylenic Sulfides and Sulfones

Stirling[40] reported the first allenic sulfone in 1964. He prepared propynyl, propargyl, and allenyl sulfides and sulfones by a series of isomerizations:

$$PhSO_2C\equiv CMe \underset{}{\overset{NEt_3}{\rightleftarrows}} PhSO_2CH=C=CH_2 \underset{Al_2O_3}{\rightleftarrows} PhSO_2CH_2C\equiv CH$$

$$\uparrow oxidation \qquad\qquad\qquad\qquad\qquad\qquad\qquad\qquad \uparrow oxidation$$

$$PhSC\equiv CMe \rightleftarrows PhSCH=C=CH_2 \underset{KOH,\ 20\%}{\overset{THF}{\rightleftarrows}} PhSCH_2C\equiv CH \overset{base}{\longleftarrow}$$
$$(85\%)$$
$$PhSH + BrCH_2C\equiv CH$$

3. ISOMERIZATIONS USING ACID CATALYSTS

Acids catalyze the allene-to-conjugated diene isomerization. Phosphorus pentachloride in ether at $-10°$ converts α-allenic secondary alcohols to the α-allenic chlorides, in 65–70% yield. Thionyl chloride or excess concentrated HCl gives 60–70% yield of a mixture of α-allenic chloride and an equal amount of 2-chloro-1,3-diene $RCH=CHC=CH_2$ (with Cl substituent). HBr gives 85–95% yield of unrearranged allenic bromide plus 2-bromo-1,3-diene.[41]

Traces of acids catalyze the rearrangement of allenyl thioethers to conjugated dienyl thioethers:

$$RCH=C=C-CHR'R'' \overset{H^+}{\longrightarrow} RCH=CHC=CR'R''$$
(with SR''' substituents)

This is a general reaction when R, R' and R'' are H or alkyl, and R''' is alkyl; the best yields (70–97%) are obtained when R, R' and R'' are alkyl.[42] Allenyl oxygen ethers rearrange much faster, indicating that the rate-determining step is addition of a proton to the β-carbon of the allenyl ether.

Phosphoric acid or p-toluenesulfonic acid catalyze the reaction of t-ethynyl-carbinols with alkyl isopropenyl (or vinyl) ethers to form β-ketoallenes. The ketoallenes are isomerized by base to form α-keto conjugated dienes.[42a] This use of acids and bases as catalysts is the reverse of the normal usages described above. The reaction of 3-methyl-1-butyn-3-ol with methyl isopropenyl ether is typical. Yields are excellent in both steps:

4. ISOMERIZATION OF METAL DERIVATIVES OF ACETYLENES, ALLENES, AND CONJUGATED DIENES

The formation and reactions of metal derivatives of propargyl, allenyl and 1,3-dienyl systems furnish many additional examples of the easy interconversion of these unsaturated systems. Primary and secondary propargyl halides and allenyl halides react with magnesium in ether to give mostly the allenic Grignard reagent. Reaction with zinc or aluminum is similar. t-Propargylic chlorides and magnesium form at least four or five Grignards, the major one being allenic. 1,3-Diene bromides give Grignards which react with aldehydes to form allenic carbinols. 1-En-3-ynes add LiR to give lithium derivatives which react with carbonyl compounds or epoxides to form allenyl alcohols, or with water to give allenic hydrocarbons.

The reaction of propargyl Grignards with carbonyl compounds to form α-allenylcarbinols, and the opposite reaction of 1-allenyl Grignards to form β-acetylenic carbinols can be represented as:[43]

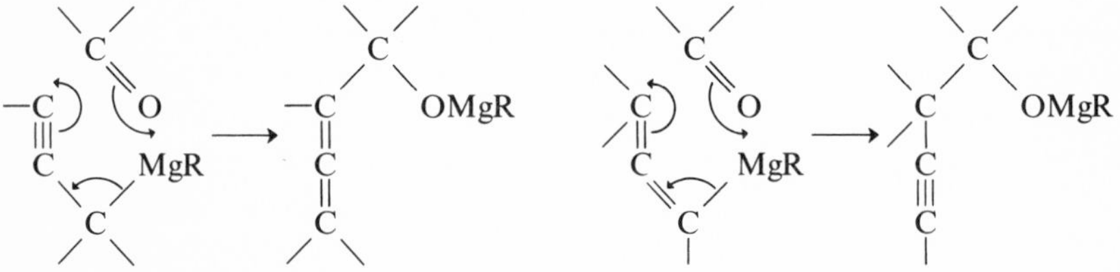

propargyl to allene   allene to β-acetylene

4.1. Grignard Reagents

Pasternak[44] reported some interesting interconversions of acetylenic, allenic and 1,3-dienic Grignard reagents. α-Acetylenic bromides or the isomeric 1-allenic bromides react with magnesium in ether to form the same allenic Grignard:

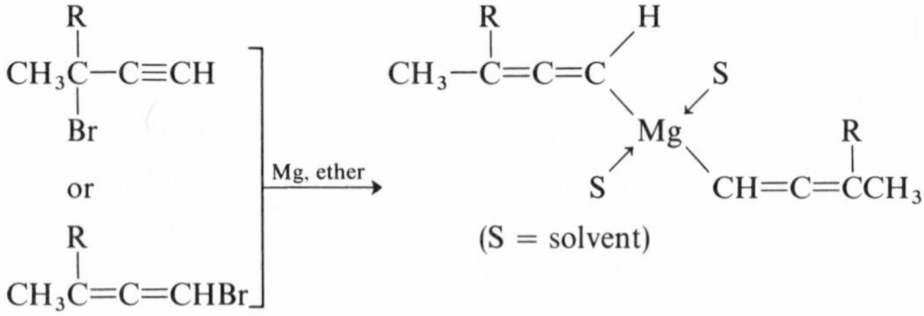

However, α-allenic bromides in *ether*, and 2-bromo-1,3-dienes in *tetrahydrofuran*, give the same 1,3-diene Grignard reagent:

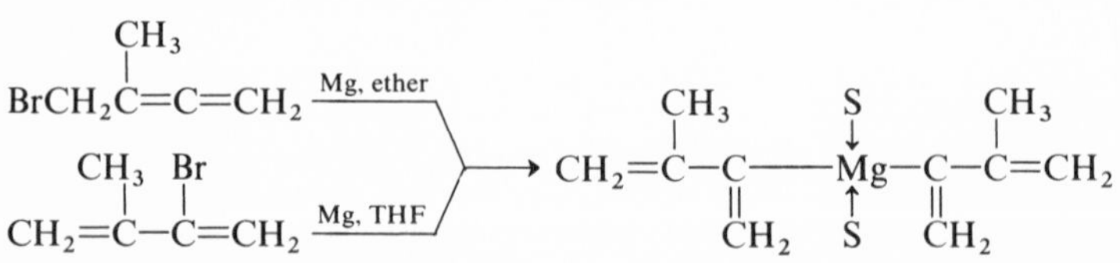

These 1,3-diene-2-Grignards react with carbonyl compounds to give 90 % yields of β-allenic carbinols:

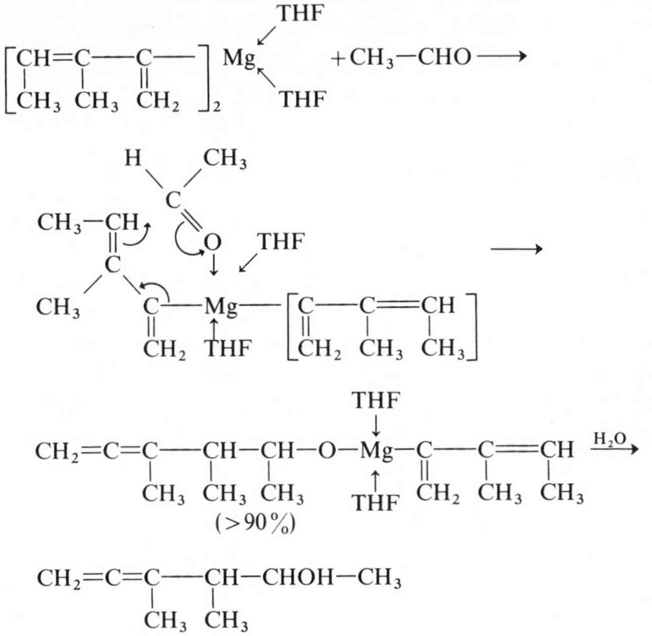

$$
\begin{bmatrix} CH=C—C— \\ |\quad |\quad || \\ CH_3\ CH_3\ CH_2 \end{bmatrix}_2 \ Mg\overset{THF}{\underset{THF}{\diagup}} \ +CH_3—CHO \longrightarrow
$$

$$
\longrightarrow
$$

$$
CH_2=C=C—CH—CH—O—Mg—C—C=CH \xrightarrow{H_2O}
$$
$$
\quad\quad |\quad\ |\quad\ |\quad\ \ \ \ \ |\ \ ||\quad\ |\quad\ |
$$
$$
\quad CH_3\ CH_3\ CH_3 \quad THF\ CH_2\ CH_3\ CH_3
$$
$$
(>90\%)
$$

$$
CH_2=C=C—CH—CHOH—CH_3
$$
$$
\quad\quad\ \ |\quad\ |
$$
$$
\quad\quad\ CH_3\ CH_3
$$
β-allenic carbinol

Allenic Grignard reacts with ethyl orthoformate to give about three times as much acetylenic acetal as allenic acetal. 1,3-Diene Grignards react to give mostly allenic acetals.

4.2. 1-Allenylboronates

Allyl bromide reacts with zinc or aluminum to form allyl—M—Br which reacts with trimethyl borate to form dimethyl allylboronate. Propargyl bromide, however, gives allenylboronate, not propargyl. Favre[45] reported the first allenylboronate, dibutyl allenylboronate:

$$
BrCH_2—C\equiv CH \xrightarrow{M} CH_2=C=CHMBr \xrightarrow{B(OCH_3)_3} CH_2=C=CH—B(OCH_3)_2
$$
$$
\xrightarrow{H_2O} CH_2=C=CH—B(OH)_2 \xrightarrow{BuOH} CH_2=C=CH—B(OBu)_2
$$

M = Al, 33 % yield
Mg, 36 % yield

Propargyl bromide reacts with aluminum or zinc to form the l-allenylmetal bromide, which reacts with α-acetylenic aldehydes to give α-allenyl-α′-acetylenic carbinols in 50–70% yields:[46]

$$HC\equiv CCH_2Br + Al \text{ (or Zn)} \longrightarrow$$

$$H_2C=C=CHAl_{2/3}Br \text{ (or } H_2C=C=CHZnBr) \xrightarrow{+RC\equiv CCHO}$$

$$\underset{\underset{RC\equiv CCHCH=C=CH_2}{|}}{OH}$$

4.3. Grignard Reagents from *t*-Propargylic Chlorides

Propargyl bromide reacts with magnesium in ether to form propargyl Grignard reagent, even though the acetylenic hydrogen ordinarily reacts with Grignard reagents to form the 1-propynyl Grignard reagent. *t*-Propargyl chlorides, however, react with magnesium only in tetrahydrofuran, and the magnesium must be activated by mercuric chloride. The products indicated that at least four or five different Grignard reagents form.[47] The major one is the allenic Grignard (2). Of the monomeric hydrocarbons formed directly (no hydrolysis), (3) is major. Coupling gives the dimers, and radical disproportionation gives (3), (4) and (9).

$$\underset{(1)}{(CH_3)_2C(Cl)-C\equiv CH} + Mg \xrightarrow{THE} (major) \underset{(2)}{(CH_3)_2C=C=CH-MgCl} \overset{?}{\rightleftharpoons}$$

$$\underset{(CH_3)_2C-C\equiv CH}{\overset{\overset{\displaystyle MgCl}{|}}{}}$$

$$(1)+(2) \longrightarrow \underset{(3)}{(CH_3)_2C=C=CH_2} + \underset{(4)}{(CH_3)_2CH-C\equiv CH}$$

$$+ \underset{(5)}{(CH_3)_2C(Cl)-C\equiv CMgCl}$$

$$(2)+(5) \longrightarrow (3)+(4)+ \underset{\underset{(6)}{\overset{\overset{\displaystyle |}{CH_3}}{}}}{CH_2=C-C\equiv CMgCl} + MgCl_2$$

$$(2)+(2) \longrightarrow (3)+(4)+ \underset{(7)}{(CH_3)_2C(MgCl)-C\equiv CMgCl}$$

$$(2)+(4) \longrightarrow (3)+ \underset{(8)}{(CH_3)_2CH-C\equiv CMgCl}$$

$$(1)+(6) \longrightarrow (5)+ \underset{\underset{(9)}{\overset{\overset{\displaystyle |}{CH_3}}{}}}{CH_2=C-C\equiv CH}$$

$$2(1)+Mg \longrightarrow MgCl_2 + [(CH_3)_2\overset{+}{C}-C\equiv CH$$

$$\longleftrightarrow (CH_3)_2C=C=\overset{+}{C}H] \begin{array}{c} \nearrow (3)+(4)+(9) \\ \searrow \end{array}$$

$$(1)+(2),(5),(6),(7),(8) \longrightarrow \text{coupling products}$$

4.4. Allenyl Grignard Reagents

1-Allenylmagnesium bromide reacts with allyl bromide to form 45% yield of a mixture of unrearranged 1-allylallene plus 1-hexen-5-yne, the rearranged product. The rearranged product possibly forms via a cyclic displacement reaction.[48] The same reaction with propargyl bromide gave three products:

$$HC{\equiv}CCH_2Br + CH_2{=}C{=}CHMgBr \longrightarrow HC{\equiv}CCH_2CH{=}C{=}CH_2 \text{ (unrearranged)}$$

$$+ HC{\equiv}CCH_2CH_2C{\equiv}CH \text{ (rearranged)}$$

$$+ H_2C{=}C{=}CH{-}CH{=}C{=}CH_2 \text{ (rearranged)}$$

About 80% of the total product was allenic.

Aufdermarsh[49] prepared cis-1,4-neoprene by an interesting series of reactions. He tried to prepare the Grignard reagent from chloroprene but obtained only polymers. He was able to make a Grignard from 1-chloro-2,3-butadiene, and he used this Grignard to prepare the polymer:

$$CH_2{=}C{=}CHCH_2Cl + Mg \xrightarrow{\text{ether}} CH_2{=}\underset{\underset{MgCl}{|}}{C}{-}CH{=}CH_2 \xrightarrow{(Bu_3Sn)_2O}$$

$$CH_2{=}\underset{\underset{SnBu_3}{|}}{C}{-}CH{=}CH_2 \xrightarrow[\text{azobisisobutyronitrile}]{\text{bulk polymerization}}$$

(76.4%)

$$\underset{\sim CH_2}{\overset{SnBu_3}{\diagdown}}C{=}C\underset{CH_2\sim}{\overset{H}{\diagup}} \xrightarrow{Cl_2}$$

$$\underset{\sim CH_2}{\overset{Cl}{\diagdown}}C{=}C\underset{CH_2\sim}{\overset{H}{\diagup}}$$

cis-1,4-Neoprene is a soft amorphous gum, with higher density and lower crystallinity than ordinary trans-1,4-neoprene.

5. ABSOLUTE CONFIGURATION OF ALLENES AND OPTICAL ACTIVITY

In 1965, Lowe[50] pointed out that van't Hoff had predicted in 1875 that unsymmetrically substituted allenes should have two enantiomeric forms. Sixty-five years later the enantiomers of 1,3-diphenylallene and 1,3-di-α-naphthylallene were synthesized. Recently, the absolute configurations of at least eight allenes were reported. Generally, absolute configurations were determined by converting an optically active molecule of known configuration to the allene, or by converting an allene into a molecule of known configuration, using sterically unambiguous reactions. The absolute configuration of molecules containing asymmetric carbon atoms can be predicted using the relative polarizability of the substituents. The same method can be used for allenes. The relative polarizabilities of substituents found in some known allenes are: Cl > Ph > COOH > Me > t-Bu > H.

Lowe described this method for predicting absolute configuration: View the allenes along their orthogonal axes and determine the handedness of the screw pattern of polarizability: "If by placing the more polarizable substituent in the vertical axis uppermost, the more polarizable substituent in the horizontal axis is to the right, then a clockwise screw pattern of polarizability will obtain and the enantiomer should be dextrorotatory; if the more polarizable substituent in the horizontal axis is to the left, then an anticlockwise screw pattern of polarizability will obtain and the enantiomer should be levorotatory."

Most of the known naturally occurring allenes are fungal metabolites. All of the dissymmetric fungal allenes have the diyne-allene system $RC{\equiv}CC{\equiv}CCH=C=CHR'$ (R = H or Me). The molecules are rigid in the portion made up of these linkages. If R' does not change conformational considerations, the rule indicates that the *dextro* enantiomers have the S-configuration and the *levo* enantiomers have the R-configuration:

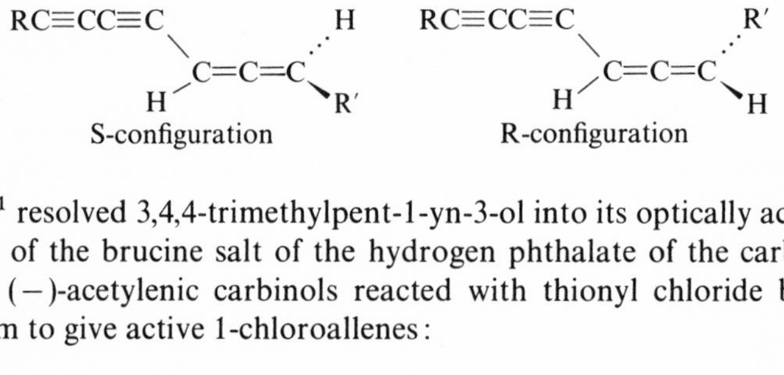

S-configuration   R-configuration

Evans[51] resolved 3,4,4-trimethylpent-1-yn-3-ol into its optically active forms by means of the brucine salt of the hydrogen phthalate of the carbinol. The (+)- and (−)-acetylenic carbinols reacted with thionyl chloride by an S_Ni' mechanism to give active 1-chloroallenes:

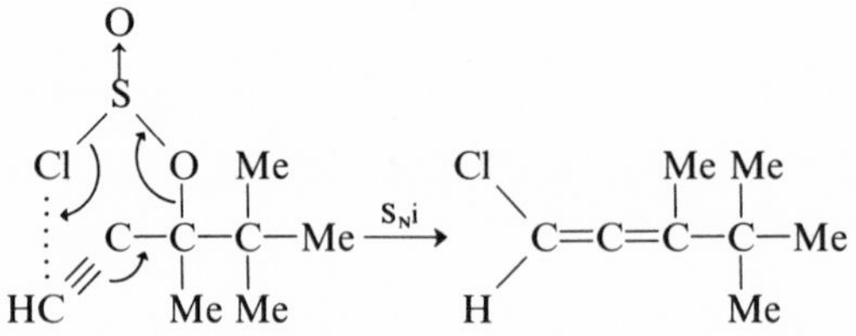

Prolonged heating during the thionyl chloride reaction, the presence of water, and the presence of acid caused racemization.

The absolute configurations of 3,4,4-trimethylpent-1-yn-3-ol (1) and 1-chloro-3,4,4-trimethylpenta-1,2-diene (2) have been determined.[52] Ordinary (1) with S-(+)-butyllactic acid resolved to (−)(1). (1) with thionyl chloride under S_Ni conditions gave (−)(2), which has the S-(−)-configuration.

6. NATURALLY-OCCURRING ALLENES, ISOMERIZATIONS

6.1. Isomerizations

In 1966, Bew[53] reported seven new allene-diacetylenes from five Basidiomycete fungi: $HC \equiv CC \equiv CCH = C = CH(CH_2)_nOH$, $n = 1, 3$ and 4;

$$HC \equiv CC \equiv CCH = C = CH\overset{\overset{OH}{|}}{C}H(CH_2)_nOH, \; n = 2 \text{ and } 3;$$

$$HC \equiv CC \equiv CCH = C = CHCH_2\overset{\overset{OH}{|}}{C}HCH_2CH_2OH;$$

and $CH_3C \equiv CC \equiv CCH = C = CHCH_2CH_2OH$. This makes 18 identified allene-poly-acetylenes derived from microorganisms. Although fungal metabolites seem to prefer to make allenes, some higher organisms also make them. For example, laballenic acid, $CH_3(CH_2)_{10}CH = C = CH(CH_2)_3CO_2H$, from *Leonotis nepetaefolia* seed oil is the first C_{18} allene isolated from higher plants.[54] Bew has found a wide variety of allene-polyacetylenes,[53] including alcohols, esters *and* diols:

	$[\alpha]_D$
$HC \equiv C-C \equiv C-CH = C = CH-CH \overset{cis}{=\!=} CH-CH \overset{trans}{=\!=\!=} CH-CH_2-CO_2H$ (1)	$-130°$
$HC \equiv C-C \equiv C-CH = C = CH-CH(OR')-CH_2-CH_2-CO_2R$	$+380^a$
$CH_3C \equiv C-C \equiv C-CH = C = CH-CH(OR')-CH_2-CH_2-CO_2R$	$+360^a$
$HC \equiv C-C \equiv C-CH = C = CH-CH_2-CH_2-OH$ (2) $[(+) \text{ and } (-)]$	±385
$HC \equiv C-C \equiv C-CH = C = CH-CH_2-CO_2R$	$+285^b$
$H_2C = C = CH-C \equiv C-C \equiv C-CH \overset{cis}{=\!=} CH-CH_2-CO_2H$	
$HC \equiv C-C \equiv C-CH = C = CH-CH_2OH$ (3)	-380
$HC \equiv C-C \equiv C-CH = C = CH-CH_2-CH_2-CH_2OH$ (4)	-290
$HC \equiv C-C \equiv C-CH = C = CH-CH_2-CH_2-CH_2-CH_2OH$ (5)	-180
$HC \equiv C-C \equiv C-CH = C = CH-CH(OH)-CH_2-CH_2OH$ (6)	-210
$HC \equiv C-C \equiv C-CH = C = CH-CH(OH)-CH_2-CH_2-CH_2OH$ (7)	
$HC \equiv C-C \equiv C-CH = C = CH-CH_2-CH(OH)-CH_2-CH_2OH$	-210^c
$CH_3-C \equiv C-C \equiv C-CH = C = CH-CH_2-CH_2OH$	$+340$
$CH_3-C \equiv C-C \equiv C-C \equiv C-CH \overset{trans}{=\!=\!=} CH-CH_2OH$	

[a] For R = R' = H.
[b] Partially synthetic sample, R = Me.
[c] For isopropylidene derivative.

Conjugated allene-polyacetylenes isomerize very easily in the presence of base. Sodium hydroxide solution at 20° produces isomerization, exemplified by:

mycomycin (1) $\xrightarrow{\text{NaOH}}$

$$CH_3-C{\equiv}C-C{\equiv}C-C{\equiv}C-CH\overset{trans}{=\!\!=\!\!=}CH-CH\overset{trans}{=\!\!=\!\!=}CH-CH_2-CO_2H$$
(isocycomycin)

marasin (2) $\xrightarrow{\text{NaOH}}$ $HC{\equiv}C-C{\equiv}C-CH{=}\overset{\overline{\quad\quad\quad\quad\quad\quad}}{C}-CH_2-CH_2-CH_2-O$
(isomarasin)

(4) or (5) $\xrightarrow{\text{NaOH}}$ $CH_3-C{\equiv}C-C{\equiv}C-C{\equiv}C-(CH_2)_n-CH_2OH$ (n = 2 or 3)
(disubstituted triyne, a new type of product)

Bew's[53] seven new allenic diacetylenes underwent some different base-catalyzed isomerizations. Some of them formed terminal triyne:

$$HC{\equiv}C-C{\equiv}C-CH{=}C{=}CH-CH(OH)- \longrightarrow$$
(3), (6), (7)

$$HC{\equiv}C-C{\equiv}C-C{\equiv}C-CH_2CH(OH)-$$

Alcohols with hydroxyl beta to the allene group behaved this way. When the hydroxyl groups are gamma or delta, the product is a nonterminal triyne:

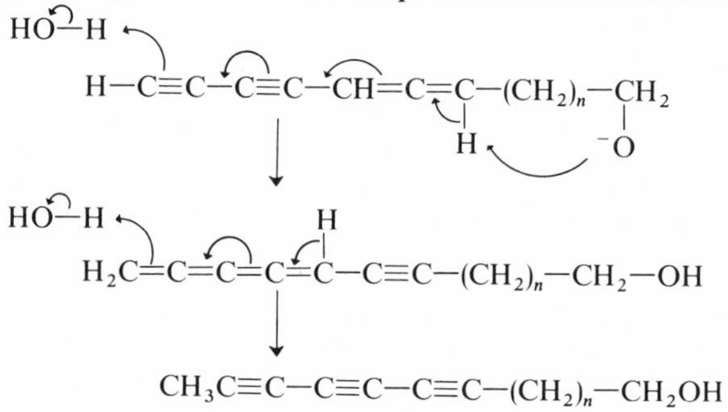

$$H-C{\equiv}C-C{\equiv}C-CH{=}C{=}C-(CH_2)_n-CH_2$$

$$H_2C{=}C{=}C{=}C{=}C-C{\equiv}C-(CH_2)_n-CH_2-OH$$

$$CH_3C{\equiv}C-C{\equiv}C-C{\equiv}C-(CH_2)_n-CH_2OH$$

where n = 2 or 3.

The high optical activity of allene-diacetylenes is due to the highly polarizable diyne grouping. The levorotatory diacetylenic allenes should have the R-configuration, and the dextrorotatory ones should have the S-configuration:

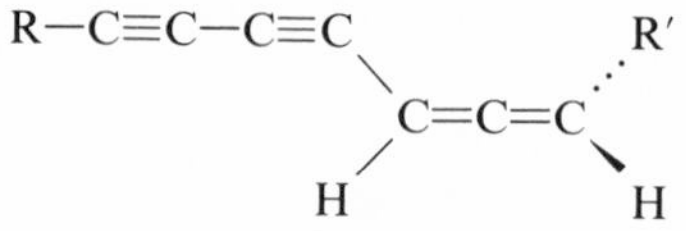

R-configuration

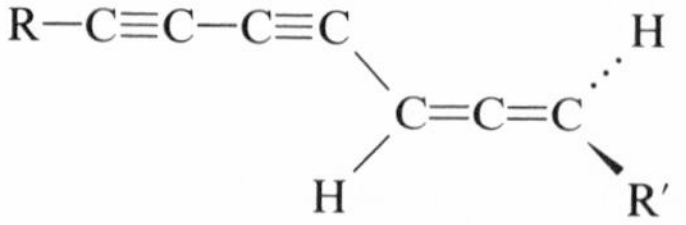

S-configuration

6.2. One-step Synthesis of Allenynes

Baker[55] noted that there are few synthetic methods available to prepare the important naturally occurring allene-polyacetylenes. He developed a one-step method which is similar to Chodkiewicz-Cadiot coupling:

$$RR'C{=}C{=}CHBr \text{ (or I)} + HC{\equiv}C(CH_2)_nR'' \xrightarrow[\text{base}]{Cu^+}$$

$$RR'C{=}C{=}CHC{\equiv}C(CH_2)_nR''$$

Simple allenynes and allenediynes do not rearrange under the mild conditions of the coupling. Yields of allenynes from allenyl iodides vary from 15–68%, and yields from allenyl bromides are 13–82% (Table 1–9).

TABLE 1-9[55]

$$RR'C{=}C{=}CHX + HC{\equiv}C(CH_2)_nR'' \longrightarrow$$

$$RR'C{=}C{=}CHC{\equiv}C(CH_2)_nR''$$

R	R′	R″	n	X	% Yield
Me	H	OH	1	I	40
i-Pr	H	OH	1	I	49
i-Pr	H	OH	2	I	36
i-Pr	H	Me	3	I	62
t-Bu	H	OH	1	I	54
Me	H	HC≡C	0	I	15
i-Pr	i-Pr	OH	1	Br	62
t-Bu	Me	OH	1	Br	62
Et	Me	OH	1	Br	51
Me	Me	OH	1	Br	33
n-Pr	H	OH	1	Br	51
Et	H	OH	1	Br	31
t-Bu	Me	Me	3	Br	82
Et	Me	HC≡C	0	Br	25

Coupling Mechanism. Blake[56] considered the yellow acetylenic copper complexes observed in the coupling reactions to be coordination polymers. The polymers have considerable back coordination from the filled metal d orbital to the antibonding orbital of at least two acetylene groups attached to metal. Although base is necessary for the coupling, it does not change the complexes. The base removes acid HX as it forms and also takes part in the coordination complex:

$$\begin{array}{c} C \\ \parallel \\ C \\ \mid \end{array}$$

$$-N \longrightarrow Cu \qquad \overset{H}{\underset{RC\equiv C}{\diagdown}} \overset{}{\underset{Br}{\diagup}} C=C=C \overset{\diagup}{\underset{\diagdown}{}} \longrightarrow -C\equiv C-\overset{}{\underset{H}{\diagup}}C=C=C\overset{\diagup}{\underset{\diagdown}{}}$$

$$+ -\overset{\mid}{\underset{\mid}{N}} \longrightarrow \overset{Br}{\underset{RC\equiv C-}{\overset{\downarrow}{Cu}}} \longleftarrow Br$$

The first step might also be elimination of a proton from the allenic carbon carrying the halogen.

Two of Baker's compounds were highly active *in vitro* against *M. tuberculosis*: 7-methylocta-4,5-dien-2-yn-1-ol and nona-4,5-dien-2-yn-1-ol.

References

 1. Bainvel, J., Wojtkowiak, B., and Romanet, R., *Bull. Soc. Chim. France*, 978 (1963).
 2. Jacobs, T. L., Akawie, R., and Cooper, R. G., *J. Am. Chem. Soc.*, **73**, 1273 (1951).
 3. Wotiz, J. H., and Parsons, C. G. (to Diamond Alkali Co.), U.S. Patent 3,166,605 (Jan. 19, 1965); *Chem. Abstr.*, **62**, 9005 (1965).
 4. Petrov, A. A., and Kormer, V. A., *Zh. Obshch. Khim.*, **34**, 1868 (1964).
 5. van Boom, J. H., *et al.*, *Rec. Trav. Chim.*, **84**, 813 (1965).
 6. Kuhn, R., and Fischer, H., *Chem. Ber.*, **92**, 1849 (1960).
 7. Craig, J. C., and Young, R. J., *J. Chem. Soc. (C)*, 578 (1966).
 8. deVries, G., *Rec. Trav. Chim.*, **84**, 1327 (1965).
 9. Blanc-Guenee, J., d'Engenieres, M. P., and Miocque, M., *Bull. Soc. Chim. France*, 603 (1964).
10. Cymerman-Craig, J., and Moyle, M., *Proc. Chem. Soc.*, 283 (1962).
11. van Daalen, J. J., Kraak, A., and Arens, J. F., *Rec. Trav. Chim.*, **80**, 810 (1961).
12. Brandsma, L., Montijn, P. P., and Arens, J. F., *Rec. Trav. Chim.*, **82**, 1115 (1963).
13. van Boom, J. H., *et al.*, *Rec. Trav. Chim.*, **84**, 31 (1965).
14. Pourcelot, G., *et al.*, *Compt. Rend.*, **252**, 1630 (1961); *Bull. Soc. Chim. France*, 1278 (1962).
15. Brandsma, L., Wijers, H. E., and Arens, J. H., *Rec. Trav. Chim.*, **82**, 1040 (1963).
15a. Hubert, A. J., and Reinlinger, H., *J. Chem. Soc. (C)*, 606 (1968).
16. Prevost, C., and Smadja, W., *Compt. Rend.*, **255**, 948 (1962).
17. Smadja, W., *Ann. Chim. (Paris)*, **10**, 105 (1965); *Chem. Abstr.*, **63**, 6834 (1965).
18. Smadja, W., *Compt. Rend.*, **256**, 2426 (1963).
19. Farmer, M. L., *et al.*, *J. Org. Chem.*, **31**, 2885 (1966).
20. Meyer, K. H., and Schuster, K., *Chem. Ber.*, **55B**, 815 (1922).
21. Jacobs, T. L., *et al.*, *J. Org. Chem.*, **17**, 475 (1952); **22**, 1424 (1957).
22. Jacobs, T. L., Dankner, D., and Singer, S., *Tetrahedron*, **20**, 2177 (1964).
23. Yen, V.-Q., *Ann. Chim. (Paris)* **7**, 785 (1962); *Chem. Abstr.*, **59**, 5043 (1963).
24. Yen, V.-Q., *Ann. Chim. (Paris)*, **7**, 799 (1962); *Chem. Abstr.*, **59**, 5044 (1963).

25. Cram, D. J., "Fundamentals of Carbanion Chemistry", p. 189, New York, Academic Press, 1965.
26. Mikolajczak, K. L., *et al.*, *J. Org. Chem.*, **29**, 318 (1964).
27. Mikolajczak, K. L., Bagby, M. D., and Wolff, I. A., *J. Am. Oil Chemists Soc.*, **42**, 243 (1965).
28. Crowley, K. J., *Proc. Chem. Soc.*, 17 (1964).
29. Cozzone, A., Grimaldi, J., and Bertrand, M., *Bull. Soc. Chim. France*, 1656 (1966).
30. Jones, E. R. H., Shaw, B. L., and Whiting, M. C., *J. Chem. Soc.*, 3212 (1954); Shaw, B. L., and Whiting, M. C., *J. Chem. Soc.*, 3217 (1954).
31. Sondheimer, F., Ben-Efraim, D. A., and Wolovsky, R., *J. Am. Chem. Soc.*, **83**, 1675 (1961).
32. Gensler, W. J., and Casella, J., Jr., *J. Am. Chem. Soc.*, **80**, 1376 (1958).
33. Mathai, I. M., Taniguchi, H., and Miller, S. I., *J. Am. Chem. Soc.*, **89**, 115 (1967).
34. Sondheimer, F., Ben-Efraim, D. A., and Gaoni, Y., *J. Am. Chem. Soc.*, **83**, 1682 (1961).
35. Sondheimer, F., *et al.*, *J. Am. Chem. Soc.*, **83**, 1675 (1961).
36. Eglinton, G., Raphael, R. A., and Willis, B. G., *Proc. Chem. Soc.*, 247 (1960).
37. Eglinton, G., *et al.*, *J. Chem. Soc.*, 2597 (1964).
38. Huntsman, W. D., DeBoer, J. A., and Woosley, M. H., *J. Am. Chem. Soc.*, **88**, 5846 (1966).
39. Bertrand, M., *Compt. Rend.*, **247**, 824 (1958).
39a. Mantione, R., *Compt. Rend.*, **264**, 1668 (1967).
40. Stirling, C. J. M., *J. Chem. Soc.*, 5856 (1964).
41. Bertrand, B., and LeGras, J., *Compt. Rend.*, **261**, 474 (1965).
42. Wijers, H. E., Brandsma, L., and Arens, J. F., *Rec. Trav. Chim.*, **85**, 601 (1966).
42a. Saucy, G., and Marbet, R., *Helv. Chim. Acta*, **50**, 1158 (1967).
43. Fischer, H., in (Patai, S., editor) "The Chemistry of Alkenes," p. 1025, New York, Interscience Publishers, 1964.
44. Pasternak, Y., and Traynard, J.-C., *Bull. Soc. Chim. France*, 356 (1966).
45. Favre, E., and Gaudemar, M., *Compt. Rend.*, **262C**, 1332 (1966).
46. Gaudemar, M., and Travers, S., *Compt. Rend.*, **262C**, 139 (1966).
47. Hennion, G. F., and DiGiovanna, C. V., *J. Org. Chem.*, **31**, 970 (1966).
48. Peiffer, G., *Bull. Soc. Chim. France*, 776 (1962).
49. Aufdermarsh, C. A., Jr., *J. Org. Chem.*, **29**, 1994 (1964).
50. Lowe, G., *Chem. Commun.*, 411 (1965).
51. Evans, R. J. D., Landor, S. R., and Smith, R. T., *J. Chem. Soc.*, 1506 (1963).
52. Evans, R. J. D., and Landor, S. R., *J. Chem. Soc.*, 2553 (1965).
53. Bew, R. E., *et al.*, *J. Chem. Soc.* (*C*), 129, 135, 139 (1966).
54. Bagby, M. D., Smith, C. R., Jr., and Wolff, I. A., *J. Org. Chem.*, **30**, 4227 (1965).
55. Baker, C. S. L., Landor, P. D., and Landor, S. R., *J. Chem. Soc.*, 4659 (1965).
56. Blake, D., Calvin, G., and Coates, G. E., *Proc. Chem. Soc.*, 396 (1959).

PART THREE

Reactions of Allenic Compounds

1. ADDITION OF RADICALS TO ALLENES

The point of addition of radicals to allenes depends on several factors. Highly electrophilic $CF_3\cdot$ adds to the terminal carbon of the allene linkage exclusively, indicating an intermediate vinylic radical,[1,2] but relatively nucleophilic methyl radical adds to the central carbon.[2] $Br\cdot$ adds to give products suggesting addition to the central carbon, which indicates an intermediate allylic radical.[3,4] At $-78°$ in pentane, γ rays initiate addition of $Br\cdot$ (from HBr) to both carbon 1 and carbon 2 of the allene linkage. Most of the bromine is on carbon 2 in the final product because the addition to carbon 1 is reversible.[5] Thiyl radicals preferentially attack the terminal carbon, but not exclusively.[6,7]

1.1. Semihydrogenation

Relatively little is known about the semihydrogenation of allenes to form monoolefins. Catalytic hydrogenation and chemical reduction usually cause complete saturation. Bulky tetraarylallenes, such as diphenyldinaphthylallene, cannot be catalytically hydrogenated at all.[8] A tetraaryldiallene added 1 mole of hydrogen, and the system rearranged to the conjugated hexatriene:[9]

$$Ph_2C{=}C{=}CH{-}CH{=}C{=}CPh_2 \xrightarrow[Pd/Pb]{H_2} Ph_2C{=}CHCH{=}CHCH{=}CPh_2$$

1,2-Butadiene hydrogenates over Pd-alumina at room temperature in a flow system to form mostly cis-2-butene, plus some trans-2-butene and 1-butene. Specific adsorption on the catalyst surface probably explains the high proportion of cis-2-butene:[10]

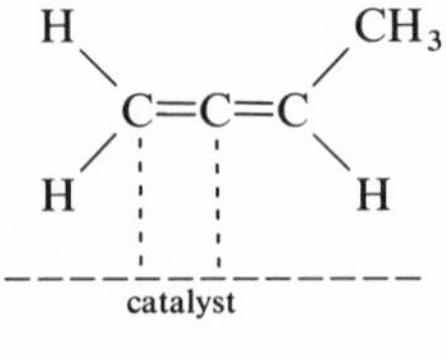

1.2. Addition of $R_3Sn\cdot$ and $RS\cdot$

Kuivila[11] studied the azobisisobutyronitrile-catalyzed addition of R_3SnH to propyne and to allenes at 100°. He compared the proportion of center attack with the center attack by EtS· obtained by Jacobs:[6]

| Allene | Addend (% to carbon↑) | |
	EtS·	Me₃Sn·
$CH_2{=}\underset{\uparrow}{C}{=}CH_2$	12.8	45
$CH_2{=}\underset{\uparrow}{C}{=}CHMe$	48.2	86.5
$CH_2{=}\underset{\uparrow}{C}{=}CMe_2$	100	100
$MeCH{=}\underset{\uparrow}{C}{=}CHMe$		100
$Me_2C{=}\underset{\uparrow}{C}{=}CHMe$		100

Propyne gave mostly product from terminal attack. Thus, allene probably does not isomerize to propyne during the addition. Since EtS· is electrophilic and Me₃Sn· is nucleophilic, polar effects are not the major factors determining the direction of addition.

On the other hand, van der Ploeg[12] concluded from his work that thiyl radicals do not show a strong preference for either position. The direction of free radical addition must be governed not only by stabilities of the initial radicals formed but also by polar factors. Griesbaum[7] believed that these conclusions are valid only if excess allene is used. He irradiated (ultraviolet) 0.1 mole thiol, 0.3 mole allene, and 0.001 mole *t*-butyl hydroperoxide. Analysis of the products by GLC, NMR and infrared showed that all of the thiol reacted except in one case (Table 1-10). The rate of reaction in the presence of peroxide

TABLE 1-10
Addition of Thiyl Radicals to Allene ($RS{\cdot} + CH_2{=}C{=}CH_2$)[7]

R	Peroxide added	Temperature (°C)	Time (hr)	% Yield of Adducts	Center attack[b] on Allene (%)
CH_3	Yes	15–17	56	88	14
CH_3	Yes	17	4.66	87	11
CH_3	Yes	17	1.5	86	10
CH_3	Yes	17	15 min[a]	54	12
CH_3	No	−45	25	—	8
CH_3	Yes	−75	12	71	6
$CH_3{-}CO$	Yes	17	15 min[a]	92	9
C_6H_5	Yes	17–18	52	—	19
C_6H_5	No	Ambient	35	75	17
C_6H_5	Yes	17	3.5	83	21

[a] 58% conversion of thiol.
[b] From amount of (6) in product, less amount of (6) formed via (3) (scheme below).

and irradiation was 75 times as fast as the uncatalyzed rates, proving that the additions were free radical in nature. The results in Table 1–10 are explained by these reaction sequences:

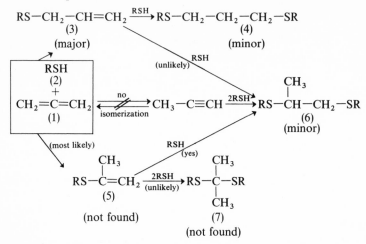

Attack of MeS· on the central carbon decreases as the temperature decreases. At least 93 % of the product from attack at carbon 2 is due to direct attack by the radical and not isomerization to propyne followed by addition. Thiolacetic acid gives 91 % terminal attack, and benzenethiol gives only 80% terminal attack. Thus, thiyl radicals generally add to the terminal carbon, and the vinylic radical formed abstracts hydrogen from thiol to give the observed allyl sulfides. Overall, this is a 1,2-addition. The yield of diadduct increases at the expense of monoadduct as the temperature decreases, since thiyl radicals add rapidly and reversibly to allyl sulfides.[13]

In the diaddition of thiyl radicals to allene, the direction of addition seems to depend on the hydrogen-donor ability of the thiol. The amount of allylic product (monoadduct) is proportional to the stability of the thiyl radicals.[13a]

Heiba[14] determined the rates of addition of benzenethiyl radicals (generated by ultraviolet irradiation) to allene, both $1M$ in various solvents. The addition products were phenylallyl sulfide, phenylisopropenyl sulfide, and 1,2- and 1,3-bisphenylthiopropane, along with some phenyl disulfide (Table 1–11). The relative rate of addition to the carbon atom in position 2 increased slowly as temperature increased. Solvent had a small effect at 80°, but not at 20 and 120°. In similar experiments, ·CCl₃ added exclusively to the carbon atom in the 2-position.

The methylene groups and the π-orbitals of allene are orthogonal, and thus the transition state for radical reaction at the 2-position is more like a primary radical than an allylic radical. Orientation of addition must be controlled by steric and orientation factors, because neither carbon offers significant energy advantages. Analysis of rate factors led Heiba to conclude that Kuivila's[11]

TABLE 1-11
Addition of PhS· to Allene in Solution[14]

$$\overset{1}{C}H_2=\overset{2}{C}=\overset{1}{C}H_2$$

Solvent	Isomer (%)[a]							
	20°		80°		120°		140°	
	Carbon 2	Carbon 1	Carbon 2	Carbon 1	Carbon 2	Carbon 1	Carbon 2	Carbon 1
None[b]	19.8	80.2	24.0	76.0	26.3	73.7	29.9	70.1
Benzene	19.8	80.2	43.0	57.0	62.9	37.1	96	4
Acetonitrile	—	—	28.6	71.4	62.3	37.7	—	—
Anisole	19.1	80.9	—	—	—	—	—	—
Benzonitrile	20.9	79.1	34.5	65.5	—	—	—	—
Chlorobenzene	—	—	37.9	62.1	67.2	32.8	—	—
Tetrachloroethylene	—	—	35.1	64.9	67.5	32.5	—	—
Trichloroethane	—	—	37.4	62.6	—	—	—	—

[a] "Carbon 2" represents mole per cent of central isomer in 1:1 and 2:2 adducts; "Carbon 1", terminal isomer in 1:1 and 2:1 adducts.
[b] Molar ratio of thiol to allene = 4.0.

proposed bridged thiyl radical (which subsequently reacts directly or opens to form a vinylic or allylic radical) is inconsistent with the data.

1.3. Addition of H₂S

Radical addition of hydrogen sulfide to allene gives products which show that HS· prefers to add to the terminal carbon of allene, just as thiols do. The products and the paths by which they probably form are represented by formulas and solid arrows; broken arrows indicate a less likely path:[15]

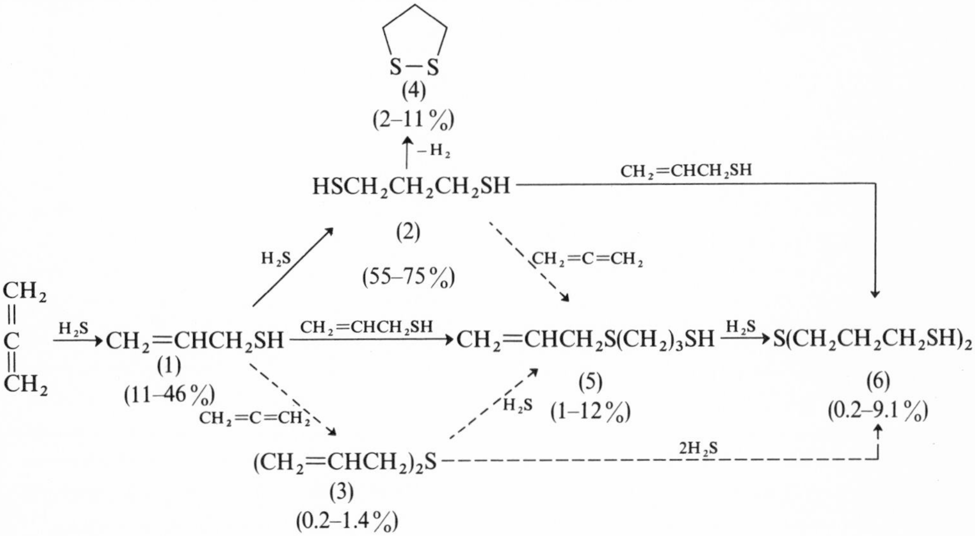

A little propane-1,2-dithiol also forms. The highest yield of diallyl sulfide, 25–30%, was obtained when excess allene was present:

$$CH_2{=}\!\!\!\!{=}C{=}\!\!\!\!{=}CH_2 \quad \begin{matrix} (CH_2{=}CHCH_2)_2S \\ (25{-}30\%) \quad +(I) \end{matrix}$$

$$\begin{matrix} CH_2 \\ \| \\ C \\ \| \\ CH_2 \end{matrix} \xrightarrow{H_2S} CH_2{=}CHCH_2SH \qquad\qquad (CH_2{=}CHCH_2SCH_2)_2CH_2$$

$$\begin{matrix} (1) & +(I) \end{matrix} \qquad CH_2{=}C{=}CH_2$$

$$CH_2{=}CHCH_2S(CH_2)_3SH$$

1.4. Addition of Phosphine

Irradiation of phosphine-allene mixtures by ultraviolet gives polymer and a product in which $-PH_2$ is attached to the central carbon $(CH_2{=}\overset{\overset{\text{Me}}{|}}{C}{-}PH_2)$. This is another example of addition to the carbon atom in the 2-position.[16]

1.5. Addition of Tetrafluorohydrazine

N_2F_4 and allene at 70° in o-dichlorobenzene react to form about 40% yield of a mixture of products. Two $\cdot NF_2$ radicals add:[16a]

$$CH_2{=}C{=}CH_2 + N_2F_4 \longrightarrow \left[CH_2{=}\overset{\overset{\text{NF}_2}{|}}{C}{-}CH_2NF_2 \right] \longrightarrow$$

$$CH_2F{-}\overset{\overset{\text{NF}}{\|}}{C}{-}CH_2NF_2 + CH_2F{-}\overset{\overset{\text{FN}}{\|}}{C}{-}CH_2NF_2$$
$$\textit{anti-} \qquad\qquad\qquad \textit{syn-}$$

1.6. Addition of Nitrogen Radicals

The aminium radical $R_2\overset{\cdot}{N}H^+$ adds to olefins, diolefins, allenes and acetylenes to give 1:1 adducts (15–68% yields). Addition to an olefinic carbon is pictured as a free radical chain reaction:[16b]

$$R_2\overset{\cdot}{N}H^+ + \underset{/}{\overset{\backslash}{{}}}C{=}C\underset{\backslash}{\overset{/}{{}}} \xrightarrow[\text{H}_2\text{SO}_4]{\text{HOAc}^-} R_2NH^+{-}\overset{|}{\underset{|}{C}}{-}\overset{|}{\underset{|}{C}}\cdot \xrightarrow{+R_2NH^+Cl}$$

$$R_2NH^+{-}\overset{|}{\underset{|}{C}}{-}\overset{|}{\underset{|}{C}}{-}Cl + R_2\overset{\cdot}{N}H^+$$

Allenes require an initiator such as ultraviolet light or ferrous ion. Allene forms the products $R_2NCH_2CCl{=}CH_2$. 1,2-Cyclononadiene undergoes ionic chlorination only. The major products are chloroacetates:

2. IONIC ADDITIONS TO ALLENES

Allenes can behave like nucleophiles in the addition of acids. The proton adds first to the more nucleophilic double bond. Which bond this is depends on the substituents, and either allyl or vinyl carbonium ion forms:

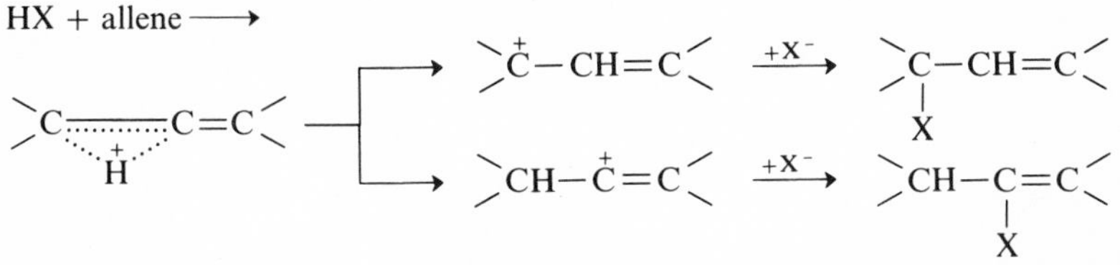

Allenes can also act as electrophiles. Amines, water (mercuric catalysts), Grignard reagents, and hydrohaloacids (metal catalyst) add by nucleophilic attack on the more electrophilic carbon. Since the central carbon of the allene bond is usually relatively positive, nucleophilic agents should add mostly to the central carbon. Thus, nucleophilic addition is more predictable than electrophilic.[16]

2.1. Allenic Hydrocarbons

2.1.1. OZONOLYSIS

Ozone is the best reagent for oxidative cleavage of allene bonds. The products are carbonyl compounds, so identification is easy. Higher cumulenes oxidize the same way to give carbonyl compounds and carbon dioxide:[16]

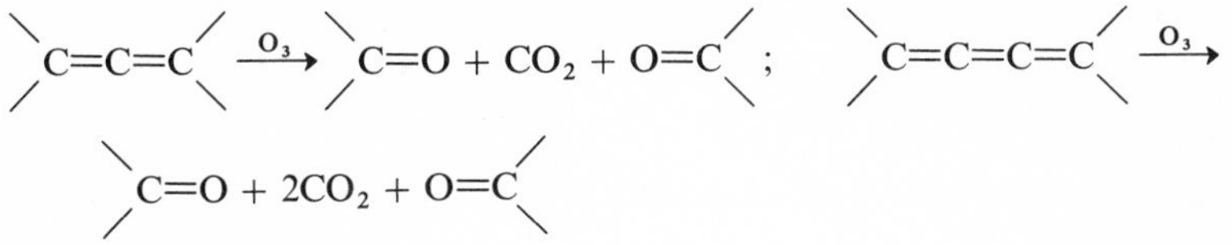

2.1.2. ADDITION OF HX

In 1965 Griesbaum[17] reviewed the early literature on ionic addition of HX to allene and propyne. Ionic additions obey Markownikoff's rule and form 2-halopropenes and/or 2,2-dihalopropanes. Water adds similarly to give acetone from both allene and propyne.

Griesbaum found that addition of HBr to allene gave cis- and trans-1,3-dibromo-1,3-dimethylcyclobutane.[18,19] He obtained the same electrophilically induced cyclodimerization from propyne:

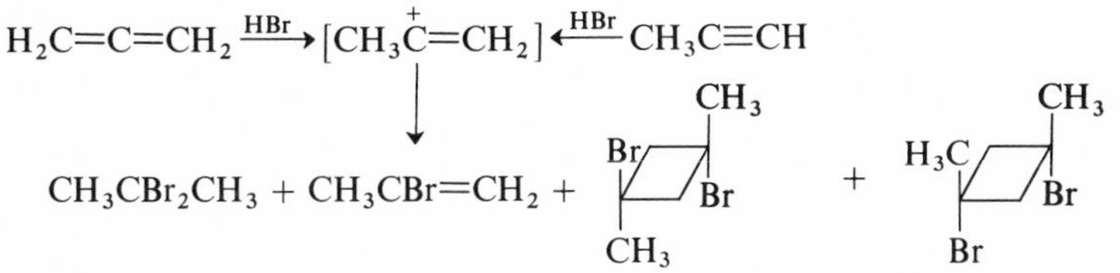

(35% of total products)

Jacobs[20] was unable to detect reaction between HCl and allene at $-78°$, but Griesbaum obtained 1% of a liquid product, the major part of which was *trans*-1,3-dichloro-1,3-dimethylcyclobutane. With bismuth chloride catalyst, *cis*- and *trans*-cyclobutanes constituted 30% of the total product. Jacobs[20] had noted that bismuth chloride catalyzed only the normal addition reaction. Under similar conditions propyne did not form cyclobutanes.

HBr adds to allene faster than HCl does.[17] Up to 52% of the products are the cyclobutanes. Under ambient conditions, HBr and allene react in the gas phase to form products containing 13% cyclobutanes. Under these conditions propyne appears to react faster, but some of the reaction is free radical addition. Inhibitors decrease the radical addition but have no effect on the ionic cyclization reaction. At $-70°$, HI adds to both allene and propyne faster than HBr does. The main product is 2,2-diiodopropane, too unstable to isolate.

Griesbaum believes his work is the first example of proton-initiated cyclo-dimerizations of allenic or acetylenic compounds to form cyclobutanes. Peterson[21] has postulated cyclobutenes as intermediates in the acid-catalyzed cyclo-trimerization of alkynes to substituted benzenes. Criegee[22] reported that the BF_3-catalyzed addition of chlorine to 2-butyne gave 1,2,3,4-tetramethyl-3,4-dichlorobutene. Griesbaum's reaction is a new one-step synthesis of four-membered rings. Cationic catalysts such as BF_3 may be useful. With propyne, acid catalysts increase both the rate and the proportion of cyclization. A reasonable mechanism for the formation of *trans*-dimethyldihalocyclobutane is:

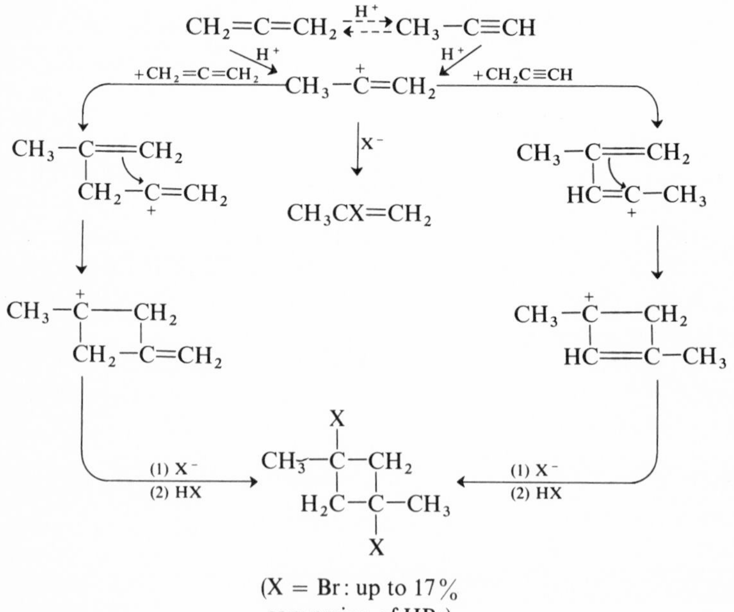

(X = Br: up to 17%
conversion of HBr)

Vinylallenes and propenylallenes add HX in concentrated aqueous solution. The product 2-halo-1,3-dienes result from 2,5-addition :[22a]

$$CH_2=C=CHCH=CHR \xrightarrow[H_2O]{12M\,HX} cis\text{- and } trans\text{-}CH_2=\underset{X}{C}-CH=\underset{H}{C}CH_2R$$

2.1.3. ADDITION OF HALOGEN

Allene usually adds only 1 mole of bromine, bromine chloride or chlorine.[23] Initial attack of X^+ is at the central carbon, to form an intermediate resonance-stabilized allyl carbonium ion: $CH_2=\overset{X}{C}-CH_2^+$. Chlorine adds in inert solvents to give equimolar amounts of propargyl chloride and 2,3-dichloropropene. Propargyl chloride may be formed by a "four-center elimination reaction":[23]

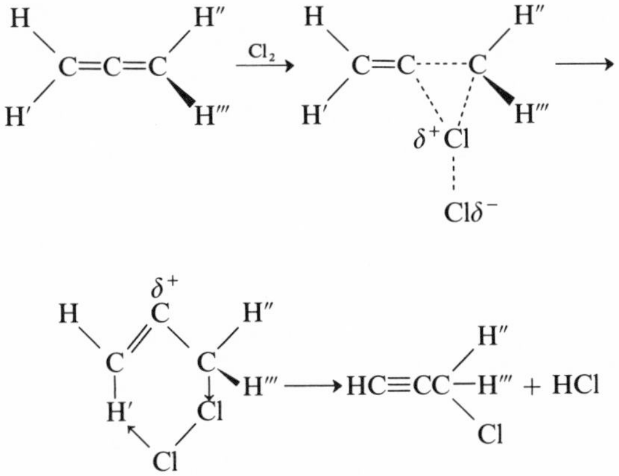

The addition of chlorine to allene in the highly ionic $NaAlCl_4-KAlCl_4$ melt at 140° gives much higher selectivity to 2,3-dichloropropene. Passing allene-Cl_2 (9.6:1) through the melt gives 77% conversion; 73% of the volatile product is 2,3-dichloropropene, and 10% is propargyl chloride (selectivity = 88:12).[24]

2.1.4. ADDITION OF CHLORINE COMPOUNDS

O,O-Dimethylphosphorylsulfenyl chloride adds to allene to give 53% yield of of $(MeO)_2P\overset{\uparrow O}{S}\overset{CH_2Cl}{C}=CH_2$.[25] Propyne also gives anti-Markownikoff products. Chlorosulfonyl isocyanate adds to substituted allenes to form β-lactams which have an exocyclic double bond; 1,3-butadiene derivatives are minor products (Table 1–12).[26]

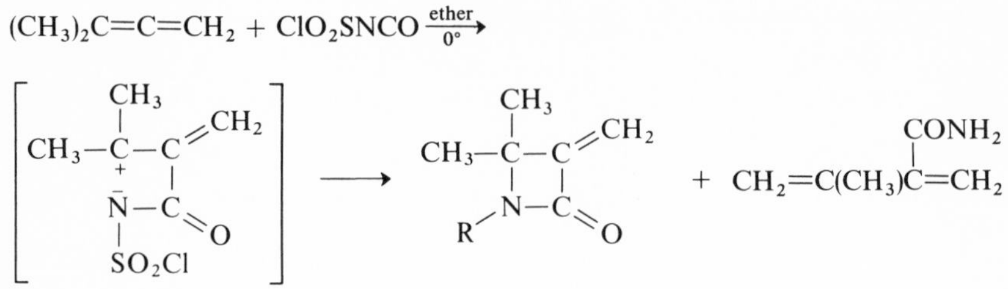

$$(CH_3)_2C{=}C{=}CH_2 + ClO_2SNCO \xrightarrow[0°]{\text{ether}}$$

TABLE 1-12
Addition of Chlorosulfonyl Isocyanates to Substituted Allenes[26]

| | | Products | | |
Allene	β-Lactam	% Yield	Diene	% Yield
$(CH_3)_2C{=}C{=}C(CH_3)_2$ 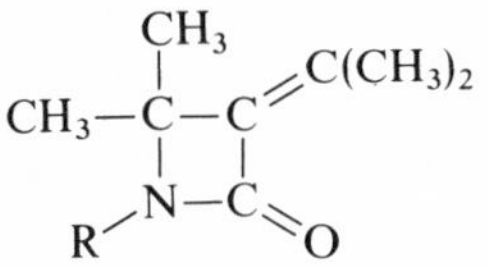		R = SO$_2$Cl 73 R = H 72		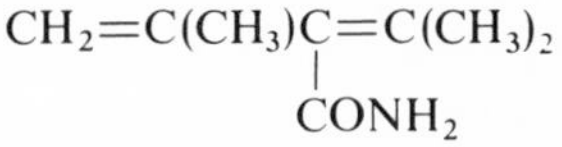 22
		R = SO$_2$Cl 38 R = H 68		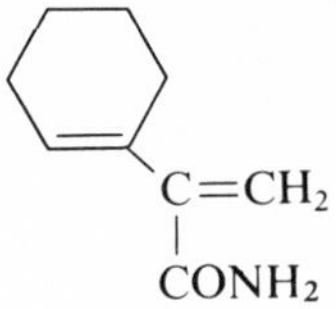 32
		R = SO$_2$Cl 89 R = H 40		

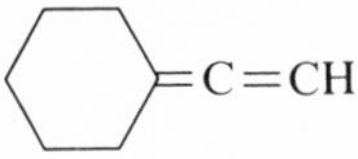

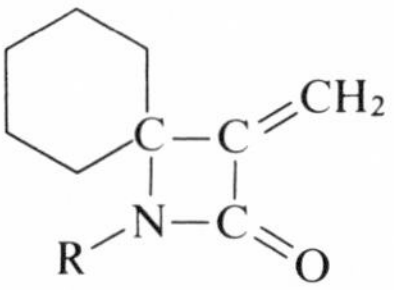

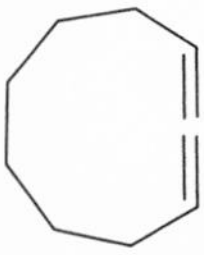

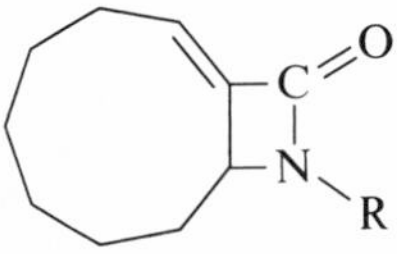

Nitrosyl chloride adds to allene in chloroform at 20° in the presence of stannous chloride to form 1,3-dichloroacetone[30a] (39% yield). In the addition step, N=O adds to the central carbon, and Cl to a terminal carbon.

Sulfenyl chlorides add to allene to form allylic chlorides $CH_2{=}\overset{\overset{\displaystyle SR}{|}}{C}{-}CH_2Cl(1)$.

Diadducts $RSCH_2{-}\overset{\overset{\displaystyle Cl}{|}}{\underset{\underset{\displaystyle SR}{|}}{C}}{-}CH_2Cl(2)$ also form. The monoadducts usually rearrange to vinylic chlorides (3):

$$
\begin{array}{c}
\text{RSCl} \\
+ \\
H_2C=C=CH_2
\end{array}
\longrightarrow
\left[
\begin{array}{c}
R \\
| \\
S \\
\diagup + \diagdown \\
H_2C\!\!-\!\!C=CH_2 \\
\downarrow \\
Cl^-
\end{array}
\right]
$$

$$\downarrow$$

$$
\underset{(2)}{\underset{\overset{|}{Cl}}{\underset{|}{ClCH_2\overset{SR}{\overset{|}{C}}CH_2SR}}}
\xleftarrow{\text{RSCl}}
\underset{(1)}{ClCH_2\overset{SR}{\overset{|}{C}}=CH_2}
\longrightarrow
\underset{\substack{(3)\\(\textit{cis} \text{ and } \textit{trans})}}{ClCH=\overset{SR}{\overset{|}{C}}CH_3}
$$

When R is CH_3, C_6H_5, or $CH_3\overset{\overset{\displaystyle O}{\parallel}}{C}S$, yields of mono adduct (1) are 67–85%. Isomerization to (3) is fairly rapid at $-20°$.[30c]

2.1.5. ADDITION OF CARBENES

In 1961, Ball[27] reported that carbenes add to allenes to give alkylidenecyclopropanes. Olefins react with CH_2I_2 and the Zn–Cu couple to form cyclopropanes.[28] This method was used to add carbene to methyl penta-3,4-dienoate:[29]

$$CH_2=C=CHCH_2COOCH_3 \xrightarrow[\text{Zn–Cu}]{CH_2I_2}$$

$$
CH_2=C
\begin{array}{c}
\diagup CH_2 \\
| \\
\diagdown CHCH_2COOCH_3
\end{array}
\quad + \quad
\begin{array}{c}
H_2C \diagdown \diagup CH_2 \\
C \\
H_2C \diagup \diagdown CHCH_2COOCH_3
\end{array}
$$

(equal amounts, total yield 35%)

Rahman[30] attempted to add carbene via the dihalide-Zn-Cu couple reaction to allenic hydrocarbons, but he obtained mostly polymers. Dibromocarbene, however, did add to 3-methyl-1,2-pentadiene, 2,3-pentadiene, and 4-methyl-2,3-pentadiene to give 75–85% yields of alkylidenecyclopropanes. This reaction was used to make 17 alkylidenecyclopropanes. The steps in the reaction are:

$$
\underset{\diagup}{\overset{\diagdown}{}}C=C=C\underset{\diagdown}{\overset{\diagup}{}} + HCBr_3 \xrightarrow[\text{pentane}]{KO-t-Bu, -25°}
\overset{\diagdown}{\underset{\diagup}{}}C\!\!-\!\!\overset{\overset{\displaystyle CBr_2}{\diagup\diagdown}}{}\!\!C=C\underset{\diagdown}{\overset{\diagup}{}} \xrightarrow{Bu_3SnH}
$$

$$
\overset{\diagdown}{\underset{\diagup}{}}C\!\!-\!\!\overset{\overset{\displaystyle CHBr}{\diagup\diagdown}}{}\!\!C=C\underset{\diagdown}{\overset{\diagup}{}} \xrightarrow{Bu_3SnH}
\triangle=C\underset{\diagdown}{\overset{\diagup}{}}
$$

Dichlorocarbene from chloroform and base reacts with allene to form only tars, but 1,3-diphenylallene or 1,1,3-triphenylallene adds two molecules of dichlorocarbene:[30b]

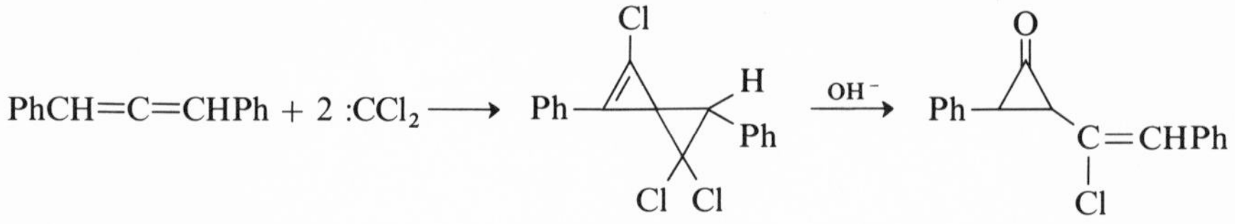

$$PhCH{=}C{=}CHPh + 2 :CCl_2 \longrightarrow$$

Dichlorocarbene prepared by heating sodium trichloroacetate to 120° in dimethoxyethane does add to allene to form the expected 1-methyl-2,2-di-chlorocyclopropane.[30a]

Alkylidenecarbenes [reaction (1)] and vinylidenecarbenes [reaction (2)] add to allenes to form 1,2-dialkylidenecyclopropanes and 1-alkylidene-2-vinylidene-cyclopropanes, respectively:[31]

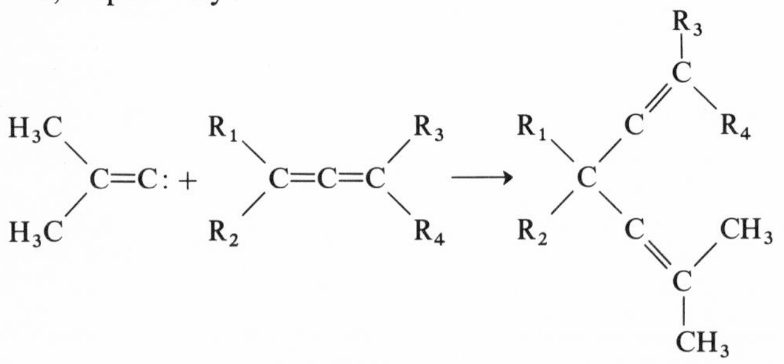

(1)

(1a) $R_1R_2R_3R_4 = CH_3$ (2a) $R_1R_2R_3R_4 = CH_3$ (12%)
(1b) $R_1R_2 = H$;   (2b) $R_1R_2 = H$;
     $R_3R_4 = CH_3$ $R_3R_4 = CH_3$

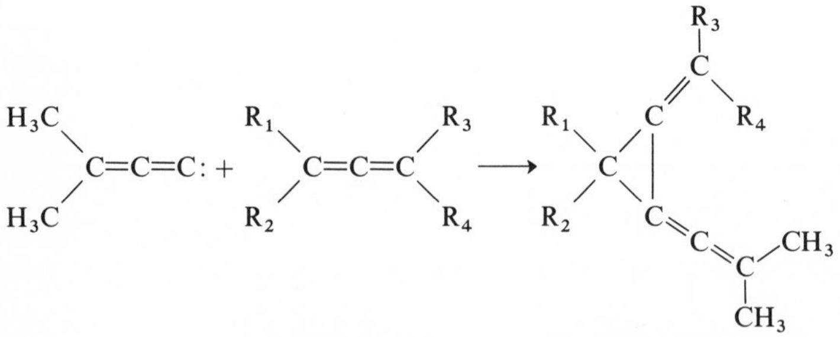

(2)

(3a) $R_1R_2R_3R_4 = CH_3$ (4a) $R_1R_2R_3R_4 = CH_3$
(3b) $R_1R_2 = H$;   (4b) $R_1R_2 = H$;
     $R_3R_4 = CH_3$ $R_3R_4 = CH_3$

At 360°, the 2-alkylidenecyclopropanes isomerize to 1,2-dimethylenecyclopropanes:[32]

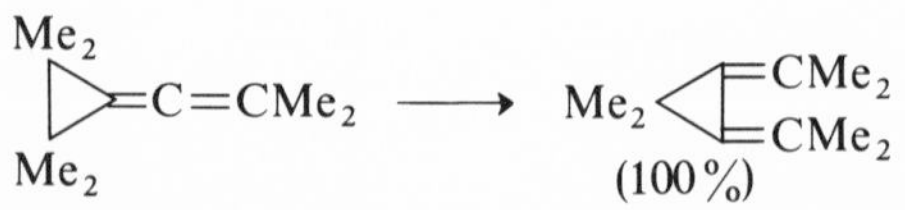

(100%)

2.1.6. ACID CATALYZED CYCLIZATION OF TETRAARYL ALLENES

The first step in this reaction is addition of a proton to the central carbon of the allene bond. Products are triarylindenes:[16,33]

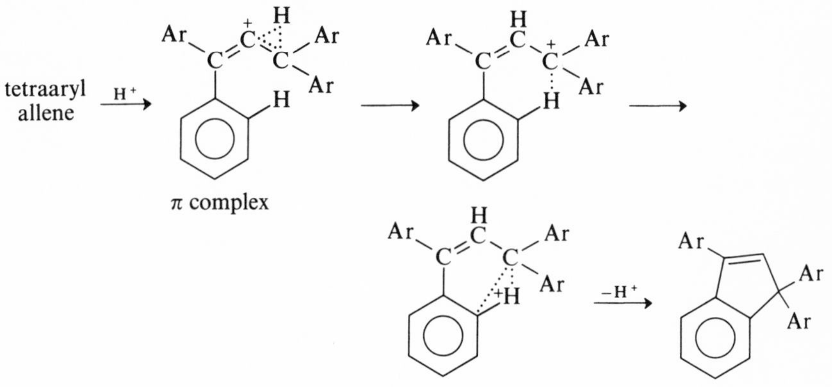

2.1.7. ADDITION OF ALKALI METALS

Tetraphenylallene with sodium forms a red crystalline disodium salt, formulated as (1) by Schlenk.[34] Dowd[35] repeated the reaction at $-78°$ and deuterolyzed the salt to form 90% of (3) and 10% of (4). At higher temperature, less than 1% of (4) formed:

$$\underset{(1)}{Ph_2\overset{-}{C}-\overset{\cdot}{C}-\overset{-}{C}Ph_2} \quad \overset{2Na^+ \quad \overline{D_2O}}{\Big\downarrow} \quad Ph_2CR^1-CR^2=CPh_2$$

(2) $R^1 = H, R^2 = H$
(3) $R^1 = D, R^2 = H$
(4) $R^1 = D, R^2 = D$

The dilithium salt hydrolyzed to form cyclic compounds:

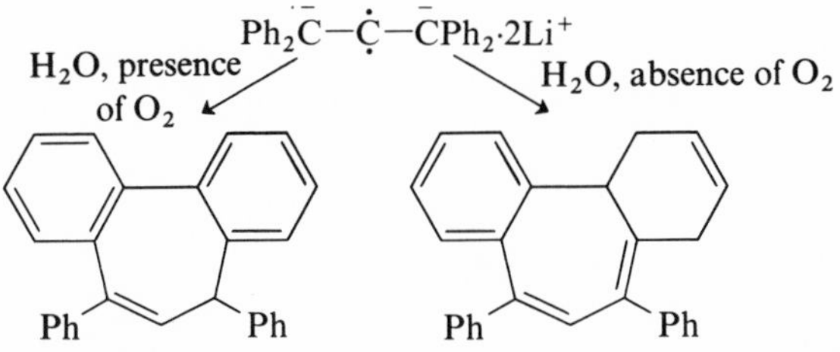

(5)
(12%)

(6)
(40%)

This is another example of an aryl ring participating in ring formation in arylallenes. Potassium salts also hydrolyzed to form the cyclic products (5) and (6).[36]

2.1.8. HYDROBORATION OF ALLENES

Hydroboration of allenes is easy, and boron goes to the central carbon of the allene bond. Diborane and 1,2-cyclononadiene give three products:[37]

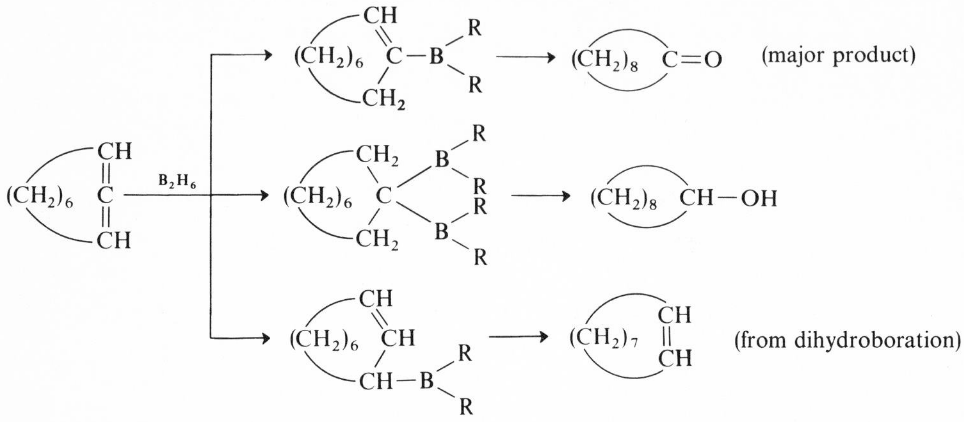

Alkaline hydrogen peroxide converts the boron adducts to the final products. Allene bonds are more reactive than olefinic, just as acetylenic bonds are:

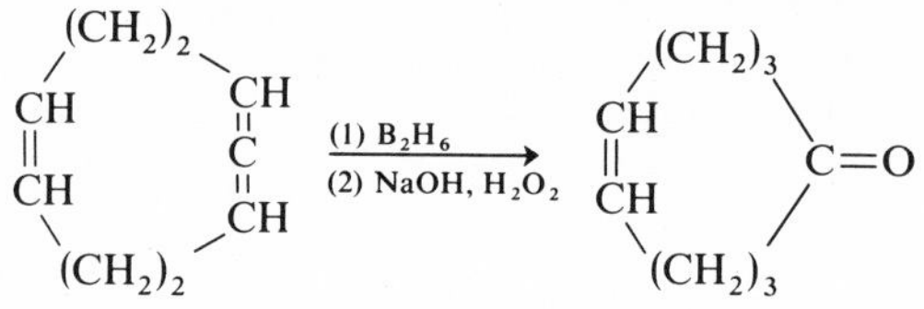

2.1.9. OXYMERCURATION OF ALLENES

1,2-Cyclononadiene reacts with ethanolic $HgCl_2$ to form a white precipitate. —HgCl adds to the central carbon of the allene bond.[38] The product is:

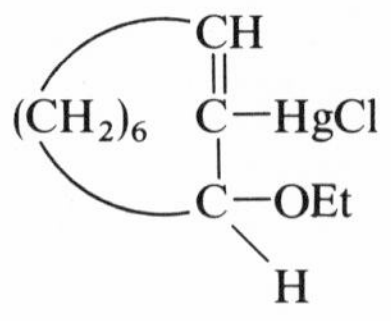

Allene bonds are more reactive than olefinic bonds:

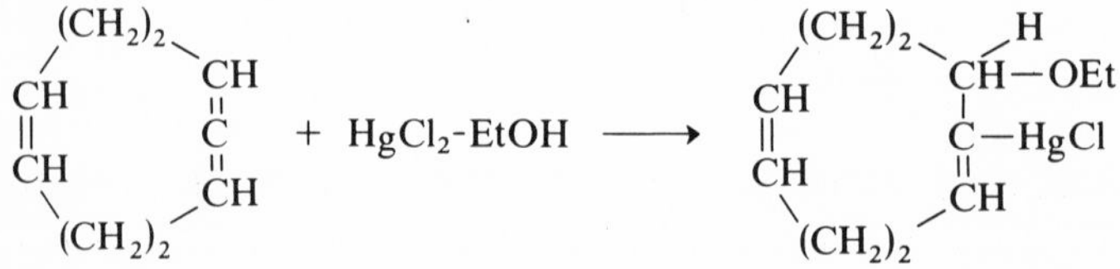

—HgCl also adds to the central carbon of aliphatic allenes:

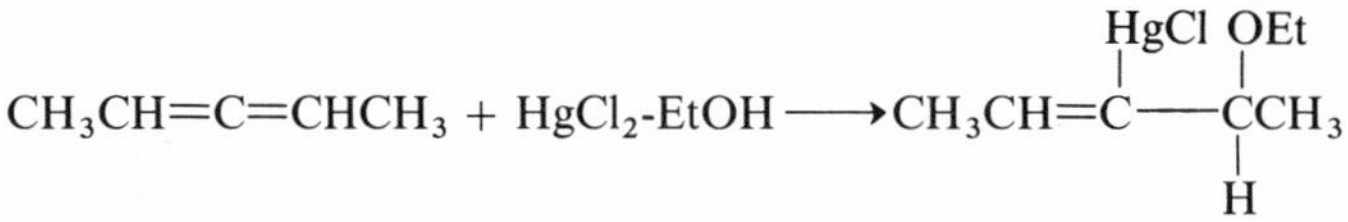

$$CH_3CH=C=CHCH_3 + HgCl_2\text{-EtOH} \longrightarrow CH_3CH=\overset{\overset{\displaystyle HgCl}{|}}{C}-\overset{\overset{\displaystyle OEt}{|}}{\underset{\underset{\displaystyle H}{|}}{C}}CH_3$$

Methoxymercuration is a good example of a clean electrophilic addition. Allene in methanolic mercuric acetate adds two —HgOAc groups, one at each terminal carbon. The resulting dimethoxy compound hydrolyzes rapidly, and the product isolated is the ketone:[38a]

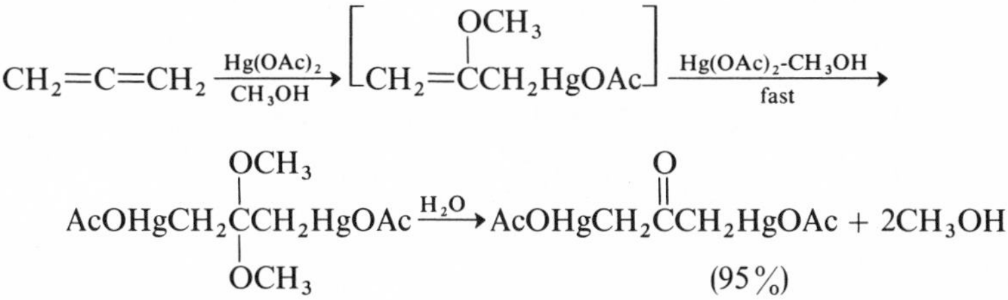

$$CH_2=C=CH_2 \xrightarrow[CH_3OH]{Hg(OAc)_2} \left[CH_2=\overset{\overset{\displaystyle OCH_3}{|}}{C}CH_2HgOAc \right] \xrightarrow[fast]{Hg(OAc)_2\text{-}CH_3OH}$$

$$AcOHgCH_2\overset{\overset{\displaystyle OCH_3}{|}}{\underset{\underset{\displaystyle OCH_3}{|}}{C}}CH_2HgOAc \xrightarrow{H_2O} AcOHgCH_2\overset{\overset{\displaystyle O}{||}}{C}CH_2HgOAc + 2CH_3OH$$

$$(95\%)$$

Substituted allenes add only one —HgOAc group, and usually to the central carbon atom. The *trans* product usually forms because the methanol attacks from the side away from the —HgOAc group in the intermediate σ-bridged mercurinium ion:

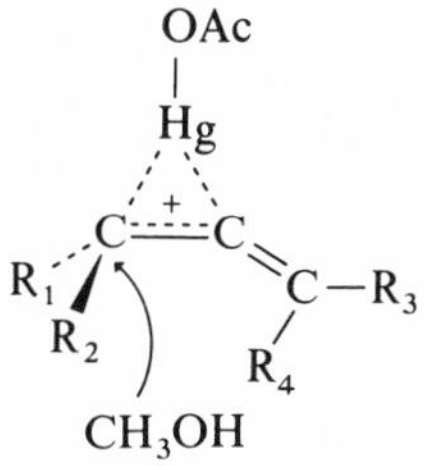

2.1.10. ADDITION OF THIOLATE ANIONS

In alcohol solution, thiolate anion adds to excess allene at 65–85°.[42] The reaction is highly selective: 90 % of the —SR groups add to the central carbon. Unreacted gas is an equilibrium mixture containing 15 % allene and 85 % propyne. Propyne gives the same addition product as allene. Tetramethylallene cannot isomerize to a propyne and does not add thiolate ions under usual conditions. Some free radical addition may occur, especially at higher temperature. Mueller and Griesbaum[42] doubt that these are purely anionic additions to allene, and believe that no completely anionic additions to allene have ever been achieved.

2.1.11. ADDITION OF FORMALDEHYDE—THE PRINS REACTION

Allene and formaldehyde react in 50% aqueous sulfuric acid. If excess formaldehyde is present, 8 moles of formaldehyde react with 1 mole of allene to give 62% yield of 2,4,10,12,15-pentaoxadispiro-[5.1.5.3]hexadecan-7-one:[42a]

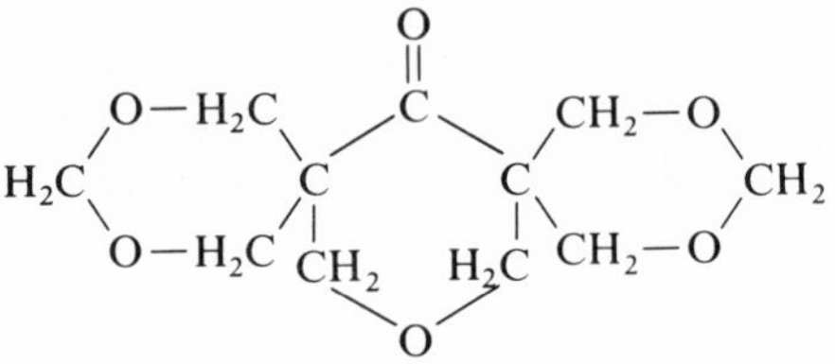

2.1.12. ADDITION OF NEGATIVELY SUBSTITUTED AZIDES

Negatively substituted azides, such as picryl azide, add to allenic hydrocarbons to form 1,2,3-triazolines.[42b] The addition of picryl azide to tetramethylallene is typical:

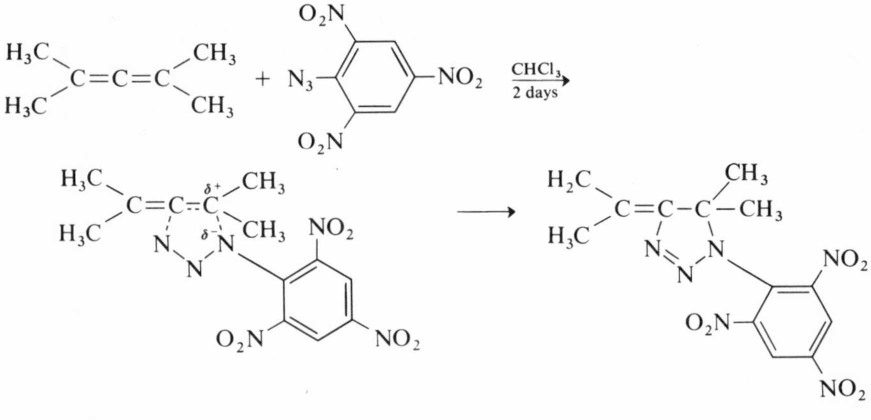

72%
4-isopropylidene-5,5-
dimethyl-1-picryl-Δ^2-
1,2,3-triazoline

2.2. Substituted Allenes

2.2.1. SOLVOLYSIS OF α-ALLENIC ALCOHOLS AND HALIDES

α-Allenic alcohols and halides solvolyze with 20% phosphoric acid in the presence of KCl or KBr. A carbonium ion is probably intermediate to the observed ketone and conjugated diene products:[39]

$$\underset{\substack{| \\ \text{RCHCH=C=CH}_2}}{\overset{\text{OH (or X)}}{}} \xrightarrow[\text{80°, 1 hr}]{\text{H}_3\text{PO}_4 + \text{H}_2\text{O} + \text{KX}} trans\text{-RCH=CH}\overset{\overset{\text{O}}{\|}}{\text{C}}\text{CH}_3 + cis\text{- and}$$

$$trans\text{-RCH=CHCX=CH}_2$$

2.2.2. ADDITIONS TO 1-CYANOALLENES

1-Cyanoallene adds dry HCl at 0°:[40]

$$\text{CH}_2\text{=C=CHCN} + \text{HCl} \longrightarrow \underset{\substack{| \\ \text{CH}_2\text{=CCH}_2\text{CN}}}{\overset{\text{Cl}}{}} \xrightarrow{\text{HCl-MeOH}} \underset{\substack{| \\ \text{CH}_3\text{C=CH}}}{\overset{\text{OMe}}{}}\overset{\overset{\text{O}}{\|}}{\text{C}}\text{NH}_2$$

1-Cyanoallenes add hydrazine or substituted hydrazines to give 3-methyl-5-aminopyrazoles:

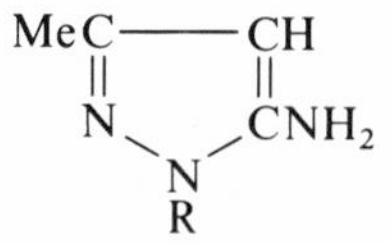

$$\begin{array}{c} \text{MeC}\text{------}\text{CH} \\ \| \qquad \| \\ \text{N} \qquad \text{CNH}_2 \\ \diagdown\text{N}\diagup \\ | \\ \text{R} \end{array}$$

Alcohols and amines also add to 1-cyanoallenes and hydroxylamine gives aminoisoxazoles.[40]

Cycloalkenamines add to cyanoallene to form four- and five-membered rings:[41]

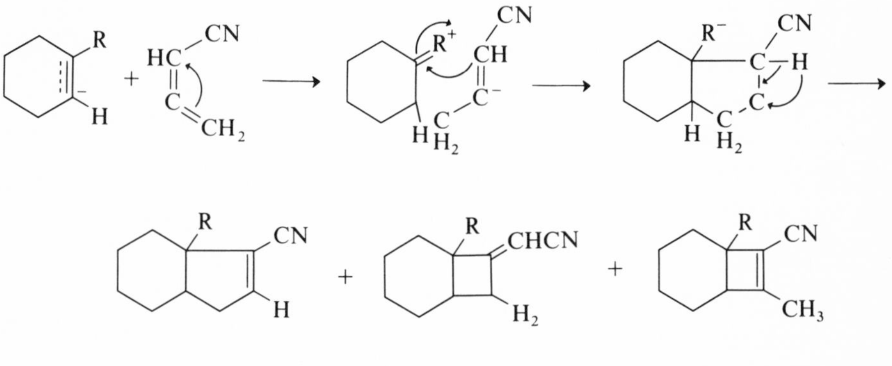

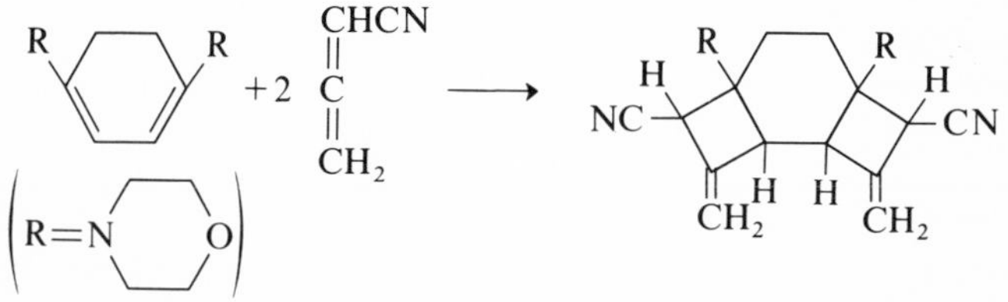

2.2.3. ADDITION TO ALLENIC ACIDS

The allenic acid $BuC(CO_2H){=}C{=}CH_2$ is stable (b.p. 102 at 2 mm) and is not poly-merized by acids, free radicals, metals, ultraviolet light or NaOH.[43] Wotiz[44] added Grignard reagents to this acid, and carbonated the adducts to make olefinic diacids:

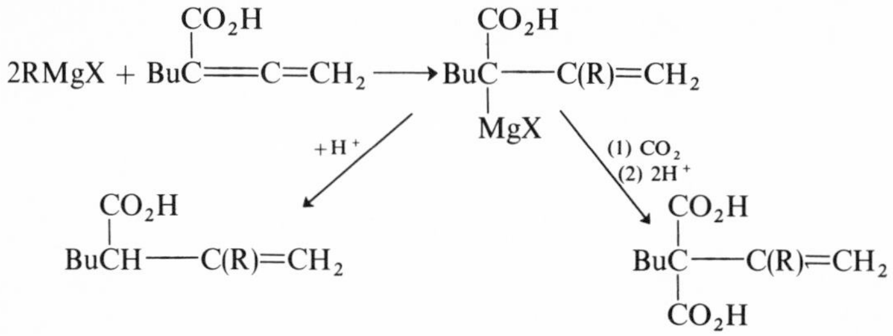

When R was ethyl, adding the Grignard to dry ice in ether gave only 9% yield. Reaction at 20° and 3 atmospheres CO_2 gave 90% yield. When R was *t*-butyl, the yield was 87%. The 1-alkylvinylmalonic acids are strong monobasic acids in 50% ethanol. (The monoanion can form strong internal hydrogen bonds, and thus the acid acts as if it were monobasic.)

α-Allenic acid esters add HN_3 to form β-azidocrotonates. This is the only convenient synthesis of these products.[45] Propargyl esters add azides to form *vic*-triazoles.

$$CH_2{=}C{=}CCO_2Et + NaN_3 \xrightarrow[H_2O]{THF-} CH_3C(N_3){=}CCO_2Et$$

(70%)

Racemic phenylallenecarboxylic acids

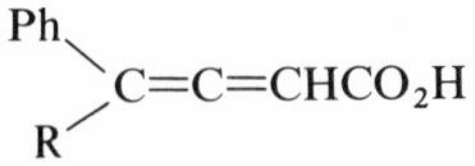

can be resolved via their cinchonidine salts. When the (+)-isomers add bromine (in carbon tetrachloride), HBr is evolved and bromocrotonolactones form in 70–90% yields:[45a]

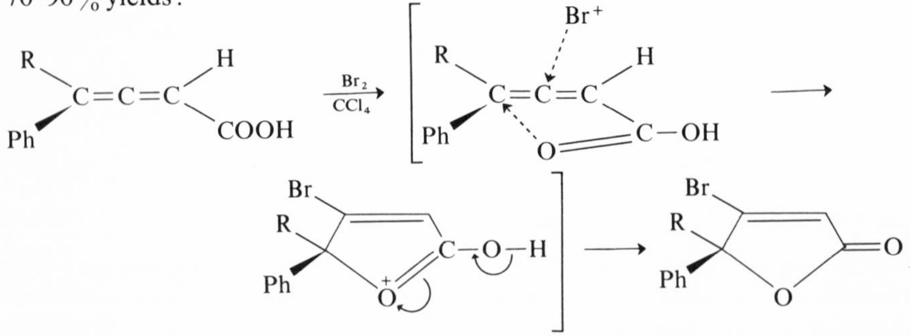

2.2.4. ADDITION TO ALLENIC KETONES

Alcohols containing aqueous 5% potassium carbonate add to α-allenic ketones to form conjugated keto ethers in 90% yield.[46] RO— adds to the central carbon

of the allene bond: $R'\overset{\overset{\displaystyle O}{\|}}{C}CH=C=CH_2 \xrightarrow{ROH} R'\overset{\overset{\displaystyle O}{\|}}{C}CH=\overset{\overset{\displaystyle OR}{|}}{C}CH_3$. α-Allenic acetals

hydrate in the presence of acid to form ketoaldehydes.[78]

$$PhCH_2\overset{}{C}=C=CH_2 \xrightarrow{H_2O, H^+} PhCH_2\overset{}{CH}-\overset{\overset{\displaystyle O}{\|}}{C}CH_3$$
$$\underset{\displaystyle CH(OEt)_2}{\big|} \qquad\qquad\qquad \underset{\displaystyle CHO}{\big|}$$

In 1958, Gaudemar-Bardone[47] reported the first symmetrical allenic diketone, 1,3-dibenzoylallene. The diketone is sensitive to heat and oxidizers, and adds nucleophiles at the central carbon. It dimerizes when acetone or ethanol solutions are allowed to evaporate. Gaudemar-Bardone thought the orange dimer was a dihydrobenzene. In 1966, Agosta[79] proved that the structure is a pyrone methide:

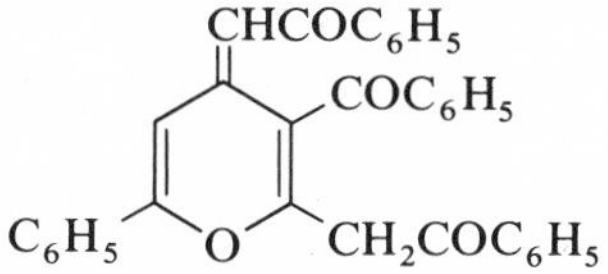

$$
\begin{array}{c}
\text{CHCOC}_6\text{H}_5 \\
\qquad \text{COC}_6\text{H}_5 \\
\\
\text{C}_6\text{H}_5 \quad \text{O} \quad \text{CH}_2\text{COC}_6\text{H}_5
\end{array}
$$

2.2.5. ADDITION TO TETRAFLUOROALLENE AND TETRACHLOROALLENE

Jacobs[48] added 2 moles of chlorine to tetrafluoroallene to form $F_2CCl-CCl_2-CF_2Cl$. He dimerized tetrafluoroallene to

$$
\begin{array}{c}
F_2C-C=CF_2 \\
\ \big|\ \ \big| \\
F_2C-C=CF_2
\end{array}
$$

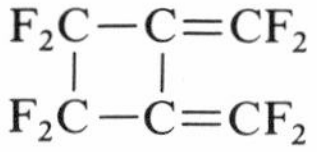

a structure confirmed by Banks.[49] Jacobs also polymerized tetrafluoroallene to a white waxy powder

$$
\left[\begin{array}{c} CF_2 \\ \| \\ C-CF_2 \end{array}\right]_n
$$

The polymer was highly crystalline and insoluble in all ordinary solvents. It melted over a 2° range between 103 and 126°, varying from one preparation to another. The polymer reacted easily with primary and secondary amines at the $=CF_2$ group. It did not depolymerize to monomer at 700°.

Tetrafluoroallene reacts much better with reagents usually regarded as electrophilic than fluoroolefins do. HX adds to give 90–99% yields of

$$CF_2XCH=CF_2.$$

Nucleophilic reagents also add very readily: neutral methanol at $-20°$ gives an ether, $MeOCF_2CH=CF_2$. Moist CeF fluorinates tetrafluoroallene to form 2H-pentafluoropropene.[49] The very easy addition of nucleophiles to tetrafluoroallene indicates that the addition of "electrophilic" HX is really a nucleophilic attack by X^- instead of electrophilic attack by H^+.

Tetrachloroallene dimerizes very easily to form

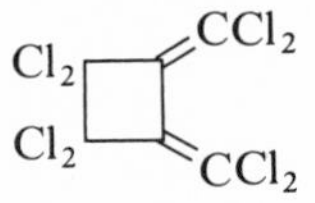

In solution at -75 to $-30°$ in the presence of $AlCl_3$, however, electrophilic additions are faster than dimerization.[49a] Chlorine, bromine, ICl and HCl add to form substituted propenes. The direction of addition of ICl and of HCl shows that the central carbon atom has the highest electron density:

$$Cl_2C=C=CCl_2 \quad \begin{cases} \xrightarrow{+ICl} Cl_2C=CI-CCl_3 \\ \xrightarrow{+HCl} Cl_2C=CH-CCl_3 \end{cases}$$

Acetyl chloride adds in the presence of $AlCl_3$ to form

2.2.6. REARRANGEMENT OF β-ALLENIC HALIDES TO CYCLO-PROPYL KETONES

This is a general synthesis of cyclopropyl ketones[50,51]

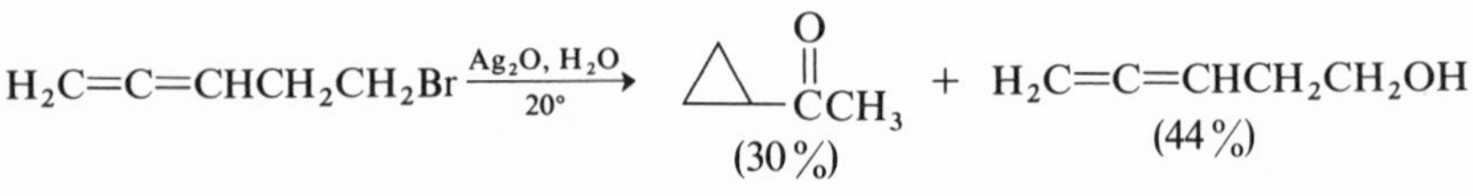

In general, the Br can be replaced by OH, NH_2, OMe or naphthylsulfonate. Water, acetic acid and methanol are good solvents, and silver oxide, calcium carbonate and sodium acetate are useful bases. The reaction has given 9–80% yields of cyclopropyl ketones.

2.2.7. ADDITION TO α-ALLENIC SULFONES

Thiophenol adds to the central carbon of the allene bond in the presence of triethylamine:[52]

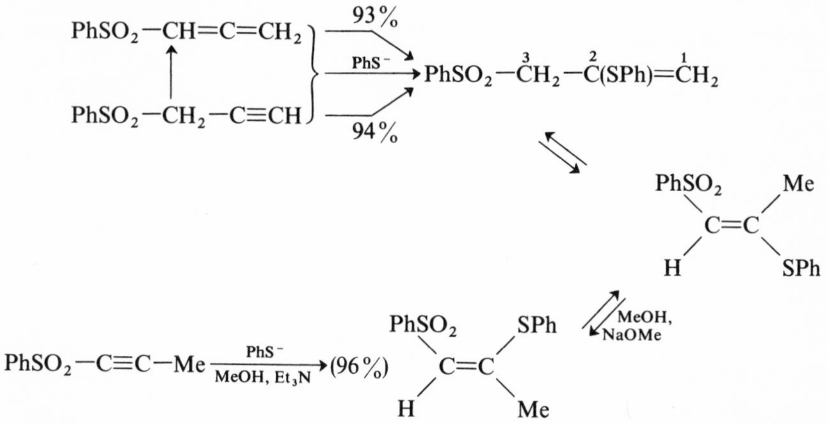

$$PhSO_2-CH=C=CH_2 \Bigg\} \xrightarrow[PhS^-]{93\%} PhSO_2-\overset{3}{C}H_2-\overset{2}{C}(SPh)=\overset{1}{C}H_2$$
$$PhSO_2-CH_2-C\equiv CH \Bigg/ \xrightarrow{94\%}$$

$$\begin{array}{cc} PhSO_2 & Me \\ \diagdown C=C \diagup \\ \diagup \qquad \diagdown \\ H & SPh \end{array}$$

$$PhSO_2-C\equiv C-Me \xrightarrow[MeOH,\ Et_3N]{PhS^-} (96\%) \begin{array}{cc} PhSO_2 & SPh \\ \diagdown C=C \diagup \\ \diagup \qquad \diagdown \\ H & Me \end{array} \nearrow \begin{array}{c} MeOH, \\ NaOMe \end{array}$$

Benzenesulfinate anion adds in a similar way. Propynyl, propargyl and allenic sulfones add carboxylate ion to form vinyl esters which react with secondary amines to form ketones: $ArSO_2CH_2COCH_3$.[51a] Yields are good.

2.2.8. ADDITION TO α-ALLENIC ALCOHOLS

Methylene iodide adds to α-allenic alcohols in the presence of zinc-copper couples to form alkylidenecyclopropylcarbinols and spiropentylcarbinols:[52a]

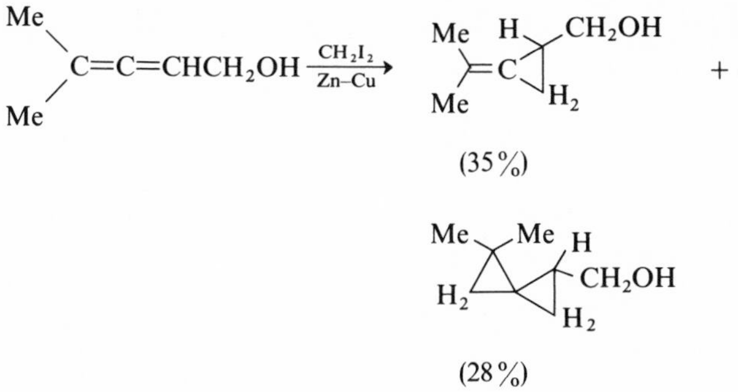

$$\begin{array}{c} Me \\ \diagdown \\ \quad C=C=CHCH_2OH \xrightarrow[Zn-Cu]{CH_2I_2} \\ \diagup \\ Me \end{array} \begin{array}{c} Me \quad H \diagdown CH_2OH \\ \diagdown \qquad \diagup C \diagdown \\ Me \diagup \qquad H_2 \end{array} \quad +$$

$$(35\%)$$

$$\begin{array}{c} Me \diagdown \diagup Me \quad H \\ \bigtriangleup \diagdown CH_2OH \\ H_2 \diagup \qquad H_2 \end{array}$$

$$(28\%)$$

2.2.9. ADDITION TO β-ALLENIC ALCOHOLS

Optically active 2,2-dimethyl-3,4-hexadien-1-ol adds 2,4-dinitrobenzenesulfenyl chloride to form optically active 3-(2,4-dinitrophenylthio)-1,5,5-trimethyl-Δ^3-dihydropyran.[52b] The hydroxyl group participates in this reaction and also in the addition of bromine:

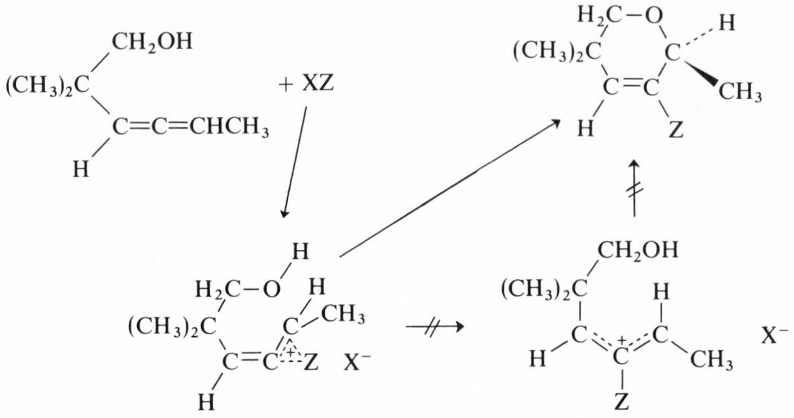

The stereospecificity of the addition reactions indicates that the open-chain carbonium ion is not involved.

3. DIELS-ALDER ADDITIONS OF ALLENES

The Diels-Alder reactions of allenes have not been studied very much, even though allenes are generally good dienophiles and give good yields.

Allene adds to cyclopentadiene in pentane at 200° to give 60% yield of 5-methylenebicyclo[2.2.1]hept-2-ene after 6 hours. The adduct polymerizes to a white solid polymer in the presence of BF_3 at $-80°$.[53]

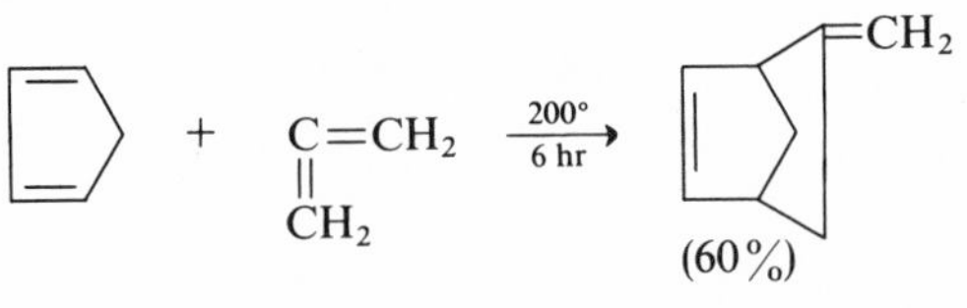

1-Cyanoallene reacts[40] with cyclopentadiene to give

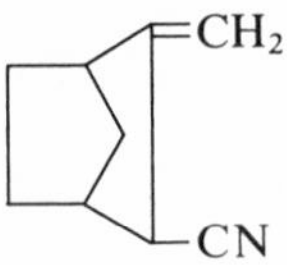

and with butadiene to form

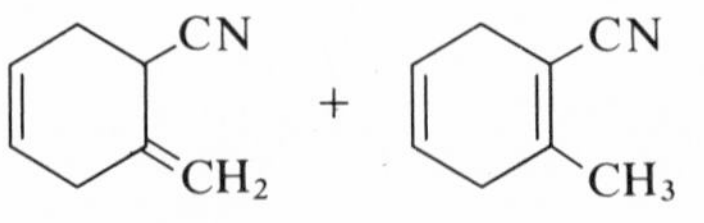

1-Carboxyallene reacts with cyclopentadiene at room temperature; isomeric 2-butynoic acid requires reflux for several hours:[54]

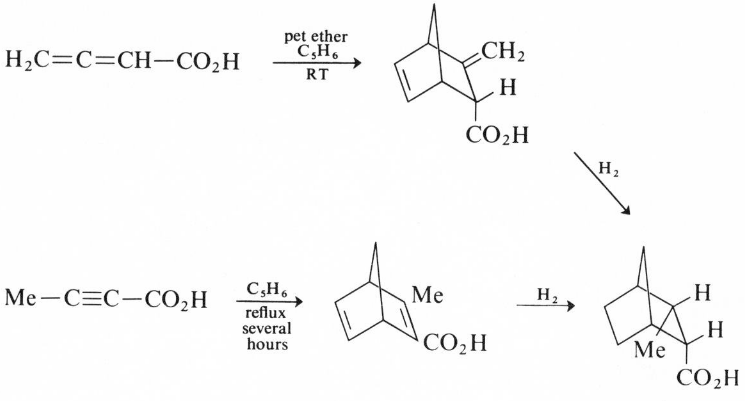

$$H_2C=C=CH-CO_2H \quad \xrightarrow[\text{RT}]{\substack{\text{pet ether} \\ C_5H_6}}$$

$$Me-C\equiv C-CO_2H \quad \xrightarrow[\substack{\text{reflux} \\ \text{several} \\ \text{hours}}]{C_5H_6}$$

(40%)

1,1-Difluoroallene is a potent dienophile. It reacts with cyclopentadiene to form the 1:1 adduct in quantitative yield.[55] In vinylallene, one of the cumulative double bonds acts as part of the diene system:[56]

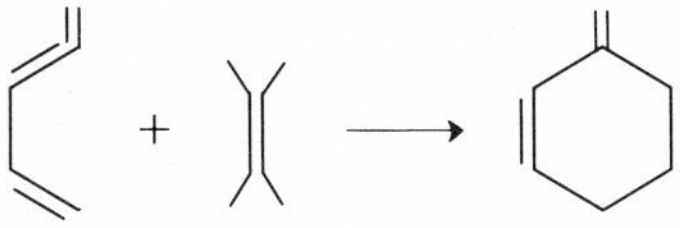

Aliphatic α-allenic ketones are also good dienophiles.[56a] The ketones

$$\underset{\displaystyle RC-CH=C=CH_2}{\overset{\displaystyle O}{\overset{\|}{}}} \quad (R = Et, \; n\text{-Pr and } i\text{-Pr})$$ react with 2,3-dimethyl-1,3-butadiene at 110° (4 hours) to give 80–85% yields of acyl methylenecyclohexenes:

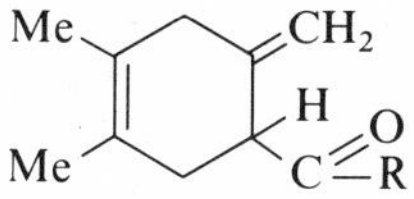

Cyclopentadiene reacts at only 50° to form 95% yields of the stereoisomers

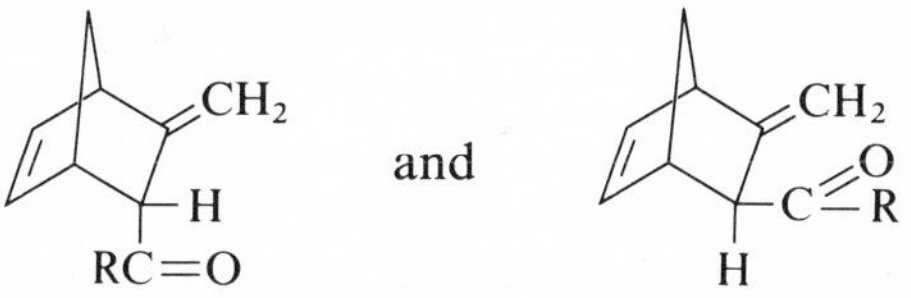

and

Furan reacts at 80–90° to give 60–65% yields of 2-acyl-3-methylene-1,4-epoxy-5-cyclohexenes:

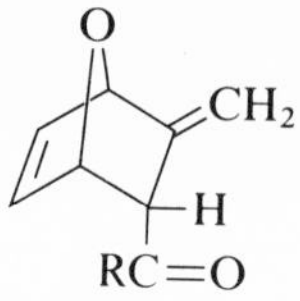

and

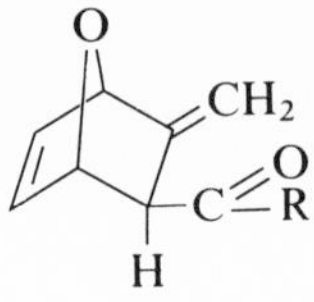

The interesting by-product from this reaction is

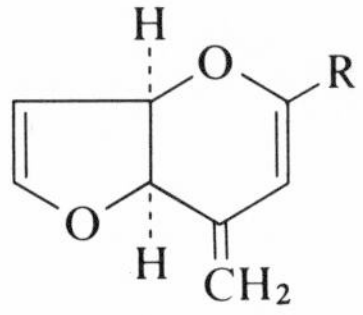

(15% yield)

4. CYCLIZATION REACTIONS OF ALLENES

4.1. Thermal Cyclodimerization

Lebedev[57] in 1913 assigned these structures to products of thermal cyclooligomerization of allene:

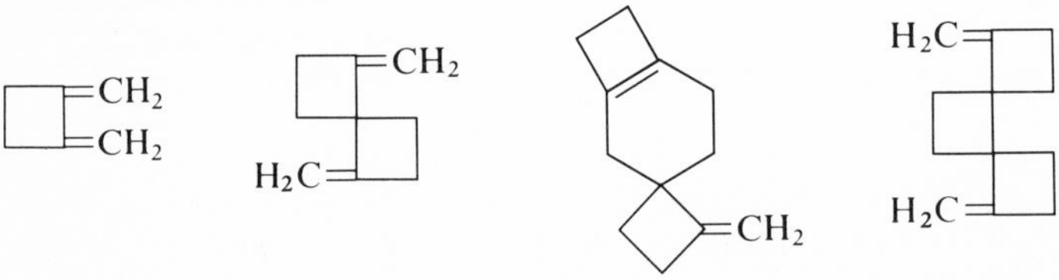

All of these have since been identified except the third one.[54a,b] Products similar to the second one have been made.[55] Almost any allene will cyclodimerize to methylenecyclobutanes under the proper conditions:[1,58]

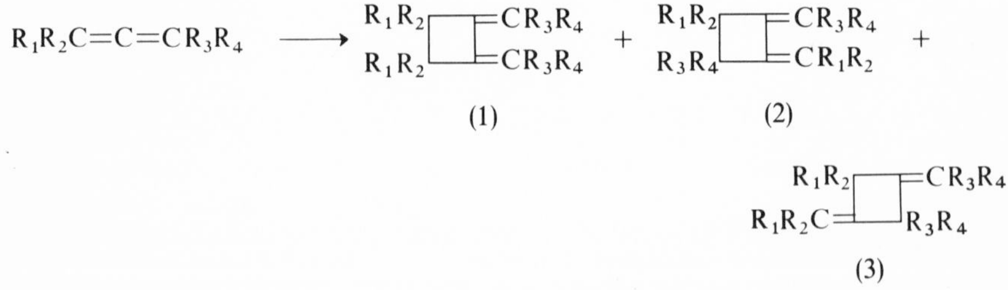

$$R_1R_2C=C=CR_3R_4 \longrightarrow$$

(1)   (2)

(3)

Dimer (1) is the major product when the R groups are H, alkyl, F or aryl. Yields of (1) are 50–90% after a few seconds at 500°, or after several days at 150°.[1] Allene itself gives 85% of (1), along with 15% yield of (3).[57] 1,1-Difluoroallene forms only (2), 35% yield, at 300°.[55] Bis(biphenylene)allene gives (3) exclusively.[33] Thermal cyclodimerization may be a radical reaction:[1]

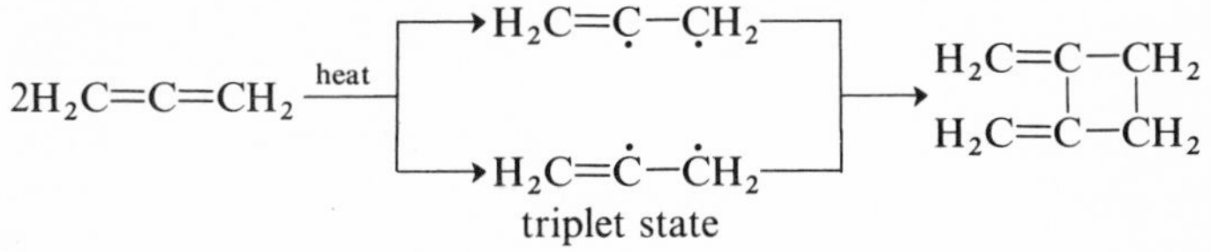

$$2H_2C{=}C{=}CH_2 \xrightarrow{\text{heat}} \begin{array}{c} \rightarrow H_2C{=}\overset{\cdot}{C}{-}\overset{\cdot}{C}H_2 \\ \\ \rightarrow H_2C{=}\overset{\cdot}{C}{-}\overset{\cdot}{C}H_2 \end{array} \longrightarrow \begin{array}{c} H_2C{=}C{-}CH_2 \\ \;\;|\quad\;\; | \\ H_2C{=}C{-}CH_2 \end{array}$$

triplet state

The direction of cyclodimerization depends on temperature in some cases. Thus, 1,1,4,4-tetraphenyl-1,2-butadiene gives good yields of the two dimers at different temperatures:[33a]

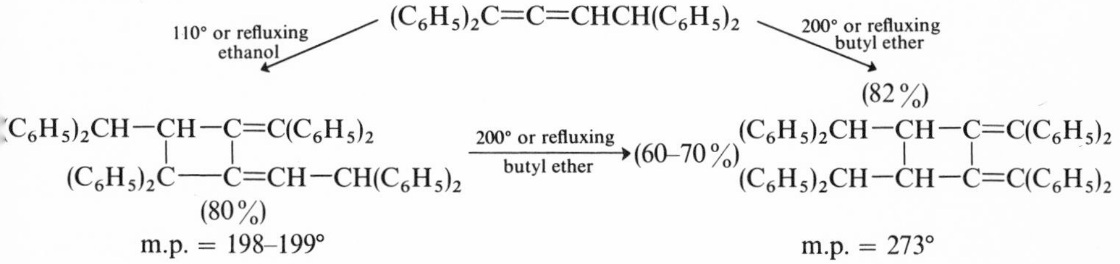

$(C_6H_5)_2C{=}C{=}CHCH(C_6H_5)_2$

110° or refluxing ethanol →

200° or refluxing butyl ether → (82%)

$(C_6H_5)_2CH{-}CH{-}C{=}C(C_6H_5)_2$
$(C_6H_5)_2C{-}\!\!-\!\!C{=}CH{-}CH(C_6H_5)_2$
(80%)
m.p. = 198–199°

$\xrightarrow[\text{butyl ether}]{200° \text{ or refluxing}}$ (60–70%)

$(C_6H_5)_2CH{-}CH{-}C{=}C(C_6H_5)_2$
$(C_6H_5)_2CH{-}CH{-}C{=}C(C_6H_5)_2$
m.p. = 273°

Initially formed dimethylenecyclobutanes can be isomerized in the presence of base to methylenecyclobutenes. Thermal dimerization of 1,3-diphenylallene in neutral medium forms *cis*-1,2-diphenyl-*anti*-3,4-dibenzylidenecyclobutane. In basic medium, the product is isomerized:[33b]

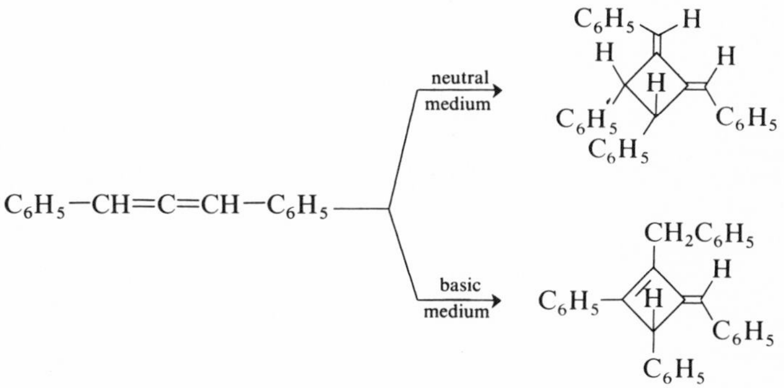

$C_6H_5{-}CH{=}C{=}CH{-}C_6H_5$

neutral medium →

basic medium →

The methylenecyclobutanes undergo Diels-Alder additions to form spiro[3.3]-heptanes. For example, tetracyanoethylene gives:

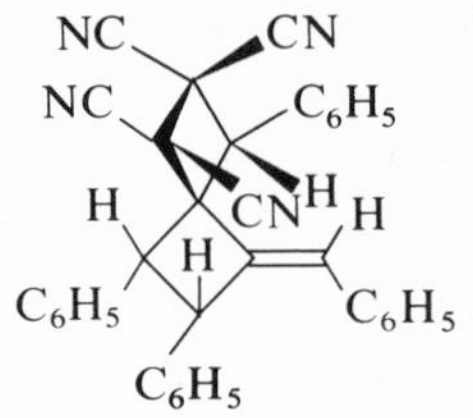

At 500° allene trimerizes to 1,5-dimethylenespiro[3.3]heptane and 2,5-dimethylenespiroheptane. Passed through a tube at 500°, allene gives 3-methylenebicyclo[4.2.0]-1-octene and 3-methylenebicyclo[4.2.1]-1(6),2-octadiene.[59] The allene diradical may have reacted with allene to form methylenecyclobutanes, the methylene groups of which reacted with another allene molecule.

Thermal dimerization and oligomerization of allene in the liquid phase at 140° gives six products.[54a] The first step may be a simple 1,2-cycloaddition. The trimers may also be formed by 1,2-cycloaddition. The cyclohexane derivatives formed as trimers probably result from a Diels-Alder reaction between allene and 1,2-dimethylenecyclobutane:

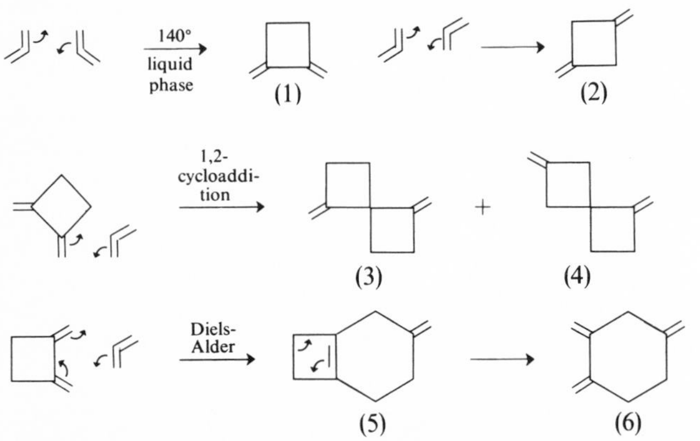

(1,2-cycloaddition reactions)

(1,2-cycloaddition)

(Diels-Alder addition)

The methylenecyclobutanes probably form by ring closure of the allylic diradicals resulting from the first interaction between two molecules of allene:

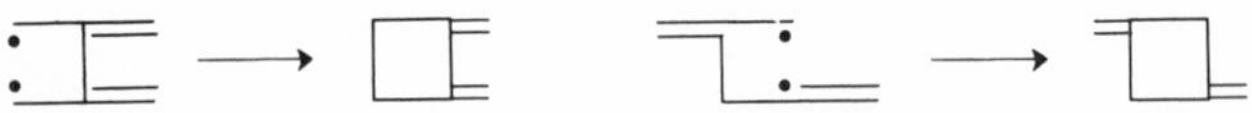

(1) (2)

Three thermal tetramers have been identified.[54b] The major tetramer is

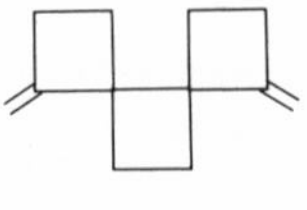

The other two are

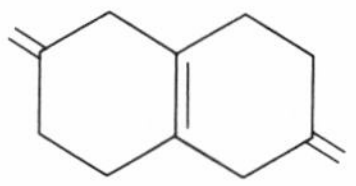

and

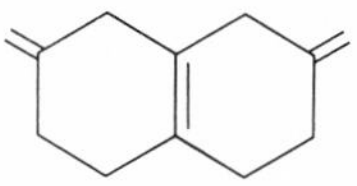

The four major thermal pentamers of allene are:[54c]

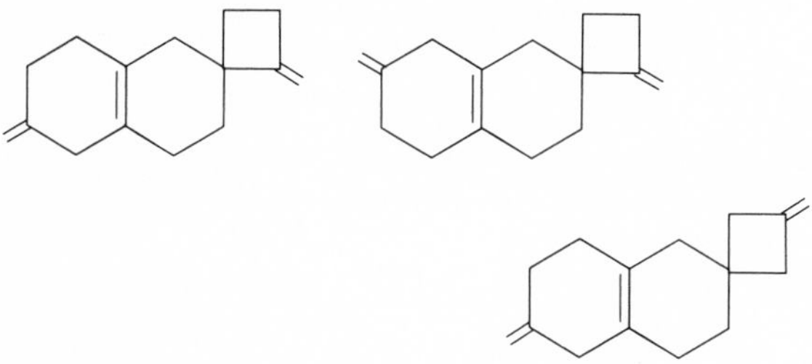

Arylallenes sometimes dimerize to indenes. One of the aryl groups participates in the formation of the indene ring.[60]

4.2. Catalytic Cyclooligomerization

Benson[61] made entirely different products by using modified nickel carbonyl catalysts. He heated allene to 105° for 5–6 hours in the presence of the catalysts and obtained two trimers, which are very reactive, and a tetramer which is surprisingly stable:

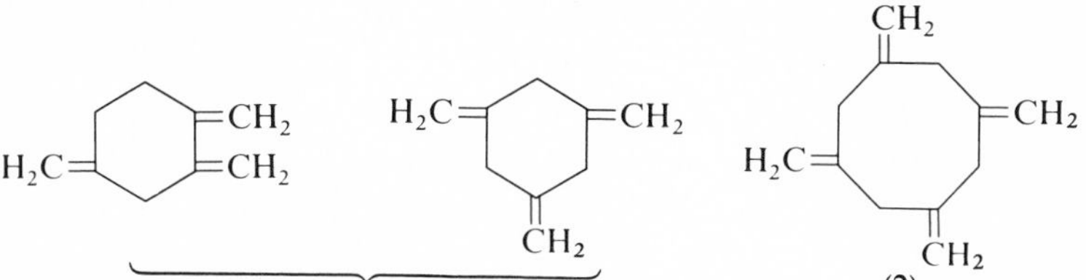

Catalyst	% Yield	
	(1)	(2)
$[(PhO)_3P]_2Ni(CO)_2$	35	6
$(PhP)_2Ni(CO)_2$	7	18
$[(PhO)_3P]_3NiCO$, and		
$(PhO)_3PNi(CO)_3$	20–25	

Similar catalysts catalyze the reaction of allene with acetylene to form exo-methylene cyclics with ring unsaturation, compounds which previously required more complex syntheses. With excess acetylene some cyclooctatetraene formed:

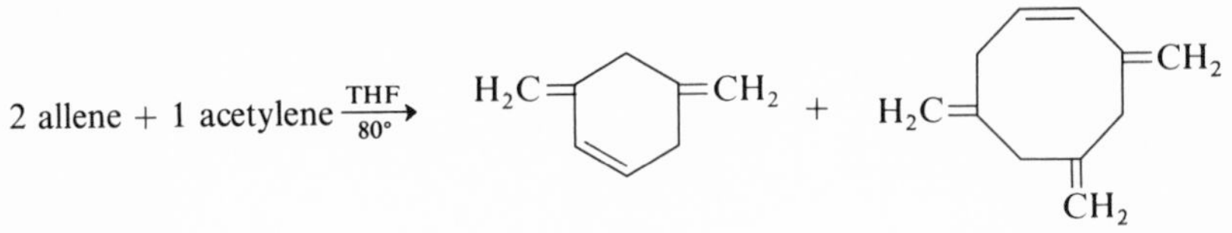

$$2 \text{ allene} + 1 \text{ acetylene} \xrightarrow[80°]{\text{THF}}$$

(3)   (4)

Catalyst	% Yield	
	(3)	(4)
Ni acetylacetonate	45	5
Ni cyanide	45	—
Ni ethylacetoacetate	37	—

Substituted acetylenes gave only the methylenecyclohexenes with nickel acetoacetate catalyst (R = Me, Ph or $CH_2=CH-$):

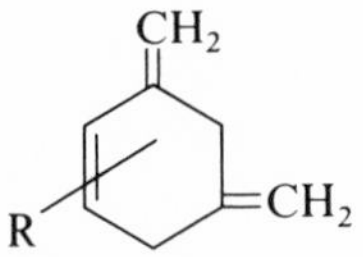

4.3. Thermal Cycloaddition of Olefins to Allenes

Cripps[62] reported a new, easy and general synthesis of cyclobutanes: allene is simply heated with a negatively substituted olefin in an autoclave:

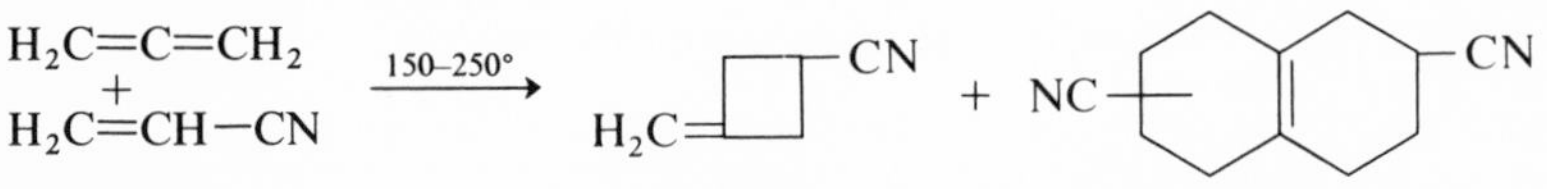

1-cyano-3-methylene-   dicyanooctahydro-
cyclobutane   naphthalene

The octahydronaphthalene probably forms via allene dimer:

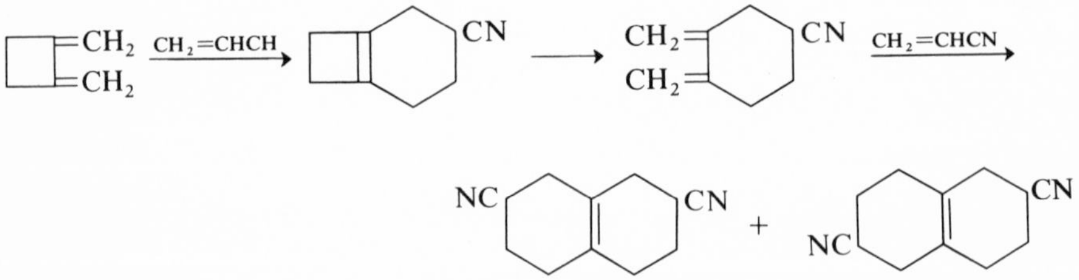

Allene and maleic anhydride give three products:

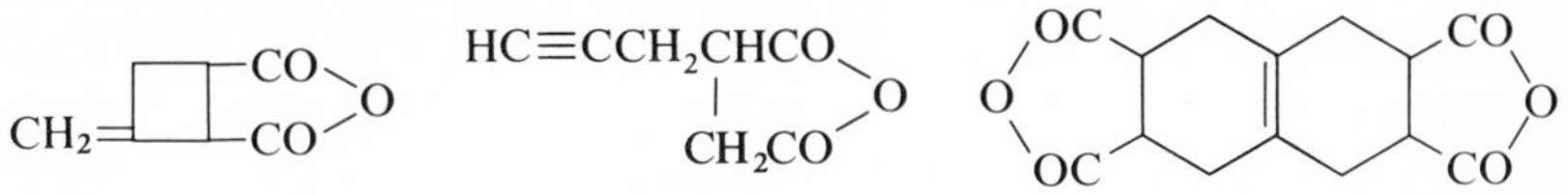

Generally, the methylenecyclobutanes

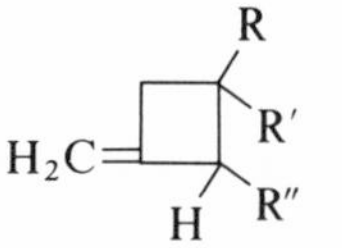

formed on 10–60% yield (R and/or R' = CN, ester, acid, aldehyde, pyridine). Methacrylonitrile gave the highest yield.

Octahydronaphthalenes are the main products from equimolar amounts of reactants under similar conditions. Acrylic acid gave only 4% yield, while acrylonitrile gave 75% yield.

4.4. Other Cycloaddition Reactions

Diphenylketene and tetramethylallene at 20° for 4 hours give 65% yield of 4-isopropylidene-3,3-dimethyl-2,2-diphenylcyclobutanone[63]

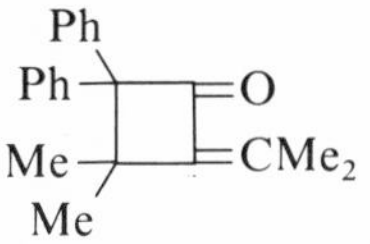

4.5. Cyclization Reactions of Fluoroallenes

Tetrafluoroallene reacts[64] with hexafluoro-2-butyne at 80° to give 50% yield of

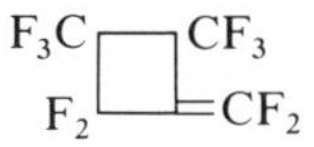

and 42% yield of

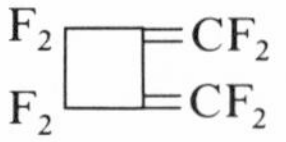

1,1-Difluoroallene shows some of the characteristics of both allene and tetrafluoroallene.[55] Passed through a quartz tube at 280°, difluoroallene gives 25–35% yield of dimer. The dimer polymerizes spontaneously within an hour at room temperature, but can be stabilized by inhibitors such as phenothiazine:

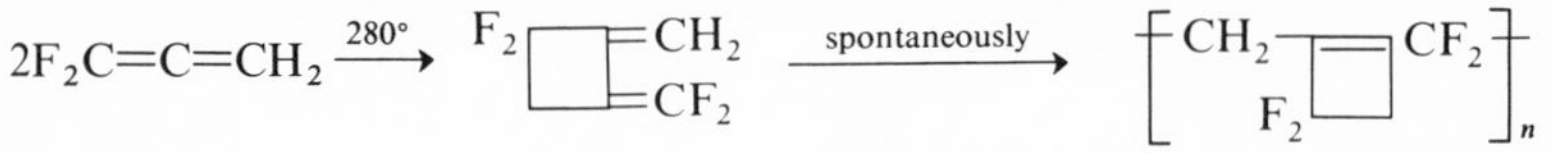

$$2F_2C{=}C{=}CH_2 \xrightarrow{280°} \quad \text{[structure: } F_2 \text{ ring } {=}CH_2, {=}CF_2\text{]} \xrightarrow{\text{spontaneously}} \quad \left[\text{structure: } CH_2, CF_2, F_2 \right]_n$$

Other addition products of 1,1-difluoroallene are:

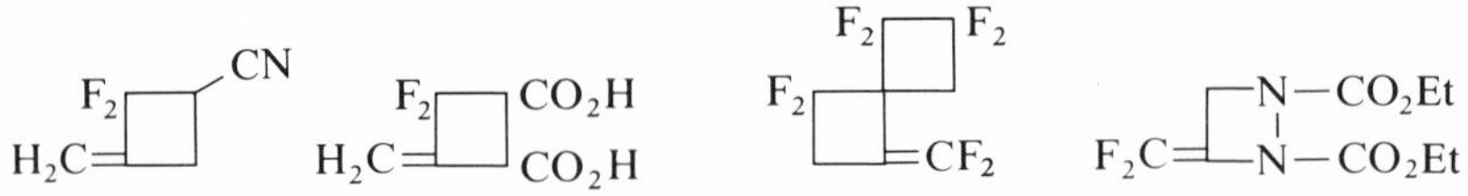

with acrylonitrile	with maleic anhydride	with tetrafluoro-	with ethyl
(43%)	(after hydrolysis)	ethylene	azidodiformate
	(16%)	(9.2%)	(6.8%)

5. METAL COMPLEXES OF ALLENE—CARBONYLATION

Carbonylation of allene was first reported in 1959.[65] Carbonylation in methanol solution containing platinum chloride-stannous chloride catalyst gave 40% yield of methyl methacrylate.

5.1. Ruthenium Catalysts

In 1961, Kealy[66] described the use of ruthenium catalysts for carbonylation. The major product is methyl methacrylate, but at higher temperatures up to 23% yield of dimethyl α,α-dimethyl-α'-methyleneglutarate forms at the expense of methyl methacrylate:

$$CH_2{=}C{=}CH_2 + CO + MeOH \xrightarrow[Ru_2(CO)_9]{140°}$$

$$\underset{(50\%)}{CH_2{=}\overset{CH_3}{\underset{}{C}}{-}CO_2Me} + \underset{(0\%)}{MeO_2C{-}\overset{CH_3}{\underset{CH_3}{C}}{-}CH_2{-}\overset{CH_2}{\underset{}{C}}{-}CO_2Me}$$

$$\xrightarrow{140{-}190°} \quad (18\%) \qquad\qquad (23\%)$$

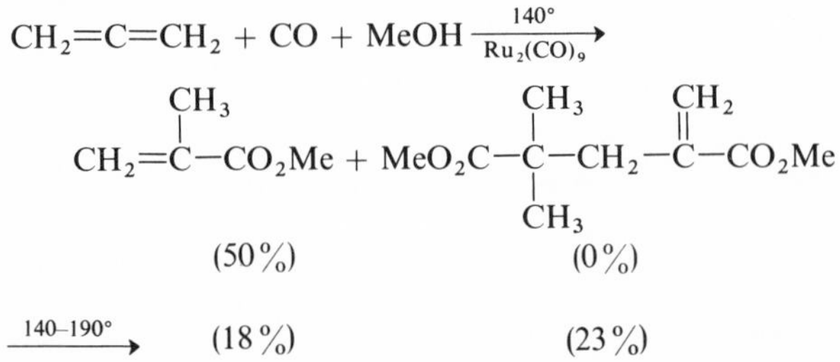

In water with ruthenium chloride-pyridine catalyst, the yield of methacrylic acid is 20%, along with two by-products:

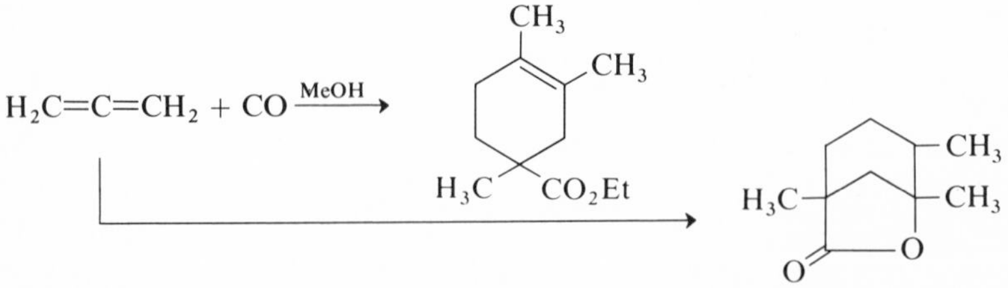

$$H_2C{=}C{=}CH_2 + CO \xrightarrow{MeOH}$$

Carbonylation at 135° with ruthenium chloride catalyst in the presence of cyclohexylamine gives 16% yield of cyclohexylacrylamide.

5.2. π-Allylic Complexes

In 1964, Schultz[67] reported the formation of π-allylic complexes directly from allene and dichlorobis(benzonitrile)palladium:

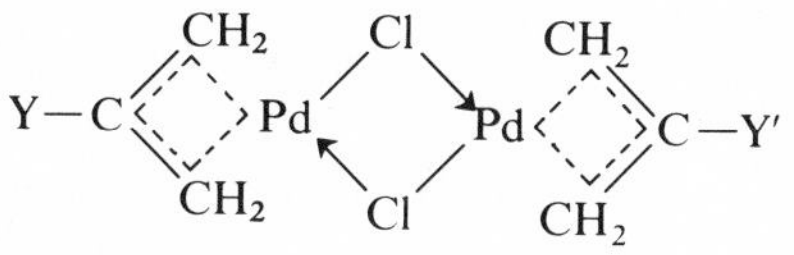

(1a) $Y = Y' = Cl$ (allene into $PdCl_2 \cdot 2PhCN$)

(1b) $Y = Cl; Y' =$ (inverse addition)

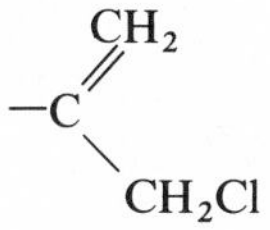

(1c) $Y = Y' =$ 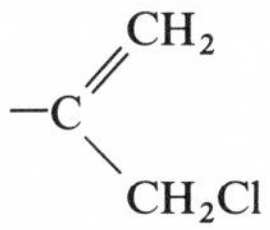   (allene into $PdCl_2 \cdot 2PhCN$ in PhCN)

(1d) $Y = Y' =$ 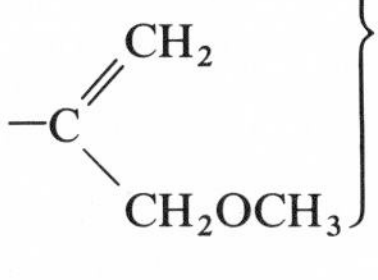   (allene into $PdCl_2 \cdot 2PhCN$ in methanol)

Complex (1a) reverts to allene and forms $PdCl_2(PMe_2Ph)_2$ when treated with phenyldimethylphosphine.[69]

Lupin[68,69] isolated only (1c) (88%) from allene and sodium chloropalladite in methanol at 25°. He identified the complex from tetramethylallene and $PdCl_2 \cdot 2PhCN$ in benzene as:

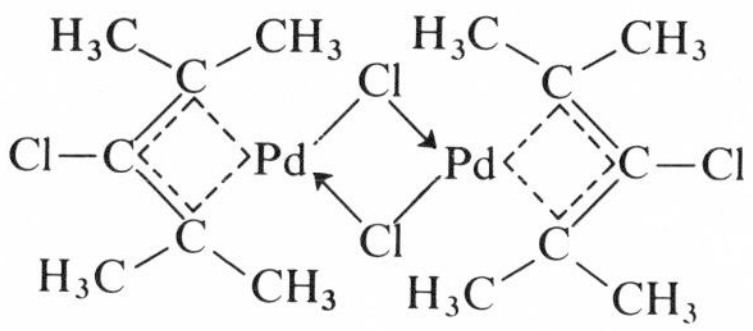

The mechanism of the allene-$PdCl_2 \cdot 2PhCN$ reaction is probably:

$$allene + PdCl_2 \cdot PhCN$$

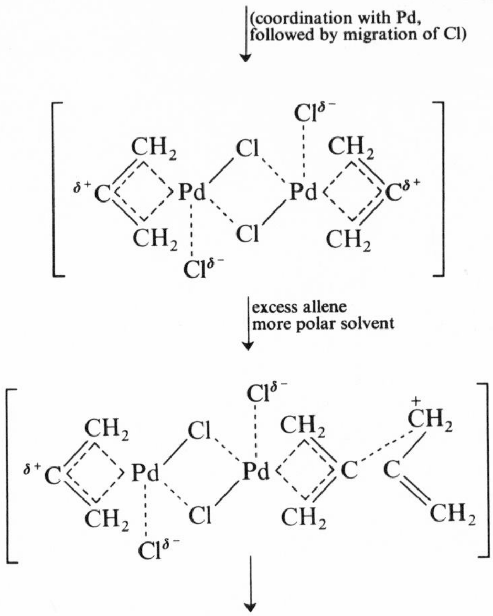

(coordination with Pd,
followed by migration of Cl)

excess allene
more polar solvent

(1b), (1c) and (1d)

5.3. Carbonylation of Allene-PdCl$_2$ Complexes

Tsuji[70] carbonylated allene-$PdCl_2$ complexes made from allene and $PdCl_2$:

$$CH_2{=}C{=}CH_2 + PdCl_2 \longrightarrow$$

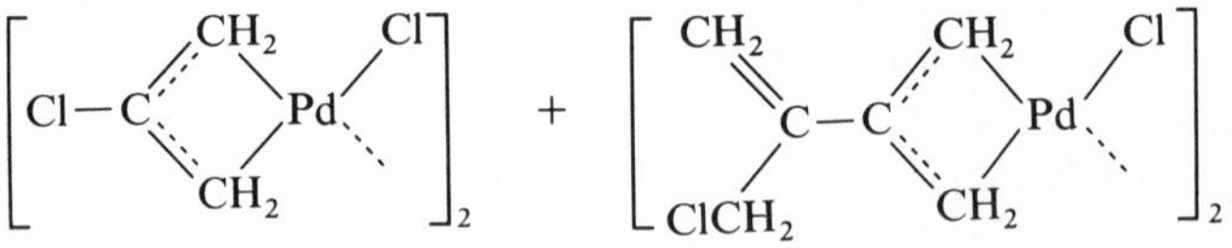

(1)   (2)

+CO, EtOH

$$\underset{\text{CH}_2{=}\text{CCH}_2\text{CO}_2\text{Et}}{\overset{\text{Cl}}{|}}$$

Complex (2) gave three products, corresponding to mono-, di- and tri-carbonylation. A possible mechanism is:

(2) + CO + EtOH $\longrightarrow$

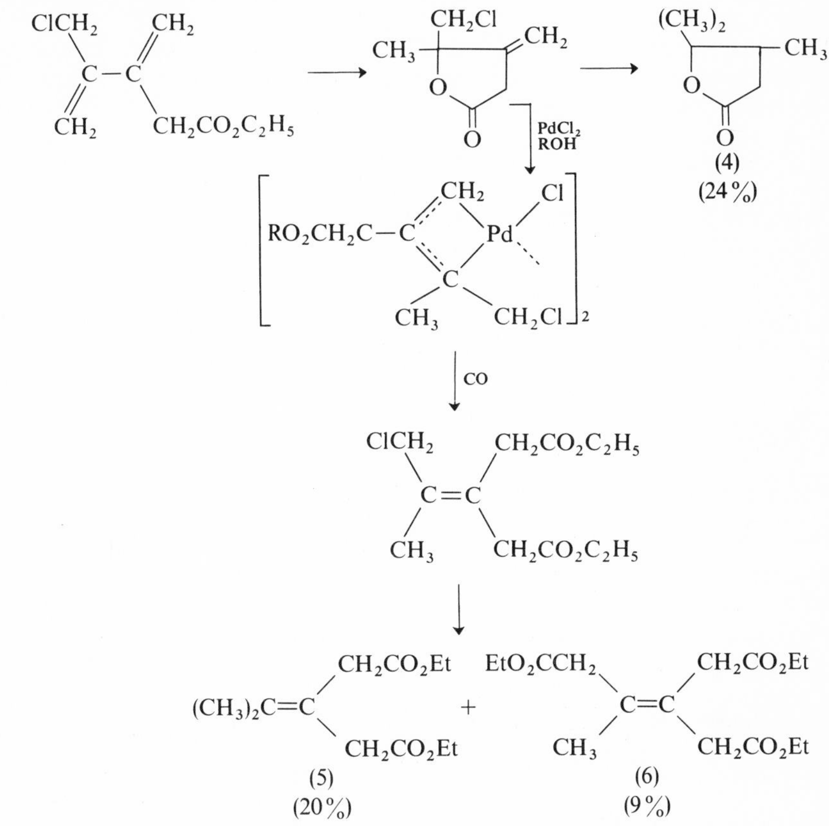

$$\underset{CH_2}{\overset{ClCH_2}{>}}C-C\underset{CH_2CO_2C_2H_5}{\overset{CH_2}{<}} \longrightarrow$$

(a lactone) $CH_3-\underset{O}{\overset{CH_2Cl}{|}}\underset{\overset{\|}{O}}{} CH_2$

$\longrightarrow$ $(CH_3)_2 \cdots CH_3$ lactone (4) (24%)

$\downarrow$ PdCl$_2$ ROH

$$\left[RO_2CH_2C-C\underset{\underset{CH_3}{C}}{\overset{CH_2}{\diagup}}Pd\underset{CH_2Cl}{\overset{Cl}{<}} \right]_2$$

$\downarrow$ CO

$$\underset{CH_3}{\overset{ClCH_2}{>}}C=C\underset{CH_2CO_2C_2H_5}{\overset{CH_2CO_2C_2H_5}{<}}$$

$\downarrow$

$(CH_3)_2C=C\underset{CH_2CO_2Et}{\overset{CH_2CO_2Et}{<}}$ + $\underset{CH_3}{\overset{EtO_2CCH_2}{>}}C=C\underset{CH_2CO_2Et}{\overset{CH_2CO_2Et}{<}}$

(5)   (6)
(20%)   (9%)

Carbonylation of allene in the presence of PdCl$_2$ without prior complex formation gave ethyl itaconate (in low yield) as the sole product:

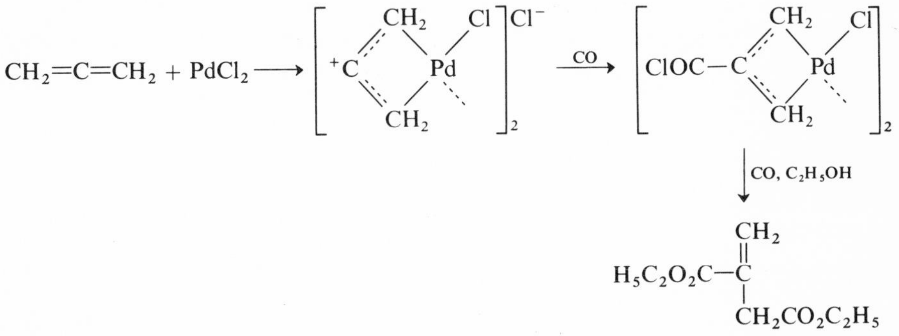

$$CH_2=C=CH_2 + PdCl_2 \longrightarrow \left[{}^+C\underset{CH_2}{\overset{CH_2}{<}}Pd\cdots Cl \right]_2 Cl^- \xrightarrow{CO} \left[ClOC-C\underset{CH_2}{\overset{CH_2}{<}}Pd\cdots Cl \right]_2$$

$\downarrow$ CO, C$_2$H$_5$OH

$$H_5C_2O_2C-\underset{CH_2CO_2C_2H_5}{\overset{CH_2}{\underset{|}{\overset{\|}{C}}}}$$

5.4. Nickel Carbonyl as Carbonylation Catalyst

Nickel carbonyl-methacrylic acid is a more effective catalyst than nickel carbonyl-water.[70a] Nickel carbonyl-HCl, commonly used for carbonylating acetylene, is inactive. The best yield of methyl methacrylate obtained with nickel carbonyl-methacrylic acid was 62%. 1,2-Hexadiene is more difficult to carbonylate because it polymerizes more readily than allene. CO attaches to the central carbon atom of the allene system exclusively.

6. POLYMERIZATION OF ALLENES

The polymerization of allenes has not been studied as much as the polymerization of acetylenes. Baker[71] reviewed allene polymerizations in 1964. Most of the recent work has involved nickel complex catalysts or Ziegler catalysts.

6.1. Allene

6.1.1. COMPLEXED METAL CATALYSTS

Otsuka[72] made crystalline, stereoregular polymers from allene by using complexes such as bis(1,5-cyclooctadiene)nickel, bis(π-allyl)nickel, π-allylnickel bromide, and bis(acrylonitrile)nickel. Zerovalent and univalent nickel complexes are active; divalent nickel complexes are inactive. Nickelous acetylacetonate reduced with $AlEt_3$ gives a gel polymer. A number of iron carbonyls were found to be inactive as catalysts.

Coordination of the monomer in such systems is usually assumed to be the first step in polymerization, but this has not often been proved. Otsuka obtained a stable allene-nickel complex by dissolving bis(1,5-cyclooctadiene)nickel in liquid allene below $-20°$. The reddish brown needles of complex were active catalysts for polymerizing allene above 10°.

Otsuka's polyallenes were colorless solids, melting around 60°, soluble in benzene, ether, carbon disulfide and carbon tetrachloride. They were poorly soluble in ethanol, benzene and cyclohexane. The polymer is a 2_1 helix, with pendant methylene groups, according to spectral data.

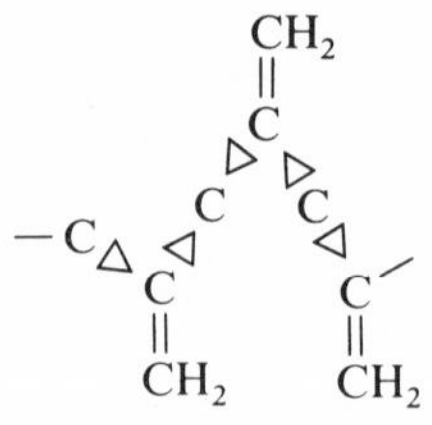

Baker[73] used transition metal catalysts to make high molecular weight crystalline polyallene. Infrared and x-ray plots suggested blocks of vinylidene

$$CH_2$$
$$\|$$

structures $+CH_2-C+_n$ and blocks of mixed vinylidene and *cis*-vinyl structures.

Low valent nickel catalysts were used to prepare polyallene with a molecular weight around 100,000.[74] From x-ray and NMR data, a purely vinylidene structure in a helix coil was apparent.

6.1.2. ZIEGLER CATALYSTS

Robinson[75] claimed that a number of Ziegler catalysts polymerized allene to a white powder. One product, formed in 16% yield, could be pressed into films. It melted above 280°, was insoluble in benzene, and had pendant methylene groups.

6.2. Allene Copolymers

Allene and ethylene copolymerize over Ziegler catalysts to form copolymers with vinylidene groups.[76] Copolymers are less crystalline than polyethylene.

6.3. Diborane Addition Polymer

Diborane adds to allene in the gas phase to form cyclic 1,2-trimethylenediborane. Polymerization occurs when trimethylenediborane is warmed quickly from $-190°$ to $+25°$. The liquid becomes more viscous and sometimes solidifies. The polymer reverts to monomer when heated to 60° under vacuum.[77] The reversible reactions may be represented by the following sequence:

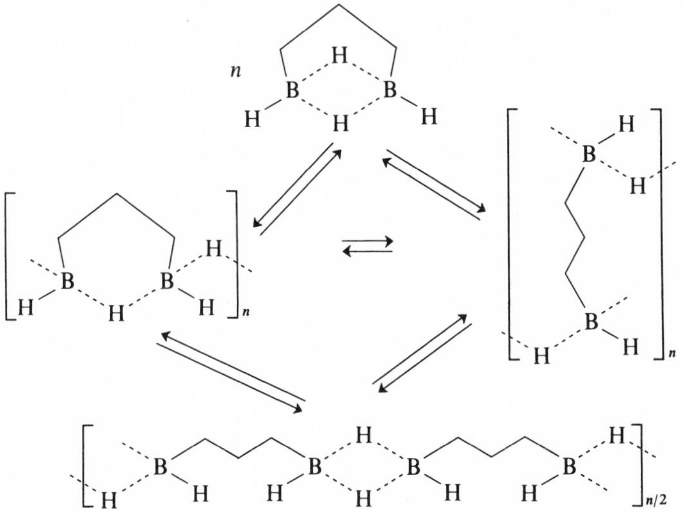

6.4. Fluoroallenes

6.4.1. 1,1-DIFLUOROALLENE

The high polymer from 1,1-difluoroallene has a cyclobutane ring:[55]

$$\left[CH_2 \square CF_2 \right]_n$$
$$F_2 \quad H_2$$

6.4.2. TETRAFLUOROALLENE

Tetrafluoroallene can polymerize by a radical process. Polymerization at 15–20° and 10 atmospheres for 54 hours gives a white, waxy solid. Allene gives only low molecular weight oligomers under more drastic radical conditions. Several structures are possible, but the preferred one involves polymerization of allylic radicals, which are the most likely radicals from addition of initiator $R\cdot$:[49]

$$R\cdot + CF_2{=}C{=}CF_2 \longrightarrow RC{-}CF_2\cdot \xrightarrow{\ nC_3F_4\ }$$
$$\overset{\|}{C}F_2$$

$$RC{-}CF_2\left[C{-}CF_2 \right]_{n-1} C{-}CF_2\cdot$$
$$\overset{\|}{C}F_2 \quad \overset{\|}{C}F_2 \qquad \overset{\|}{C}F_2$$

References

1. Haszeldine, R. N., Leedham, K., and Steele, B. R., *J. Chem. Soc.*, 2040 (1954).
2. Rajbenbach, A., and Szwarc, M., *Proc. Roy. Soc. (London), Ser. A*, **251**, 1266 (1959).
3. Griesbaum, K., *et al.*, *J. Org. Chem.*, **29**, 2404 (1964).
4. Kovacic, D., and Leitch, L., *Can. J. Chem.*, **44**, 1239 (1966).
5. Heiba, E.-H. I. and Haag, W. D., *J. Org. Chem.*, **31**, 3814 (1966).
6. Jacobs, T. L., and Illingworth, G. E., *J. Org. Chem.*, **28**, 2692 (1963).
7. Griesbaum, K., *et al.*, *J. Org. Chem.*, **28**, 1952 (1963).
8. Maitland, P., and Mills, W. H., *J. Chem. Soc.*, 987 (1936).
9. Kuhn, R., and Fischer, H., *Chem. Ber.*, **94**, 3060 (1961).
10. Meyer, E. F., and Burwell, R. L., Jr., *J. Am. Chem. Soc.*, **85**, 2877, 2881 (1963).
11. Kuivila, H. G., Rahman, W., and Fish, R. H., *J. Am. Chem. Soc.*, **87**, 2835 (1965).
12. van der Ploeg, H. J., Knotnerus, J., and Bickel, A. F., *Rec. Trav. Chim.*, **81**, 775 (1962).
13. Hall, D. N., Oswald, A. A., and Griesbaum, K., *J. Org. Chem.*, **30**, 3829 (1965).
13a. Oswald, A. A., *et al.*, *Can. J. Chem.*, **45**, 1173 (1967).
14. Heiba, E.-H. I., *J. Org. Chem.*, **31**, 776 (1966).
15. Griesbaum, K., *et al.*, *J. Org. Chem.*, **30**, 261 (1965).
16. Fischer, H., in (Patai, S., editor) "The Chemistry of Alkenes", p. 1025, New York, Interscience Publishers, 1964.
16a. Sausen, G. N., and Logothetis, A. L., *J. Org. Chem.*, **32**, 226 (1967).
16b. Neale, R. S., *J. Org. Chem.*, **32**, 3263 (1967).

17. Griesbaum, K., Naegele, W., and Wanless, G. G., *J. Am. Chem. Soc.*, **87**, 3151 (1965).
18. Griesbaum, K., *J. Am. Chem. Soc.*, **86**, 2301 (1964).
19. Griesbaum, K., *Angew. Chem.*, **76**, 782 (1964).
20. Jacobs, T. L., and Johnson, R. L., *J. Am. Chem. Soc.*, **82**, 6397 (1960).
21. Peterson, P. E., and Duddey, J. E., *J. Am. Chem. Soc.*, **85**, 2865 (1963).
22. Criegee, R., and Moschel, A., *Chem. Ber.*, **92**, 2181 (1959); *Org. Syn.*, **46**, 34 (1966).
22a. Grimaldi, J., Cozzone, A., and Bertrand, M., *Bull. Soc. Chim. France*, 2723 (1927).
23. Peer, H. G., *Rec. Trav. Chim.*, **81**, 113 (1962).
24. Mueller, W. H., Butler, P. E., and Griesbaum, K., *J. Org. Chem.*, **32**, 2651 (1967).
25. Mueller, W. H., Rubin, R. M., and Butler, P. E., *J. Org. Chem.*, **31**, 3537 (1966).
26. Moriconi, E. J., and Kelly, J. F., *J. Am. Chem. Soc.*, **88**, 3657 (1966).
26a. Peer, H. G., and Schors, A., *Rec. Trav. Chim.*, **86**, 167 (1967).
27. Ball, W. J., and Landor, S. R., *Proc. Chem. Soc.*, 246 (1961).
28. Simmons, H. E., and Smith, R. D., *J. Am. Chem. Soc.*, **81**, 4256 (1959).
29. Ullman, E. F., and Franshawe, W. J., *J. Am. Chem. Soc.*, **83**, 2379 (1961).
30. Rahman, W., and Kuivila, H. G., *J. Org. Chem.*, **31**, 772 (1966).
30a. Peer, H. G., and Schors, A., *Rec. Trav. Chim.*, **86**, 161 (1967).
30b. Dehmlow, E. V., *Chem. Ber.*, **100**, 2779 (1967).
30c. Mueller, W. H., and Butler, P. E., *J. Org. Chem.*, **33**, 1533 (1968).
31. Bleiholder, R. F., and Shechter, H., *J. Am. Chem. Soc.*, **86**, 5032 (1964).
32. Crandall, J. K., and Paulson, D. R., *J. Am. Chem. Soc.*, **88**, 4302 (1966).
33. Fischer, H., and Fischer, H., *Chem. Ber.*, **97**, 2975 (1964).
33a. Sisenwine, S. F., and Day, A. R., *J. Org. Chem.*, **32**, 1770 (1967).
33b. Dehmlow, E. V., *Chem. Ber.*, **100**, 3260 (1967).
34. Schlenk, W., and Bergmann, E., *Ann. Chem.*, **463**, 228 (1928).
35. Dowd, P., *Chem. Commun.*, 568 (1965).
36. Hoffmann, A. K., and Zweig, A., *J. Am. Chem. Soc.*, **84**, 3278 (1962).
37. Devaprabhakara, D., and Gardner, P. D., *J. Am. Chem. Soc.*, **85**, 1458 (1963).
38. Sharma, R. K., Shoulders, B. A., and Gardner, P. D., *J. Org. Chem.*, **32**, 241 (1967).
38a. Waters, W. L., and Kiefer, E. F., *J. Am. Chem. Soc.*, **89**, 6261 (1967).
39. Bertrand, M., and Le Gras, J., *Compt. Rend.*, **261**, 762 (1965).
40. Kurtz, P., Gold, H., and Disselnkötter, H., *Ann. Chem.*, **624**, 1 (1959).
41. Ried, W., and Käppeler, W., *Ann. Chem.*, **687**, 183 (1965).
42. Mueller, W. H., and Griesbaum, K., *J. Org. Chem.*, **32**, 856 (1967).
42a. Peer, H. G., and Schors, A., *Rec. Trav. Chim.*, **86**, 597 (1967).
42b. Bleiholder, R. F., and Shechter, H., *J. Am. Chem. Soc.*, **90**, 2131 (1968).
43. Wotiz, J. H., and Bletsoe, N. C., *J. Org. Chem.*, **19**, 403 (1954).
44. Wotiz, J. H., and Merrill, H. E., *J. Am. Chem. Soc.*, **80**, 866 (1958).
45. Harvey, G. R., and Ratts, K. W., *J. Org. Chem.*, **31**, 3907 (1966).
45a. Shingu, K., Hagishita, S., and Nakagawa, M., *Tetrahedron Letters*, 4371 (1967).
46. Bertrand, M., and Le Gras, J., *Compt. Rend.*, **260**, 6926 (1965).
47. Gaudemar-Bardone, F., *Ann. Chim. Paris*, **XIII3**, 52 (1958).
48. Jacobs, T. L., and Bauer, R. S., *J. Am. Chem. Soc.*, **81**, 606 (1959).
49. Banks, R. E., Haszeldine, R. N., and Taylor, D. R., *J. Chem. Soc.*, 978 (1965).
49a. Roedig, A., and Heinrich, B., *Chem. Ber.*, **100**, 3716 (1968).
50. Hanack, M., and Haeffner, J., *Tetrahedron Letters*, 2191 (1964).
51. Hanack, M., and Haeffner, J., *Chem. Ber.* **99**, 1077 (1966).

51a. Appleyard, G. P., and Stirling, C. J. M., *J. Chem. Soc. (C)*, 2686 (1967).

52. Stirling, C. J. M., *J. Chem. Soc.,* 5856 (1964).

52a. Bertrand, M., and Maurin, R., *Bull. Soc. Chim. France*, 2779 (1967).

52b. Jacobs, T. L., Macomber, R., and Zunker, D., *J. Am. Chem. Soc.,* **89**, 7001 (1967).

53. Cripps, H. N. (to E. I. du Pont de Nemours & Co.), U.S. Patent 3,214,483 (Oct. 26, 1965); *Chem. Abstr.*, **64**, 3380 (1966).

54. Jones, E. R. H., Mansfield, G. H., and Whiting, M. C., *J. Chem. Soc.,* 4073 (1956).

54a. Weinstein, B., and Fenselau, A. H., *J. Chem. Soc. (C)*, 368 (1967).

54b. Weinstein, B., and Fenselau, A. H., *J. Org. Chem.,* **32**, 2278 (1967).

54c. *Ibid.*, 2988 (1967).

55. Knoth, W. H., and Coffman, D. D., *J. Am. Chem. Soc.,* **82**, 3873 (1960).

56. Jones, E. R. H., Lee, H. H., and Whiting, M. C., *J. Chem. Soc.,* 341 (1960).

56a. Bertrand, M., and Le Gras, J., *Bull. Soc. Chim. France*, 4336 (1967).

57. Lebedev, S. V., and Merezhkowskii, B. K., *J. Russ. Phys. Chem. Soc.,* **45**, 1249 (1913); *Chem. Zentr.*, **85**, 1410 (1914).

58. Bertrand, M., Reggio, H., and Leandri, G., *Compt. Rend.,* **259**, 827 (1964).

59. Slobodin, Ya. M., and Khitrov, A. D., *Zh. Org. Khim.,* **1**, 1531 (1965); *Chem. Abstr.*, 613 (1966).

60. Strausz, F., and Ehrenstein, M., *Ann. Chem.,* **442**, 93 (1925).

61. Benson, R. E., and Lindsey, R. V., Jr., *J. Am. Chem. Soc.,* **81**, 4247, 4250 (1959).

62. Cripps, H. N., Williams, J. K., and Sharkey, W. H., *J. Am. Chem. Soc.,* **80**, 751 (1958); **81,** 2723 (1959).

63. Gott, P. G. (to Eastman Kodak Co.), French Patent 1,414,456 (Oct. 15, 1965); *Chem. Abstr.*, **64**, 6523 (1966).

64. Banks, R. E., *et al., J. Chem. Soc. (C)*, 2051 (1966).

65. Jenner, E. L., and Lindsey, R. V., Jr. (to E. I. du Pont de Nemours & Co.), U.S. Patent 2,876,254 (March 3, 1959).

66. Kealy, T. J., and Benson, R. E., *J. Org. Chem.,* **26**, 3126 (1961).

67. Schultz, R. G., *Tetrahedron*, **20**, 2809 (1964).

68. Lupin, M. S., Powell, J., and Shaw, B. L., *J. Chem. Soc. (A)*, 1687 (1966).

69. Lupin, M. S., and Shaw, B. L., *Tetrahedron Letters*, 883 (1964).

70. Tsuji, J., and Susuki, T., *Tetrahedron Letters*, 3027 (1965).

70a. Kunichika, S., Sakakibara, Y., and Okamoto, T., *Bull. Chem. Soc. Japan*, **40**, 885 (1967).

71. Baker, W. P., Jr., *Encycl. Polymer Sci. Technol.,* **1**, 746 (1964).

72. Otsuka, S., Mori, K., and Imaizumi, F., *J. Am. Chem. Soc.,* **87**, 3017 (1965).

73. Baker, W. P., Jr., *J. Polymer Sci.,* **A1**, 655 (1963).

74. Tadokoro, H., and Takahashi, Y., *J. Polymer Sci. (Part B)*, **3**, 9 (1965).

75. Robinson, I. M. (to E. I. du Pont de Nemours & Co.), U.S. Patent 3,151,104 (Sept. 29, 1964); *Chem. Abstr.,* **62**, 1760 (1965).

76. Masuda, Y., *Asahi Garasu Kenkyu Hokoku*, **15**, 73 (1965); *Chem. Abstr.,* **63**, 14986 (1965).

77. Lindner, H. L., and Onak, T., *J. Am. Chem. Soc.,* **88**, 1886 (1966).

78. Gelin, R., Gelin, S., and Arcis, A., *Compt. Rend. (Ser. C)*, **263**, 499 (1966).

79. Agosta, W. C., *Tetrahedron*, **22**, 1195 (1966).

PART FOUR

Higher Cumulenes

Introduction

Allene is the first and simplest cumulene (compounds which have cumulative carbon–carbon double bonds). Allene is 2-cumulene, by common nomenclature. Higher cumulenes are similar to allenes in their reactions. Most of the cumulenes are synthesized from acetylenic compounds, hence their interest in connection with acetylene-allene chemistry. Several recent reviews are available: Cadiot, Chodkiewicz and Rauss (1961),[1] Kuhn and Shulz (1963);[2] Fischer (1964).[3] The last is the most comprehensive review published.

1. BUTATRIENES $\diagdown$C=C=C=C$\diagup$ (3-Cumulenes)

1.1. Preparation

1.1.1. AROMATIC BUTATRIENES

Most known butatrienes have only aromatic substituents.[3–5] The best preparative method is reduction of butynediols with stannous chloride-ether-HCl, with KI-H_2SO_4-EtOH,[4,6] or with PBr_3-pyridine:

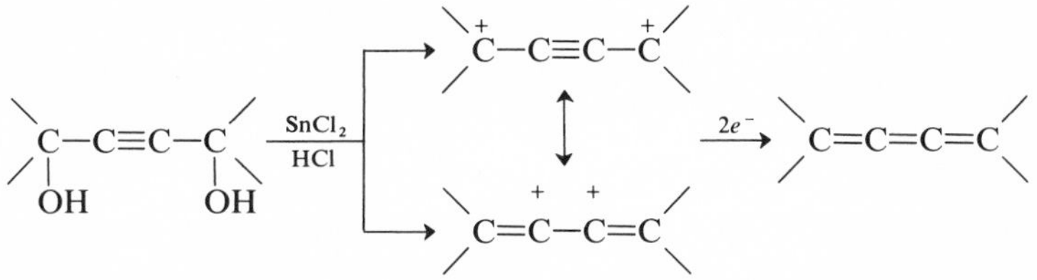

Tertiary butynediols are required. Secondary butynediols give diynes. Fischer[3] listed all the butatrienes known in 1964. The butatrienes are usually stable solids, with very high melting points. Tetrafluorobutatriene is an explosive gas. Butatrienes show *cis-trans* isomerism in a few cases.[7] Several bis(butatrienes) have been made[8] (50–90% yields).

1.1.2. ALIPHATIC BUTATRIENES—ACETYLENE-ALLENE-CUMULENE REARRANGEMENT

In 1964, Vartanyan[5] reported aliphatic butatrienes made by reacting ethylmethyl(vinylethynyl)chloromethane with dimethylamine:

$$\underset{(1)}{\overset{\overset{\text{Cl}}{|}}{\text{EtMeCC}}\equiv\text{CCH}=\text{CH}_2} \xrightarrow[\text{6 days}]{\text{HNMe}_2(2),\ 1\ \text{mole}} \underset{(3)}{\overset{\overset{\text{NMe}_2}{|}}{\text{EtMeCC}}\equiv\text{CCH}=\text{CH}_2}\ (\text{enyne, unrearranged})$$

$$(1) + 2\ \text{moles}\ (2) \longrightarrow \underset{\underset{(8.3\%)}{(4)}}{\overset{\overset{\text{NMe}_2}{|}}{\text{EtMeC}}=\text{C}=\text{C}-\text{CH}=\text{CH}_2}\ (\text{ene-allene})$$

+

$$\underset{\underset{(40\%)}{(5)}}{\text{EtMeC}=\text{C}=\text{C}=\text{CHCH}_2\text{NMe}_2}\ (\text{3-cumulene})$$

$$(1) + 3\ \text{moles}\ (2) \longrightarrow \underset{(1.3\%)}{(4)}\quad \underset{(35\%)}{+\ (5)}$$

Dimethyl(vinylethynyl)chloromethane gave similar reactions.[9]

A new method for preparing 3-cumulenes was reported in 1967.[25] 1,4-Dibromo-2-butyne, 1,2,3,4-tetrabromo-2-butene, or 1-bromo-4-phenoxy-2-butyne at 80° in aprotic solvents (dimethyl sulfoxide, dimethylformamide, or sulfolane) reacted with zinc dust to give butatriene. A further requirement (in addition to aprotic solvent) was that the reaction flask be evacuated so the unstable, volatile product is removed immediately into a cold trap. In yet another system, 1,4-dihaloalkynes reacted with sodium iodide in dimethyl-sulfoxide. 1,2,3-Butatriene formed in 75% yield, and higher homologs in 55% yield. The reaction requires 3 moles of sodium iodide per mole of dihalide.

1-Cumulenyl ethers were also reported in 1967.[26] Bis (propargylic) ethers react with sodamide in ammonia to form the new ethers. In the reaction, neither R′ nor R″ can be hydrogen:

$$C_2H_5O\text{—}\overset{\overset{\text{H}}{|}}{\underset{\underset{\text{CH}_3}{|}}{\text{C}}}\text{—O—}\overset{\overset{\text{R}'}{|}}{\underset{\underset{\text{R}''}{|}}{\text{C}}}\text{—C}\equiv\text{C—}\overset{}{\underset{\underset{\text{NH}_2}{\overset{|}{\text{H}}}}{\text{CHOCH}_3}} \xrightarrow{\text{liquid NH}_3}$$

$$\overset{\text{R}'}{\underset{\text{R}''}{>}}\text{C}=\text{C}=\text{C}=\text{CHOCH}_3 + C_2H_5O^- + CH_3\overset{\overset{O}{\parallel}}{\underset{\underset{\text{H}}{}}{\text{C}}}$$

$$(\sim65\%)$$

R′	R″	b.p. (°C @ 12 mm)
Me	Me	43
Me	Et	59
Et	Et	73

The ethers are relatively stable to heat but are very easily polymerized by oxygen. NMR indicates *cis* and *trans* isomers.

Grignard reagents react with *t*-monoacetylenic dihalides to form 90% yields of 3-cumulenes:[10]

$$\underset{\displaystyle Me_2C-C\equiv C-CMe_2}{\overset{\displaystyle Cl \qquad\quad Cl}{|\qquad\qquad|}} \xrightarrow{MeMgBr} Me_2C=C=C=CMe_2$$

Tetramethylbutatriene absorbed oxygen from the air very rapidly to form polymeric peroxide.

Dichloro- and dibromocarbenes add to allenes to give 35–70% yields of 1,1-dihalo-2-methylenecyclopropanes, which react with *n*-butyllithium at −60° to form 3-cumulenes in 75% yields.[11]

1.2. Reactions of Butatrienes

1.2.1. REDUCTION

Catalytic hydrogenation gives saturated hydrocarbons and conjugated dienes. Chemical reduction gives saturated, conjugated, diene, acetylene and allene hydrocarbons:[3]

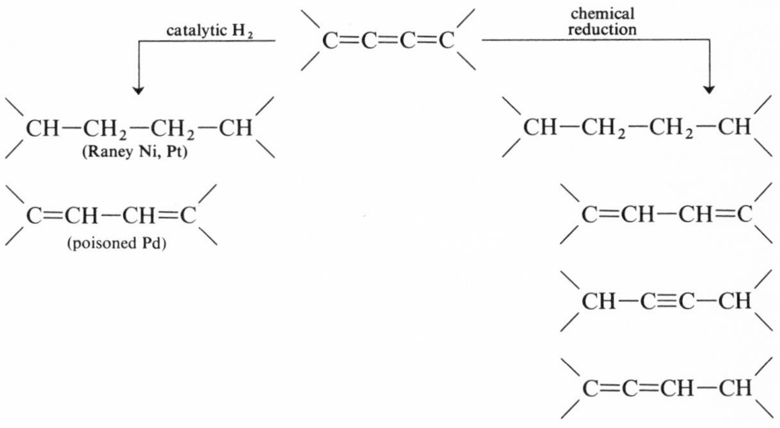

1.2.2. OXIDATION

Ozone cleaves butatrienes to two molecules of ketone and two molecules of CO_2.[3]

1.2.3. HALOGENATION

Halogen usually results in 2,3-addition which yields 2,3-dihalo-1,3-buta-dienes.[12,13]

1.2.4. ADDITION OF ALKALI METALS—FORMATION OF DIAMAGNETIC DIANIONS

In 1962, Zweig[6] reviewed the literature on diamagnetic dianions made by adding alkali metals to aromatic olefins and other electron-accepting unsaturates. Very little is known about the chemical nature of the dianions. They can (1) act as strong bases and remove protons from weaker bases, (2) react in nucleophilic displacements, and (3) transfer electrons to an acceptor, and revert to starting hydrocarbon and metal cation. Zweig made dianions from either tetraarylbutatrienes or dimethyl ethers of tetraarylbutynediols, and observed all three possible reactions:

$$\underset{(6)}{(C_6H_5)_2\overset{\overset{OCH_3}{|}}{C}C\equiv C\overset{\overset{OCH_3}{|}}{C}(C_6H_5)_2} \xrightarrow{K\cdot} \underset{K^+}{(C_6H_5)_2\overset{\cdot\cdot}{C}-C\equiv C-\overset{OCH_3}{C}(C_6H_5)_2} \longrightarrow$$

$$\xrightarrow[\text{reaction (1)}]{\text{Na, K, DME}} \underset{(2)}{(C_6H_5)_2\overset{-}{C}C\equiv C\overset{\cdot}{C}(C_6H_5)_2} \xrightarrow{CH_3I}$$

$$\underset{(1)}{(C_6H_5)_2C=C=C=C(C_6H_5)_2} + I^- + CH_3\cdot$$

$$\underset{(1)}{(C_6H_5)_2C=C=C=C(C_6H_5)_2}$$

$$\xrightarrow[\text{reaction (2)}]{\text{Na, K, ether}} \underset{(3)}{(C_6H_5)_2\overset{-}{C}C\equiv C\overset{-}{C}(C_6H_5)_2} \xrightarrow{CH_3I}$$

$$\underset{(5)}{(C_6H_5)_2\overset{\overset{CH_3}{|}}{C}C\equiv C\overset{\overset{CH_3}{|}}{C}(C_6H_5)_2}$$

Reactions (1) and (2) usually amounted to around 90% of the total. Solvent cleavage was observed only when the dianion (3), with potassium cation, was heated to 85° in dimethoxyethane.

1.2.5. IRON CARBONYL–CUMULENE COMPLEXES

Metal carbonyls remove halogen from polyhalogen-carbon compounds. Products are sometimes metal carbonyl-ene complexes, and sometimes are olefins. Joshi[14] prepared some hexacarbonyldiironbutatriene complexes by reacting dodecacarbonyltriiron with halomethylene compounds or with propargylic chlorides. 9-Dibromomethylenefluorene at 120° gave dark red complex (2). Under the same conditions, reaction with 1,1-dibromo-2,2-diphenylethylene, (3) was more complex:

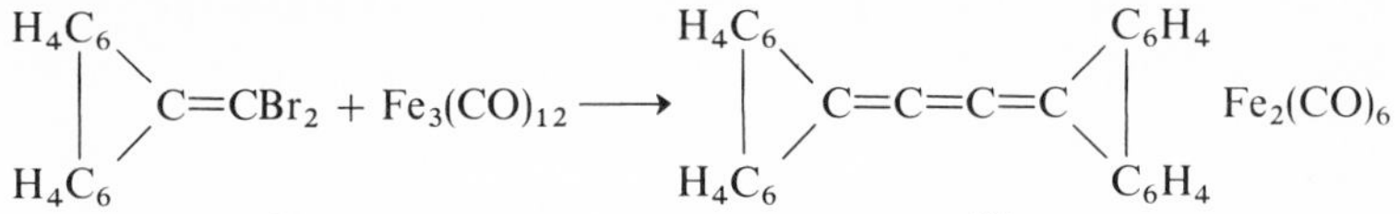

$$\underset{(1)}{\overset{H_4C_6}{\underset{H_4C_6}{>}}C=CBr_2} + Fe_3(CO)_{12} \longrightarrow \underset{(2)}{\overset{H_4C_6}{\underset{H_4C_6}{>}}C=C=C=C\overset{C_6H_4}{\underset{C_6H_4}{<}}Fe_2(CO)_6}$$

$$\underset{(3)}{Ph_2C=CBr_2 + Fe_3(CO)_{12}} \longrightarrow$$

$$\left.\begin{array}{l} Ph_2C=C=C=CPh_2Fe_2(CO)_6 \\ C_{14}H_{10}Fe_2(CO)_6 \\ C_{34}H_{20}Fe_2O_6 \end{array}\right\} \begin{array}{l} \text{This is 1,1,1-tricarbonyl-2-phenylferra-} \\ \text{indane, the same as the complex from} \\ \text{diphenylacetylene and dodecacarbonyl-} \\ \text{triiron.}[14a] \end{array}$$

Propargyl chlorides reacted best in tetrahydrofuran with Na_2 (or $K_2)Fe_2(CO)_6$:

$$ClR_2C-C\equiv C-CR_2Cl + Fe_3(CO)_{12} \longrightarrow$$

$$R_2C=C=C=CR_2Fe_2(CO)_6 \text{ (poor yield)}$$

$$Na_2Fe_2(CO)_6 + ClR_2C-C\equiv C-CR_2Cl \overset{THF}{\longrightarrow}$$

$$R_2C=C=C=CR_2Fe_2(CO)_6 \text{ (higher yield)}$$

$$K_2Fe_2(CO)_8 + ClH_2C-C\equiv C-CH_2Cl \overset{MeOH}{\longrightarrow}$$

$$H_2C=C=C=CH_2Fe_2(CO_6)$$

Complexes were also made by refluxing butatrienes with $Fe_3(CO)_{12}$ in toluene:

$$R_2C=C=C=CR_2 + Fe_3(CO)_{12} \longrightarrow R_2C=C=C=CR_2Fe_2(CO)_6$$
$$(60\% \text{ yield in less than 1 hr})$$

(where $R_2 = Me_2, Ph_2$ and biphenyl-2,2'-). Nakamura[15] recently prepared similar complexes, but believed they were $Fe_2(CO)_5$ complexes. Joshi proved that they were the $Fe_2(CO)_6$ complexes he identified.

The butatriene complexes are diamagnetic and do not oxidize in air. Infrared analysis indicated coordinated double bonds and no bridging carbonyls. Cumulene C=C absorption was masked. The structure of the complex from 9-dibromofluorene is:

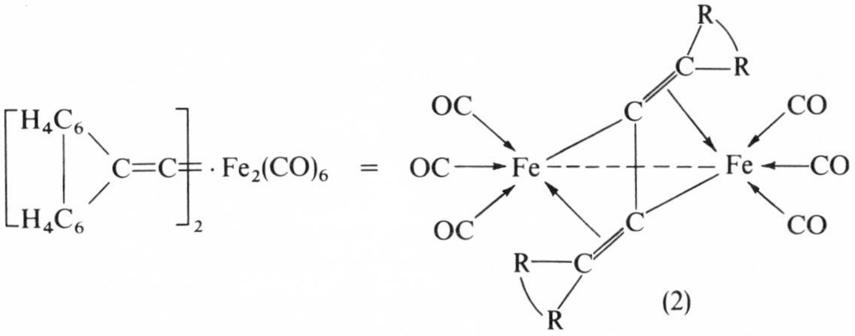

Joshi also described several other butatriene complexes with iron carbonyls. Reactions of the complexes and their use as catalysts have not been reported.

1.2.6. ISOMERIZATION

Butatrienes which have at least one hydrogen on terminal carbon are isomerized by base to enynes in an irreversible retropropargylic rearrangement:[3]

$$\begin{array}{c}\diagdown\\ \diagup\end{array}C{=}C{=}C{=}C\begin{array}{c}\diagup^H\\ \diagdown\end{array} \longrightarrow \begin{array}{c}\diagdown\\ \diagup\end{array}C{=}CHC{\equiv}C-$$

Acids isomerize tetraarylbutatrienes to methyleneindenes. This is similar to the acid isomerization-cyclization of tetraarylallenes.

Tetrasubstituted butatrienes are isomerized to enallenes by sodamide in ammonia, but the butatrienes with two terminal hydrogen atoms are isomerized to 3,1-enynes:[25]

$$R(R')C{=}C{=}C{=}CH_2 + NH_2^- \xrightarrow{\text{liquid }NH_3} R(R')C{=}CH{-}C{\equiv}C^- + NH_3$$
$$\downarrow{\scriptstyle H_2O}$$
$$R(R')C{=}CH{-}C{\equiv}CH$$

Acetylenic ethers form enynes when treated with sodamide in ammonia. These reactions support the hypothesis that cumulenes are intermediate in the acetylenic ether-enyne isomerization. The terminal 3-cumulenes add MeO^- slowly to form 2-butynyl ethers. This reaction was best for butatriene itself. Thiolate anion adds to the carbon atom in position 4 of 1,2,3-butatrienes to form butynyl thioethers and allenyl thioethers.

2. PENTATETRAENES $\begin{array}{c}\diagdown\\ \diagup\end{array}C{=}C{=}C{=}C{=}C\begin{array}{c}\diagup\\ \diagdown\end{array}$

2.1. Preparation

Very few 4-cumulenes have been prepared, and little is known about their properties.[3] In 1964, Fischer[16] and Kuhn[2,17] reported the first tetraarylpentatraenes:

$$\overset{Br}{\underset{|}{Ar_2C}}{=}\overset{|}{C}{-}CH_2{-}\overset{Br}{\underset{|}{C}}{=}CAr_2 \xrightarrow[\text{DMF-EtOH}]{KOH} Ar_2C{=}C{=}C{=}C{=}CAr_2$$

(Ar = Ph, p-MeOPh, 90% yield)

If an acetylene-allene rearrangement occurred during the reaction of a Grignard reagent with a t-diacetylenic monohalide, one product should be a

4-cumulene. Skattebøl[10] tried such reactions. Although cumulene did not form, an allene rearrangement did take place. Products were conjugated allenynes, and the reaction is a general synthesis for allenynes.[18] With primary and secondary halides, more normal acetylenic product forms:[19]

$$Me_2C(OH)C{\equiv}CC{\equiv}CMe + conc.\ HCl \xrightarrow{CaCl_2} Me_2C(Cl)C{\equiv}CC{\equiv}CMe \xrightarrow{RMgX}$$
$$(90\%)$$

$$[MeC{\equiv}CC{\equiv}CCMe_2]_2 + Me_2C{=}C{=}CHC{\equiv}CMe + Me_2CHC{\equiv}CC{\equiv}CMe$$
$$(major)$$

$$+ CH_2{=}\overset{\overset{\displaystyle Me}{|}}{C}C{\equiv}CC{\equiv}CMe$$

In 1965, Skattebøl reported a new synthesis for cumulenes.[20] He added 2 equivalents of dibromocarbene to tetramethylallene to form the cyclopropane (1) and the spiropentane (2). The cyclopropane cleaved (lithium in ether) to give a 3-cumulene, and the spiro-pentane cleaved to a 4-cumulene:

$$Me_2C{=}C{=}CMe_2 + 2Br_2C{:} \longrightarrow$$

(1)
(67%)

(2)
(6%)

(1) + MeLi $\xrightarrow[-78°]{ether}$ $Me_2C{=}C{=}C{=}CMe_2$, crystals, highly sensitive to oxygen
(100%)

(2) + MeLi $\xrightarrow[-78°]{ether}$ $Me_2C{=}C{=}C{=}C{=}CMe_2$, the first even cumulene with all aliphatic substituents

2.2. Reactions

Fischer[16] compared the properties of these odd-carbon cumulenes with the properties of known even-carbon cumulenes. The odd-carbon cumulenes are less stable and have different absorption spectra. The pentatetraenes have high electron density on the sp-hybridized carbons and so they react slowly in electrophilic additions. Sodium adds slowly, alkyllithium adds faster, and both give 1,4-addition. Conversely, pentatetraenes add nucleophilic fragments much better than even-carbon cumulenes do. Protons add to form salts. Salts are unstable, but if HX is the proton source, X^- also adds and stable halogenated salts form. 4-Cumulenes add mercuric chloride to form salts:

$$Ar_2\overset{+}{C}C{=}C{=}C{=}CAr_2{\cdot}HgCl_3^-$$
$$\underset{HgCl}{|}$$

Iodine adds to form periodide salts. Pentatetraenes do not dimerize thermally.

3. HEXAPENTAENES $\diagup$ C=C=C=C=C=C $\diagdown$

3.1. Preparation

Hexapentaenes (5-cumulenes) are usually made by the reduction of diacetylenic diols.[8] Symmetrical diols are made from disodiodiacetylene and carbonyl compounds. Symmetrical or unsymmetrical diols are made from two molecules of acetylenic carbinol by Chodkiewicz-Cadiot coupling. Acetates of acetylenic carbinols react with allene carbenes to form 5-cumulenes.[21]

Nearly all of the known 5-cumulenes are tetraaryl substituted. A few exceptions are known, such as 1,6-diphenyl-1,6-di-t-butyl-hexapentaene[22] and tetramethylhexapentaene[20] (from tetramethylallene and dibromocarbene, via the spiropentane; see Section 2.1). Some bis(hexapentaenes) have been made by stannous chloride reduction of bis(diacetylenic) carbinols.[8]

3.2. Reactions

Catalytic hydrogenation gives mixed dienes of uncertain structure. Aluminum amalgam reduction saturates the central double bond first, to form a bis(allene) which rearranges to a vinylbutatriene in the presence of base:

$$Ar_2C=C=C=C=C=CAr_2 \xrightarrow{Al/Hg} Ar_2C=C=CHCH=C=CAr_2 \xrightarrow{OH^-}$$

$$Ar_2C=CHCH=C=C=CAr_2.$$

Bromine adds to the middle two carbons to form bis(allenes).[23]

4. HIGHER CUMULENES

Three 7-cumulenes and two 9-cumulenes have been made, but they are too unstable to isolate.[3] The only known preparative method is reduction of the proper polyynediols.[24] Several bis(octaheptaenes) were reported in 1966.[8]

References

1. Cadiot, P., Chodkiewicz, W., and Rauss-Godineau, J., *Bull. Chim. Soc. France,* 2176 (1961).
2. Kuhn, R., and Schulz, B., *Chem. Ber.,* **96**, 3200 (1963).
3. Fischer, H., in (Patai, S., editor) "The Chemistry of Alkenes," p. 1025, New York, Interscience Publishers, 1964.
4. Jasiobedzki, W., *Roczniki Chem.,* **39**, 763 (1965); *Chem. Abstr.,* **63**, 9839 (1965).
5. Vartanyan, S. A., Badanyan, Sh. O., and Mushegyan, A. V., *Izv. Akad. Nauk. Arm. SSSR Khim. Nauk.,* **17**, 505 (1964); *Chem. Abstr.,* **62**, 11672 (1965).
6. Zweig, A., and Hoffmann, A. K., *J. Am. Chem. Soc.,* **84**, 3278 (1962).
7. Cadiot, P., Chodkiewicz, W., and Rauss-Godineau, J., *Bull. Soc. Chim. France,* 2176 (1961).

8. Rauss-Godineau, J., Chodkiewicz, W., and Cadiot, P., *Bull. Soc. Chem. France*, 2877 (1966); Skowronski, R., Chodkiewicz, W., and Cadiot, P., *ibid.*, 4235 (1967).
9. Vartanyan, S. A., Badanyan, Sh. O., and Muchegyan, A. V., *Izv. Akad. Nauk. Arm. SSSR Khim. Nauk.*, **16**, 547 (1963); *Chem. Abstr.*, **61**, 1745 (1964).
10. Skattebøl, L., *Tetrahedron*, **21**, 1357 (1965).
11. Ball, W. J., Landor, S. R., and Punja, N., *J. Chem. Soc. (C)*, 194 (1967).
12. Ried, W., and Dankert, G., *Chem. Ber.*, **92**, 1223 (1959).
13. Wolinski, J., *Roczniki Chem.*, **31**, 1189 (1957).
14. Joshi, K. K., *J. Chem. Soc. (A)*, 594, 598 (1966).
14a. Braye, E. H., and Hübel, W., *J. Organometall. Chem.*, **3**, 25 (1965).
15. Nakamura, A., *et al.*, *Bull. Chem. Soc. Japan.*, **37**, 292 (1964); *J. Organometall. Chem.*, **3**, 7 (1965).
16. Fischer, H., and Fischer, H., *Chem. Ber.*, **97**, 2959 (1964).
17. Kuhn, R., Fischer, H., and Fischer, H., *Chem. Ber.*, **97**, 1760 (1964).
18. Hennion, G. F., and Sheehan, J. J., *J. Am. Chem. Soc.*, **71**, 1964 (1959).
19. Jacobs, T. L., Teach, E. G., and Weiss, D., *J. Am. Chem. Soc.*, **77**, 6254 (1955).
20. Skattebøl, L., *Tetrahedron Letters*, 2175 (1965).
21. Hartzler, H. D., *J. Am. Chem. Soc.*, **83**, 4990, 4997 (1961).
22. Kuhn, R., Schulz, B., and Jochims, J. C., *Angew. Chem., Intern. Ed.*, **5**, 420 (1966).
23. Bohlmann, F., and Kieslich, K., *Chem. Ber.*, **87**, 1363 (1954).
24. Kuhn, R., and Zahn, H., *Chem. Ber.*, **84**, 566 (1951).
25. Montijn, P. P., Brandsma, L., and Arens, J. F., *Rec. Trav. Chim.*, **86**, 129 (1967).
26. Montijn, P. P., *et al.*, *Rec. Trav. Chim.*, **86**, 115 (1967).

Chapter Two

ADDITION REACTIONS OF ACETYLENIC COMPOUNDS

Introduction

Additions to the triple bond are among the most important reactions of acetylenic compounds. Although acetylenes have been available to chemists almost as long as olefins have, addition to acetylenes has not been studied nearly as extensively or thoroughly as addition to olefins. Nevertheless, the recent literature contains hundreds of references to addition reactions of acetylenic compounds. All of these cannot be included in this compilation. Some reactions in each class of addition are covered to outline the scope and mechanisms where possible.

The presentation is organized partly by reaction and partly by type of acetylenic compound. Organization by kind of acetylene is the clearest and most convenient way to cover the compounds which have been most frequently used for addition studies. The parts of Chapter 2 deal with:

(1) Hydrogenation and reduction of acetylenes,
(2) Hydration—addition of water to the acetylenic bond,
(3) Free radical addition,
(4) Oxidation of acetylenic compounds,
(5) Addition reactions of "nonactivated" acetylenes—hydrocarbons and alcohols,
(6) Addition to enynes and polyynes,

(7) Addition to acetylenic ethers,

(8) Addition to negatively substituted "activated" acetylenes—acetylenic acids, esters, ketones, nitriles, sulfones and sulfoxides,

(9) Addition to polyfluoroacetylenes.

Most of the reactions discussed in this chapter are known to be either ionic or free radical.

Ionic Addition Reactions. These represent the most extensively studied addition reactions of acetylenic compounds. Acetylenes undergo many of the same addition reactions observed with olefins, but 2 moles of reagent can add instead of just one. The bonding pairs of electrons in acetylenic bonds are closer to the carbon nuclei than in olefinic bonds because of the high degree of s character of the sp σ orbitals of the carbons. Thus, the acetylenic hydrogens are more acidic than ethylenic hydrogens and are more easily removed as protons. The greater afinity of the acetylenic carbons for electrons causes the rates of electrophilic additions to the triple bond to be slower than electrophilic additions to the double bond. Conversely, the rates of many nucleophilic additions to the triple bond are faster than those to the double bond.

In polyynes, the π-electrons tend to concentrate in the middle of the conjugated system to give mesomeric cumulene-like structures:

$$-C{\equiv}C\overset{\frown}{-}C{\equiv}C\overset{\frown}{-}\overset{\delta^+}{C{\equiv}C}-\ \longleftrightarrow\ \overset{\delta^-}{-(C{=}C)_n}{=}\overset{\delta^+}{C}-$$

An electron donor, such as C=C, decreases this effect. Thus, dimethyltriacetylene adds methanol five times as fast as the mono-addition product does. The acidity of polyynes is also explained by this picture. Diacetylene is so acidic that aqueous KOH can be used in alkynylating ketones with diacetylene.[1]

(1) *Nucleophilic Addition.* Nucleophiles add only to olefinic bonds which are polarized by an electronegative group. The acetylenic bond does not need this inductive polarization:

(a) Acetylene requires a base catalyst for addition of alcohols. Diacetylene reacts more easily, and tri- and tetraacetylenes react even faster. Monoethers are easily isolated because of the great difference in reactivity of polyynes and enynes. Amines add most easily to the polarized triple bonds as in acetylenic ketones and acids:

$$\overset{\delta^+}{-C}{\equiv}\overset{}{C}{-}\overset{\delta^-}{C}{=}\underline{\underline{O}}|\ +\ \overset{\diagdown}{\underset{\diagup}{N}}H\ \longrightarrow\ -C(N\overset{\diagup}{\underset{\diagdown}{}})=CH-C=O$$

(b) Diazomethane and hydrazoic acid react only with polarized olefins, but they react easily with ordinary acetylenes. These are 1,3-dipolar additions.

(c) Acetylenic acids and polyynes are sufficiently polarized to enter the Michael addition reaction.

(d) Lithium alkyls add rapidly at 20° to polyynes with at least three conjugated triple bonds, but Grignard reagents are not active enough to add. Lithium aluminum hydride is similar to lithium alkyls; tetraynes react more rapidly than triynes, etc. The addition is stepwise and reduces triple bonds to double bonds. Reduction starts at the end of the conjugated system.

(2) *Electrophilic Addition.* Burnelle[2] investigated the possibility that electrophilic additions might involve a distorted acetylenic bond, and not the usual equilibrium configuration. Electrophilic additions are usually *trans.* At some stage of the reaction the acetylene molecule can be described by a wave function which has a significant contribution from the

bent excited state. The approach of the adding reagent causes distortion of the acetylene molecule, through an excited state. This makes the acetylene more accessible to the adding reagent. The second step, addition of the anion, must occur very quickly after the initial attack, for otherwise rearrangement would be likely.

Electrophilic reagents add much more slowly to triple bonds than to double bonds. The reaction rates with olefinic bonds are often greater by powers of 10, so double bonds can react selectively in the presence of triple bonds.

(a) Oxidation: Peracids add much faster to olefinic bonds. Thus, enynes can be converted into acetylenic epoxides. The triple bond reacts only if large excess of peracid is used and if reaction times are very long. Ozone in air does not attack the triple bond, but ozone in oxygen cleaves the acetylenic bond nonquantitatively. Chromic acid also selectively oxidizes double bonds.

(b) Polyynes add halogen acid even more slowly than monoacetylenes do. Tetraynes do not add HX at all. Water adds easily to acetylenes, but the rate decreases as the number of conjugated triple bonds in the molecule increases. Tetraynes do not add water. Di-*t*-butyl-pentaacetylene can be heated with 50% sulfuric acid for hours without any reaction.

(c) Halogen addition: Halogen adds to olefins electrophilically. The reaction is fast even in the dark. Acetylenes react very slowly in the dark, and the rate increases as the number of conjugated triple bonds increases. This indicates that halogen addition to polyynes may be nucleophilic.

Radical Addition: (a) Addition of halogen: Light increases the rate of addition of halogen. 5-Decyne reacts with bromine 10^4 times faster in light than in the dark. Halogen usually adds to the ends of polyyne chains first to form cumulenes. NO_2 also adds to the ends first.

(b) Catalytic hydrogenation: As the number of conjugated triple bonds increases, the hydrogenation rate increases only linearly. Hydrogenation must involve addition of dissociated hydrogen as radicals. Adsorption complexes are usually oriented on catalyst surfaces, so the addition of hydrogen is *cis*.

(c) Addition of thiols: Thiols add easily if a radical catalyst is present. Thiol acids add to terminal acetylenes to form $RCH{=}CH{-}SCH_2CO_2R'$, which reacts with carbonyl reagents to give RCH_2CHO. This is a preparative method for "hydrating" terminal acetylenes to aldehydes.

(d) Polymerization: Polyynes polymerize more easily than monoacetylenes when irradiated, even in longer-wavelength light. When tetraynes are irradiated, they form blue-black insoluble polymers in a few seconds. The products from radical polymerization of polyynes are crystalline, so the polyyne chains must align themselves parallel. Energy from the light uncouples the π-electron pairs.

References

1. Bohlmann, F., *Angew. Chem.,* **69**, 82 (1957).
2. Burnelle, L., *Tetrahedron*, **20**, 2403 (1964).

1. BACKGROUND

It is often easier to introduce an acetylene group into a molecule and hydrogenate it to an olefinic or saturated group than to introduce these groups directly. Hydrogenation can be directed to either olefin group or saturated group by choice of catalyst and conditions. Hydrogenation and semihydrogenation can be done in heterogeneous and in homogeneous systems. Bond[1] published a review in 1955 covering hydrogenation of acetylenes.

1.1. Heterogeneous Catalysis

Semihydrogenation is possible because the acetylene group adsorbs very strongly on the catalyst surface, and not because acetylenes hydrogenate more quickly than olefins. In fact, olefins frequently hydrogenate faster than acetylenes. In competitive hydrogenation, the acetylene group adsorbs so strongly that it excludes all other groups from the catalyst surface.[2] The chemisorbed acetylene group is probably very similar to the metal-acetylene compounds found in complexes of metals (such as platinum) and acetylene in solution.[3] Catalytic semihydrogenation of internal acetylenes usually gives *cis*-olefins. Internal acetylene groups must adsorb on the surface of the metal so the $C \equiv C$ group is parallel to the surface. Substituents must be bent upward from the surface. If the substituents are large, the acetylene bond cannot be as close to the surface, and the hydrogenation rate may be faster because more hydrogen can chemisorb.[4]

The usual heterogeneous catalysts are palladium, platinum, nickel and other transition group metals. The acetylenic group hydrogenates on nickel or palladium preferentially to all other groups. With small amounts of catalyst, semihydrogenation is usually maximum. Allenes also adsorb strongly, do not isomerize to acetylenes, and hydrogenate to *cis*-olefins.[2] Lindlar catalyst is widely used,[5] but palladium on barium sulfate deactivated by an equal weight of quinoline is easier to make, and gives better selectivity and rates.[2]

1.2. Homogeneous Catalysis and Reduction

Catalytic hydrogenations can be carried out in homogeneous solution. Chemical reduction in homogeneous solution is also a useful procedure, especially when chromous sulfate or boranes are the adding-reducing reagents. For many laboratory reactions, homogeneous systems are easier and more convenient.

2. HYDROGENATIONS IN HETEROGENEOUS SYSTEMS

2.1. Vapor Phase

Vapor phase hydrogenation of acetylenes to olefins is important industrially. Ethylene from cracking contains a little acetylene. Semihydrogenation to olefin is the best purification method. Acetylenes are strong poisons for the catalysts generally used for polymerization of olefins. Acetylene must be removed from ethylene to be polymerized over chromia-silica-alumina catalysts or Ziegler catalysts.

Ruthenium or osmium on alumina selectively hydrogenate 2-butyne to cis-2-butene at 80–150°.[6] Palladium on alumina also gives cis-2-butene.[7] Deuteration gives cis-2-butene-2,3-d_2 as the major product.[8] Under similar conditions, the acetylenic hydrogen of 1-butyne exchanges with deuterium. The product is 1-butene containing deuterium. Allenes give deuterated olefins from 1,2-addition to one of the double bonds. When 1,3-diynes are hydrogenated over palladium, the first hydrogen adds at the monosubstituted acetylene bond.[9]

2.2. Liquid Phase

When acetylene is hydrogenated over Group VIII metals, some polymer often forms.[10] Iron, cobalt and nickel give more polymer than palladium or platinum. Substituted acetylenes tend to polymerize less, and acetylenes above propyne usually dimerize.

Some of the metals cause reactions of acetylenes in the absence of hydrogen.[11] Some acetylenes react on palladium-charcoal catalysts in nitrogen. Phenylacetylene reacts rapidly at 20°, and violently at higher temperatures, to give a complex mixture of solid yellow linear polymers. Dimethyl acetylenedicarboxylate refluxed in benzene with palladium-charcoal gives 93% yield of trimer, hexamethyl mellitate. The catalyst made by reducing palladous chloride with sodium borohydride slowly dimerizes phenylacetylene to 1-phenylnaphthalene at 100°. Note that chromium bis(arenes) are also able to abstract a proton from a complexed benzene ring to give naphthalenes.[12] Palladium is the most active polymerizing-cyclizing catalyst. Platinum, iridium, ruthenium, rhodium and nickel are less active. Although polymerization reactions are less important in the presence of hydrogen, it is wise to check the effect of a catalyst on an acetylenic compound in nitrogen before using the catalyst for a hydrogenation reaction.

Palladium-charcoal catalyst has been used to study hydrogenation of 1,7-undecadiyne.[13] One mole of hydrogen gives 1-undecen-7-yne, 2 moles give

1,7-undecadiene, and 3 moles give 4-undecene. All of these are clean reactions. Thus, the order of addition of hydrogen is: (1) terminal acetylene group, (2) internal acetylene group, and (3) terminal olefin group. The 4-undecene is 1:1 cis:trans. Lindlar catalyst gives only cis-4-undecene. The amount of trans-olefin sometimes depends on pH and sometimes on the catalyst. It is not predictable and is not related to the reaction rate. Lindlar and Raney nickel catalysts are usually stereoselective for forming cis-olefins.

The amount of trans-olefin from 1,7-octadiyne is changed by substances added to the palladium catalyst:

	% trans-Olefin
Pd-CaCO$_3$-Pb(OAc)$_2$-quinoline (Lindlar)	4
Pd-CaCO$_3$	64
Pd-BaSO$_4$	40

The amount of Pd-C catalyst also has a large effect on the amount of trans-olefin formed:[13]

% Catalyst	% trans-Olefin	
9.8	32	
10.00	68	(this is an amazing change for such a small increase in the amount of catalyst)
17.4	31	

Sodium in ammonia cannot be used to semihydrogenate ω-fluoroalkynes because it removes all the fluorine. Lindlar catalyst in hydrocarbon solvent is useful, and the rate curves break nicely after 1 mole of hydrogen adds. Oxygen-containing solvents cause complete hydrogenation to 1-fluoroalkanes.[14]

The relative rates of hydrogen uptake by several acetylenes over Pd-C in ethanol are:[15]

Propargyl alcohol	0.3
Butynediol	0.2
2-Pentyne-1,4-diol	0.2
1-Ethynylcyclohexanol	1
Phenylacetylene	0.7
Monopotassium acetylenedicarboxylate	0.4

Butynediol gives cis-butenol, but when alkali is present, much of the product is trans. The catalyst and base do not isomerize cis- to trans-butenediol, so the trans product must be a direct hydrogenation product. Acetylenedicarboxylic acid usually adds hydrogen trans, as it does over Pd-C catalyst.

The semihydrogenation of the central triple bond in triynes which have two terminal triple bonds can be accomplished if the terminal groups are first converted to trimethylsilyl derivatives.[15a] Hydrogenation over Pd-BaSO$_4$ reduces the central acetylenic group to the ethylenic group, after which the free acetylene groups are restored to the terminal positions by treatment with silver nitrate followed by KCN in water:

$$HC{\equiv}C{-}CH{=}C(CH_3){-}CH_2{-}C{\equiv}C{-}CH_2{-}C(CH_3){=}CH{-}C{\equiv}CH$$

$$\downarrow \begin{array}{l}\text{(1) EtMgBr}\\ \text{(2) Me}_3\text{SiCl}\end{array}$$

$$Me_3Si{-}C{\equiv}C{-}CH{=}C(CH_3){-}CH_2{-}C{\equiv}C{-}CH_2{-}C(CH_3){=}CH{-}C{\equiv}C{-}SiMe_3$$

$$\downarrow \text{H}_2/\text{Pd-BaSO}_4$$

$$Me_3Si{-}C{\equiv}C{-}CH{=}C(CH_3){-}CH_2{-}CH{=}CH{-}CH_2{-}C(CH_3){=}CH{-}C{\equiv}C{-}SiMe_3$$

$$\downarrow \begin{array}{l}\text{(1) AgNO}_3\\ \text{(2) KCN/H}_2\text{O}\end{array}$$

$$HC{\equiv}C{-}CH{=}C(CH_3){-}CH_2{-}CH{=}CH{-}CH_2{-}C(CH_3){=}CH{-}C{\equiv}CH$$
(80–90% yield in the cleavage reaction)

This is a simple and useful laboratory procedure.

Tedeschi[16] reviewed the literature on selective hydrogenation of tertiary ethynylcarbinols in 1962, and reported his own work. He could not get high-purity ethylenic carbinols from tertiary ethynylcarbinols using Lindlar or any other palladium catalysts. However, alkali metal hydroxide, preferably KOH, added to Pd-C catalysts (KOH:catalyst weight ratio 0.5:1) gave 77–92% yield of 95–97% pure ethylenic alcohols from several acetylenic carbinols. Without base, the purity is 82–98%. The alkali hydroxide slows the hydrogenation of the olefinic group more than the hydrogenation of the acetylenic group. As the chain length of the carbinol increases, the effect of base on product purity is more pronounced.

Completely selective semihydrogenation of 3-methyl-1-butyn-3-ol requires a rhodium-charcoal catalyst and sodium methoxide. Palladium or platinum catalysts are unsatisfactory, and so is rhodium catalyst if the sodium methoxide is omitted. Potassium metal in the ethynylcarbinol gives a mixture of potassium alcoholate and potassium acetylide, and has a different effect than KOH. KOH-ethynylcarbinol complex[17] has the same effect as KOH.

The inhibitory and selectivity of bases is more pronounced with catalysts less active than Pd-C (e.g., Pd-BaCO$_3$), and hydrogenation stops after 1 mole of hydrogen is used. The same catalysts are used for semihydrogenation of t-acetylenic glycols.[18]

Brown[19] developed a very convenient laboratory hydrogenation system. Sodium borohydride is added to a stirred suspension of carbon in a solution of a metal salt. The active catalyst forms *in situ*. The acetylenic compound is then added, followed by hydrogen generated in an attached vessel by addition of standard acid to a solution of sodium borohydride of known concentration. The vessels are arranged so the hydrogen pressure is constant (slightly above atmospheric), and generation of hydrogen stops when hydrogen consumption stops. The amount of hydrogen consumed is calculated from the volume of standard acid consumed. Palladium catalyst is better than platinum, nickel, cobalt or iron for semihydrogenation of acetylenes in this system. Addition of an amine such as quinoline doubles the hydrogenation rate. Ethyl propiolate, 1-butyn-3-ol, phenylacetylene and ethynylcyclohexanol give 74–97% yield of olefinic products. In the absence of the amine, hydrogenolysis of carbinols is a competing reaction. Quinoline completely prevents hydrogenolysis. Disubstituted acetylenes give high yields of *cis*-olefins. Amine has no effect on the rate of hydrogenation of disubstituted acetylenes.

Lindlar catalyst at 30° in pyridine is used to hydrogenate α-acetylenic acids to *cis*-olefinic acids. Yields are 90–95%.[20] Lindlar catalyst also gives *cis*-dienols from enynols.[21] Palladium-CaCO$_3$ catalyst reduces only the acetylenic bond in vinyl 1-propynyl ketone, but reduces the double bond, triple bond, and some of the carbonyl group in 1-propenyl ethynyl ketone and 1-propenyl 1-propynyl ketone.[22]

Raney zinc is used to semihydrogenate acetylene, tolan, methylbutynol and butynediol. Raney zinc reduces vinylacetylene to butadiene.[23] Copper-zinc couples usually give *cis*-olefins from acetylenes. Stearolic acid gives 78% yield of oleic acid, and 3-hexyne-1,6-dioic acid gives the *cis*-olefinic diacid.[24] The reduction is not stereospecific with dimethyl acetylenedicarboxylate and phenylpropiolic acid.[15]

The conclusion from the literature data is that certain systems are known which usually give stereospecific semihydrogenation of acetylenes to olefins, but no single system is universally applicable. For hydrogenation of acetylenes not previously studied, some search for suitable catalyst, additives, solvents and conditions may be necessary.

3. HOMOGENEOUS SYSTEMS

3.1. Catalytic Hydrogenation

In 1956, Miller[25] reviewed earlier reports of semihydrogenation of acetylenes in homogeneous systems.

A solution of stannous chloride and chloroplatinic acid (10:1 mole ratio) in methanol mixed with 1:1 acetylene-hydrogen rapidly reduces the acetylene to a 3:1 mixture of ethane and ethylene. Thus, acetylene hydrogenates more slowly than ethylene in the presence of this catalyst.[26]

Tris(triphenylphosphine)rhodium chloride is easy to make, and in benzene-ethanol solution at 20° it is an active catalyst for full hydrogenation of acetylenes. 1-Hexyne gives quantitative yield of n-hexane.[27] 2-Hexyne is hydrogenated to hexane via 2-hexene.[28]

Dicyclopentadienyltitanium dicarbonyl is soluble in benzene, heptane or toluene, and is an excellent catalyst for semi-hydrogenation of acetylenes at 50° and 50 atmospheres pressure.[29] The complex formed by two molecules of acetylene and one molecule of the catalyst is not a hydrogenation catalyst.

3.2. Chemical Reduction

3.2.1. CHROMOUS SULFATE

In 1909, Berthelot[30] reported the reduction of acetylene by aqueous ammoniacal chromous sulfate solution. In 1964, Castro[31] noted several more recent reports, and also described a detailed study of the rate and mechanism of the reaction. Various acetylenes (0.1–$0.3M$) were reduced in homogeneous solution by chromous sulfate (0.4–$0.7M$), at room temperature in water at pH around 4, or in aqueous dimethylformamide, under nitrogen. The reaction is stereospecific; internal acetylenes are reduced to *trans*-ethylenes. The stoichiometry is:

$$2H^+ + 2Cr^{++} + RC{\equiv}CR' \longrightarrow 2Cr^{+++} + \begin{matrix} R & & H \\ & C{=}C & \\ H & & R' \end{matrix}$$

Yields are usually 85–95%.

Acetylenes are classified according to reactivity:[31]

(1) Very fast: essentially complete reaction in 5–15 minutes at room temperature; propargyl alcohol, 1-butyn-3-ol, acetylenedicarboxylic acid, phenylpropiolic acid, butynediol.

(2) Fast: Complete reaction in 2–3 hours; 2-butyn-1-ol, phenylacetylene, 1-hexyne.

(3) Slow: Complete reaction in one day; 2-carboxydiphenylacetylene.

(4) Inert: Not reacted in one day; diphenylacetylene, 2-pentyne, 2,5,5-trimethyl-3-hexyn-2-ol, and various 2- and 4-substituted diphenylacetylenes.

If no hydroxylic substituents are present, terminal acetylenes are more reactive than internal, but if hydroxyl is present, the availability of the triple bond for coordination is more important. With large excesses of acetylenes, kinetics are pseudo-second order. With stoichiometric reactants, the reductions are third order. Coordination is essential: o-carboxytolan reacts, but p-carboxytolan does not. When chromous ion is complexed with ethylenediaminetetraacetic acid, the reaction rate decreases drastically.

The functional groups adjacent to the triple bond obviously have a strong influence on Cr^{++} reductions. Reaction of Cr^{++} with acetylenedicarboxylic acid in dilute perchloric acid gives fumaric acid as the only organic product:[31a]

$$2Cr^{++} + HO_2CC{\equiv}CCO_2H + 2H^+ \longrightarrow 2Cr^{+++} + \text{fumaric acid, } 94\%$$

A careful rate-mechanism study has shown that the reaction takes place in four consecutive steps:[31a]

(1) $2Cr^{++} + HO_2CC{\equiv}CCO_2H \longrightarrow [Cr_2 \cdot HO_2C{\equiv}CCO_2^-] + H^+$
    species 1

(2) Species $1 + H^+ \rightleftharpoons$ species $1\text{-}H^+ \longrightarrow$ species 2 (Cr^{++}:ADA $= 2$)

(3) Species $2 + H_2O \longrightarrow Cr^{+++} +$ species 3 (Cr^{++}:fumaric acid $= 1$)

(4) Species $3 + H_2O \longrightarrow Cr^{+++} +$ fumaric acid

As indicated, in species 2 the Cr^{++}: acetylenedicarboxylic acid ratio is 2. Species 3 is a $1:1$ Cr^{++} fumaric acid complex.

A $1:1$ complex forms first, and a second chromous ion probably approaches the complex from the opposite side:[31]

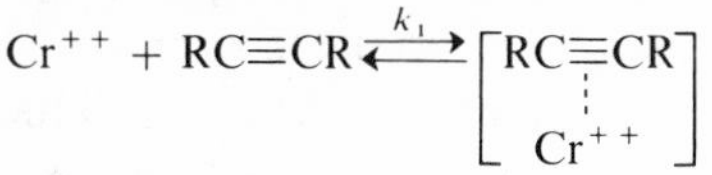

$$Cr^{++} + RC{\equiv}CR \underset{}{\overset{k_1}{\rightleftharpoons}} \begin{bmatrix} RC{\equiv}CR \\ \vdots \\ Cr^{++} \end{bmatrix} \tag{1}$$

$$\begin{matrix} Cr^{++} \\ \downarrow \\ \begin{bmatrix} RC{\equiv}CR \\ \vdots \\ Cr^{++} \end{bmatrix} \end{matrix} \xrightarrow[\text{slow}]{k_2} 2Cr^{+++} + \begin{matrix} R \\ \diagdown \\ C{=}C \\ \diagup \quad \diagdown \\ H \qquad R' \end{matrix} \qquad \begin{matrix} H \\ \diagup \\ \\ \end{matrix} \tag{2}$$

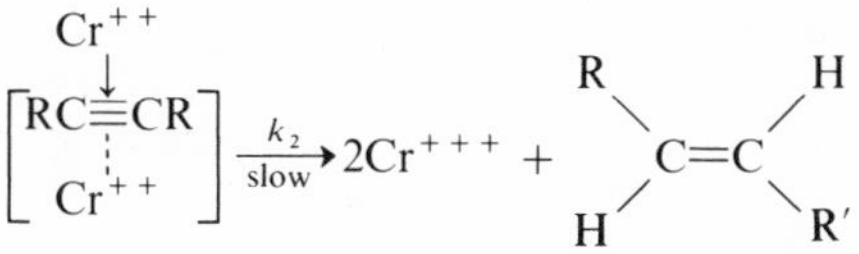

This suggests that the transition state is:

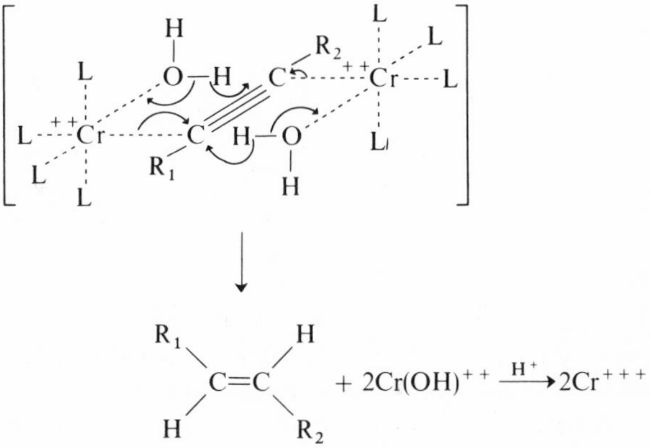

In the formulas, L denotes solvent ligands. The large entropy loss calculated for the propargyl alcohol reduction is what is expected for the charged, strongly solvated, ordered transition state. Proton transfer must be involved, since propargyl alcohol reacts 3.5 times as fast in water as in D_2O. Castro observed that a complex as pictured for this reduction could also be used to explain the mechanism of the Birch reduction of acetylenes. In this sodium-ammonia reduction, another mechanism has been proposed by Greenlee;[32] it involves a solvated electron addition and protonation sequence. If a 2:1 sodium-ammonia:acetylene complex collapsed, *trans*-olefin and sodamide should form directly.

3.2.2. HYDROBORATION

The diborane-olefin addition reaction has been studied extensively, but relatively little work on acetylenes has been reported.[33] Curiously, the first diborane-unsaturate reaction used acetylene. Stock[34] wrote in 1923 that acetylene and diborane reacted explosively at 100° to form condensation products with an aromatic odor. No further work was reported until 1948, when Hurd began to publish his studies on the diborane-olefin reactions.

Hydroboration is an easy, fast reaction. Unsaturate and sodium borohydride are mixed in an ether solvent, usually diglyme, and BF_3 etherate is added. Organoborane forms very rapidly at room temperature.[33] Hydrolysis of the organoborane by dilute acid gives *cis*-olefinic products, and oxidation-hydrolysis by alkaline H_2O_2 gives carbonyl compounds. Terminal acetylenes are thus converted into aldehydes.

(1) *Monohydroboration.* Internal acetylenes, such as 3-hexyne, add one $\diagdown\!\!BH\!\diagup$

rapidly, and the second $\diagdown\!\!BH\!\diagup$ slowly, so monohydroboration is the major

reaction:[35]

(84%)

Addition of B—H is *cis*, via a 4-center transition state. The monohydroboration product is hydrolyzed by acid to a *cis*-olefin, and is hydrolyzed-oxidized by basic hydrogen peroxide to a ketone. Under the same conditions, 1-hexyne gives mostly dihydroborated product. Hydroboration of terminal acetylenes by diisoamylborane gives only monohydroboration. The size of the diisoamylborane molecule apparently prevents addition of a second molecule (Table 2–1).[33]

TABLE 2-1
Monohydroboration of Internal and Terminal
Alkynes to *cis*-Olefins[33]

Acetylene	Hydroborating Agent	% Acetylene Reacted	% Monohydroboration
3-Hexyne	$NaBH_4$-BF_3 in diglyme	84	66
3-Hexyne	Diisoamylborane in THF	100	100
1-Hexyne	$NaBH_4$-BF_3 in diglyme	56	12
1-Hexyne	Diisoamylborane in THF	100	100

Trimethylamine-*t*-butylborane also monohydroborates terminal alkynes.[36] A fair number of acetylenes have been monohydroborated.[36-39] The products oxidize to aldehydes in high yields. Protonolysis gives the *cis*-olefinic products. The effect of position of the triple bond and the hydroborating agent on the yield of olefin is illustrated by data in Table 2–2.

TABLE 2-2
Monohydroboration of Internal and Terminal
Alkynes to *cis*-Olefins[33]

Acetylene	Hydroborating Agent	% Yield of Olefin
1-Hexyne	$NaBH_4$-BF_3, diglyme	7
1-Hexyne	Diisoamylborane in diglyme	92[a]
2-Pentyne	$NaBH_4$-BF_3-diglyme	60
3-Hexyne	$NaBH_4$-BF_3-diglyme	68
3-Hexyne	$NaBH_4$-BF_3-diglyme	80[a]
3-Hexyne	Diisoamylborane-diglyme	90
Diphenylacetylene	Diisoamylborane-diglyme	69

[a] Small excess of hydroborating reagent used. In all other cases, theoretical amount used.

Semihydrogenation over Lindlar catalyst is sometimes tricky, and catalyst preparation is critical. Olefins from Lindlar semihydrogenations are sometimes difficult to obtain in purities above 95 %. On the other hand, hydroboration is very simple, yields of olefins are good, and olefins which are 98–99 % pure are easily obtained.

Oxidation of monohydroboration products from terminal acetylenics is one of the best and occasionally the only way to go directly to aldehyde. Hydration to form aldehyde is sometimes possible, but the yields are usually low (see Part 2 of this chapter).

(2) *Dihydroboration*. 2,2-Dimethyl-2-butylborane and cyclohexylborane add to the terminal carbon of 1-alkynes:

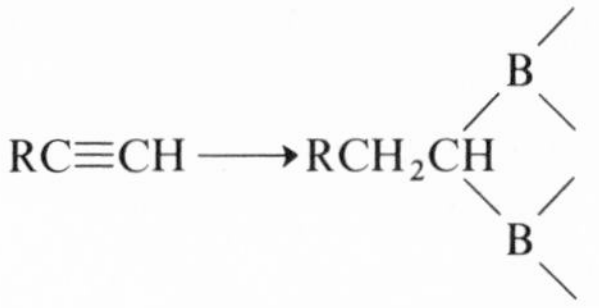

$$RC\equiv CH \longrightarrow RCH_2CH \begin{matrix} \diagup B \diagdown \\ \diagdown B \diagup \end{matrix}$$

Hydrolysis-oxidation by alkaline hydrogen peroxide gives primary alcohols, RCH_2CH_2OH. Oxidation by anhydrous *m*-chloroperbenzoic acid gives carboxylic acids: $RCH_2CH_2CO_2H$.[40]

Hydroboration reactions can thus be used to convert internal acetylenes into *cis*-olefins or ketones, and to convert terminal acetylenes into olefins, aldehydes, alcohols or carboxylic acids. All of these are high-yield reactions under proper conditions.

3.2.3. REDUCTION BY TRIALKYLBORANES

Olefins exchange alkyl groups with trialkylboranes, but very slowly, and reversibly. Hubert[41] was the first to study the reaction of acetylenic compounds

TABLE 2-3
Reduction of Acetylenic Hydrocarbons by Triisobutylborane[41]

Acetylenic Compound	Isobutene Evolved (% of Theory)	Product	% Yield
Dec-5-yne	91	*cis*-Dec-5-ene	88
Cyclotetradeca-1,8-diyne	100	*cis,cis*-Diene	67
		(+ impure crop)	(+28)
Diphenylacetylene	100	*cis*-Stilbene	80
Hexa-2,4-diyne	—	*cis,cis*-Diene	~50
Dodeca-5,7-diyne	99	*cis,cis*-Diene	42
Hexadeca-7,9-diyne	91	*cis,cis*-Diene	64–75[a]
2,2,7,7-Tetramethylocta-3,5-diyne	—	Mixture of unsaturated hydrocarbons	—
Dicyclohexylbuta-1,3-diyne	120	*cis,cis*-Diene	50
Diphenylbuta-1,3-diyne	75	*cis,cis*-Diene	36
1,14-Diphenyltetradeca-6,8-diyne	92	*cis,cis*-Diene	13
Cyclooctadeca-1,3,10,12-tetrayne	100	all-*cis*-Tetraene	33

[a] With no hydroquinone present, the yield is 64 %. With 1, 5, 10 and 20 % hydroquinone present, the yields are 58, 71, 74 and 75 %, respectively.

with trialkylboranes. At 160–200° a fast irreversible reaction liberates olefin:

$$3RC\equiv CR + B(CH_2CH_3)_3 \longrightarrow B(RC=CHR)_3 + 3CH_2=CH_2$$

The products are trialkenylboranes, which are hydrolyzed to *cis*-olefins, and are oxidized to ketones. The displacement is a good method for making pure terminal olefins from olefin mixtures. Mixed olefins are hydroborated, isomerized and then the terminal olefin is liberated by treating with an acetylene. This is easier than liberation by exchange with a higher-boiling olefin.

Triisobutylborane is good and is usually added in excess. Disubstituted acetylenes and disubstituted diacetylenes are the only acetylenes which give good reactions.

The reaction gives high yields of *cis*-olefins and is much better than Lindlar semihydrogenation, especially for diacetylenes (Table 2–3).

This reaction was used as one step in the preparation of civetone homologs:[42]

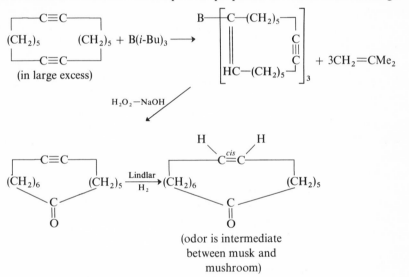

(odor is intermediate
between musk and
mushroom)

3.2.4. REDUCTION BY ALUMINUM HYDRIDES

Diisobutyl aluminumhydride reduces acetylenes to *cis*-olefins.[25a] Lithium aluminum hydride reduces propargylic alcohols to *trans*-olefinic alcohols. When Miller[25] reduced 1-ethynyl-1-cyclohexanol with diisobutyl aluminumhydride, he obtained 48% conversion and 100% yield of the olefinic alcohol. The reduction was not selective with butynediol, however.

Lithium aluminum hydride reduces acetylenic hydrocarbons to *trans*-olefins if an ether solvent is used. In toluene solvent, the major product is saturated hydrocarbon, but the olefin also formed is largely *cis*. Thus, the solvent controls the direction and extent of addition.[25a]

The lithium aluminum hydride reductions are done at atmospheric pressure under nitrogen. Hydrolysis liberates the *trans*-olefins. This is a convenient procedure. Arylacetylenes reduce faster than alkylacetylenes do:

$$PhC{\equiv}CCH_3 \xrightarrow[\text{reflux, 12 hr}]{\text{LiAlH}_4, \text{ THF}} \textit{trans}\text{-l-phenylpropene}$$
$$(98\text{--}99\%)$$

$$PhC{\equiv}CPh \xrightarrow[\text{reflux, <2 hr}]{\text{LiAlH}_4, \text{ THF}} \textit{trans}\text{-stilbene}$$
$$(100\%)$$

Deuterolysis shows the point of addition of $LiAlH_4$:[25a]

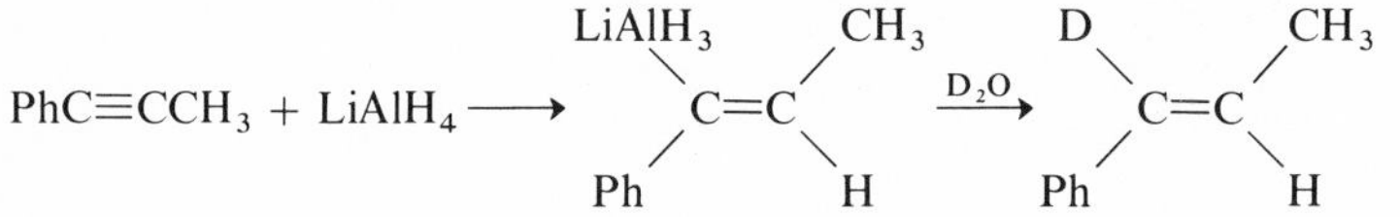

Diisobutyl aluminum hydride adds the same way, but the product is the *cis*-olefin.

3.2.5. MECHANISM OF LITHIUM ALUMINUM HYDRIDE REDUCTION OF UNSATURATED ALCOHOLS

Bohlmann[43] observed that the rate-determining step in lithium aluminum hydride reduction is nucleophilic addition of hydride ion to the unsaturated alcohol. Enynols are reduced to β-allenic alcohols. The rate is slow if the hydroxyl group of the alcohol is alpha to a carbon bearing an induced negative charge. Diacetylenic alcohols reduce faster than monoacetylenic alcohols, consistent with the greater nucleophilic activity of the diacetylenes (Table 2.4). For acetylenic alcohols, the phenyl group activates the triple bond more

TABLE 2-4

Times for Half-reduction of Acetylenic Alcohols[43]

(20°, concentrations of lithium aluminum hydride and alcohol are $7 \cdot 55 \times 10^{-3}$ mole/liter)

Alcohol	$T_{1/2}$ (min)
$C_6H_5{-}C{\equiv}C{-}CH_2OH$	5
$(p)H_3CO{-}C_6H_4{-}C{\equiv}C{-}CH_2OH$	32
$(p)Cl{-}C_6H_4{-}C{\equiv}C{-}CH_2OH$	1.1
$H_3C{-}C{\equiv}C{-}C{\equiv}C{-}CH_2OH$	0.46
$H_3C{-}CH{=}CH{-}C{\equiv}C{-}CH_2OH$	37
$C_6H_5{-}C{\equiv}C{-}C{\equiv}C{-}CH_2OH$	0.022

than the olefinic group does, just as with olefinic alcohols. The reduction of pentenynols to allenes is represented as:

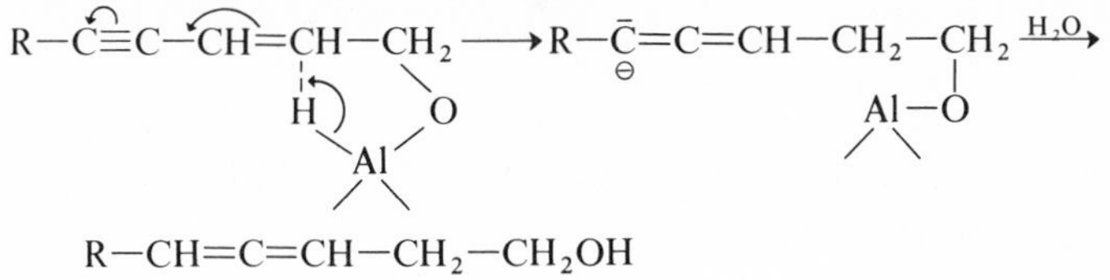

$$R-C\equiv C-CH=CH-CH_2 \longrightarrow R-\overset{\ominus}{C}=C=CH-CH_2-CH_2 \xrightarrow{H_2O}$$
$$\overset{}{\underset{\overset{\displaystyle H}{}}{}} \quad O \qquad\qquad\qquad\qquad Al-O$$
$$\overset{}{Al}$$

$$R-CH=C=CH-CH_2-CH_2OH$$

Reduction with $LiAlD_4$ and hydrolysis with water, and reduction with $LiAlH_4$ and deuterolysis with D_2O, showed where the Li and H atoms add:

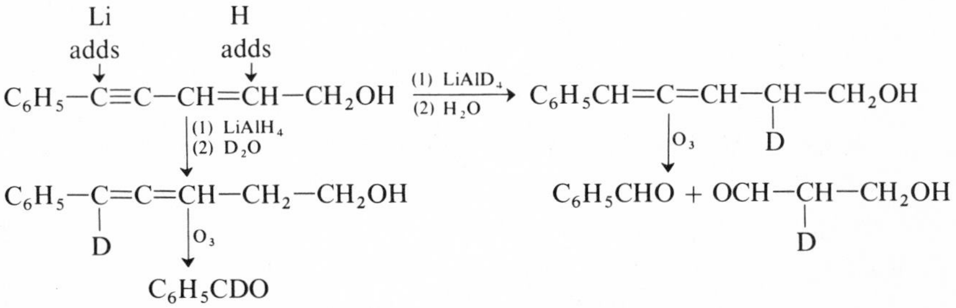

$$\begin{array}{cc} \text{Li} & \text{H} \\ \text{adds} & \text{adds} \\ \downarrow & \downarrow \end{array}$$

$$C_6H_5-C\equiv C-CH=CH-CH_2OH \xrightarrow[\text{(2) } H_2O]{\text{(1) } LiAlD_4} C_6H_5CH=C=CH-CH-CH_2OH$$

$$\downarrow \begin{array}{l}\text{(1) } LiAlH_4 \\ \text{(2) } D_2O\end{array} \qquad\qquad\qquad\qquad\qquad \downarrow O_3 \qquad D$$

$$C_6H_5-C=C=CH-CH_2-CH_2OH \qquad\qquad C_6H_5CHO + OCH-CH-CH_2OH$$
$$\overset{|}{D} \quad \downarrow O_3 \qquad\qquad\qquad\qquad\qquad\qquad\qquad \overset{|}{D}$$

$$C_6H_5CDO$$

Some additional studies showed that the rate of reduction of α-unsaturated alcohols increases as the number of conjugated acetylenic or olefinic bonds increases, and nonallylic and nonpropargylic alcohols reduce more slowly. Some alcohols such as $PhC\equiv CCH_2CH_2OH$ do not reduce at all.

Bohlmann demonstrated the utility of lithium aluminum hydride reduction in preparing a naturally occurring thiophene compound:

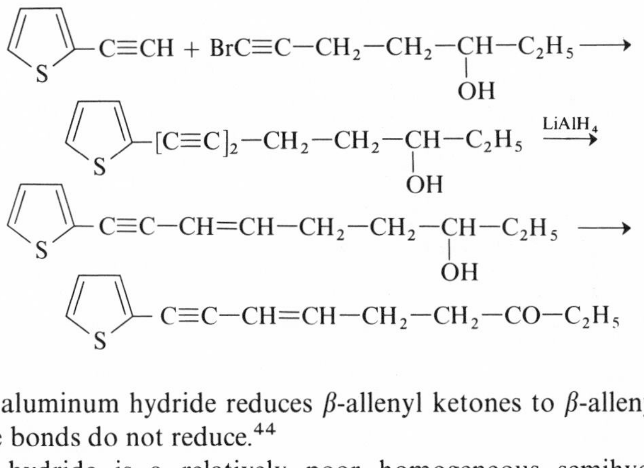

Lithium aluminum hydride reduces β-allenyl ketones to β-allenyl alcohols. The double bonds do not reduce.[44]

Sodium hydride is a relatively poor homogeneous semihydrogenation reagent.[45]

3.2.6. LITHIUM-AMMONIA REDUCTION

Chemical reduction of γ-ethynyl ketones by lithium metal in ammonia-tetrahydrofuran gives 2-methylenecyclopentanols via an ion radical intermediate:[46]

$$\begin{matrix} CH_2\!-\!\!-\!\!-O \\ | \qquad\quad \diagdown \\ | \qquad\quad Me\!-\!C\!-\!CH_2CH_2CH_2Br + NaC\!\equiv\!CH, NH_3 \xrightarrow{hydrolysis} \\ CH_2\!-\!O \diagup \end{matrix}$$

$$\underset{NH_3-THF}{\overset{O}{\overset{\|}{Me\ddot{C}(CH_2)_3C\equiv CH}}\xrightarrow{Li,}}$$

$$[\overset{O}{\overset{\|}{Me\ddot{C}(CH_2)_3\bar{C}=\ddot{C}H}}]\longrightarrow$$

(Me–ring with O⁻, •CH) $\longrightarrow$ (Me–ring with OH, CH₂)

3.3. Reductive Dimerization of α-Acetylenic Carbonyl Compounds

Durand[47] in 1961 reviewed the sparse literature on reductive dimerization of acetylenic aldehydes to glycols. Chemical reduction of aliphatic aldehydes gives mostly alcohol, with little glycol. Ethylenic aldehydes give mostly glycol, with some alcohol. 1,1-Diethoxybutyne-2 with zinc in acetic acid at 0° does not react. Phenylpropynal gives diphenylhexadiynediol. Polarographic reduction of phenylpropynal shows two breaks, one for alcohol and one for glycol.

Durand prepared butynal, hexynal, heptynal, octynal and phenylpropynal by the classic acetylenic Grignard-ethyl orthoformate reaction. He hydrolyzed the acetals and used the liberated aldehyde without purification. Reduction with zinc-acetic acid gave: 5% alcohol ($RC\!\equiv\!CCH_2OH$), 30% glycol

$$\underset{RC\equiv CCHCHC\equiv CR}{\overset{HO\ \ OH}{\overset{|\ \ \ \ |}{}}}$$

and about 5–10% unreacted aldehyde. If the reaction is a radical coupling, 50% *meso* and 50% *racemic* product should form if the radicals are equally reactive. Ketones do not react like aldehydes.

α-Acetylenic ketones in isopropanol, however, are reductively dimerized upon irradiation by U.V. light:[47a]

$$(CH_3)_3CC\equiv C\overset{O}{\overset{\|}{-}C}-CH_3 \xrightarrow[isopropanol]{U.V.} 15\% (CH_3)_3CC\equiv C\overset{OH}{\overset{|}{\underset{\underset{CH_3}{|}}{C}}}\overset{OH}{\overset{|}{\underset{\underset{CH_3}{|}}{C}}}-CH_3 +$$

$$28\% \left((CH_3)_3C-C\equiv C-\overset{OH}{\overset{|}{\underset{\underset{CH_3}{|}}{C}}}\!\!-\right)_2$$

This is an example of a photochemical reaction in which the excitation energy is in an acetylenic molecule.

References

1. Bond, G. C., in (Emmet, P. H., editor) "Catalysis," Third ed., New York, Reinhold Publishing Corp., 1955.
2. Augustine, R. L., "Catalytic Hydrogenation," pp. 62, 69–71, 152, New York, M. Dekker, Inc., 1965.
3. McQuillin, F. J., Ord, W. O., and Simpson, P. L., *J. Chem. Soc.,* 5996 (1963).
4. Jardine, I., and McQuillin, F. J., *J. Chem. Soc. (C),* 458 (1966).
5. Lindlar, H., *Helv. Chim. Acta,* 35, 446 (1952); Lindlar, H., and Dubuis, R., *Org. Syn.,* 46, 89 (1966).
6. Webb, G., and Webster, P. B., *Trans. Faraday Soc.,* 61, 1232 (1965).
7. Siegel, S., in *Advan. Catalysis,* 16, 160, 164 (1966).
8. Meyer, E. F., and Burwell, R. L., Jr., *J. Am. Chem. Soc.,* 85, 2877 2881 (1963).
9. Sokolov, L. B., *et al., Zh. Organ. Khim.,* 1, 1544 (1965); *Chem. Abstr.,* 64, 577 (1966).
10. Bond, G. C., *J. Chem. Soc.,* 2705 (1958).
11. Bryce-Smith, D., *Chem. Ind. (London),* 239 (1964).
12. Zeiss, H., in "Organometallic Chemistry," pp. 406–421, New York, Reinhold Publishing Corp., 1960.
13. Dobson, N. A., *et al., Tetrahedron,* 16, 16 (1961).
14. Pattison, F. L. M., and Dear, R. E. A., *Can. J. Chem.,* 41, 2600 (1963).
15. McQuillin, F. J., and Ord, W. A., *J. Chem. Soc.,* 2902 (1959).
15a. Schmidt, H. M., and Arens, J. F., *Rec. Trav. Chim.,* 86, 1138 (1967).
16. Tedeschi, R. J., and Clark, G., Jr., *J. Org. Chem.,* 27, 4323 (1962).
17. Tedeschi, R. J., *et al., J. Org. Chem.,* 27, 1740 (1963).
18. *Ibid.,* 27, 2398 (1962).
19. Brown, C. A., and Brown, H. C., *J. Org. Chem.,* 31, 3989 (1966).
20. Grimmer, G., and Hildebrandt, A., *Ann. Chem.,* 685, 154 (1965).
21. Spangler, C. W., and Woods, G. F., *J. Org. Chem.,* 30, 2218 (1965).
22. Buckley, G. D., and Levy, W. J., *J. Chem. Soc.,* 3016 (1951).
23. Freidlin, L. Kh., and Gorshkov, V. I., *Dokl. Acad. Nauk USSR,* 131, 1109 (1960); *Chem. Abstr.,* 54, 20983 (1960).
24. Clarke, A. J., and Crombie, L., *Chem. Ind. (London),* 143 (1957).
25. Miller, A. E. G., Biss, J. W., and Schwartzman, L. H., *J. Org. Chem.,* 24, 627 (1959).
25a. Magoon, E. F., and Slaugh, L. H., *Tetrahedron,* 23, 4509 (1967).
26. Cramer, R. D., *et al., J. Am. Chem. Soc.,* 85, 1691 (1963).
27. Jardine, F. H., *et al., Chem. Ind. (London),* 560 (1965).
28. Osborn, J. A., *et al., J. Chem. Soc. (A),* 1711 (1966).
29. Sonogashira, K., and Hagihara, N., *Bull. Chem. Soc. Japan,* 39, 1178 (1966).
30. Berthelot, M., *Ann. Chim. Phys.,* 9, 401 (1909).
31. Castro, C. E., and Stephens, R. D., *J. Am. Chem. Soc.,* 86, 4358 (1964).
31a. Bottei, R. S., and Joern, W. A., *J. Am. Chem. Soc.,* 90, 297 (1968).
32. Greenlee, K. W., and Fernelius, W. C., *J. Am. Chem. Soc.,* 64, 2505 (1942).
33. Brown, H. C., "Hydroboration," pp. 227, New York, W. A. Benjamin, Inc., 1962.
34. Stock, A., and Kuss, E., *Chem. Ber.,* 56, 789 (1923).

35. Brown, H. C., and Zweifel, G., *J. Am. Chem. Soc.,* **81**, 1512 (1959).
36. Hawthorne, M. F., *J. Am. Chem. Soc.,* **83**, 2541 (1961).
37. Cywinski, N. F., *J. Org. Chem.,* **30**, 361 (1965).
38. Cope, A. C., *et al., J. Am. Chem. Soc.,* **82**, 6370 (1960).
39. Logan, T. J., and Flautt, T. J., *J. Am. Chem. Soc.,* **82**, 3446 (1960).
40. Zweifel, G., and Arzoumanian, H., *J. Am. Chem. Soc.,* **89**, 291 (1967).
41. Hubert, A. J., *J. Chem. Soc.,* 6669 (1965).
42. *Ibid.,* 6679 (1965).
43. Bohlmann, F., Enkelmann, R., and Plettner, W., *Chem. Ber.,* **97**, 2118 (1964).
44. Bertrand, M., and Santelli, M., *Compt. Rend.,* **262C**, 1601 (1966).
45. Slaugh, L. H., *J. Org. Chem.,* **32**, 108 (1967).
46. Stork, G., *et al., J. Am. Chem. Soc.,* **87**, 1148 (1965).
47. Durand, M. H., *Bull. Soc. Chim. France,* 2387, 2393 (1961).
47a. Wilson, J. W., and Stubblefield, V. S., *J. Am. Chem. Soc.,* **90**, 3423 (1968).

PART TWO

Hydration—Addition of Water to the Acetylenic Bond

1. BACKGROUND

Hydration of triple bonds usually gives ketones. The structure of the acetylenic compound determines which carbon adds —OH to form the intermediate vinyl alcohol that rearranges to form the carbonyl compound:

$$-CH=\overset{|}{C}-OH \longrightarrow -CH_2\overset{|}{C}=O$$

Some activated acetylenes (arylacetylenes, α-acetylenic acids, alkoxyalkynes) add water in the presence of aqueous acid only, but most acetylene compounds require acids and catalysts. Mercuric salt, usually mercuric sulfate, is the most common catalyst. Some acetylenes also hydrate in the presence of bases. Acetylene itself hydrates in gas or liquid phase to acetaldehyde, acetic acid, or acetone, depending on the catalyst and the conditions. Most substituted acetylenes hydrate best in the liquid phase, usually at atmospheric pressure. Most mechanism studies have been done in mercury-catalyzed liquid phase systems.

2. VAPOR PHASE HYDRATION OF ACETYLENE

Acetylene hydrates to acetaldehyde over carbon-phosphoric acid catalyst at 280°.[1] The rate varies with acetylene concentration and with catalyst acidity.

Acetylene may form a π complex with a proton. With zinc phosphate catalyst, the rate-determining step is formation of $H\overset{+}{C}{=}CHZn^+$. Hydration of acetylene is no longer the major industrial source of acetaldehyde, acetic acid and acetone. Olefinic hydrocarbons have assumed the lead in this area. A few examples of recently reported catalysts for vapor phase hydration of acetylene are given in Table 2-5.

TABLE 2-5
Vapor Phase Hydration of Acetylene

Catalyst	Temperature (°C)	Product	% Yield	Reference
Cd-Ca phosphate	200–300	Acetaldehyde	High	2, 3
Cd-Zn molybdate	350	Acetaldehyde	72	4
Ni molybdate	350	Acetic acid	45	4
ZrO_2	400–500	Acetaldehyde	High	5
	400–500, $+NH_3$	Acetonitrile	99	5

3. LIQUID PHASE HYDRATION OF ACETYLENE

Ban[6] suggested that the function of zinc and cadmium salts in acetylene hydration is to polarize the water, not the acetylene. The polarization is by coordination, while aqueous acids or cations are formed. Polarized water then attacks acetylene nucleophilically to form acetaldehyde. When acetylene is "hydrated" by D_2O at 75°, using mercuric and ferric sulfates and sulfuric acid, all of the D goes to the methyl group of the product acetaldehyde: $C_2H_2 + D_2O \longrightarrow CD_2HCHO$.[7] Acetylene passed into cuprous sulfate-sulfuric acid causes variation of potential of a copper electrode (under nitrogen).[8] The formation of acetylides stops and the maximum concentration of acetaldehyde occurs at a definite ratio between electrode potential and acidity. The higher the acidity, the higher is the potential at maximum concentration of acetaldehyde. Thus complex formation, a reaction associated with the ionization of acetylenic hydrogen, and the equilibrium constant increase as acidity increases.[9]

In 1961, Halpern[10] reported a new hydration catalyst: $RuCl_3$ in water. The reaction proceeds under comparatively mild and apparently homogeneous conditions. Acetylene at atmospheric pressure and 50° in $5M$ HCl containing $0.1M$ $RuCl_3$ formed 2×10^{-3} mole/liter of acetaldehyde. Propyne gave acetone, and 1-butyne gave methyl ethyl ketone. Phenylpropiolic acid gave acetophenone and CO_2 from decarboxylation of the intermediate β-keto acid. The following mechanism is proposed:[10]

$$-\overset{|}{\underset{|}{Ru}}{}^{+++}\cdots OH_2 \xrightarrow{\;+RC\equiv CR'\;} -\overset{|}{\underset{|}{Ru}}{}^{+++}\cdots OH_2 \longrightarrow -\overset{|}{\underset{|}{Ru}}{}^{+++} \quad OH + H^+ \longrightarrow$$

$$\underset{\uparrow}{R-C\equiv C-R'}$$

$$\overset{\diagdown}{\underset{\diagup}{C}}=\overset{\diagup}{\underset{\diagdown}{C}}$$

$$R \qquad R'$$

$$-\overset{|}{\underset{|}{Ru}}{}^{+++} + RCH_2\overset{O}{\overset{\|}{C}}R'$$

This sequence accounts for all the facts and explains the dependence of the rate on Cl^-. The requirement of a ligand water molecule explains the decrease in activity at high Cl^- concentration because stable $RuCl_6^{-3}$ is formed. The decrease in activity at low Cl^- concentration is probably caused by the high concentration of the lower neutral and cationic chloro complexes, which do not react with acetylene.

Cuprous chloride-ammonium chloride, acid, and a metal sulfide hydrate acetylene to acetaldehyde (90% yield).[11-13] The metal sulfide is a promoter. Zinc sulfate in water requires 22 atmospheres acetylene pressure and 140°, but leads to 99% yield of acetaldehyde.[14]

Thallic chloride in acetic acid is more active for hydrating acetylenes than mercuric salts in acetic acid. Thallic chloride is also more active than mineral acids alone, at equivalent concentrations. Phenylacetylene is hydrated to acetophenone (84% yield) in only 3 hours at 100° in the presence of thallic chloride and 90% acetic acid.[14a]

4. MECHANISM OF MERCURY-CATALYZED HYDRATIONS

Hess and Munderloh[15] reported the first mercury-catalyzed hydration of a triple bond in 1918. Mercury-catalyzed additions to triple bonds have become commonplace since then. Budde and Dessy[16] noted that hydration of acetylene has been studied extensively because of its former importance commercially, and the role of the mercury catalyst has been widely speculated on. Many workers have suggested that acetylene-mercuric π complexes are intermediates, but "kinetic studies of the hydration of acetylene in aqueous sulfuric acid solutions of mercuric sulfate have been cumbersome, inconclusive, lacking in detail, and contradictory."[16]

Lemaire[17] presented evidence that a complex between 3-hexyne and acetoxy-mercuric ion is intermediate in mercuric acetate-perchloric acid catalyzed addition of acetic acid. This does not necessarily apply to hydration in aqueous systems, however. Dessy[18] has reported reaction of organomercuric halides with simple acetylenes in basic aqueous dioxane and has established the conditions which favor formation of C—Hg bonds. Budde and Dessy[16,19] obtained complex insoluble intermediates when they used mercuric chloride-HCl. They later used mercuric perchlorate because perchlorate has less tendency to complex than chloride does. Phenylacetylene hydrates very slowly in perchloric acid-dioxane-water, and acetophenone is inert in this mixture.

Ultraviolet spectra suggest that a dioxane-mercuric ion complex is the active catalyst. Phenylacetylene added to the active catalyst forms a yellow complex instantly. The rate of formation of the yellow complex depends on catalyst activity. The rate of disappearance of color is related to the structure of the acetylene, and ketone forms while the color disappears. Spectral data indicate that a 2:1 acetylene-mercuric ion complex reacts with water to form ketone. This is the only complex which adds water rapidly. Similar complexes form with 1-hexyne and with diphenylacetylene. The complex might be a substituted cyclobutadiene-mercuric ion, similar to complexes of acetylenes with palladium chloride[20] and with cobaltocene.[21] The complex might also be a bis(alkyne) π complex which retains the character of the alkyne. In 1921, Vogt and Niewland[22] noted that mercuric perchlorates, fluoroborates, etc., are effective hydration catalysts, but halides, acetates and phosphates are ineffective. It is interesting that Budde and Dessy were able to identify a bis(alkyne) complex intermediate with a mercuric salt of a strong acid, the anion of which is a weak complexer. Weak complexing ability of the anion allows the mercuric ion to complex the acetylene more completely.

Acetylenic carboxylic acids can hydrate via an intramolecular reaction in which the carboxyl group adds across the triple bond and ultimately furnishes the carbonyl oxygen:[23]

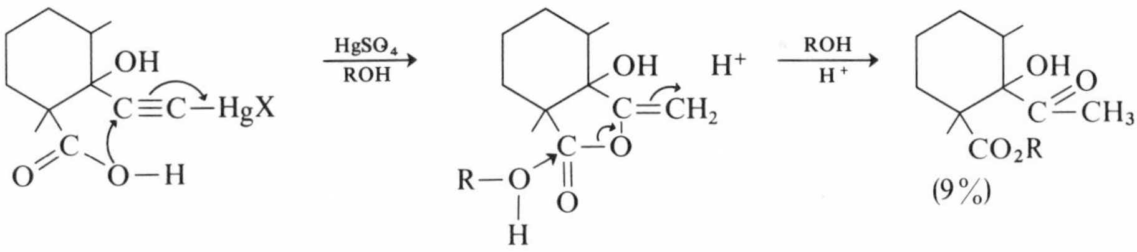

5. MECHANISM OF HYDRATIONS CATALYZED BY ACIDS ONLY

Jacobs and Searles[24] concluded from their study of acid-catalyzed hydration of alkoxy and aryloxyacetylenes that hydronium ion is a specific catalyst:

$$\mathrm{ROC{\equiv}CH + H_3O^+ \longrightarrow [ROC{=}CH_2]^+ + H_2O}$$

$$\downarrow +\mathrm{H_2O}$$

$$\mathrm{[ROC(OH_2){=}CH_2]^+ \xrightarrow{+H_2O} \left[ROC(OH){=}CH_2\right] + H_3O^+}$$

$$\downarrow$$

$$\mathrm{ROC(=O)CH_3}$$

Stamhuis and Drenth[25] later studied the hydration of alkoxyalkynes in detail and observed general acid catalysis. Their mechanism is essentially the same, except any acid (HA) adds first instead of H_3O^+. Hydration of acetylenic thioethers is also subject to general acid catalysis, and products are oriented the same way. Carbonyl forms on the carbon connected to thioether sulfur:

$$RCH_2\overset{\overset{\displaystyle O}{\|}}{C}SEt.^{26}$$

Acid-catalyzed hydration of phenylpropiolic acids and phenylacetylenes in 65% H_2SO_4 involves rate-determining addition of proton to form vinylic cation, $Ph\overset{+}{C}=CHCO_2H$.[27] Rate constants are pseudo-first order. Phenylacetylene hydrates 28 times as fast as phenylmethylacetylene. Phenylpropiolic acid hydrates 19 times faster than cinnamic acid; phenylacetylene is 2.3 times as fast as styrene. Thus, vinylic cations form more easily in moderately acid solutions than saturated carbonium ions do.

α-Olefinic ketones undergo acid-catalyzed hydration more easily than the analogous olefinic acids. Phenylbenzoylacetylene hydrates at nearly the same rate as phenylpropiolic acid does. Thus, the mechanism of hydration of the acetylenes is different. Phenylbenzoylacetylene is hydrated rapidly in aqueous 80% sulfuric acid. The rate-determining step is addition of a proton to the triple bond to form a vinylic cation, which reacts with water to form dibenzoylmethane:[27a]

$$PhC\equiv\overset{\overset{\displaystyle O}{\|}}{C}Ph + H^+ \xrightarrow{\text{slow}} Ph\overset{+}{C}=CH\overset{\overset{\displaystyle O}{\|}}{C}Ph \xrightarrow[-H^+]{H_2O} Ph\overset{\overset{\displaystyle O}{\|}}{C}CH_2\overset{\overset{\displaystyle O}{\|}}{C}Ph$$

6. HYDRATION OF PROPARGYLIC AMINES

Attempts to hydrate N,N,1,1-tetramethyl-2-butynylamine by the usual mercuric sulfate-sulfuric acid method gave mesityl oxide and not the expected 2-dimethylamino-2-methyl-3-pentanone.[28] Hydration of N,N-diethyl-1,1-dimethyl-2-propynylamine gave 3-diethylamino-3-methyl-2-butanone as expected.[29] This difference must be due to an effect of the terminal methyl group on the neighboring acetylenic bond, and not to the amine character.

Easton[30] found that hydration of 2-propynylamines as amido derivatives involved a novel neighboring group effect, and extended the study to include hydration of 3-alkyl-2-propynylamines as amido derivatives:

$$\underset{\displaystyle Me_2C-C\equiv CMe}{\overset{\displaystyle NHMe}{|}} \xrightarrow[\text{water}]{HCl} \text{no reaction}$$

$$\underset{\displaystyle Me_2C-C\equiv CMe}{\overset{\displaystyle MeNAc}{|}} \xrightarrow{HCl,\ water} \underset{\displaystyle CH_2=C-C\equiv CMe}{\overset{\displaystyle Me}{|}}$$
$$(73\%)$$

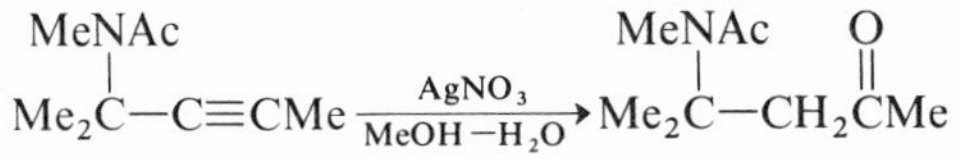

$$\underset{\underset{\displaystyle Me_2C-C\equiv CMe}{|}}{MeNAc} \xrightarrow[\text{MeOH}-\text{H}_2\text{O}]{\text{AgNO}_3} \underset{\underset{\displaystyle Me_2C-CH_2CMe}{|\qquad\quad||}}{MeNAc\;\;\;O} \qquad \begin{array}{l}\text{(this is a general method for}\\ \text{hydrating propynylamines)}\end{array}$$

$$\underset{\underset{\displaystyle Me_2C-C\equiv CMe}{|}}{MeNAc} \xrightarrow{\text{dry HCl}} \underset{\underset{\displaystyle\underset{Me_2C-CH=CHMe}{|}}{O}}{\overset{+}{MeN}=\overset{\overset{\displaystyle Cl^-}{}}{C}-Me} \xrightarrow[\text{base}]{\text{H}_2\text{O}} \underset{\underset{\displaystyle Me_2C-----CCH_2Me^*}{|\qquad\qquad||}}{MeNAc\;\;\;O} + \text{oxazolines}$$

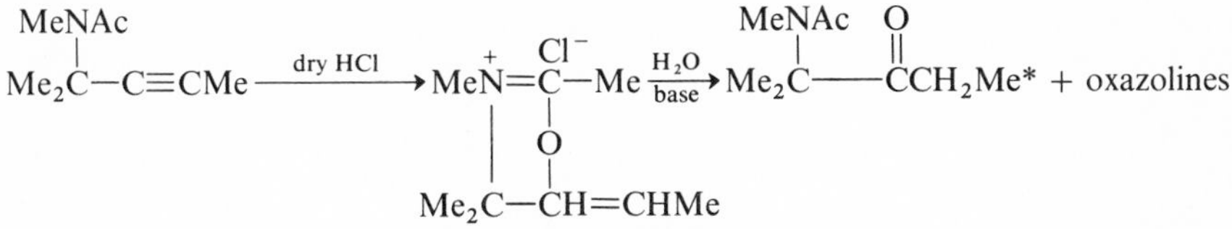

(indicated by IR, NMR)

The new propargyl *gem*-multiamines, first reported in 1968, add acetic acid to form α-olefinic ketones in a reaction which bears some similarity to catalyzed hydration. N,N,N′,N′-Tetramethyl-3-phenyl-2-propynylidenediamine (1) and N,N,N′,N′,N″,N″-hexamethyl-3-phenyl-2-propynylidenetriamine (2) are obtained by reacting phenylacetylene with tris(dimethylamino)methane and tetrakis(dimethylamino)methane, respectively, at 80° for 18 hours. Yields are 87–88%.[30a] The reaction of (2) with acetic acid in benzene at room temperature gives 80% yield of 3,3-bis(dimethylamino)acrylophenone (3) and 10% yield of tris(dimethylamino)-1-phenylallylium acetate (4):

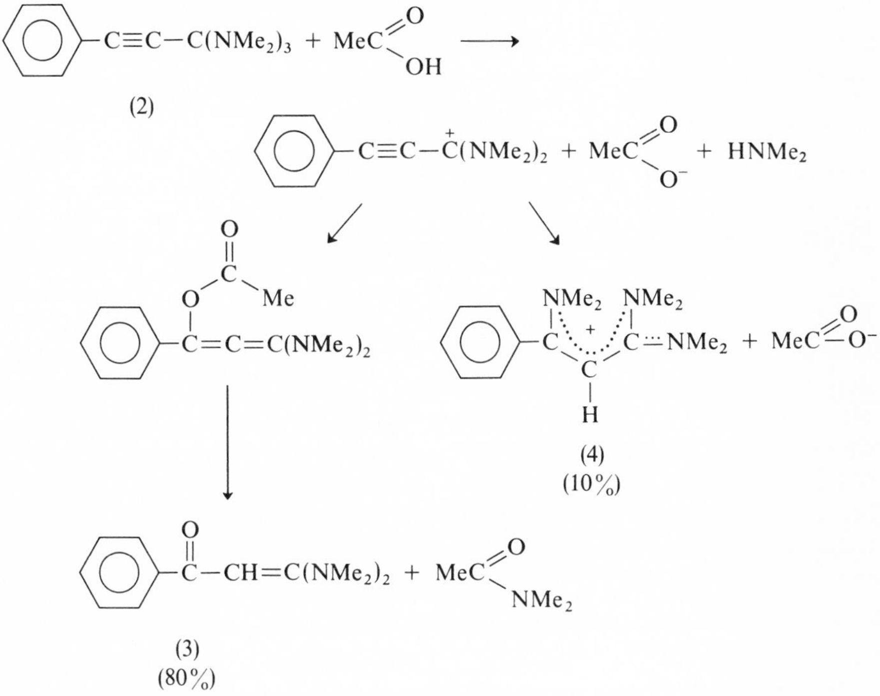

(2)

(4)
(10%)

(3)
(80%)

* This is hydration in the other direction.

Acetic acid also adds easily to (1), giving 1,3-bis(dimethylamino)-1-phenyl-allylium acetate as the only isolable product:

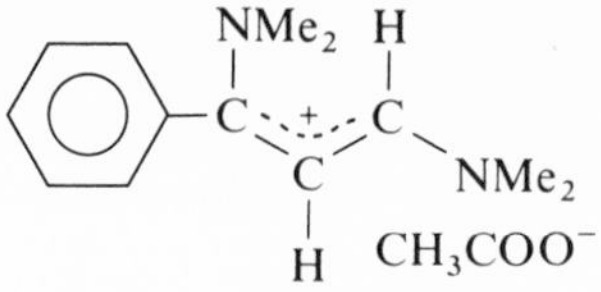

When methanol reacts with (1), one dimethylamino group is replaced by a methoxyl group. The product is 1-methoxy-N,N-dimethyl-3-phenyl-2-propynylamine:

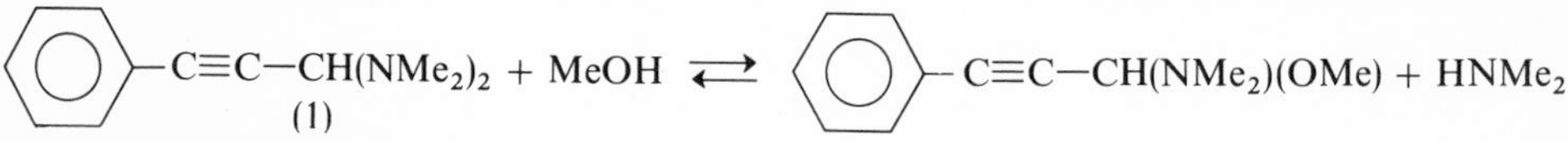

7. DIRECTION OF ADDITION OF WATER TO SUBSTITUTED ACETYLENES

The direction of addition of water depends mainly on the structure of the acetylene, and to a lesser extent on the hydrating reagent. The examples shown in Tables 2-6 through 2-12 lead to some general rules:

(1) Terminal alkynes almost invariably form methyl ketones by adding OH at carbon 2. Nonterminal alkynes usually add OH to the acetylenic carbon connected to the less branched substituent.

(2) Arylacetylenes add water to form α-aryl ketones:

$$ArC\equiv C-\longrightarrow Ar\overset{\overset{\displaystyle O}{\|}}{C}CH_2-$$

(3) Arylacetylenic acids form aryl-α-keto-β-carboxylic acids.

(4) Vinylacetylenes form vinyl alkyl ketones. Aryl is a stronger directing group than vinyl: $ArC\equiv CCH=CH_2 \longrightarrow Ar\overset{\overset{\displaystyle O}{\|}}{C}CH_2CH=CH_2$.

(5) Acetylenic ethers, thioethers and amines add OH to the carbon connected to the hetero atom.

(6) Nonconjugated diacetylenes add 2 moles of water to form diketones. Conjugated diacetylenes add 1 mole of water to a carbon of the acetylenic group connected to the more branched alkyl group (R'):

$$RC\equiv CC\equiv CR' \longrightarrow RC\equiv C\overset{\overset{\displaystyle O}{\|}}{C}CH_2R'$$

(7) Secondary and tertiary α-ethynylcarbinols add OH to carbon 2 to form 2-carbonyl-3-hydroxy compounds. Primary alcohols do not hydrate easily.[31]

(8) α-Acetylenic glycols add water and then usually cyclize to dihydro-furanones. Acetylenic γ,γ-dicarboxylic acids hydrate to the ketone which cyclizes to dilactone:[32]

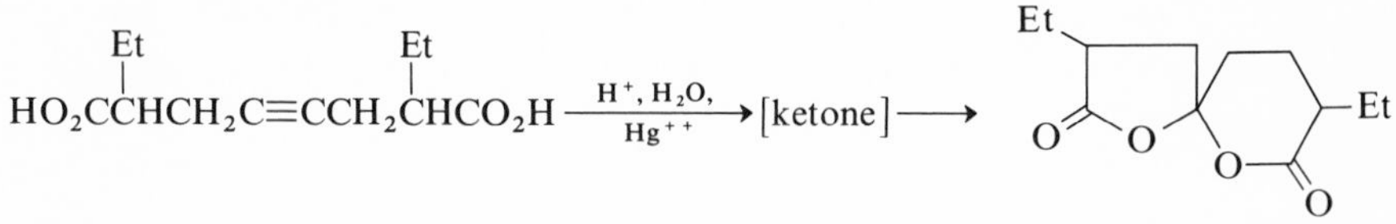

TABLE 2-6
Direction of Addition of Water to Nonterminal Acetylenic Hydrocarbons

$R-\overset{1}{C}\equiv\overset{2}{C}-R'$		Reagent, Conditions	% Addition OH to Carbon		Reference
R	R'		1	2	
Me	Pr	85% HCO$_2$H, 145°, 7 hr	50	50	33
Me	t-Bu	85% HCO$_2$H, 145°, 7 hr	70	30	33
Me	Ph	85% HCO$_2$H, 145°, 7 hr	0	100	33
t-Bu	Ph	85% HCO$_2$H, 145°, 7 hr	0	100	33
Me	Ph	Sulfonated polystyrene, 60°	0	100	34
Me	t-Bu	80% H$_2$SO$_4$	25	75	35
Me	Ph	ROH-NaOMe, then H$_2$O	0	100	34
Me	Et	Acid, Hg^{++}	42	58	36
Me	Pr	HgO, H$_2$SO$_4$	44	56	37
Me	i-Pr	Fe$_2$(SO$_4$)$_3$	54	46	37
Me	t-Bu	Fe$_2$(SO$_4$)$_3$	35	65	37
Me	1-Cyclohexenyl	HgO, H$^+$, H$_2$O, Fe$_2$(SO$_4$)$_3$	0	100	38
R	H	Cd or Ca phosphate, 375°	100	0	39
R	H	HOBr	100(R$\overset{\text{O}}{\overset{\|}{C}}CH_2$Br)		40

TABLE 2-7
Direction of Addition of Water to Conjugated Diacetylenic Hydrocarbons

$RC\overset{1}{\equiv}\overset{2}{C}C\overset{3}{\equiv}CR'$		Reagent, Conditions	% Addition to Carbon		Reference
R	R'		1	2	
Me	Me	Conc. H$_2$SO$_4$	100 (49)[a]	0	41
Me	i-Pr	Conc. H$_2$SO$_4$	0	100 (89)[a]	41
Me	t-Bu	Conc. H$_2$SO$_4$	0	100 (100)[a]	41
H	Et	HgSO$_4$, 20% H$_2$SO$_4$	To 1 and 3 →diketone		42
Ph	Ph	HgSO$_4$, H$_2$SO$_4$, MeOH, H$_2$O	To 1 and 3 →furan		43

[a] Total yield.

TABLE 2-8
Direction of Addition of Water to α-Ethynylcarbinols

$RR'C(OH)\overset{1}{C}\equiv CH$		Reagent, Conditions	% Addition to Carbon 1	Reference
R	R'			
Ph	Ph	H_2SO_4, HgO, H_2O	100 (19)[a]	44
$CH_2=\overset{Me}{\underset{|}{C}}-$	Me	HgO, HOAc, 90°	100 (76)[a]	45
Me	Me	$PhCH_2SH$, $HgSO_4$, H_2SO_4, H_2O	100 (80,[a] thiobenzyl ether)	46

[a] Total yield.

TABLE 2-9

Direction of Addition of Water to Enyne Hydrocarbons

$RCH=CH-\overset{1}{C}\equiv\overset{2}{C}R'$		Reagent, Conditions	% Addition to Carbon		Reference
R	R'		1	2	
H	H	Cd or Ca phosphate, 350°	100		47
Et	H	Cd or Ca phosphate, 350°	100		39
H	1-Cyclo-hexenyl	$HgSO_4$, H_2O, MeOH		100	48

Other Enynes

$\overset{1}{R}C\equiv\overset{2}{C}R'$					
R	R'				
Vinyl	Allyl	H_2SO_4, HgO, $Fe_2(SO_4)_3$	100		49
Phenyl	Allyl	H_2SO_4, HgO, $Fe_2(SO_4)_3$	100		49
Allyl	Me	$HgSO_4$, H_2SO_4	100		50
Allyl	Phenyl	$HgSO_4$, H_2SO_4 + MeOH		100 (90)[a]	51

TABLE 2-10
Direction of Addition of Water to Nonconjugated Diacetylenic Hydrocarbons

$HC\equiv\overset{1}{C}-X-\overset{2}{C}\equiv CH$	Reagent, Conditions	% Addition to Carbon		Reference
X		1	2	
$(CH_2)_{4,5}$	HgO, 10% H_2SO_4, MeOH	100 (90)[a]		43
1,4-Phenylene	HgO, 10% H_2SO_4, MeOH	(85)[a]		
9, 10-Anthracenyl	HgO, 10% H_2SO_4, MeOH	(55)[a]		

[a] Total yield.

TABLE 2-11
Direction of Addition of Water to Conjugated Diacetylenic Glycols

$$R-C{\equiv}C-C{\equiv}C-R'$$

R	R'	Reagent, Conditions	Product	% Yield	Reference
Me₂C(OH)—	Me₂C(OH)—	Aqueous Me₂NH, 3 hr, 100°	(structure with Me₂, Me, O, O)	52	52
		HgSO₄-H₂SO₄, 80°			
		HgSO₄, H₂SO₄	(structure with Me₂, Me₂C=, H, O, O)		53
		HgSO₄, H₂SO₄, heat	(furan structure with Me₂, Me₂)		54
(1-hydroxycyclohexyl, OH)	(1-hydroxycyclohexyl, OH)	Aqueous Me₂NH, heat, or alcoholic Et₂NH, −20°	(spirocyclic structure with O, O)		52

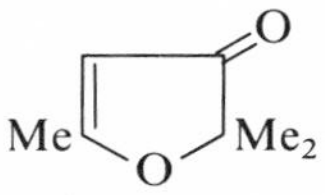

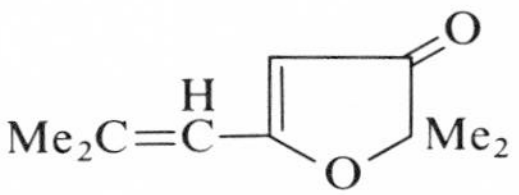

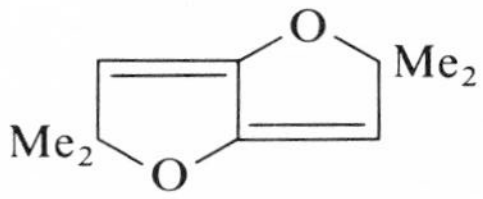

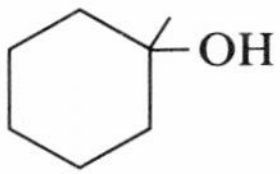

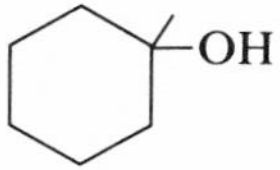

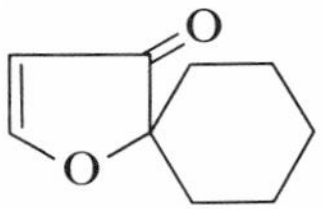

TABLE 2-12
Direction of Addition of Water in Hydrations
Catalyzed by Acid Only

$R-\overset{1}{C}\equiv\overset{2}{C}-R'$		Reagents,	% Addition to Carbon		
R	R'	Conditions	1	2	Reference
Et$_2$N	Et	1N H$_2$SO$_4$, very fast	100 (70)[a]		55
Ar	CO$_2$H	H$^+$	100		27
RO	H	5N HCl, fast	100		24, 25
Bu	OPh	dil. HCl		100	56
CH$_2$=CHO—	H	6N H$_2$SO$_4$	100 (goes to HOAc + HAc)		57
$\overset{\text{OH}}{\underset{\vert}{\text{Me}_2\text{C}-}}$	OCH=CH$_2$	1N H$_2$SO$_4$	100 (to Me$_2$C=CH— $\overset{}{\underset{\Vert O}{\text{COCH=CH}_2}}$ by rearrangement)		57
$\overset{\text{O}}{\underset{\Vert}{\text{R}'\text{CS}-}}$	R	Water, neutral		100 (nucleophilic addition)	58
CH$_2$=CH—	SMe	H$_2$SO$_4$		100 (53)[a]	59

[a] Total yield.

8. REARRANGEMENTS RELATED TO HYDRATION

8.1. Meyer-Schuster and Rupe Rearrangements—Mechanism

Acids rearrange ethynylcarbinols to α,β-unsaturated carbonyl compounds.[60,61] These, the Meyer–Schuster and Rupe rearrangements, are unimolecular, and probably go through the intermediate mesomeric carbonium ion (1) which reacts in either of two ways, depending on its structure:

$$
\begin{array}{c}
\text{R}-\text{CH}_2 \\
\diagdown \\
\text{C}-\text{C}\equiv\text{C}-\text{R}_2 \\
\diagup \quad | \\
\text{R}_1 \quad \text{OH}
\end{array}
\quad
\underset{-\text{H}^+}{\overset{+\text{H}^+}{\rightleftharpoons}}
\quad
\begin{array}{c}
\text{R}-\text{CH}_2 \\
\diagdown \\
\text{C}-\text{C}\equiv\text{CR}_2 \\
\diagup \quad | \\
\text{R}_1 \quad {}^+\text{OH}_2
\end{array}
$$

$$-\text{H}_2\text{O} \updownarrow +\text{H}_2\text{O}$$

$$
\left[
\begin{array}{c}
\text{R}-\text{CH}_2 \\
\diagdown \\
\text{C}=\text{C}=\overset{+}{\text{C}}-\text{R}_2 \\
\diagup \\
\text{R}_1
\end{array}
\quad \rightleftharpoons \quad
\begin{array}{c}
\text{R}-\text{CH}_2 \\
\diagdown \overset{+}{} \\
\text{C}-\text{C}\equiv\text{C}-\text{R}_2 \\
\diagup \\
\text{R}_1
\end{array}
\right]
$$

$$\Big\downarrow {\text{H}_2\text{O} \atop -\text{H}^+} \qquad (1) \qquad \Big\downarrow -\text{H}^+$$

$$
\begin{array}{c}
\text{R}-\text{CH}_2 \\
\diagdown \\
\text{C}=\text{C}=\text{C}-\text{R}_2 \\
\diagup \quad\quad | \\
\text{R}_1 \quad\quad \text{OH}
\end{array}
\qquad\qquad
\begin{array}{c}
\text{R}-\text{CH} \\
\diagdown\!\!\diagdown \\
\text{C}-\text{C}\equiv\text{C}-\text{R}_2 \\
\diagup \\
\text{R}_1
\end{array}
$$

$$\downarrow \qquad\qquad\qquad\qquad \Big\downarrow {+\text{H}_2\text{O}, \atop \text{H}^+}$$

$$
\begin{array}{c}
\text{R}-\text{CH}_2 \\
\diagdown \quad\quad\quad \text{O} \\
\quad\quad\quad\quad \|\\
\text{C}=\text{CH}-\text{C}-\text{R}_2 \\
\diagup \\
\text{R}_1
\end{array}
\qquad\quad
\begin{array}{c}
\text{R}-\text{CH} \\
\diagdown\!\!\diagdown \\
\text{C}-\text{C}=\text{CH}-\text{R}_2 \\
\diagup \quad | \\
\text{R}_1 \quad \text{OH}
\end{array}
$$

$$\downarrow$$

$$
\begin{array}{c}
\text{R}-\text{CH} \quad\quad \text{O} \\
\diagdown\!\!\diagdown \quad\quad \| \\
\text{C}-\text{C}-\text{CH}_2\text{R}_2 \\
\diagup \\
\text{R}_1
\end{array}
$$

Meyer–Schuster
rearrangement

Rupe rearrangement

8.2. Hydrations and Rearrangements in Ethynyl Steroids

When 17-ethynyl-17-hydroxysteroids are hydrated, five-membered rings expand into six-membered rings.[60,62] Hennion and Fleck[63] used 9-ethynyl-9-fluorenol as a model compound to see if hydration was accompanied by expansion of the five-membered ring. Hydration by the methods used for steroids gave polymer. Hydration by aqueous methanol with a little mercuric sulfate-sulfuric acid catalyst gave 9-acetyl-9-fluorenol in 84% yield, with no evidence of ring expansion. 9-Acetylfluorenol also failed to undergo D-homoannulation under conditions which expand the ring in 17β-hydroxy-20-ketosteroids. Unrearranged starting material was recovered. Dilute solutions of 9-ethynyl-9-fluorenol in refluxing acidic ethanol underwent the Meyer–Schuster rearrangement to give 60% yield of fluoroenylidene acetaldehyde.

The course of hydration reactions has always been subject to controversy because two products are frequently possible. Meyer–Schuster rearrangement can form aldehyde, and Rupe reaction forms ketone. Ketone is usually the major product, but the ratio of aldehyde to ketone depends on the structure of the acetylenic carbinol. Neither 9-ethynyl-9-fluorenol nor 1,1,3-triphenyl-2-propyn-1-ol can undergo the Rupe reaction, so the Meyer–Schuster is the only reaction observed.

Ghaudhuri and Gut[60] studied the acid-catalyzed rearrangement of a 20-ethynyl-20-hydroxysteroid to see if the Rupe rearrangement would take place and form a $\Delta^{17(20)}$-22-ketosteroid. They recognized that the intermediate propargyl cation,[64] perhaps in equilibrium with the allenic cation, might undergo a Wagner-Meerwein rearrangement to form a D-homosteroid to relieve the steric strain of the *trans*-hydrindane system. D-Homoannulation is a general reaction of steroids which react through a carbonium ion at carbon 20.[62]

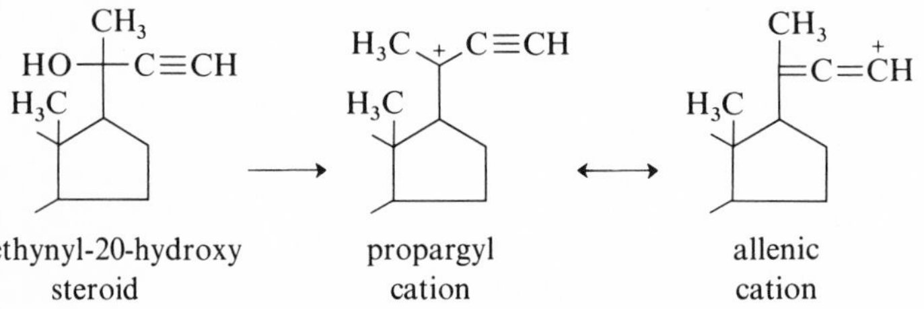

20-ethynyl-20-hydroxy   propargyl   allenic
steroid   cation   cation

When 20α-ethynylpregn-5-ene-3β,20β-diol 3-acetate was heated in 97% formic acid for 3–5 minutes on a steam bath, products from both the Wagner–Meerwein (2) and Rupe (3, 4) rearrangements were isolated (25 and 35% yields, respectively).

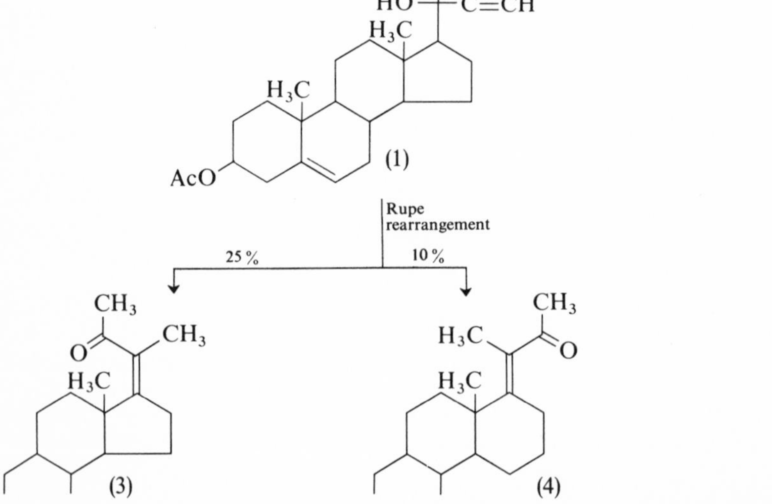

Consideration of steric factors led to the conclusion that D-homoannulation took place by movement of the 16,17-bond. Thus, the rearrangement of (1) to (2) can take place either in a concerted manner, or stepwise through (5) and (6):

(1) (5) (6)

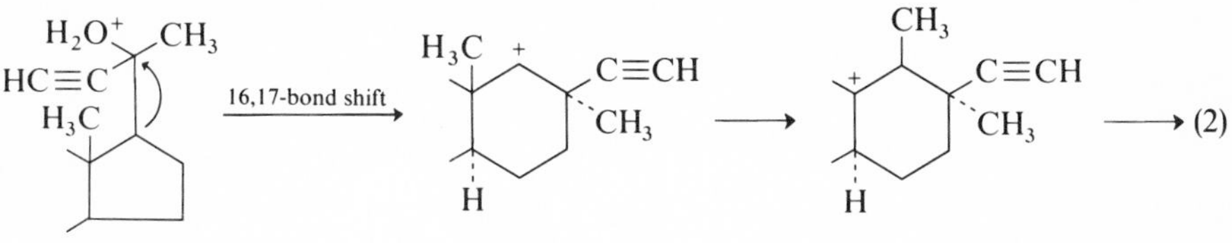

8.3. Acid-catalyzed Rearrangement of Ethoxyethynyl- and Thioethynylcarbinols

Reaction of ethoxyethynyl ether or ethylthioethynyl ether as the lithium or sodium derivative with carbonyl compounds gives the corresponding carbinols. These carbinols are rearranged by acids to α,β-unsaturated esters and aldehydes. Esters are usually the main products, but under some reaction conditions the product is a mixture of unsaturated and β-hydroxy esters:

$$>C(OH)-C\equiv COEt \xrightarrow{H^+,\ H_2O} >C=CHCO_2Et + >C(OH)-CH_2CO_2Et$$

The rearrangements are surprisingly fast compared to hydration of the parent ethynyl ethers under similar conditions. The preferred procedures are: (a) reaction of solutions of carbinols in alcohol, dioxane or tetrahydrofuran with a little 10% aqueous sulfuric acid at room temperature for $\frac{1}{4}$–2 hours, or (b) shaking an ether solution with 10% sulfuric acid at room temperature for $\frac{1}{4}$ hour.[65] Some carbinols are so sensitive to traces of acids that they rearrange when distilled from "Pyrex" flasks which have been cleaned with nitric or chromic acid. In 1960, Arens[65] stated that the unsaturated esters probably form by a rearrangement, not a hydration followed by loss of water, because reaction conditions are too mild for loss of water. Co-product β-hydroxy esters probably form by hydration. The ratio is determined by the relative rates of the two reactions.

Another explanation of the unsaturated ester formation is: A proton adds to the hydroxyl group, and the oxonium ion rearranges and loses water to form carbonium ion. The carbonium ion adds water and loses a proton:[66]

$$>C(OH)-C\equiv COEt \xrightarrow{H^+} >C(\overset{+}{O}H_2)-C\equiv COEt \xrightarrow{-H_2O} \left[>\overset{+}{C}-C\equiv COEt \rightleftarrows >C=C=\overset{+}{C}OEt \right] \xrightarrow[-H^+]{+H_2O}$$

$$>C=C=C(OH)(OEt) \longrightarrow >C=CHCO_2Et$$

It is also possible that both products form from the carbonium ion which results from the addition of a proton to the triple bond (not addition to the hydroxyl group). Either the carbonium ion adds water to form hydroxy ester, or a 1,3-shift occurs to form the unsaturated ester:[67]

$$\begin{array}{c}\diagdown \\ \diagup \end{array}C-C\equiv COEt \xrightarrow{+H^+} \begin{array}{c}\diagdown \\ \diagup \end{array}C-CH=\overset{+}{C}OEt \xrightarrow{-H_2O} \begin{array}{c}\diagdown \\ \diagup \end{array}C-CH_2CO_2Et + H^+$$
with OH groups, leading to
$$\longrightarrow \begin{array}{c}\diagdown \\ \diagup \end{array}C=CHCO_2Et + H^+$$

In 1961, Hekkert and Drenth[68] reported a careful study of rearrangement rates and product ratios. The rate constant decreases by a factor of 1.4 when ethyl groups are substituted for methyl in $Me_2C(OH)C\equiv CSEt$. The mechanism involving addition of a proton to hydroxyl[66] therefore is probably not correct, since in this scheme the rate-determining step is probably carbonium ion formation, and substitution of ethyl groups for methyl should increase the rate. A mechanism with a common rate-determining step[67] is more reasonable.

One explanation of the faster rate of rearrangement of the carbinols relative to the rate of hydration of the parent ethers is anchimeric acceleration by the hydroxyl group in an intermediate in which the hydroxyl group shows a neighboring group participation:

$$\left[\begin{array}{c} Me \diagdown \quad \overset{H}{\underset{}{O}} \\ \qquad C \diagdown \diagup \diagdown CSEt \\ Me \diagup \quad C \\ \qquad \underset{H}{\overset{}{}} \end{array} \right]^+$$

An equation was derived for predicting the ratio of hydroxy ester to unsaturated ester.[68] These products are not formed in parallel reactions. They are also not the result of subsequent reactions because they are not converted into each other under the reaction conditions. The rate-determining step is common to both products. Since the reaction is general acid catalyzed, the rate-determining step is proton transfer to carbon. The proton attaches to the β-carbon, because of resonance between the π electrons of the triple bond and the unshared pairs on the oxygen:

$$\begin{array}{c}\diagdown \\ \diagup \end{array}C-C\equiv COR \rightleftharpoons \begin{array}{c}\diagdown \\ \diagup \end{array}C-\overset{-}{C}=C=\overset{+}{O}R \xrightarrow{+H^+} \begin{array}{c}\diagdown \\ \diagup \end{array}C-CH=\overset{+}{C}OR$$

The slow step in the rearrangement is analogous to the slow step in hydration of the parent ethynyl ethers. Thus, the overall mechanism is:

β-hydroxy ester α,β-unsaturated ester

The β-lactone (1) is probably the second intermediate. The direction of ring opening depends on pH. Under strongly acidic or basic conditions, acyl-oxygen fission is major, as it is in ester hydrolysis. This gives hydroxy ester. At pH 1–7, alkyl-oxygen fission is the major cleavage, and the product is unsaturated ester.

References

1. Tsybina, E. N., et al., Zh. Fiz. Khim., **32**, 865, 1002 (1958); Chem. Abstr., **52**, 19370 (1958).
2. Kalinin, A. A., Kabanova, G. B., and Kirillov, I. P., Izv. Vysshykh Uchebn. Zavedenii Khim. i Khim. Tekhnol., **8**, 88 (1965); Chem. Abstr., **63**, 2427 (1965).
3. Gorin, Yu. A., Gorn, I. K., and Makashina, A. N., Zh. Organ. Khim., **1**, 852 (1965); Chem. Abstr., **63**, 6833 (1965).
4. Sarbaev, A. N., and Kirillov, I. P., Izv. Vysshykh Uchebn. Zavedenii Khim. i Khim. Tekhnol., **7**, 948 (1964); Chem. Abstr., **63**, 2893 (1965).
5. Berkmann, S., et al., "Catalysis," p. 440, New York, Reinhold Publishing Corp., 1940.
6. Ban, M., Kogyo Kagaku Zasshi, **61**, 1166 (1958); Chem. Abstr., **56**, 49 (1962).
7. Rekasheva, A. F., and Samchenko, I. P., Dokl. Akad. Nauk SSSR, **133**, 1340 (1960); Chem. Abstr., **54**, 23643 (1960).
8. Temkin, O. N., German, E. D., and Flid, M., Zh. Obshch. Khim., **30**, 699 (1960); Chem. Abstr., **54**, 24357 (1960).
9. Temkin, O. N., et al., Kinetika i Kataliz, **2**, 205 (1961); Chem. Abstr., **55**, 22088 (1961).
10. Halpern, J., James, B. R., and Kemp, A. L. W., J. Am. Chem. Soc., **83**, 4097 (1961).
11. Vartanyan, S. A., et al., Izv. Akad. Nauk Arm. SSR Khim. Nauk, **14**, 565 (1961); Chem. Abstr., **58**, 4412 (1963).
12. Vartanyan, S. A., et al., Izv. Akad. Nauk Arm. SSR Khim. Nauk, **12**, 345 (1959); Chem. Abstr., **54**, 10841 (1960).
13. Vartanyan, S. A., Pirenyan, S. K., and Manasyan, N. G., Zh. Obshch. Khim., **31**, 2436 (1961); Chem. Abstr., **56**, 4599 (1962).
14. Ban, M., and Ida, F., U.S. Patent 2,791,614 (May 7, 1957); Chem. Abstr., **51**, 15552 (1957).
14a. Vemura, S., et al., Bull. Chem. Soc. Japan, **40**, 1499 (1967) (in English).
15. Hess, K., and Munderloh, H., Chem. Ber., **51**, 377 (1918).

16. Budde, W. L., and Dessy, R. E., *J. Am. Chem. Soc.*, **85**, 3964 (1963); *Tetrahedron Letters*, 651 (1963).
17. Letsinger, R. L., and Nazy, J. R., *J. Am. Chem. Soc.*, **81**, 3013 (1959).
18. Dessy, R. E., Budde, W. L., and Woodruff, C., *J. Am. Chem. Soc.*, **84**, 1172 (1962).
19. Budde, W. L., and Dessy, R. E., *Chem. Ind. (London)*, 735 (1963).
20. Blomquist, A. T., and Maitlis, P. M., *J. Am. Chem. Soc.*, **84**, 2329 (1962).
21. Boston, J. L., Sharp, D. W. A., and Wilkinson, G. A., *J. Chem. Soc.*, 3488 (1962).
22. Vogt, R. R., and Nieuwland, J. A., *J. Am. Chem. Soc.*, **43**, 2071 (1921).
23. Riemer, A. C., and Rigby, W., *J. Chem. Soc.* (C), 764 (1966).
24. Jacobs, T. L., and Searles, S., Jr., *J. Am. Chem. Soc.*, **66**, 686 (1944).
25. Stamhuis, E. J., and Drenth, W., *Rec. Trav. Chim.*, **80**, 797 (1961).
26. Drenth, W., and Hogeveen, H., *Rec. Trav. Chim.*, **79**, 1002 (1960).
27. Noyce, D. S., *et al.*, *J. Am. Chem. Soc.*, **87**, 2295 (1965); **89**, 6225 (1967); *J. Org. Chem.*, **33**, 845 (1968).
27a. Noyce, D. S. and De Bruin, K. E., *J. Am. Chem. Soc.*, **90**, 372 (1968).
28. Kruse, C. W., and Kleinschmidt, R. F., *J. Am. Chem. Soc.*, **83**, 216 (1961).
29. Rose, J. D., and Weedon, B. C. L., *J. Chem. Soc.*, 782 (1949).
30. Easton, N. R., Cassady, D. R., and Dillard, R. D., *J. Org. Chem.*, **30**, 3084 (1965).
30a. Weingarten, H., *Tetrahedron*, **24**, 2767 (1968).
31. Jadot, J., and Mullers, S., *Bull. Soc. Roy. Sci. Liege*, **29**, 203 (1960).
32. Akhnazaryan, A. A., *et al.*, *Izv. Akad. Nauk Arm. SSR Khim. Nauk*, **17**, 660 (1964); *Chem. Abstr.*, **63**, 2894 (1965).
33. Kupin, B. S., and Petrov, A. A., *Izv. Vysshykh Uchebn. Zavedenii Khim. i Khim. Tekhnol.*, **6**, 75 (1963); *Chem. Abstr.*, **59**, 5042 (1963).
34. Kupin, B. S., *Zh. Organ. Khim.*, **1**, 1206 (1965); *Chem. Abstr.*, **63**, 12995 (1965).
35. Kupin, B. S., and Petrov, A. A., *Izv. Vysshykh Uchebn. Zavedenii Khim. i Khim. Tekhnol.*, **5**, 439 (1962); *Chem. Abstr.*, **58**, 436 (1963).
36. Petrov, A. A., and Kupin, B. S., *Zh. Obshch. Khim.*, **29**, 3153 (1959); *Chem. Abstr.*, **54**, 11968 (1960).
37. Kupin, B. S., *et al.*, *Tr. Leningr. Tekhnol. Inst. im. Lensoveta*, **60**, 63 (1960); *Chem. Abstr.*, **55**, 20929 (1961).
38. Favorskaya, I. A., and Auvinen, E. M., *Zh. Organ. Khim.*, **1**, 486 (1965); *Chem. Abstr.*, **63**, 1712 (1965).
39. Gorin, Yu. A., and Bogdanova, L. P., *Zh. Obshch. Khim.*, **28**, 1144 (1958); *Chem. Abstr.*, **53**, 275 (1959).
40. Kennedy, J., *et al.*, *Proc. Chem. Soc.*, 148 (1964).
41. Turbanova, E. S., Porfir'eva, Yu. I., and Petrov, A. A., *Zh. Organ. Khim.*, **2**, 772 (1966); *Chem. Abstr.*, **65**, 10480 (1966).
42. Herbertz, T., *Chem. Ber.*, **85**, 475 (1952).
43. Yen, V.-Q., *Ann. Chim. (Paris)*, **7**, 799 (1962); *Chem. Abstr.*, **59**, 5044 (1963).
44. Venus-Danilova, E. D., Serkova, V. I., and El'tsov, A. V., *Zh. Obshch. Khim.*, **27**, 334 (1957); *Chem. Abstr.*, **51**, 15467 (1957).
45. Mavrov, M. V., and Kucherov, V. F., *Izv. Akad. Nauk SSSR Otd. Khim. Nauk*, 1267 (1962); *Chem. Abstr.*, **58**, 1338 (1963).
46. Stacy, G. W., Barnett, B. F., and Strong, P. L., *J. Org. Chem.*, **30**, 592 (1965).
47. Gorin, Y. A., and Bodganova, L. P., *Zh. Obshch. Khim.*, **28**, 657 (1958); *Chem. Abstr.*, **52**, 17095 (1958).

48. Nazarov, I. N., and Zaretskaya, I. I., *Zh. Obshch. Khim.*, **27**, 624 (1957); *Chem. Abstr.*, **51**, 16316 (1957).
49. Kupin, B. S., Petrov, A. A., and Koptev, D. A., *Zh. Obshch. Khim.*, **32**, 1758 (1962); *Chem. Abstr.*, **58**, 7858 (1963).
50. Kupin, B. S., and Petrov, A. A., *Zh. Obshch. Khim.*, **29**, 3999 (1959); *Chem. Abstr.*, **54**, 20823 (1960).
51. Kurtz, P., *Ann. Chem.*, **658**, 6 (1962).
52. Gusev, B. P., Nazarova, I. I., and Kucherov, V. F., *Izv. Akad. Nauk SSSR Ser. Khim.*, 688 (1965); *Chem. Abstr.*, **63**, 6939 (1965).
53. Bohlmann, F., *Chem. Ber.*, **94**, 1104 (1961).
54. Audier, L., *Ann. Chim. (Paris)*, 105 (1957).
55. Montijn, P. P., Harryvan, E., and Brandsma, L., *Rec. Trav. Chim.*, **83**, 1211 (1964).
56. Jacobs, T. L., Cramer, R., and Hanson, J. E., *J. Am. Chem. Soc.*, **64**, 223 (1942).
57. Brandsma, L., and Arens, J. F., *Rec. Trav. Chim.*, **81**, 539 (1962).
58. Wijers, H. E., *Rec. Trav. Chim.*, **84**, 1284 (1965).
59. Boiko, Yu. A., Kupin, B. S., and Petrov, A. A., *Zh. Organ. Khim.*, **2**, 1008 (1966); *Chem. Abstr.*, **65**, 16849 (1966).
60. Cherbuliez, E., *et al.*, *Helv. Chim. Acta*, **48**, 632 (1965).
61. de la Mare, P. B. D., in "Molecular Rearrangements," Part I, p. 27, New York, Interscience Publishers, 1963.
62. Wendler, N. L., in "Molecular Rearrangements," Part II, p. 1019, New York, Interscience Publishers, 1964.
63. Hennion, G. F., and Fleck, B. R., *J. Am. Chem. Soc.*, **77**, 3253 (1955).
64. Richey, H. G., Phillips, J. C., and Rennick, L. E., *J. Am. Chem. Soc.*, **87**, 1381 (1965).
65. Arens, J. F., in "Advances in Organic Chemistry: Methods and Results," Vol. 2, p. 117, New York, Interscience Publishers, 1960.
66. Shchukina, M. N., and Rubtsov, I. A., *J. Gen. Chem. USSR*, **18**, 1645 (1948).
67. Sarrett, L. H., *et al.*, *J. Am. Chem. Soc.*, **76**, 1715 (1954).
68. Hekkert, G. L., and Drenth, W., *Rec. Trav. Chim.*, **80**, 1285 (1961).

PART THREE

Free Radical Addition to Acetylenes

1. BACKGROUND

Radicals often do not orient the same as ions when they add to triple bonds. Radical addition is rarely stereospecific, but ionic addition frequently is. Compared to ionic addition, radical addition has not been studied very much. More is known about addition of hetero atom radicals than about addition of carbon radicals. In 1963 an *Organic Reactions* chapter on free radical additions to acetylenes and olefins was published.[1] Sosnovsky's[2] book on free radical reactions in preparative organic chemistry appeared in 1964. These comprehensive surveys show an enormous amount of work on addition of radicals to olefins, but relatively little on addition to acetylenes.

2. ADDITION OF HETERO ATOM RADICALS

2.1. Halogen

Radical chlorination of acetylene gives vinyl chloride, 1-chloro-1,3-butadiene, benzene and traces of chloroacetylene.[3] Dimethyl acetylenedicarboxylate adds bromine as radicals, even in the dark.[4] Light increases the rate and the extent of reaction. Illumination of bromine in acetic acid before adding the dimethyl acetylenedicarboxylate speeds up the subsequent dark reaction. This effect lasts for about 3 hours. Chlorine adds only to the central carbons of tetraphenylbutatriene,[2] but halogen usually adds at the end carbons of polyyne systems to give cumulenes.[5] Propyne and allene react with t-butyl hypochlorite in light to form propargyl chloride by a substitution rather than addition. Propargyl radical is the intermediate from both propyne and allene.[6]

Nazarov[7] studied the radical addition of bromine to acetylenes and observed some stereospecific additions: In 1-alkynes, the tendency to *cis* products increases as the bulk of the alkyl group increases (16% from propyne, 90% *cis* product from t-butylacetylene). Propargyl alcohol gives all *trans*-dibromoallyl alcohol, but t-acetylenic alcohols give 90% *cis* products. Dimethylhexynediol gives all *cis*-olefinic product. Polar solvents sometimes increase the amount of *cis* addition to form *trans* product (Table 2-13).

TABLE 2-13
Effect of Solvent on Addition of Bromine to Acetylenes[7]

Acetylene	Solvent	% *cis* Product	% *trans* Product
Acetylene	Hexane	55	45
	Acetic acid	0	100
Acetylenedicarboxylic acid	Ether	100	0
	Methanol	0	100

2.2. Halogen Acids

HBr adds to terminal acetylenes to give 1-bromoolefins. HBr adds the other way to 1-bromoacetylenes to give 1,2-dibromoolefins. Yields are usually high. Acetylene gives 85% yield of ethylene bromide or 80% yield of vinyl bromide, depending on conditions.[2] Skell[8] added HBr to liquid propyne. After 10 minutes at $-70°$, the product was 70% *cis*-1-bromopropene, formed by stereospecific *trans* addition. Ionic addition gives 2-bromopropene. Radical hydrobromination is a typical chain reaction.

2.3. Carbon Polyhalides

Carbon tetrachloride adds to 1-heptyne in the presence of benzoyl peroxide at 77° to form 40% yield of normal addition product, 6% yield of chlorination product and 20% yield of a cyclopentane. The cyclopentane forms by an intermolecular 1,5-hydrogen migration—the first example of this often postulated reaction.[9]

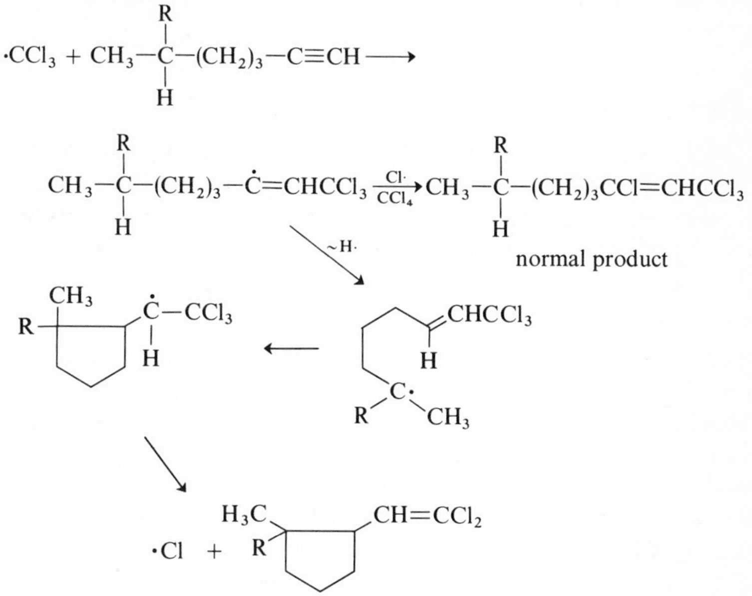

$$\cdot CCl_3 + CH_3 - \overset{\overset{\displaystyle R}{|}}{\underset{\underset{\displaystyle H}{|}}{C}} - (CH_2)_3 - C \equiv CH \longrightarrow$$

$$CH_3 - \overset{\overset{\displaystyle R}{|}}{\underset{\underset{\displaystyle H}{|}}{C}} - (CH_2)_3 - \overset{\displaystyle \cdot}{C} = CHCCl_3 \xrightarrow[CCl_4]{Cl\cdot} CH_3 - \overset{\overset{\displaystyle R}{|}}{\underset{\underset{\displaystyle H}{|}}{C}} - (CH_2)_3 CCl = CHCCl_3$$

normal product

$\sim H\cdot$

CH=CCl₂

$\cdot Cl +$

(20% yield, R = H)
cis and *trans*

Peroxide-initiated addition of CCl_4 to propargyl esters gives the normal 1:1 addition products and also gives 20–25% yields of substituted γ-butyrolactones.[11] An intramolecular 1,5-hydrogen abstraction is the key step:

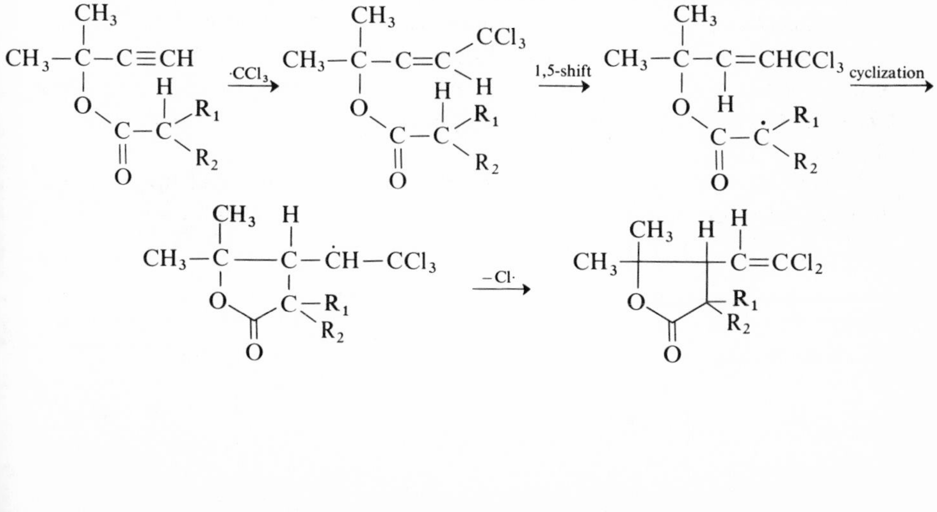

The relative yield of lactone is highest if the hydrogen atom abstracted from the ester is tertiary or benzylic. Optically active lactones form if the asymmetric center of the ester involves the carbon to which abstractable hydrogen is attached. Thus, free radical reaction at an asymmetric center is possible without complete racemization.

BrCCl$_3$ adds to terminal acetylenes to give 1,1,1-trichloro-3-bromo-2-olefins.[2] Octyne-1 gives 80% yield; phenylacetylene gives 9% yield. CF$_3$I is similar. Propyne and CF$_3$I give 91% yield of CH$_3$CI=CHCF$_3$.[10]

2.4. Thiols and Disulfides

Thiol addition is the most studied radical addition to acetylenes. One or two radicals add, depending on the proportions of reactants. The first thiol adds faster than the second. The first molecule of thiol adds anti-Markownikoff, and the second usually adds so that the thiol groups are on adjacent carbons. Acetylene gives RSCH$_2$CH$_2$SR, and ethynyl ethers give ROCH=CHSR'

$$\underset{\underset{|}{SR'}}{}$$

(nucleophilic addition[12] gives H$_2$C=COEt). Phenylacetylene is abnormal; the product is PhCH$_2$CH(SR)$_2$. Some acetylenes give only monoaddition: dimethyl acetylenedicarboxylate with ethanethiol, phenylacetylene with thiolacetic acid, 3-hydroxy-1-butynyl ethyl ether with ethanethiol, and 4-hydroxy-1-pentyne with thiolacetic acid.[13-15]

Blomquist[16] mixed 2 moles of ethanethiol with 1 mole of acetylenic compound, added benzoyl peroxide, and irradiated the solutions for 2–4 weeks. Propargyl alcohol gave 95% yield of 2,3-bis(ethylmercapto)-1-propanol. Propargyl acetate, butynediol diacetate and 1-hexyne gave quantitative yield of the symmetrical bis(ethylmercapto) adducts. Butynediol diacetate and ethanedithiol gave 43% yield of 1,2-bis(acetoxymethyl)-1,4-dithiane. Methylbutynol (a) and phenylacetylene (b) gave some "abnormal" products*:

(a) Me$_2$C(OH)CH=CHSEt, Me$_2$C=CHCH(SEt)$_2^*$, +symmetrical diadduct
(b) PhCH=CHSEt, PhCH$_2$CH(SEt)$_2^*$

Vinylacetylene and excess ethanethiol react at 100° to form four products in the presence of benzoyl peroxide. EtSH adds (1) to the triple bond, (2) to the double bond, and (3) 1,4, followed by rearrangement to allenic product CH$_2$=C=CHCH$_2$SEt.[17] p-Methylthiophenol adds only to the olefinic bonds of divinylacetylene.[2]

Sauer[18] injected acetylene into a bomb containing thiol and initiator, and then heated to decompose the initiator. He obtained 1,2-disulfides in 63–94% yield. Ethanethiol adds by a radical chain reaction:

$$EtSH \longrightarrow EtS\cdot \xrightarrow{+C_2H_2} EtSCH=\dot{C}H \xrightarrow{+EtSH} EtSCH=CH_2 + EtS\cdot$$

$$EtSCH=CH_2 + EtS\cdot \longrightarrow EtS\dot{C}HCH_2SEt \xrightarrow{+EtSH} EtSCH_2CH_2SEt + EtS\cdot$$

In the presence of CO (1000–3000 atmospheres), 3-alkylthiopropenal was one product:

$$BuS\cdot + C_2H_2 \longrightarrow BuSCH=\dot{C}H \xrightarrow{+CO} BuSCH=CH\dot{C}=O \xrightarrow{+BuSH}$$

$$BuS\cdot + BuSCH=CHCHO$$
$$(17\%)$$

This is similar to the formation of 3-alkylthiopropanols from thiol, ethylene and CO.[19]

Ethanethiol adds to ethoxyethyne to give EtSCH=CHOET and

$$\overset{\displaystyle SEt}{\underset{\displaystyle |}{EtSCH_2CHOEt.}}$$

Symmetrical product forms from EtSH and thioethoxyethyne.[20] EtSeH adds the same way. Acid hydrolysis of EtSeCH=CHOEt or EtSeCH=CHSEt gives EtSeCH₂CHO. Thiolacetic acid adds to ethynyl

$$\overset{\displaystyle O}{\overset{\displaystyle ||}{}}$$

thioethyl ether to give $CH_3CSCH=CHSEt$. Ethanethiol adds to similar acetylenes to form symmetrical products:

$$RO_2CC\equiv CSEt \longrightarrow \overset{\displaystyle SEt}{\underset{\displaystyle |}{RO_2CC}}=CHSEt^{21}$$

$$Me_2C(OH)C\equiv COEt \longrightarrow \overset{\displaystyle SEt}{\underset{\displaystyle |}{Me_2C(OH)C}}=CHOEt^{22}$$

$$Bu_2PC\equiv CH \longrightarrow Bu_2PCH=CHSEt^{23}$$

Organic disulfides add to acetylene bonds in hydrocarbons to form 1:1 products in which the *trans/cis* ratio is constant at various conversions:[23a]

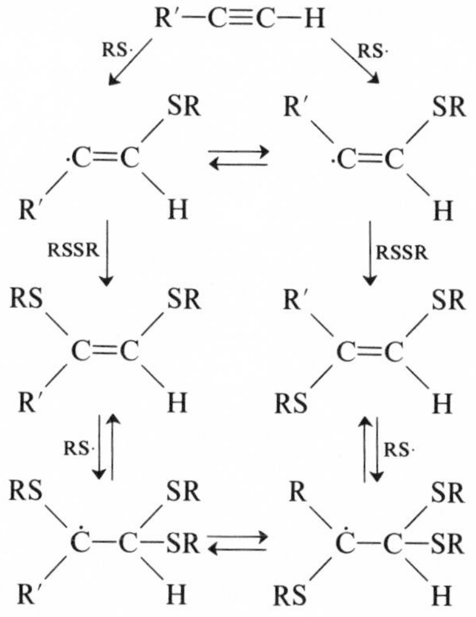

Propyne and 1-heptyne give 80% yields of the olefinic products. No saturated products are found, presumably because the addition of RS· to the olefinic product is an equilibrium which contains mostly the vinyl product. This is a radical chain addition, initiated by ultraviolet light at 20° or by decomposing *t*-butyl peroxide at 120°. The corresponding reaction with olefins is slower and gives poor yields.

2.5. Sulfur

Krespan[24] made 3,4-bis(trifluoromethyl)-1,2-dithietene (1) from hexafluoro-2-butyne and sulfur. Iodine was a good catalyst. In batch reactions, he obtained some other products. In a flow system at 445°, the dithietene was the only product. The dithietene is the first member of a new class of compounds containing the unusual dithietene ring. Other fluorinated acetylenes gave the same ring system. Sulfur adds as the diradical $\cdot SS(S)_x S \cdot$:

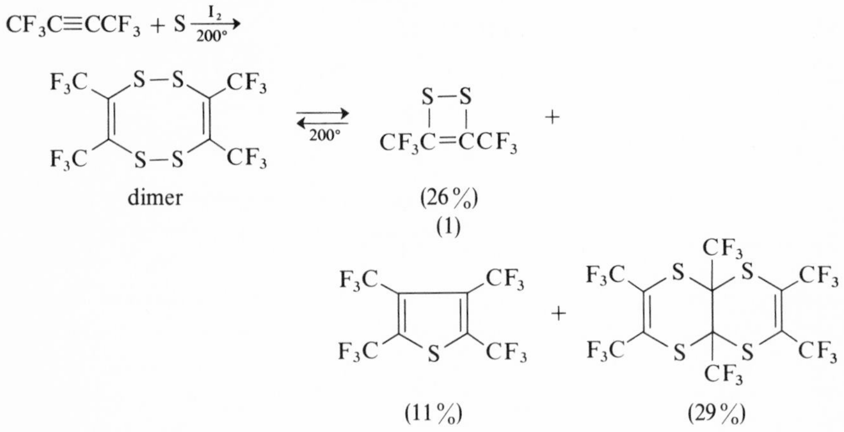

$$CF_3C{\equiv}CCF_3 + S \xrightarrow[200°]{I_2}$$

dimer

$$\underset{200°}{\rightleftharpoons} \quad \begin{array}{c} S{-}S \\ | \quad | \\ CF_3C{=}CCF_3 \end{array} \quad +$$

(26%)
(1)

(11%)   (29%)

2.6. Mono- and Disubstituted Phosphines

Radicals generated from mono- and di-β-cyanoethylphosphine by decomposing azobisisobutyronitrile add to 1-alkynes at the terminal carbon:[25]

$$RC{\equiv}CH + R'_2PH \longrightarrow R'_2PCH{=}CHR$$
(30%)

$$RC{\equiv}CH + R'PH_2 \longrightarrow R'P(CH{=}CHR)_2$$
(26%)

Acetylene gives no definite product.

(Triphenylstannyl)diphenylphosphine adds to acetylenes.[26] The bulky triphenylstannyl group probably adds to the least hindered position of the acetylenic compound. Phenylacetylene gives an olefinic product assumed to be

$$HC\!\!\underset{\underset{SnPh_3}{|}}{=}\!\!\underset{\underset{PPh_2}{|}}{C}Ph.$$ The yield without catalyst is 21%, and 78% with azobisiso-butyronitrile catalyst.

2.7. Nitrogen Oxides

N_2O_4 adds to 3-hexyne in a radical mechanism:[27]

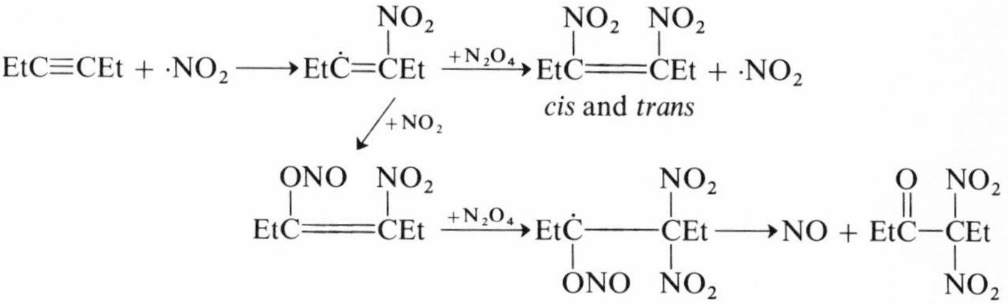

$$EtC\equiv CEt + \cdot NO_2 \longrightarrow Et\overset{\overset{NO_2}{|}}{C}=\overset{}{C}Et \xrightarrow{+N_2O_4} Et\overset{\overset{NO_2}{|}}{C}=\!\!=\overset{\overset{NO_2}{|}}{C}Et + \cdot NO_2$$

cis and trans

$$\Big\downarrow{+NO_2}$$

$$Et\overset{\overset{ONO}{|}}{C}=\!\!=\overset{\overset{NO_2}{|}}{C}Et \xrightarrow{+N_2O_4} Et\overset{\overset{NO_2}{|}}{\underset{\underset{ONO}{|}}{C}}\!\!-\!\!-\overset{\overset{}{}}{\underset{\underset{NO_2}{|}}{C}}Et \longrightarrow NO + Et\overset{\overset{O}{||}}{C}-\overset{\overset{NO_2}{|}}{\underset{\underset{NO_2}{|}}{C}}Et$$

N_2O_4 and iodine add to tolan to give 15% yield of 1-nitro-2-iodostilbene. Phenylacetylene gives α-iodo-β-nitrostyrene, 86% yield.

Diphenyldiacetylene adds N_2O_4 at the 1,4-positions to form the dinitro-cumulene[1]

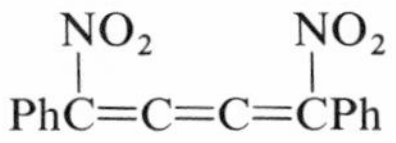

$$Ph\overset{\overset{NO_2}{|}}{C}=C=C=\overset{\overset{NO_2}{|}}{C}Ph$$

Some unusual products are made by adding NO plus NO_2 to acetylene at 60° and 28 atmospheres.[28] 3,3'-Biisoxazole forms in 65% yield, along with minor amounts of two other isoxazoles. The key intermediate may be ONC—CNO, and the intermediate to the aldehyde may be oxycyanoform-aldehyde:

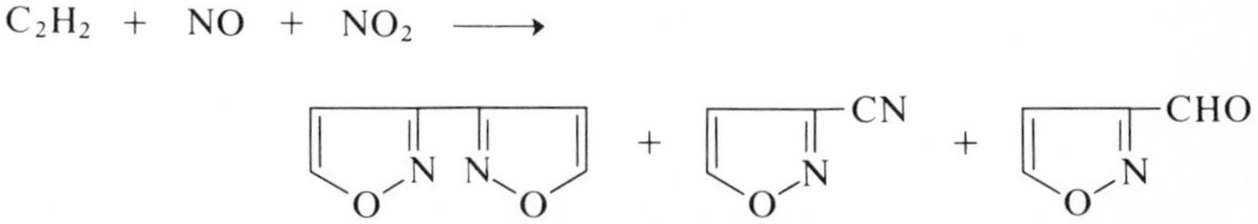

$$C_2H_2 + NO + NO_2 \longrightarrow$$

2.8. Other Hetero Atom Additions

Stacy[1] listed a few examples of some other hetero atom additions. 1-Acetylenes add alkali metal bisulfites to form olefin 1-sulfonates; diaddition gives sym-metrical products. Propyne adds SF_5Cl to form $SF_5CH\!\!=\!\!\overset{\overset{Cl}{|}}{C}CH_3$. H_2S_2 reacts with $PhC{\equiv}C\overset{Cl}{\underset{|}{C}}{=}NPh$ to form

$$\underset{\underset{S-S}{\displaystyle|}}{PhC{=}CH}\diagdown\atop C{=}NPh$$

The platinum-catalyzed addition of silicochloroform has been studied more than the radical addition. $\cdot SiCl_3$ adds to the terminal carbon of 1-acetylenes. $HGeCl_3$ gives the symmetrical product $Cl_3GeCH_2CH_2GeCl_3$ from acetylene. Alkynes react with PCl_3 and oxygen to form $RCCl{=}CHPOCl_2$.

N_2F_4 adds across the acetylene bond:[28a]

$$N_2F_4 + RC{\equiv}CR' \xrightarrow{170°} RC\underset{\underset{NF_2}{\displaystyle|}}{\overset{\overset{NF_2}{\displaystyle|}}{=\!\!=\!\!=}}CR'$$

In some cases, the products rearrange on further heating to give

$$RC\underset{\underset{NF}{\displaystyle\|}}{\text{---}}\overset{}{C}FR' \quad \text{and} \quad RC\underset{\underset{NF_2}{\displaystyle|}}{\text{---}}\overset{\overset{F}{\displaystyle|}}{C}R' \atop {}\underset{NF}{\displaystyle\|}$$

Diynes add N_2F_4 across only one triple bond to give products such as

$$MeC{\equiv}CC\underset{\underset{NF_2}{\displaystyle|}}{=\!\!=\!\!=}\overset{}{C}\underset{\underset{NF_2}{\displaystyle|}}{}{-}Me$$

Allenes give mostly the fluoroimino rearranged products.

Dialkyl-N-chloramines add to acetylenes. This free radical addition involves a chain reaction in which $R_2\overset{\cdot}{N}H^+$ adds to the acetylenic bond.[28b] Products are β-chloroenamines which hydrolyze to α-chloroaldehydes during work-up:

$$Et_2NCl + EtC{\equiv}CEt \xrightarrow[\text{(2) water}]{\text{(1) acid}} Et\overset{\overset{O}{\displaystyle\|}}{C}\underset{\underset{Cl}{\displaystyle|}}{C}HEt + EtC{\equiv}C\underset{\underset{Cl}{\displaystyle|}}{C}HMe$$
$$\phantom{Et_2NCl + EtC{\equiv}CEt \xrightarrow}(59\%) \qquad\quad (4\%)$$

3. ADDITION OF CARBON RADICALS

In these additions, yields are often high, but conversions are usually uneconomically low.

3.1. Acyl Radicals

Schlubach[29] added acyl radicals, generated from acetaldehyde and benzoyl peroxide, to acetylene at 65°. He obtained a little crotonaldehyde and 2,5-

hexanedione. Later, Wiley[30] compared γ-induced addition of aldehydes to maleate, fumarate and acetylenedicarboxylate esters. Acyl radical added to dimethyl acetylenedicarboxylate by a radical chain process to form bis(acyl)-succinates. Yields and G values are:

Aldehyde	Diethyl Maleate % Yield	Diethyl Maleate G Value	Dimethyl Fumarate % Yield	Dimethyl Fumarate G Value	DMAD % Yield	DMAD G Value
Isobutyraldehyde	27	35			9	11
n-Butyraldehyde	84	70	36	23	23	54
Acetaldehyde					31	54

Elad[31] irradiated formamide and dimethyl acetylenedicarboxylate. Two moles of formamide added to DMAD. The product was

$$\underset{\text{MeO}_2\text{CH}}{\overset{\text{CONH}_2}{|}}\text{———}\underset{\text{CHCO}_2\text{Me}}{\overset{\text{CONH}_2}{|}}$$

which hydrolyzed to 1,1,2,2-ethanetetracarboxylic acid.

3.2. Carboxylic Acids

Very little is known about this potentially valuable method of synthesis of unsaturated acids and saturated diacids.

The carboxymethyl radical adds to acetylene. Acetylene and acetic acid at 110° in the presence of benzoyl peroxide gave 50% conversion of acetylene and 26% yield of adipic acid. One intermediate was vinylacetic acid. 1-Hexyne gave 24% yield of 3-octenoic acid.[32] Ogibin and Nikishin[33] used di-t-butyl peroxide to initiate addition of radicals from acids to 1-hexyne and other acetylenes:

$$\text{RC}\equiv\text{CH} + \text{R}'\text{CH}_2\text{CO}_2\text{R}'' \longrightarrow \text{RCH}=\text{CH}-\overset{\text{R}'}{\underset{|}{\text{C}}}\text{HCO}_2\text{R}'' + \text{RCH}_2\text{CH}=\overset{\text{R}'}{\underset{|}{\text{C}}}\text{CO}_2\text{R}''$$

The charge was usually 2 moles of acid per mole of acetylenic compound, with 15 mole % peroxide catalyst. Reactions were carried out at 135–150°. The yields were usually low, and significant amounts of high-boiling residues formed. The residues were not identified, but they may contain products of addition of 2 moles of acid to the triple bond. The results are summarized in Table 2-14.

3.3. Hydrocarbon Radicals

"Polymerization" of ethylene, propylene, acetylene and butadiene can be initiated in the gas phase by decomposing di-t-butyl peroxide.[35] The "polymers"

TABLE 2-14
Addition of Radicals from Carboxylic Acids and Esters to Acetylenes[33]

Acid or Ester (1)	Acetylene (2)	Moles Charged (1)	Moles Charged (2)	Peroxide	Temperature (°C)	% Unreacted (2)	% Yield (1:1 Adduct)
$CH_2(COOCH_3)_2$	$C_4H_9C{\equiv}CH$	4.0	0.20	0.03	148–153	—	30.5
$CH_2(COOCH_3)_2$	$C_4H_9C{\equiv}CH$	2.9	0.14	0.02	145–148	—	30.0
CH_3CH_2COOH	$CH_3COOCH_2C{\equiv}CH$	6.0	0.30	0.03	135–138	—	25.0
CH_3CH_2COOH	$CH_3COOCH_2C{\equiv}CH$	6.0	0.30	0.03	137–138	—	29.0
$CH_2(COOCH_3)_2$	$CH_3COOCH_2C{\equiv}CH$	3.0	0.15	0.02	145–160	32	9.0
$CH_2(COOCH_3)_2$	$CH_3COOCH_2C{\equiv}CH$	4.2	0.21	0.03	143–150	33	11.0
$CH_2(COOCH_3)_2$	$CH_3COOC(CH_3)_2C{\equiv}CH$	2.2	0.11	0.02	145–150	44	6.0
CH_3CH_2COOH	$CH_3COOC(CH_3)_2C{\equiv}CH$	4.0	0.20	0.03	136–137	—	—
$C_5H_{11}COOCH_3$	$CH_3COOC(CH_3)_2C{\equiv}CH$	4.0	0.20	0.03	135–140	45	2.0
$C_5H_{11}COOCH_3$	$CH_3COOCH(CH_3)C{\equiv}CH$	3.0	0.15	0.02	135–140	70	3.5
$C_9H_{19}COOCH_3$	$CH_3COOCH_2C{\equiv}CH$	2.2	0.22	0.02	145–150	60	9.0

were not identified, but the polymerization rate constants and activation energies were determined. "Polymerizations" of acetylene, butadiene and propylene have nearly the same total activation energy, 24.1–24.3 kcal/mole. Acetylene polymerizes fastest. With the proper reactant ratios and conditions, acetylene might react with olefins to form oligomers with double or triple bonds, interesting as chemical intermediates and monomers.

The rates of addition of methyl radicals to acetylenes and olefins have been compared.[36] Methyl radical adds much faster than it abstracts a hydrogen atom. The radical adds and forms a new C—C σ bond, using one of the π electrons in the multiple bond. Where the interaction of the π electrons is stronger, the activation energy of addition should be larger. Triple bonds are shorter than double bonds, and interaction between their π electrons is stronger. The activation energy of addition to triple bonds is thus larger than that to double bonds. This is a measure of the ease of propagation of radical addition and does not necessarily reflect the total activation energy relationships.

Acetylene adds methyl radicals faster than ethylene does, but the other acetylenic compounds add radicals more slowly than the corresponding ethylenic compounds. For example, the rates of addition of radicals to styrene and phenylacetylene are compared:

$\dfrac{k \text{ (addition to styrene)}}{k \text{ (addition to PhC}\equiv\text{CH)}}$	Methyl Acrylate	Acrylonitrile	Methyl
	6	3.5	4.5

Methyl radicals add to 1-alkynes and also abstract α-hydrogen. The rate of addition is independent of the length of R in RC≡CH. The rate of abstraction of α-hydrogen is slightly higher in acetylenes than in olefins.

Photolysis of ketene gives methylene radicals, which add to acetylene to form propyne and allene. The propyne:allene ratio is 1.5 and is independent of pressure and time.[37]

Thermal and acid-catalyzed alkylations of olefins with paraffins are well-understood processes in the petroleum industry, but little is known about the acetylene-paraffin alkylation. One of the reasons is that acetylene polymerizes and decomposes under thermal alkylation conditions. Radiation is an ideal way to initiate temperature-sensitive reactions, because radical processes can be started at relatively low temperature. Although some ions are generated at the same time, most radiation-induced paraffin-olefin alkylations are best explained by assuming that the radical process predominates. Bartok[38] irradiated acetylene-propane mixtures at 10–15 atmospheres and 250–400°, using a large excess of propane to minimize explosive decomposition of acetylene. Neutron flux in the reactor was around 10^6 rad/hr of thermal and fast neutrons.

The thermal alkylation is negligible between 200 and 400°. At 320° with irradiation, 20–30% of the acetylene gave addition product, identified as 3-methyl-1-butene, the only monoadduct. Selectivity was around 50%. The only other products were typical radiolysis products of propane. Although this is a new reaction which has no thermal counterpart, it is probably a radical chain reaction in which isopropyl radical adds to acetylene.[38]

3.4. Carbon Radicals from Hydrocarbons, Ethers, Ketones, Amines and Alcohols

Cywinski[39] generated radicals from saturates by decomposing di-t-butyl peroxide in the presence of acetylene. The products indicated very little attack by primary radicals on acetylene, some attack by secondary, and a high selectivity for addition of tertiary radicals. Under his conditions, secondary radicals seemed borderline, with good yields from cycloparaffins and some isoparaffins.

The reaction is very general, and allows formation of a wide variety of 1-olefins or α-substituted allylic compounds with the original functional group. Yields depend on the generation of a free radical in the right stability range under the reaction conditions. The reactions in Table 2-15 constituted a survey, and no attempt was made to optimize yields.

Acetylene reacts with isopropyl alcohol at 150° in the presence of di-t-butyl peroxide to give 2,5-dimethylhexane-2,5-diol:[34]

$$C_2H_2 + H\overset{\displaystyle Me}{\underset{\displaystyle Me}{\overset{|}{\underset{|}{C}}}}OH \longrightarrow Me\overset{\displaystyle Me}{\underset{\displaystyle OH}{\overset{|}{\underset{|}{C}}}}-CH_2CH_2-\overset{\displaystyle Me}{\underset{\displaystyle OH}{\overset{|}{\underset{|}{C}}}}Me$$

(76% yield on C_2H_2, 90% on isopropyl alcohol)

3-Methyl-1-butyn-3-ol adds isopropyl alcohol to form 2,5-dimethyl-3-hexene-2,5-diol in 90% yield. Either peroxide or ultraviolet irradiation can be used to initiate this reaction.

Ethanol, cyclohexane, 2-methyltetrahydrofuran and diethyl ether add to acetylene. Ultraviolet light of wave length 2537 Å is effective at 25° when acetone is used as the absorber.[34a] The product from ethanol is $CH_3\overset{\displaystyle OH}{\overset{|}{C}}H-CH=CH_2$. Cyclohexane gives vinylcyclohexane. Diethyl ether forms

$$CH_3CH_2O\overset{\displaystyle }{\underset{\displaystyle CH=CH_2}{\overset{}{\underset{|}{C}}}}HCH_3$$

and 2-methyltetrahydrofuran gives 2-methyl-2-vinyltetrahydrofuran.
Rates of reaction are:

Addend	Hours	Formation of RCH=CH$_2$ (moles/ml min)	Formation of RCH=CH$_2$ per "active" H Atom (moles/ml min)
Ethanol	1.5	0.86	0.43
Cyclohexane	1.5	0.19	0.02
2-MethylTHF	2	0.10	0.10
Diethyl ether	2	0.03	0.01

TABLE 2-15
Addition of Carbon Radicals to Acetylene[39]

Addend	Addend (g)	DTBP (g)	N_2 (psi)	Total pressure (psig)	Heating schedule (hr)	Heating schedule (°C)	Monoadduct	Weight (g)	Moles per Mole of DTBP
2,3-Dimethylbutane	425	58.8	75	300	a		3,3,4-Trimethyl-1-pentene	35.7	0.80
	439	39.2	75	300	a		3,3,4-Trimethyl-1-pentene	26.7	0.89
Cyclopentane	480	59.0	0	260	22	140	Vinylcyclopentane	35.3	0.93
	514	19.8	0	280	21	140	Vinylcyclopentane	30.4	2.50
Cyclohexane	461	48.5[b]	100	300	4	130	Vinylcyclohexane	35.8	1.20
					4	140			
Ethyl ether	454	58.8	0	300	9[c]	140	3-Ethoxy-1-butene	8.0	0.20
Methyl isopropyl ketone	521	38.9	100	250	10[c]	148	3,3-Dimethyl-1-penten-4-one	15.5	0.52
Methyl isobutyrate	499	19.6	100	240	8[c]	145	Methyl dimethylvinylacetate	22.0	1.30
n-Butylamine	477	39.5	0	230	5.5	140	3-Amino-1-hexene	20.2	0.74
Methyl alcohol	513	60.5	0	300	5[c]	140	Allyl alcohol	6.4	0.27
Ethyl alcohol	509	58.7	0	300	5[c]	140	3-Buten-2-ol	16.6	0.57
	546	19.6	0	300	5[c]	140	3-Buten-2-ol	20.4	2.10
Isopropyl alcohol	507	58.6	0	300	5[c]	140	2-Methyl-3-buten-2-ol	52.0	1.50
	545	18.7	0	300	5[c]	140	2-Methyl-3-buten-2-ol	37.0	3.70
	561	7.8	0	300	2.3[c]	140	2-Methyl-3-buten-2-ol	35.0	7.70
	562	2.2	0	290	5[c]	140	2-Methyl-3-buten-2-ol	15.9	12.00

a 3 hr at 125°, 4 hr at 135°, and 6 hr at 145°.
b 7.1 g of di-t-butyl peroxide was recovered from the products.
c The acetylene valve was closed and heating was continued for about 17 hr to decompose residual peroxide.

The products form by a radical chain addition sequence:

$$R\cdot\ +\ HC\equiv CH\longrightarrow RCH=CH\cdot\xrightarrow{\ +RH\ }R\cdot\ +\ RCH=CH_2$$

When diphenylacetylene in excess 2,3-dihydropyran is irradiated with 2537 Å ultraviolet light, a different kind of addition occurs.[34b] 7,8-Diphenyl-2-oxabicyclo[4.2.0]oct-7-ene is formed in 81% yield:

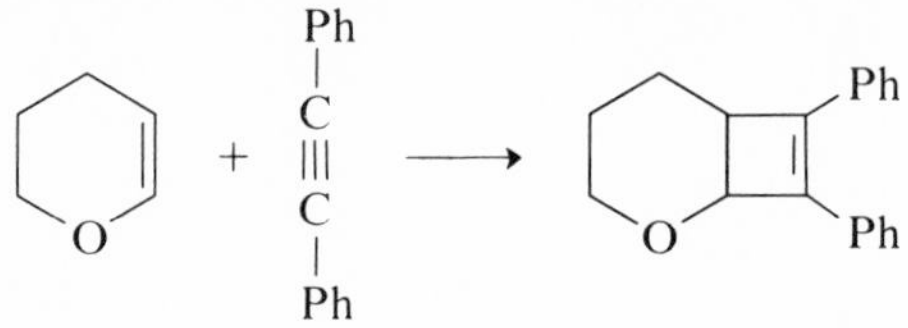

This is probably not a free radical addition, but rather is the addition of the dihydropyran to the first excited triplet state of diphenylacetylene. Diphenylacetylene also adds 2-methoxy-1,4-naphthoquinone when irradiated with ultraviolet light in acetonitrile. The product is

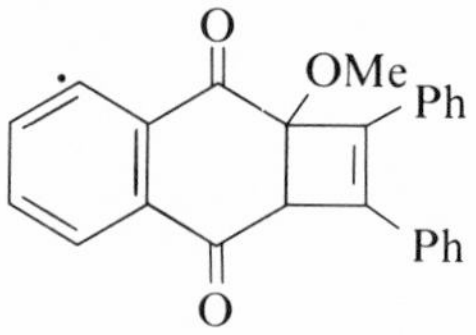

Other naphthoquinones form similar products. This is a general synthesis for tricyclo[6.4.0.0³,⁶]dodeca-4,8,10,12-tetraene-2,7-diones.[34c] Anthraquinone adds to diphenylacetylene under photolysis conditions to give yet another kind of product:[34d]

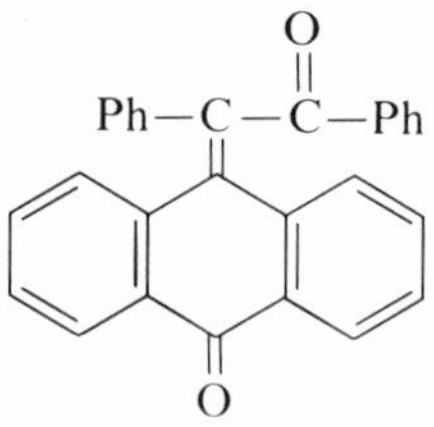

Irradiation of isopropyl alcohol solutions of dimethyl acetylenedicarboxylate with ultraviolet light (benzophenone activator) gives lactones. Lactones form by addition of secondary carbon radical, followed by lactonization of carboxyl and hydroxyl (from isopropanol added):[40]

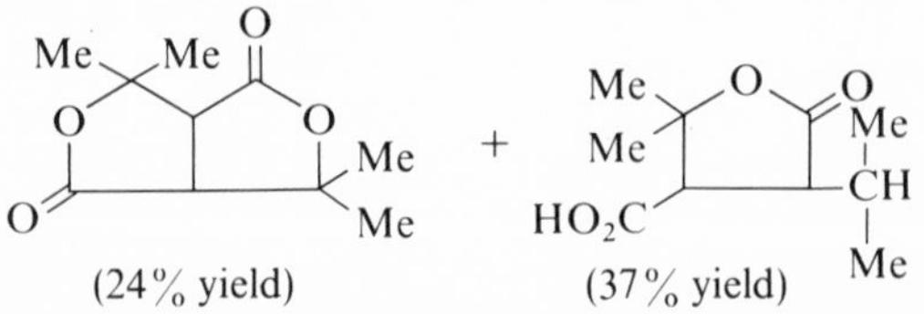

Tertiary carbon radical from 1,4-diaminocyclohexane adds to acetylene at 150° (di-*t*-butyl peroxide) to give low yields of three products.[41] Two of the products are identified as:

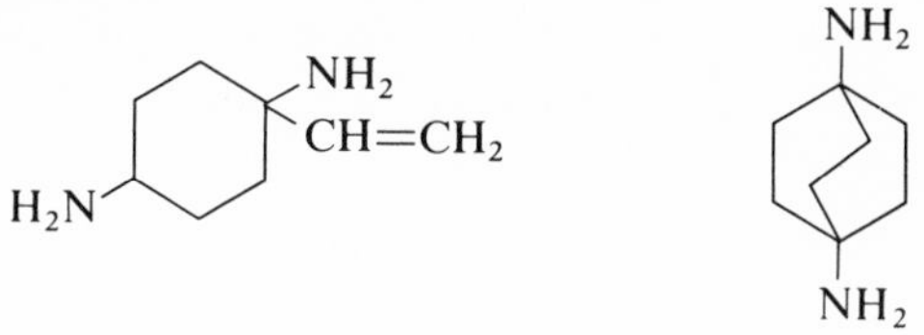

Isopropylamine gives 6% yield of *t*-pentenylamine, $CH_2{=}CHCNH_2$,

and a little diamine from 1,2-diaddition. The yield of *t*-pentenylamine is increased to 60% if a glass lined vessel and specially pretreated isopropylamine are used.[41]

Olefins and benzonitrile irradiated by ultraviolet light give cyclobutane derivatives

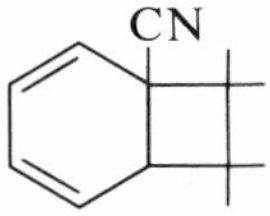

but acetylenes ($RC{\equiv}CR$) give cyclooctatetraenes by an addition rearrangement:[42]

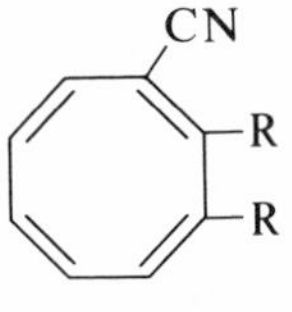

Methyl 2-butynoate and benzene give a similar reaction:[43]

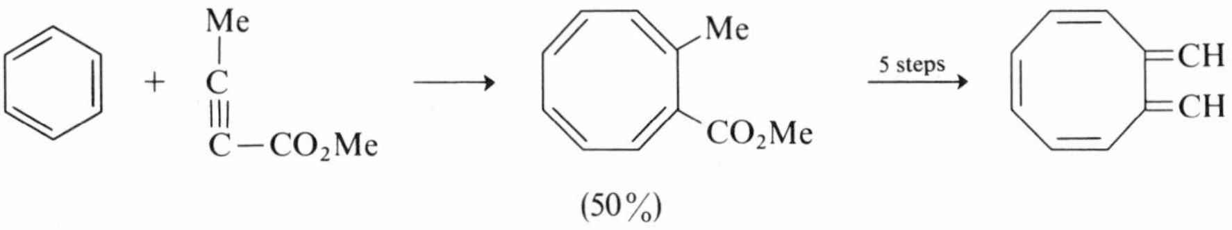

(50%)

Methyl propiolate and dimethyl acetylenedicarboxylate also react with benzene to give cyclooctatetraene derivatives under similar conditions, but phenylacetylene gives 1-phenylazulene and 1-phenylnaphthalene in addition to phenylcyclooctatetraene.[44] Maleic anhydride and acetylene irradiated in acetone solution form cyclobutene and cyclopropane derivatives:[45]

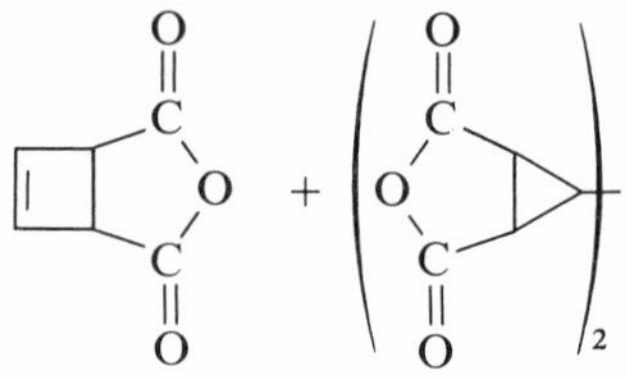

(10–20%)　　　　　　(90–80%)

4. THERMAL ALKYLATION

A recent patent[46] claims that acetylene reacts with isobutylene at 300–450° and 4 atmospheres to form 2-methyl-1,4-pentadiene in 6% yield. This is an example of thermal addition of an allylic radical across the triple bond:

$$H_2C{=}\overset{\displaystyle Me}{\underset{|}{C}}{-}CH_2\cdot + C_2H_2 \longrightarrow H_2C{=}\overset{\displaystyle Me}{\underset{|}{C}}{-}CH_2CH{=}CH\cdot$$

5. REACTIONS OF THE ETHYNYL RADICAL

Tarr[47] in 1965 reviewed the reactions of ethynyl radicals and described the results of his own extensive studies. He generated ethynyl radicals by photolysis of either chloroacetylene or bromoacetylene. Bromoacetylene is more convenient because it is higher boiling and less flammable in air. Nitric oxide is a good polymerization inhibitor. A mixture of nitric oxide and bromoacetylene was stable for at least three months in a dark storage vessel. Irradiation of bromoacetylene-NO mixtures gave CO and BrCN as the major products, and HCN, acetylene and diacetylene as minor products. Tarr did a series of experiments with constant pressure of bromoacetylene and NO, and variable pressures of added alkanes. The relative rate constants for hydrogen abstraction by ethynyl radicals can be determined by the rate of formation of diacetylene. Alkyl radicals can be scavenged by NO, and vinyl radicals undergo π-bond fission, but ethynyl radical cannot be scavenged by NO. Thus, the ethynyl radical is not a simple one-form radical. It is probably stabilized by delocalization of the free electron by the triple bond:

$$HC{\equiv}C\cdot \rightleftarrows H\dot{C}{=}C:$$

in which the left-hand form predominates.

The ethynyl radical shows less discrimination between secondary and tertiary C—H bonds than the methyl radical does. The relative rates for

abstracting primary, secondary and tertiary H by ethynyl radicals are about
$1:2:8$, and by methyl radicals $1:7:50$.

In similar studies with ethynyl radical and olefins, substantial quantities of
1-alkynes formed only if NO was absent. Some enynes also formed, probably
by addition of the ethynyl radical to the olefinic bond, followed by an elimination:[48]

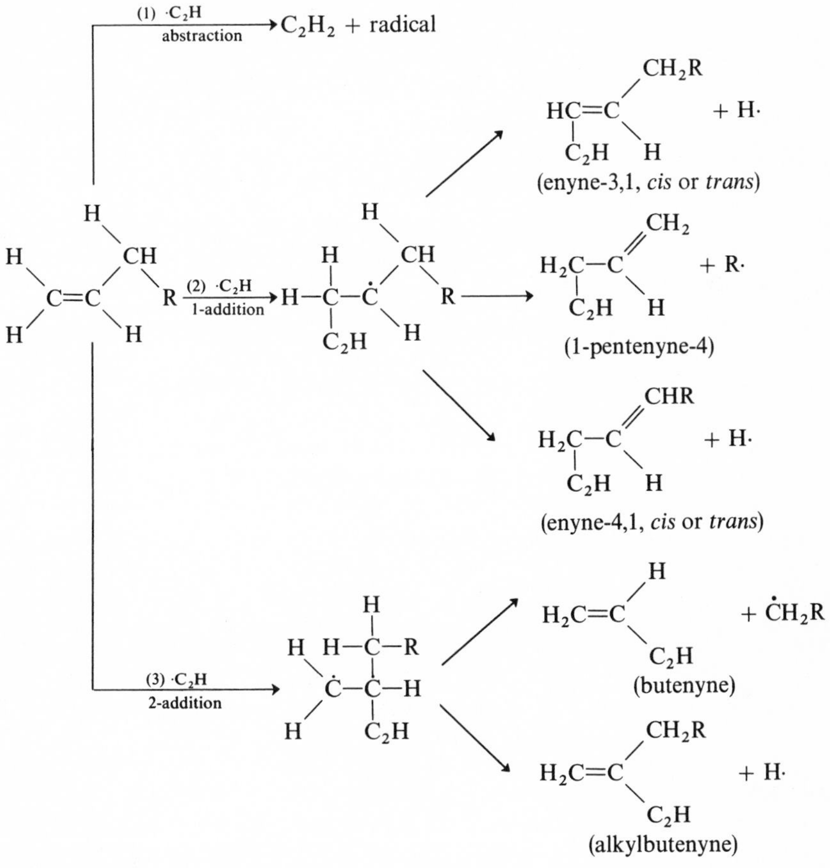

References

1. Stacey, F. W., and Harris, J. F., Jr., *Org. Reactions*, **13**, 150 (1963).
2. Sosnovsky, G., "Free Radical Reactions in Preparative Organic Chemistry," New
   York, The Macmillan Co., 1964.
3. Wijnen, M. H. J., *J. Chem. Phys.*, **36**, 1672 (1962).
4. Smirnov-Zamkov, I. V., and Piskovitina, G. A., *Ukr. Khim. Zh.*, **23**, 208 (1957);
   Chem. Abstr., **51**, 14557 (1957).

5. Bohlmann, F., *Angew. Chem.*, **69**, 82 (1957).
6. Caserio, M. C., and Pratt, R. E., *Tetrahedron Letters*, 91 (1967).
7. Nazarov, I. N., and Bergel'son, L. D., *Izv. Akad. Nauk SSSR Otd. Khim. Nauk*, 887, 896, 1066, 1073, 1235 (1960); *Chem. Abstr.*, **54**, 24332 (1960).
8. Skell, P. S., and Allen, R. G., *J. Am. Chem. Soc.*, **80**, 5997 (1958).
9. Heiba, E.-A. I., and Dessau, R. M., *J. Am. Chem. Soc.*, **88**, 1589 (1965); **89**, 3772 (1967).
10. Haszeldine, R. N., and Leedham, K., *J. Chem. Soc.*, 1261 (1954).
11. Heiba, E.-A. I., and Dessau, R. M., *J. Am. Chem. Soc.*, **89**, 2238 (1967).
12. Alkema, H. J., and Arens, J. F., *Rec. Trav. Chim.*, **79**, 1257 (1960).
13. Stacey, F. W., and Harris, J. F., Jr., *Org. Reactions*, **13**, 150 (1963).
14. Bader, H., *et al.*, *J. Chem. Soc.*, 619 (1949).
15. Behringer, H., *Ann. Chem.*, **564**, 219 (1949).
16. Blomquist, A. T., and Wolinsky, J., *J. Org. Chem.*, **23**, 551 (1958).
17. Sulimov, I. G., and Petrov, A. A., *Zh. Organ. Khim.*, **2**, 767 (1966); *Chem. Abstr.*, **65**, 12099 (1966).
18. Sauer, J. C., *J. Am. Chem. Soc.*, **79**, 5314 (1957).
19. Foster, R. E., *et al.*, *J. Am. Chem. Soc.*, **78**, 5606 (1956).
20. Arens, J. F., "Advances in Organic Chemistry: Methods and Results," Vol. 2, p. 117, New York, Interscience Publishers, 1960.
21. Bonnema, J., and Arens, J. F., *Rec. Trav. Chim.*, **79**, 1137 (1960).
22. Wieland, J. H. S., and Arens, J. F., *Rec. Trav. Chim.*, **75**, 1358 (1956).
23. Voskuil, W., and Arens, J. F., *Rec. Trav. Chim.*, **81**, 993 (1962).
23a. Heiba, E. I., and Dessau, R. M., *J. Org. Chem.*, **32**, 3837 (1967).
24. Krespan, C. G., *J. Am. Chem. Soc.*, **83**, 3434 (1961).
25. Rauchut, M. M., *et al.*, *J. Org. Chem.*, **26**, 5138 (1961).
26. Schumann, H., Jutzi, P., and Schmidt, M., *Angew. Chem.*, **77**, 912 (1965).
27. Freeman, J. P., and Emmons, W. D., *J. Am. Chem. Soc.*, **79**, 1712 (1957).
28. Cramer, R., and McClellan, W. R., *J. Org. Chem.*, **26**, 2976 (1961).
28a. Sausen, G. N., and Logothetis, A. L., *J. Org. Chem.*, **32**, 2261 (1967).
28b. Neale, R. S., *J. Org. Chem.*, **32**, 3263 (1967).
29. Schlubach, H. H., *et al.*, *Ann. Chem.*, **587**, 124 (1954).
30. Wiley, R. H., and Harrell, J. R., *J. Org. Chem.*, **25**, 903 (1960).
31. Elad, D., *Proc. Chem. Soc.*, 225 (1962).
32. Roberts, W. J., and DiPietro, J., *Angew. Chem. Intern. Ed.*, **5**, 415 (1966); *Can. J. Chem.*, **44**, 2241 (1966).
33. Ogibin, Yu. N., and Nikishin, G. I., *Zh. Organ. Khim.*, **2**, 1565 (1966).
34. Hoffmann, J. K. (to Cumberland Chem. Corp.), U.S. Patent 3,304,247 (Feb. 14, 1967); *Chem. Abstr.*, **66**, 75668 (1967).
34a. Srinivasan, R., and Carlough, K. H., *Can. J. Chem.*, **45**, 3209 (1967).
34b. Rosenberg, H. M., and Serve, P., *J. Org. Chem.*, **33**, 1653 (1968).
34c. Pappas, S. P., and Portnoy, N. A., *J. Org. Chem.*, **33**, 2200 (1968).
34d. Bryce-Smith, D., Gilbert, A., and Johnson, M. G., *Tetrahedron Letters*, 2863 (1968).
35. Landers, L. C., and Volman, D. H., *J. Am. Chem. Soc.*, **79**, 2966 (1957).
36. Gazith, M., and Szwarc, M., *J. Am. Chem. Soc.*, **79**, 3339 (1957).
37. Tereo, T., Sakai, N., and Shida, S., *J. Am. Chem. Soc.*, **85**, 3919 (1963).
38. Bartok, W., and Lucchesi, P. J., *J. Am. Chem. Soc.*, **81**, 5918 (1959).
39. Cywinski, N. F., and Hepp, H. J., *J. Org. Chem.*, **30**, 3814 (1965).

40. Schenck, G. D., and Steinmetz, R., *Naturwissenschaften*, **47**, 514 (1960); *Chem. Abstr.*, **55**, 11386 (1961).
41. David, I. A., (to E. I. du Pont de Nemours) U.S. Patent 3,151,159 (Sept. 29, 1964); *Chem. Abstr.*, **61**, 14554 (1964).
42. Atkinson, J. G., *et al.*, *J. Am. Chem. Soc.*, **85**, 2257 (1963).
43. Anet, F. A. L., and Gregorovich, B., *Tetrahedron Letters*, 5961 (1966).
44. Bryce-Smith, D., and Lodge, J. E., *J. Chem. Soc.*, 695 (1963).
45. Koltzenburg, G., Fuss, P. G., and Leitich, J., *Tetrahedron Letters*, 3409 (1966).
46. SNAM, S.P.A., Netherlands Patent Application 6,408,696 (Feb. 2, 1965); *Chem. Abstr.*, **62**, 16049 (1965).
47. Tarr, A. M., Strausz, O. P., and Gunning, G. E., *Trans. Faraday Soc.*, **61**, 1946 (1965).
48. *Ibid.*, **62**, 1221 (1966).

PART FOUR

Oxidation of Acetylenic Compounds

1. BACKGROUND

Oxidation can be a free radical or an electrophilic addition reaction. Most oxidizing agents cleave olefinic bonds much more easily than acetylenic bonds.

Alkaline $KMnO_4$ oxidizes diphenylacetylene to benzil. The α-diketones formed in acetylene oxidations usually cleave to 2 moles of acid. This is a commonly used method for establishing the position of a triple bond.[1] No one has reported the relative rates of oxidation of triple and double bonds by permanganate. Chromic acid-acetic acid oxidizes *o,o'*-dinitrotolan to the benzil, 60% yield.[2] Selenium dioxide at 280° oxidizes tolan to benzil, 35% yield. Stilbene gives only 17% yield.[3] Nitrous oxide at 300° and high pressure oxidizes 1-alkynes to ketenes. In ethanol, the ketene reacts to form the ethyl ester of the acid: Acetylene gives 41% yield of ethyl acetate; 1-heptyne gives 8% yield of ethyl heptanoate. N_2O oxidizes tolan in cyclohexane to diphenylketene dimer, 87% yield.[4] Ketenes are frequently postulated but seldom isolated intermediates in oxidations of acetylenes. One patent claims that silver nitrate-calcium zinc salt-silica gel is a catalyst for oxidizing acetylene to ketene.[5]

2. OXIDATION BY OZONE

Hurd[6] briefly reviewed the literature before 1936 on ozonolysis of acetylenes. Air containing less than 1% ozone will not cleave acetylenic bonds, but ozone in oxygen will. Hurd ozonized actylenic compounds in carbon tetrachloride, using an excess of 5–10% ozone in oxygen. The ozonides, of uncertain structure, cleaved during steam distillation to give the expected acids. Yields were only 40–60%. Acetylene under the same conditions did not cleave. The products were glyoxal and hydrogen peroxide, probably formed by hydrolysis of an ozonide such as HC=CH.

$$\begin{array}{ccc} HC & = & CH \\ | & & | \\ \end{array}$$
$$O-O-O$$

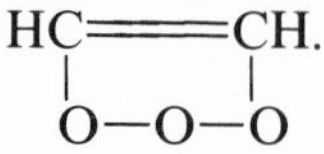

Clay[7] irradiated aqueous acetylene with γ rays from ^{60}Co. Without oxygen, the products were a solid yellow white polymer, formaldehyde, glycolaldehyde, and larger amounts of acetaldehyde and crotonaldehyde. With oxygen, the main products were glyoxal and hydrogen peroxide. With oxygen and ferrous sulfate, the products were glyoxal, hydrogen peroxide and glycolaldehyde. These products form directly by a radical chain mechanism.

The most significant ozonolysis work with acetylenes prior to 1962 was done by Criegee[8] in 1953. Ozonolysis in acetic acid gives an acetoxy hydroperoxide containing all the atoms of the original acetylene. Reduction gives the 1,2-dicarbonyl derivative, and decomposition gives anhydride or acid. Bailey[9] ozonolyzed unsymmetrical acetylenes to determine which half of the molecule produced by the Criegee zwitterion. For example, phenylalkyl acetylenes gave 90% yield of peroxide:

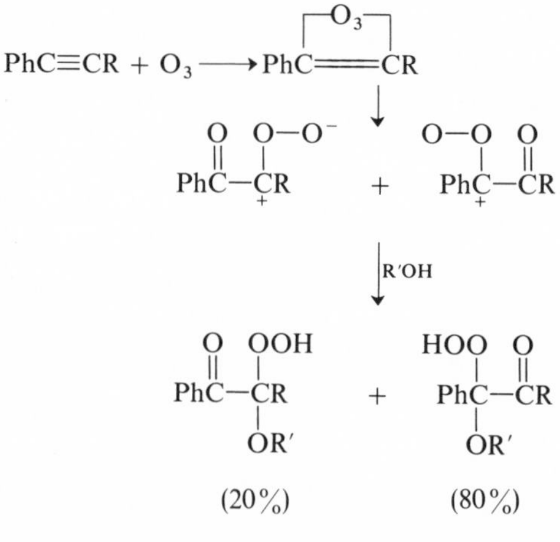

Thus, the zwitterion formed on the acetylenic carbon alpha to phenyl.

Ozonolysis of dimethyl acetylenedicarboxylate in formamide gives oxalic acid and 75% yield of hydroperoxymalonic acid, by an unknown mechanism.[10]

Ozone attacks olefinic bonds first in enynes. Bohlmann[11] ozonized poly-enynes in acetic acid at 0° and then cleaved the ozonides with dioxane-hydrogen peroxide to form acids and aldehydes. Only the olefinic bonds oxidized.

3. RADICAL-INDUCED CLEAVAGE

Sherwood found a new free radical reaction which cleaved acetylenic bonds.[12] He generated "inducer" radicals in an alkyne which contained 1 mole % NO. The "inducer" radical apparently adds to the alkyne to form a substituted vinyl radical which then adds NO. The cyclic intermediate decomposes to the products:

$$X-C{\equiv}C-Z + Y\cdot \longrightarrow \quad \begin{array}{c} X \\ \diagdown \\ \\ Y \diagup \end{array} C{=}\overset{\diagup Z}{\underset{\cdot}{C}}$$

$$\begin{array}{c} X \\ \diagdown \\ \\ Y \diagup \end{array} C{=}\overset{\diagup Z}{\underset{\cdot}{C}} + NO \rightleftharpoons \left[\begin{array}{c} X \\ \diagdown \\ \\ Y \diagup \end{array} C{=}C \begin{array}{c} \diagup Z \\ \\ \diagdown NO \end{array}\right] \longrightarrow$$

$$\left[\begin{array}{c} X \\ \diagdown \\ \\ Y \diagup \end{array} \underset{O-N}{\overset{}{C}}{-}C \begin{array}{c} \diagup Z \\ \\ \diagdown \end{array}\right] \longrightarrow \begin{array}{c} X \\ \diagdown \\ \\ Y \diagup \end{array} C{=}O + ZCN$$

Sherwood used several initiators, including photolysis by ultraviolet light. H_2S adds H·, not HS·, and the resulting vinyl radical adds NO. The intermediate decomposes to formaldehyde and HCN. Cl· from the photolysis of phosgene adds to chloroacetylene at the ≡CH. The NO adduct decomposes to form CO, HCl and CNCl.

Photochemically excited nitrobenzene can cleave an olefinic bond like ozone does. Scheinbaum[13] found that excited nitrobenzene also cleaves acetylenic bonds. Tolan and nitrobenzene in petroleum ether irradiated for three days under nitrogen give benzophenone-anil, CO_2, nitrosobenzene, dibenzanilide, 2-hydroxyazobenzene, and the β-lactam of N-phenyl-β-amino-tetraphenyl-propionic acid. 100 mmoles of PhC≡CR give 34.5 mmoles of cleavage products.

$$\underset{\text{0.1 mmole}}{PhNO_2\cdot} + \underset{\text{100 mmole}}{PhC{\equiv}CPh} \xrightarrow[\text{3 days}]{\text{ultraviolet}} \underset{\text{12 mmoles}}{Ph_2C{=}NPh} + \underset{\text{20 mmoles}}{CO_2}$$

$$+ \underset{\text{3.2 mmoles}}{PhNO} + \underset{\text{18 mmoles}}{(PhCO)_2NPh}$$

0.4 mmole

$$\begin{array}{c} Ph_2C{-}NPh \\ | \quad\ | \\ Ph_2C{-}C{=}O \end{array}$$

0.9 mmole

In aqueous dioxane, 100 mmoles of tolan give 72 mmoles of cleavage products:

$$PhNO_2 \cdot \ + \ PhC{\equiv}CPh \xrightarrow{+H_2O} Ph_2CHCO_2H \ + \ PhN(COPh)_2 \ + \ PhCOCOPh$$

0.1 mmole   100 mmoles   24 mmoles   40 mmoles   12.6 mmoles

$+ \ PhN{=}N(O)Ph \ +$

11 mmoles

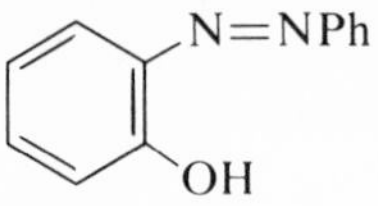

1.0 mmole

The reaction intermediate is probably diphenylketene.

4. PEROXIDATION

4.1. Addition of Peracids

Peracids react with olefins faster than with acetylenes. Peracetic acid reacts at these relative rates:[14]

Ethylene,	1
Acetylene,	too slow to measure
1-Acetylenes,	0.1
Internal acetylenes,	0.5

Polyenynes react with peracids to give mono- and diepoxides by exclusive reaction at the olefinic bonds.[11] The epoxyacetylenes are valuable as starting materials for the "Perveev" reaction to form thiophenes:[15]

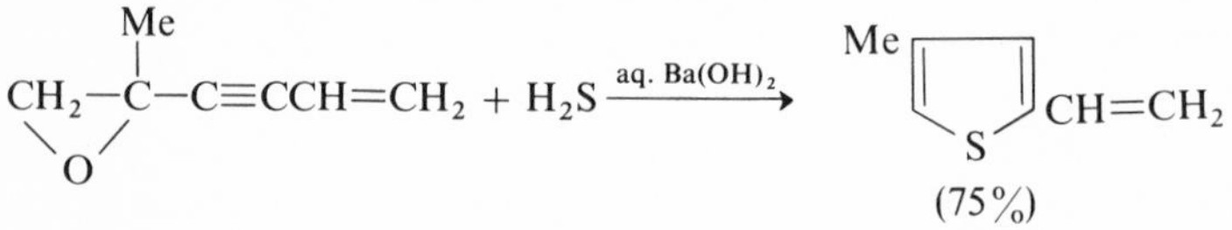

Phthaloyl peroxide adds 30–300 times faster to diarylethylenes than to diaryl-acetylenes. An interesting addition sequence has been proposed:[16]

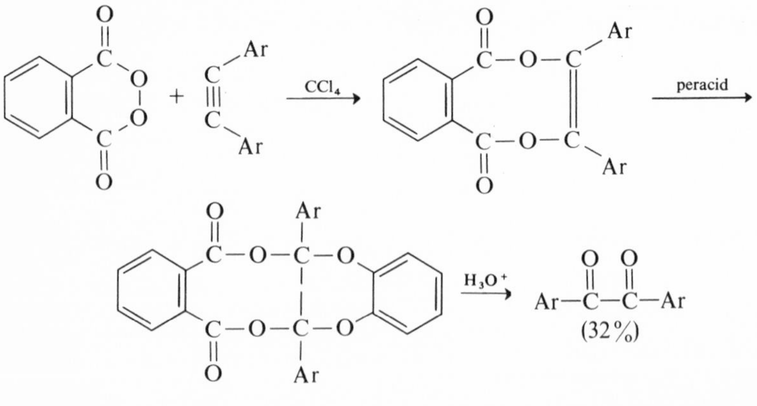

4.2. Mechanism of Peroxidation

Peroxidation of olefins forms the oxirane ring easily. By the same reaction,

acetylenes would give the oxirene ring $HC{\overset{\displaystyle O}{\triangle}}CH$. Oxirene is isoelectronic with cyclobutadiene and is probably very unstable. Some evidence indicates that the oxirene ring is intermediate in acetylenic bond peroxidation. The best chance for a stable oxirene ring is from acetylenes whose electrons are delocalized as in diphenylacetylene. Peroxidation of diphenylacetylene gives several products, depending on conditions. The oxirene is not isolable. One common product is benzil, which is also formed in other oxidations of tolan (Table 2-16). With trifluoroperacetic acid, phenylacetylene gives 38% yield of phenylacetic acid and 25% yield of benzoic acid.

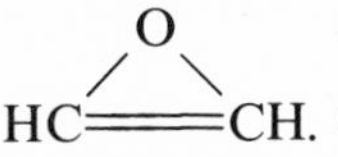

TABLE 2-16
Peroxidation of Tolan

Peracid	Conditions	Product	% Yield	Reference
Trifluoroperacetic	CH_2Cl_2, Na_2HPO_4, reflux	Benzil	76	17
		Benzoic acid	17	
Peroxybenzoic	Benzene-ether, 20° dark	Benzil	9.6	18
		Diphenylacetic ester	7.8	
		Benzyl benzoate	9	
m-Chloroperoxy-benzoic	Ethanol	Benzil	3.2	18
		Diphenylacetic ester	18.3	
		Benzoic acid	66.2	
		Ethyl benzoate	5.3	

The first product is probably the oxirene from addition of "singlet" oxygen to the triple bond. The oxirene can add another oxygen to form a "bisoxirane" or can rearrange to diphenylketene. These two intermediates account for all the products:[17,18]

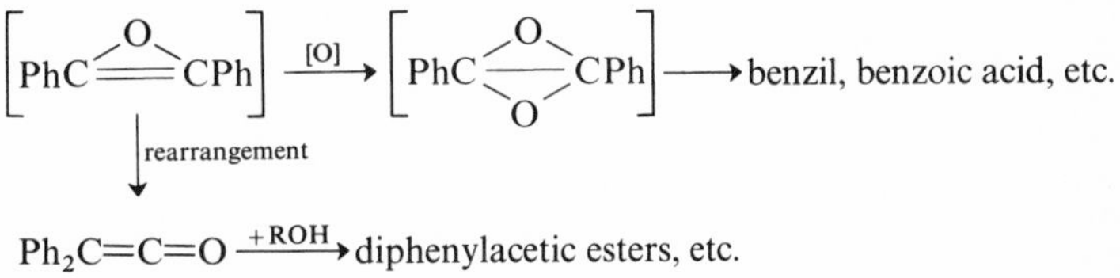

$$\left[PhC{\overset{\displaystyle O}{\triangle}}CPh\right] \xrightarrow{[O]} \left[PhC{\overset{\displaystyle O}{\underset{\displaystyle O}{\diamond}}}CPh\right] \longrightarrow \text{benzil, benzoic acid, etc.}$$

(rearrangement)

$$Ph_2C=C=O \xrightarrow{+ROH} \text{diphenylacetic esters, etc.}$$

Thus, peroxidation of acetylenes is much more complex than peroxidation of olefins because the intermediate oxirene is so reactive it rapidly forms the secondary oxidation products.

5. OXIDATION BY LEAD TETRAACETATE

Acetylenic acids might be less reactive toward lead tetraacetate than other acetylenes because of the mesomerism to an allenic structure:[19]

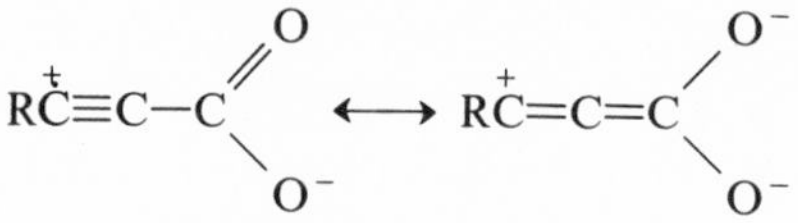

The acetylenic acids are less reactive than other acetylenes. Phenylpropiolic acid and lead tetraacetate in acetic acid at 80° do not react unless a large excess of $Pb(OAc)_4$ is present. Acetylenedicarboxylic acid is even less reactive. o-Methoxyphenylpropiolic acid oxidizes rapidly, the first mole of oxidant adding faster than the second. Carbon dioxide and methane are formed. The enhanced reactivity of the triple bond in o-methoxyphenylpropiolic acid is accounted for:

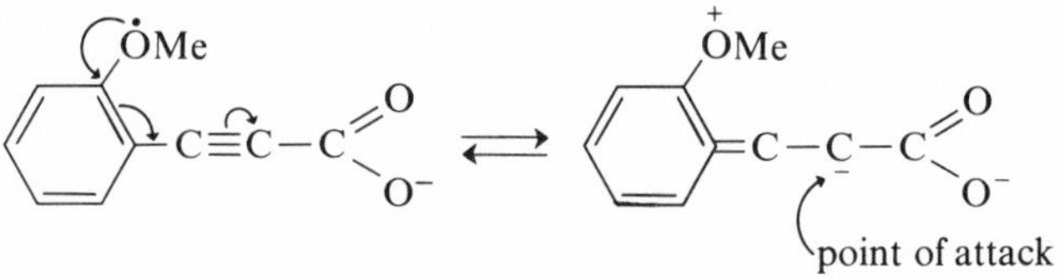

Phenylacetylene refluxed with lead tetraacetate in acetic acid for an hour gives some unusual products:[20] phenylmethylacetylene, 24%; phenyl isopropyl ketone, 4%; $PhC(OAc)=CHMe$, 19%, formed by addition of methyl acetate. Formation of phenylmethylacetylene is the first known replacement of acetylenic hydrogen by methyl from lead tetraacetate.

6. OXIDATION BY OF_2

Acetylenes are less reactive in addition of OF_2 than olefins, convenient rates being obtained at $-40°$ instead of $-78°$.[21] Products are α,α-difluoroketones. The effect of substituents on the direction of addition to triple bonds is illustrated in Table 2-17.

TABLE 2-11[21]

$$RC{\equiv}CR' \quad + OF_2 \longrightarrow R\overset{O}{\overset{\|}{C}}CF_2R' + \; RCF_2\overset{O}{\overset{\|}{C}}R'$$

R	R'	% Yield	
Ph	Ph	82	—
Ph	Me	69	0
Et	Me	27	42
Ph	H	33	—

This is often a more convenient synthesis of difluoroketones than standard methods are. OF_2 is a treacherous oxidizer, and adequate shielding is necessary.

References

1. Raphael, R. A., "Acetylenic Compounds in Organic Synthesis," p. 31, London, Butterworths Scientific Publications, 1955.
2. Reisch, J., and Walker, H., *Deutsche Apotheker-Z.*, **103**, 1139 (1964); *Chem. Abstr.*, **60**, 373 (1964).
3. Postowsky, J. J., and Lugowkin, B. P., *Chem. Ber.*, **68**, 852 (1935).
4. Buckley, G. D., and Levy, W. J., *J. Chem. Soc.*, 3016 (1951).
5. Ciocchetti, J. E. (to Acidos Grosos Limitadas), U.S. Patent 3,193,512 (July 6, 1965); *Chem. Abstr.*, **63**, 13073 (1965).
6. Hurd, C. D., and Christ, R. E., *J. Org. Chem.*, **1**, 141 (1936).
7. Clay, P. G., Johnson, G. R. A., and Weiss, J., *J. Phys. Chem.*, **63**, 862 (1959).
8. Criegee, R., and Lederer, M., *Ann. Chem.*, **583**, 29 (1953).
9. Bailey, P. S., Chang, Y.-G., and Kwie, W. W. L., *J. Org. Chem.*, **27**, 1198 (1962).
10. Bernatek, E., Ledaal, T., and Asen, S., *Acta Chem. Scand.*, **18**, 1317 (1964); *Chem. Abstr.*, **61**, 13180 (1964).
11. Bohlmann, F., and Sinn, H., *Chem. Ber.*, **88**, 1869 (1965).
12. Sherwood, A. G., and Gunning, H. E., *J. Am. Chem. Soc.*, **85**, 3506 (1963).
13. Scheinbaum, M. L., *J. Org. Chem.*, **29**, 2200 (1964).
14. Schlubach, H., and Franzen, V., *Ann. Chem.*, **577**, 60 (1952).
15. Gronowitz, S., "Advances in Heterocyclic Chemistry," Vol. 1, p. 1, New York, Academic Press, 1963.
16. Greene, F. D., and Rees, W. W., *J. Am. Chem. Soc.*, **82**, 893 (1960).
17. McDonald, R. N., and Schwab, P. A., *J. Am. Chem. Soc.*, **86**, 4866 (1964).
18. Stille, J. K., and Whitehurst, D. D., *J. Am. Chem. Soc.*, **86**, 4871 (1964).
19. Jones, E. R. H., Whitham, G. H., and Whiting, M. C., *J. Chem. Soc.*, 4628 (1957).
20. Moon, S., and Campbell, W. J., *Chem. Commun.*, 470 (1966).
21. Merritt, R. F., and Ruff, J. K., *J. Org. Chem.*, **30**, 328 (1965).

PART FIVE

Addition Reactions of Acetylenic Hydrocarbons and Acetylenic Alcohols

Introduction

Acetylenic alcohols and hydrocarbons are typical "unactivated" acetylenes. Addition reactions usually require more severe conditions than with "activated" acetylenes. The ionic reactions discussed in this part include reactions of arylacetylenes, which can be regarded as only moderately "activated."

1. ADDITION OF HALOGEN AND PSEUDO-HALOGENS

1.1. Elemental Halogen

Halogens usually react electrophilically, first adding X^+ to one carbon of the triple bond, then adding X^- to the other carbon to form dihaloethylenes. Halogen adds selectively to the olefinic bond in enynes, unless cuprous catalyst is used.[1] Dibromides and diiodides made by addition are frequently used as crystalline derivatives of acetylenes. Addition-substitution reactions of chlorine and acetylene give the commercially important chloroethylenes, widely used as solvents. Although addition of chlorine to acetylenic bonds is a textbook reaction, very few quantitative data exist for substituted acetylenes.[2] 1-Butyne will not react with chlorine at $-9°$ unless irradiated. The product is 90% *trans*-1,2-dichloro-1-butene.

Addition of fluorine to acetylenes was first reported in 1967.[2a] Acetylenic hydrocarbons in CCl_3F or methanol add F_2 at $-78°$. With CCl_3F, the major product is the tetrafluoride:

$$PhC{\equiv}CR + 2F_2 \xrightarrow{CCl_3F} PhCF_2CF_2R$$

In methanol, the methanol also adds and the products also contain *gem*-fluoroethers and dimethyl ketals. The composition of the product varies with R:

$$C_6H_5C{\equiv}CR + 2F_2\text{-}CH_3OH \longrightarrow \underset{(1)}{C_6H_5CF_2CF_2R} + \underset{\underset{OCH_3}{|}}{\underset{(2)}{C_6H_5CFCF_2R}} + \underset{\underset{OCH_3}{|}}{\overset{\overset{OCH_3}{|}}{\underset{(3)}{C_6H_5CCF_2R}}}$$

		%	
R	(1)	(2)	(3)
C_6H_5	23	57	20
CH_3	19	50	31
H	13	35	52

The products are accounted for equally well by assuming initial addition of either F^+ or methanol. The fluoroethers and the ketals are easily hydrolyzed to form α-difluoroketones. This reaction has synthetic utility, provided facilities for handling fluorine are available.

1.2. Halogenation by Cupric Halides

Cupric halides halogenate aromatic, aliphatic unsaturated, and alicyclic hydrocarbons at high temperature in the gas phase. Relatively little has been reported on low-temperature liquid phase cupric halide halogenations. Castro[3] reviewed earlier work and described his reactions with acetylenes. He used 0.2–$0.3M$ substrate, 0.8–$1.0M$ cupric halide, and refluxed in methanol. Terminal acetylenes give some trihalogenation by addition-elimination:

$$RC\equiv CH + CuBr_2 \longrightarrow RCBr{=}CHBr \longrightarrow RCBr_2CHBr_2 \longrightarrow HBr + RCBr{=}CBr_2$$

Internal acetylenes give *trans*-dihaloethylenes. Trihalogenation is minimum when excess acetylenic is present. Yields generally are good. With equimolar reagents, propargyl alcohol gives 93% yield of tribromoallyl alcohol within a half hour; cupric chloride is slower. The yield of trichloroallyl alcohol is only 31% after 22 hours. Propargyl aldehyde is slower than propargyl alcohol, and acrolein is slightly faster than propargyl aldehyde. Methyl propiolate and phenylacetylene are even less reactive.

Reactions of allyl alcohol and propargyl alcohol are overall second order, but have different rate expressions. For allyl alcohol, the rate is independent of alcohol concentration. Rate = $k_2(CuBr_2)^2$. For propargyl alcohol, rate = $k_2(CuBr_2)(\text{alcohol})$. Bromine will distill from cupric bromide in refluxing acetonitrile, indicating that bromine may be the active addend. On the other hand, the reaction may involve a copper complex within which bromine transfers from copper to carbon.

The acetylenic bonds in *o*-bis(phenylethynyl)benzene do not interact during reaction with iron pentacarbonyl or during radiation. Much interaction occurs when the compound reacts with electrophilic, nucleophilic or radical reagents, however.[4] Apparently electrophilic reagents like HBr, Br_2 or water add to form a cyclic intermediate by a concerted ring closure, followed by addition of X^-:

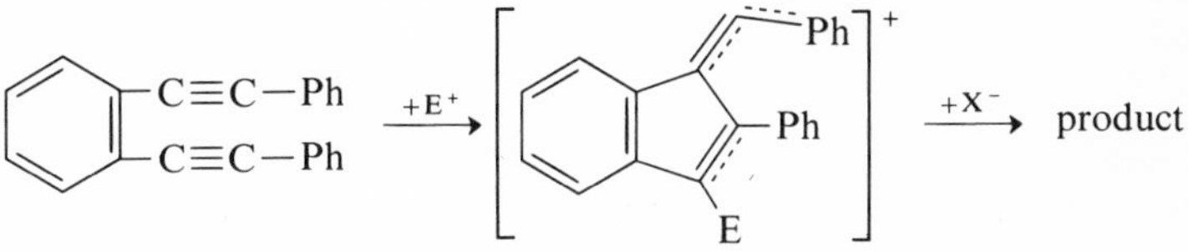

When the electrophilic reagent is HBr, H^+ is E^+, and Br^- is X^-, and the product is

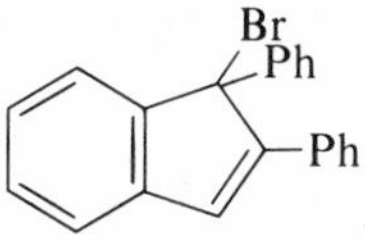

1.3. Positive Halogen

Chlorination of 1-hexyne in methanol gives some 1,1-dichlorohexanone-2.[5] Vinylacetylene gives 1,1,4-trichlorobutanone-2.[6] Alkynes are chlorinated by N-chlorosuccinimide or by *t*-butyl hypochlorite in methanol.[7] The β-chlorovinyl ether is intermediate:

$$RC\equiv CR' + Cl^+ + MeOH \longrightarrow R\overset{\overset{\displaystyle OMe}{|}}{C}=CR'Cl \longrightarrow R\overset{\overset{\displaystyle OMe}{|}}{\underset{\underset{\displaystyle OMe}{|}}{C}}-C(Cl_2)R' \xrightarrow[\text{acid}]{\text{dilute}} R\overset{\overset{\displaystyle O}{||}}{C}CCl_2R'$$

60–80% yield from N-chlorosuccinimide, 50–70% yield from t-butyl hypochlorite. Only one product forms, except when unsymmetrically disubstituted alkynes react. Then both of the possible dichloroacetals form in equal amount. α,α-Dichloroacetophenone was reported earlier[6] to be the main product from the reaction of phenylacetylene in carbon tetrachloride with ethyl hypochlorite.

Walling[9] reacted t-butyl hypochlorite with some alkynes: "Preliminary deductions as to the existence of a 'spontaneous' reaction were based upon the detonation of sealed tubes containing t-butyl hypochlorite and 2-butyne on warming from a liquid N_2 bath." 2-Butyne neat, or in toluene solution, in light or dark, gives 1-chloro-2-butyne by substitution. This probably is the result of a radical attack on propargylic hydrogen. This reaction should be valuable for preparing propargylic chlorides because the products are essentially free of chloroallenes. Yields can probably be improved by using a higher acetylene:hypochlorite ratio. Caution must be exercised when hypohalites are used as halogenating agents in unexplored systems; violent reactions can occur.

The work of Reed[10] and of Walling[9] serves as a good example of the same reactants giving completely different products by different reaction mechanisms in different solvents.

N-Chloramines add to butadiene to give 1-chloro-4-dialkylamino-2-butenes. N-Chloroamines also add to acetylenes and to allenes.[11] Acetylenes react spontaneously to form the α-chlorocarbonyl compounds:

$$R_2NCl + R'C\equiv CR'' \longrightarrow \left[R_2N\overset{\overset{\displaystyle R'}{|}}{C}=\overset{\overset{\displaystyle R''}{|}}{C}Cl \right] \xrightarrow{H_2O} R'\overset{\overset{\displaystyle O}{||}}{C}CHR''Cl$$
$$\text{enamine}$$

If secondary C—H is available, intramolecular hydrogen abstraction predominates over addition. Allene reacts like a terminal olefin and gives a 1,2-addition product. 1,1-Dimethylallene gives partly the same reaction, but the major product is from an ionic reaction; the product is $Et_2NCH_2\overset{\overset{\displaystyle Me}{|}}{C}=C(Cl)CH_2Cl$, 34% yield.

1.4. Addition of Iodine Isocyanate (INCO)

Very little has been reported on the addition of iodine isocyanate to acetylenes. Phenylacetylene, diphenylacetylene, 4-octyne and 2-octyne add 1 mole of INCO very rapidly at −30°. Stearolic acid does not react. Yields of carbamates (hydration of the isocyanate) are 20–40% if THF is used as solvent instead of ether.[11a]

2. ADDITION OF HALOGEN ACIDS

1-Alkynes add halogen acids to give 2-haloethylenes.[12] Arylacetylenes give α-halostyrenes.[13] The addition of halogen acids to acetylene is discussed as a special addition reaction, vinylation, in Chapter 3.

HCl adds to acetylenes in acetic acid solution in the presence of tetramethyl-ammonium chloride catalyst.[13a] 1-Phenylpropyne gives mainly the *cis*-hydrochloride, while 3-hexyne gives equal amounts of ketone and *trans*-hydrochloride. The ketone probably forms via the addition of acetic acid to the triple bond, followed by hydrolysis during the work-up. A mechanism accounting for the products from 3-hexyne and the observed kinetics is:

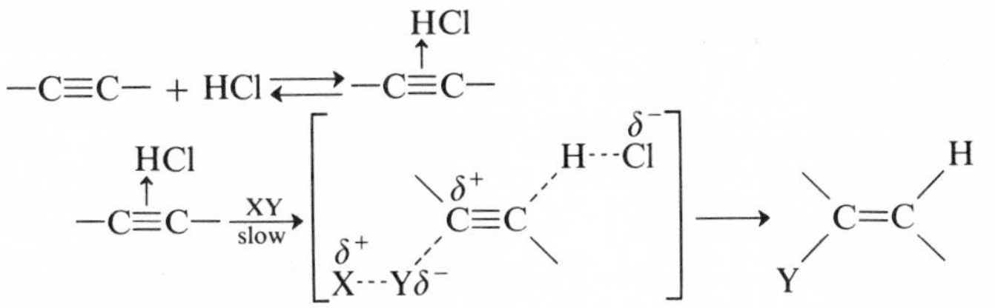

$$-C\equiv C- + HCl \rightleftharpoons -\overset{\overset{\displaystyle HCl}{\uparrow}}{C}\equiv C-$$

$$-\overset{\overset{\displaystyle HCl}{\uparrow}}{C}\equiv C- \xrightarrow[\text{slow}]{XY} \left[\begin{array}{c} \text{C}\equiv\text{C} \\ \text{X}\cdots\text{Y} \end{array} \right] \longrightarrow \text{C}=\text{C}$$

X = H; Y = OAc or Cl
X = (CH$_3$)$_4$N; Y = Cl

3. ADDITION OF NOCl AND NO$_2$Cl TO PHENYLACETYLENES

The main product from addition of NOCl or NO$_2$Cl to phenylacetylenes is the α-chloro-β-nitrostyrene.[14] NOCl reacts in chlorinated solvent at $-50°$, and NO$_2$Cl reacts in dry ether at $10°$. In a few reactions with NO$_2$Cl, significant amounts of dichlorostyrenes form (Table 2-18).

TABLE 2-18
Addition of NOCl and NO$_2$Cl to Phenylacetylenes[14]

Phenylacetylene	Addend	Product	% Yield
PhC≡CH	NOCl	*trans*-PhCCl=CHNO$_2$	30
		cis-PhCCl=CHNO$_2$	5
	NO$_2$Cl	*trans*-PhCCl=CHNO$_2$	40
PhC≡CPh	NOCl	PhCCl=C(NO$_2$)Ph	16
	NO$_2$Cl	*trans*-PhCCl=CClPh	17
		cis-PhCNO$_2$=C(NO$_2$)Ph	5
PhC≡CCO$_2$H	NOCl	PhCCl=C(NO$_2$)CO$_2$H	18
	NO$_2$Cl	*cis*-PhCCl=CClCO$_2$H	42
		PhCCl=C(NO$_2$)CO$_2$H	4

4. ADDITION OF SO_3

One mole of acetylenic compound can react with 1, 2 or 4 moles of SO_3.[15] Acetylene, 1-hexyne and phenylacetylene react similarly.

$$acetylene + 1SO_3 \longrightarrow HC\equiv CSO_3H$$

$$acetylene + 2SO_3 \longrightarrow HO_3SCH_2CHO \text{ (after hydrolysis)}$$

$$acetylene + 4SO_3 \longrightarrow (HO_3S)_2CHCHO \text{ (after hydrolysis)}[16,17]$$

Acetaldehyde disulfonic acid is a valuable sulfoalkylating agent.

5. ADDITION OF HALOMETHYL ALKYL ETHERS

Aluminum chloride catalyzes the low-temperature addition of halomethyl alkyl ethers to acetylenes:[18]

$$C_2H_2 + ClCH_2OMe \longrightarrow MeOCH_2CH=CHCl$$
$$(60\% \text{ yield})$$

Halogen in the aluminum halide exchanges with ether halogen, and enters the final product:

(1) $EtOCH_2Cl + AlBr_3$ ⎤
                    $\xrightarrow{+C_2H_2}$ $EtOCH_2CH=CHCl + EtOCH_2CH=CHBr$
(2) $EtOCH_2Br + AlCl_3$ ⎦

ratio	from (1)	1	4.5
	from (2)	2.4	1

In a similar reaction, t-butyl chloride adds to acetylene to give 10–15% yield of t-BuCH=CHCl.[19]

6. ADDITION OF ACYL CHLORIDES— FORMATION OF β-CHLOROVINYL KETONES

Pohland[20] published a comprehensive review of synthesis and reactions of β-chlorovinyl ketones, covering the literature through mid-1964. Best yields from acyl chlorides are obtained by making a solution of catalyst (usually $AlCl_3$) and acyl chloride in CCl_4, and then adding acetylene at 0–10° over several hours. For aroyl chlorides, it is best to make the complex in 1,2-dichloroethane at 10°, warm to 45°, and then add the acetylene. α,β-Unsaturated acyl halides react. α- and β-Haloacyl halides add, and products can be dehydrohalogenated to give the same alkenyl β-chlorovinyl ketones. β-Chlorovinyl ketones from acetylene are mostly *trans-transoid*:

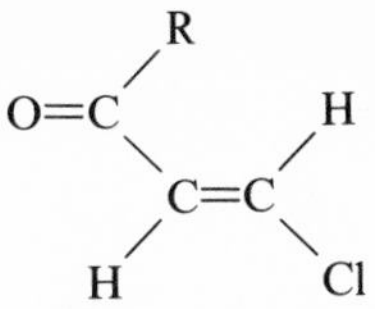

trans-transoid 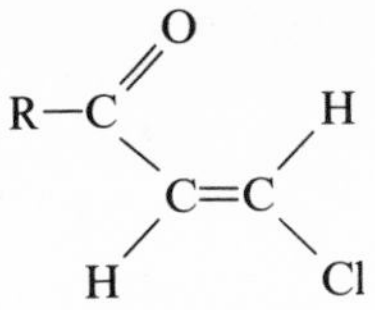　　　　　　　*trans-cisoid*

Aromatic β-chlorovinyl ketones are more stable than the aliphatic ketones.

The β-chlorovinyl ketones have three active sites, and reactions at all of them are well known. A few examples illustrate the preparative scope of β-chlorovinyl ketones:

(1) Reaction at carbonyl: Carbonyl reagents react as expected. Hydrazines give 3-alkylpyrazoles:

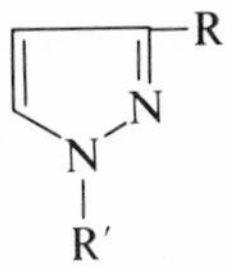

Hydroxylamine gives isoxazoles:

3-alkyl- and 5-alkylisoxazoles

(2) Reaction at the double bond: The Diels–Alder reaction gives high yields:

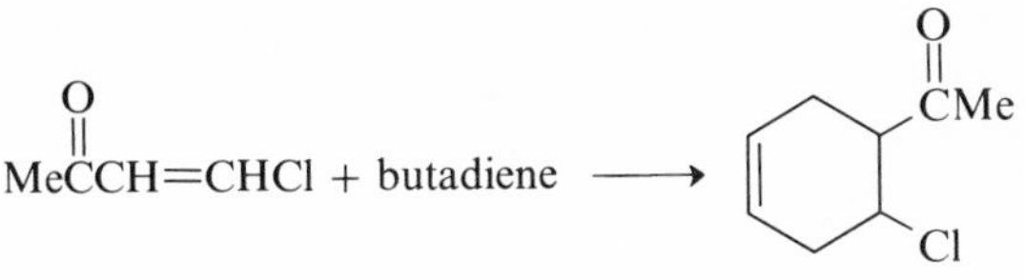

MeCCH=CHCl + butadiene ⟶

(60%)

Azo compounds add to form pyrazoles:

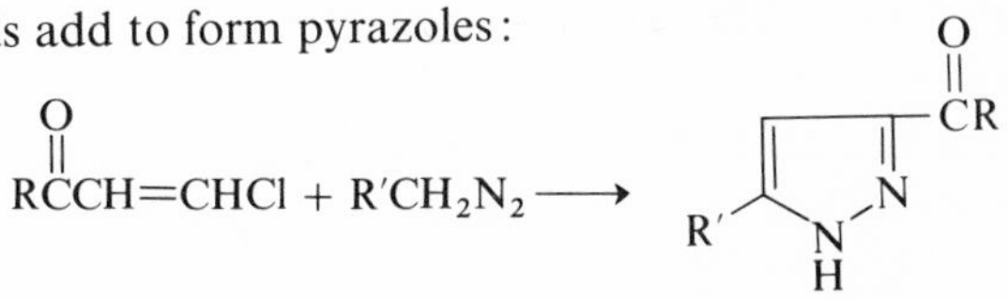

RCCH=CHCl + R'CH$_2$N$_2$ ⟶

(30–90%)

Azides give triazoles:

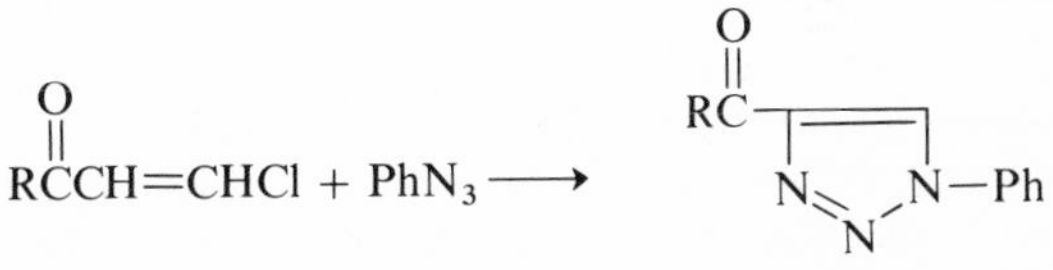

RCCH=CHCl + PhN$_3$ ⟶

Cyanates give isoxazoles:

$$\underset{\text{O}}{\overset{\text{O}}{\text{RCCH}}}\text{=CHCl} + \text{R'CNO} \longrightarrow$$

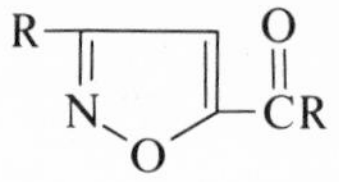

3-alkyl-5-isoxazoles

(3) Reaction at the Cl: Amines and ammonia react to give pyridines:

$$\underset{\text{O}}{\overset{\text{O}}{\text{RCCH}}}\text{=CHCl} + \text{R'NH}_2 \longrightarrow \text{RC(O)CH=CHNHR'} \xrightarrow{+\text{RC(O)CH=CHCl}}$$

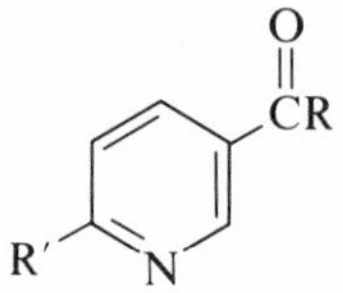

2-alkyl-5-acylpyridines

Alcohols, thiols, and organic or inorganic anions easily replace Cl:

$$\text{ROH} \longrightarrow \text{ROCH=CH}\overset{\text{O}}{\overset{\|}{\text{C}}}\text{R'}$$

$$\text{RSH} \longrightarrow \text{RSCH=CH}\overset{\text{O}}{\overset{\|}{\text{C}}}\text{R'}$$

$$\text{acetoacetic ester} \longrightarrow \text{EtO}_2\text{C}\overset{\text{R''}}{\underset{\text{C(O)R}}{\overset{|}{\text{C}}}}\text{CH=CH}\overset{\text{O}}{\overset{\|}{\text{C}}}\text{R'}$$

7. NUCLEOPHILIC ADDITION OF THIOLS

Sodium *p*-toluenethiolate adds to phenylacetylene in refluxing alcohol. The only product is *cis*-ω-styryl-*p*-tolyl sulfide, in good yield:[21]

$$\text{PhC}\equiv\text{CH} + p\text{-MePhSNa} \longrightarrow \underset{\text{Ph}}{\overset{\text{H}}{>}}\text{C=C}\underset{\text{SPh}-p\text{-Me}}{\overset{\text{H}}{<}}$$

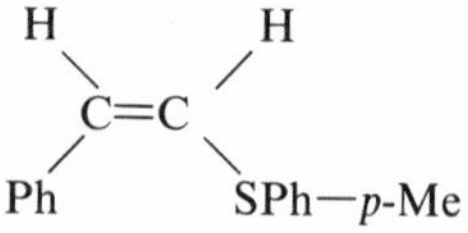

2-Butyne in ethanol reacts to form 2-*p*-tolylmercapto-*trans*-2-butene. In both of these additions, only the isomers from *trans* addition form. This orientation is consistent with the nucleophilic nature of thiolate anion and the electronegative character of phenyl compared to methyl. The intermediate should be the configurationally stable carbanion

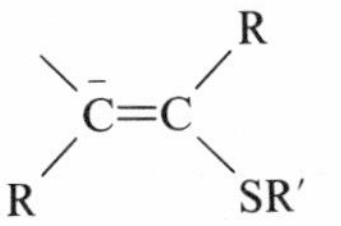

Other examples are known. Ethyl mercaptan adds to t-butylacetylene in the presence of KOH at 180° to form 85% yield of t-BuCH=CHSEt.[22] Chloroacetylene adds aryl mercaptans without catalyst to form ArSCH=CHCl in good yields.[23] Selenophenol adds *trans* to phenylacetylene in the presence of sodium methoxide.[24]

8. ADDITION OF ARYLSULFENYL HALIDES

Arylsulfenyl halides add across the triple bonds of acetylenes as ArS and X to give ethylenes (Table 2-19).

TABLE 2-19
Addition of Arylsulfenyl Halides to Acetylenes

Acetylene	Addend	Product	% Yield			Reference
Acetylene	ArSCl	ArCH=CHCl	43–90			25
Acetylene	ArSeBr	*trans*-ArSeCH=CHBr	—			27
				Ratio		
PhC≡CH	p-NO$_2$PhSCl (ArSCl)		Solvent	(1)	(2)	26
			EtOAc	85	15	
			CHCl$_3$	65	35	
			HOAc	20	80	

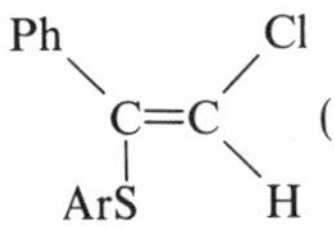

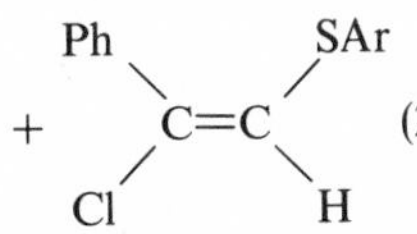

9. ADDITION OF AMINES

Amines add across nonactivated triple bonds. Initial products frequently cyclize, particularly at high temperatures (Table 2-20). For a recent review of these reactions, see reference 34.

TABLE 2-20
Addition of Amines to Acetylenes

Acetylene	Amine	Conditions	Product	% Yield	Reference
$RC\equiv CR'$	$R''NH_2$	CuCl, 170°	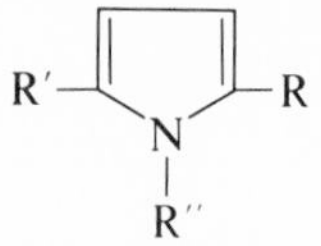	40–80	28
3-Methyl-1-butyn-3-ol	NH_3, MEK	$Cd_3(PO_4)_2$-Al_2O_3, 400°	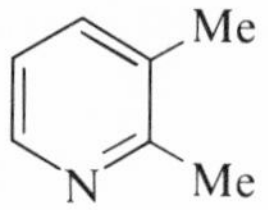	15	29
Acetylene	Aniline	$Zn_3(PO_4)_2$, 300°	Quinaldine	22	30
Acetylene	Aniline	$PbEt_4$, 400°	Indole	—	31
Acetylene	Aniline-$HgCl_2$	75°	3-Phenylbenzo[f] quinoline	—	33
Acetylene	Benzalaniline	75°	3-Phenylbenzo[f] quinoline	61	32

10. 1,3-DIPOLAR ADDITION OF DIAZO COMPOUNDS
(see part 8 for a discussion of the mechanism of 1,3-dipolar additions)

10.1. Diazoalkanes

Diazo compounds in ether react with acetylene at 12–15 atmospheres pressure and 20°.[35,36] In this system, good yields of pyrazoles form after several hours. Diazomethane gives pyrazole, 95% yield; diazoethane gives 77% yield of 3-methylpyrazole. Diazoacetic ester requires 100° to form ethyl pyrazole-3-carboxylate (72% yield). Bis(diazoalkanes) give bis(pyrazolyl)alkanes:

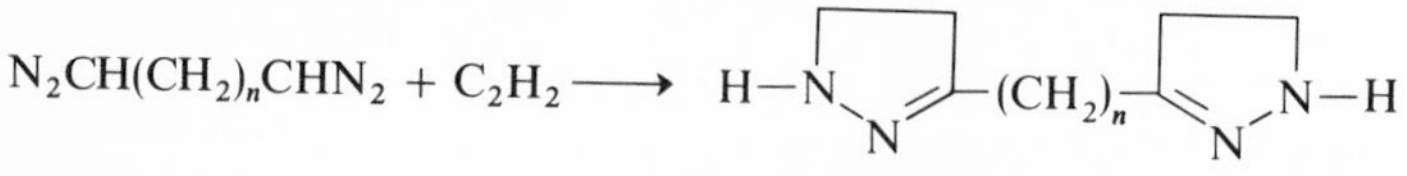

$$N_2CH(CH_2)_nCHN_2 + C_2H_2 \longrightarrow$$

$$n = 3, 85\%$$
$$4, 90\%$$
$$5, 90\%$$
$$10, 93\%$$

Diazoacetic ester and $(BuO)_2BC\equiv CH$ react and give (after hydrolysis) the pyrazoline

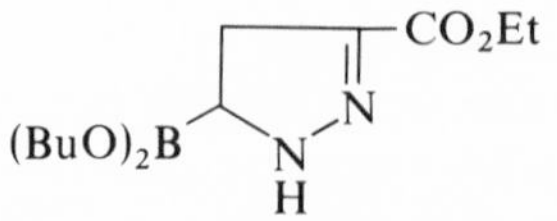

75% yield.[37] With diazoacetic ester, 5-decyne adds carbene in the presence of cupric sulfate or benzoyl peroxide; the product is the cyclopropene. With a small amount of cupric sulfate, the product is 2,3-dibutyl-4-ethoxyfuran (84% yield).[38] Disubstituted diazomethanes add to acetylenes to form 3,3-disubstituted pyrazolines.[39]

Diazofluorene reacts with 2 moles of acetylene to form 2H-phenanthro-[9.10-c]pyrazol and 9-[pyrazolyl-(3)-methylene]-fluorene. The first product results from a ring enlargement.[38a]

10.2. Azidoalkanes

Durden[40] reported 36 new triazoles from alkyl and aralkyl azido compounds and alkynes. He described an easy method for making the azidoalkanes: Alkyl chlorides react with sodium azide in dimethylformamide. He refluxed the azido compound with the acetylene in a solvent, such as toluene, alcohol or acetone. For propyne and butyne, a pressure bomb was necessary. Yields were 20–95%.

Lithium acetylides react with toluene-p-sulfonyl azide in the presence of β-naphthol.[41] Two moles of azide react, and then 2 moles of naphthol, to give

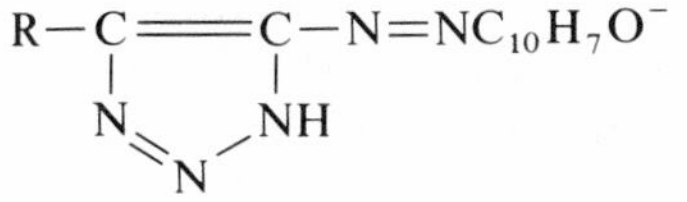

$$\begin{array}{c} R-C=\!\!=\!\!=C-N=\!\!=NC_{10}H_7O^- \\ | \qquad | \\ N \qquad NH \\ \diagdown_{N}\diagup \end{array}$$

Organic azides and acetylenic compounds give 1,2,3-triazoles.[42-46] Acetylene and alkynes add cyanogen azide (N_3CN) to form 1-cyano-1,2,3-triazoles, in equilibrium with α-diazo-N-cyanoethylideneimines:[46a]

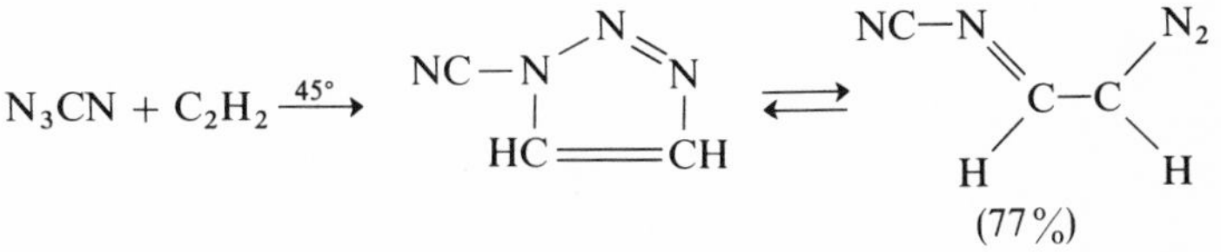

$$N_3CN + C_2H_2 \xrightarrow{45°} \begin{array}{c} NC-N \diagup^{N}\diagdown_{N} \\ | \qquad | \\ HC=\!\!=\!\!=CH \end{array} \rightleftharpoons \begin{array}{c} NC-N \diagdown \qquad N_2 \\ C-C \\ \diagup \quad \diagdown \\ H \qquad H \end{array}$$

$$(77\%)$$

Ethyl azidoformate gives oxazoles and other products.[47]

11. ADDITION OF CARBENES

11.1. Carbenes

The first example of addition of a carbene to an acetylene was reported in 1956. D'yakonov[48] reacted ethyl dizaoacetate with phenylpropyne in cyclohexane (copper sulfate catalyst):

$$EtO_2CCHN_2 + PhC\equiv CMe \longrightarrow \underset{H\quad CO_2H}{\overset{Ph-C=CMe}{\diagdown C \diagup}} + \underset{\underset{H\quad H}{MeC-C-CO_2Et}}{\overset{CO_2Et}{\underset{|}{\overset{|}{C=C-Ph}}}}$$

(42%)

Similar reactions have since been done with many acetylenes and carbenes. Rozantsev[49] reviewed the literature in 1965.

Double bonds react faster than triple bonds; the olefinic bonds in enynes react first and fastest. Products are acetylenic cyclopropanes. Products from addition to the triple bond are cyclopropenes. Gaseous diazomethane and acetylene formed propyne and allene, not the expected cyclopropene.[50] Doering[51] prepared cyclopropenes from carbene and 2-butyne in solution. He added hot N-nitrosomethylurea-KOH in decalin to 2-butyne in methanol. After irradiation with ultraviolet light, the yield of dimethylcyclopropene was 17%. Unsubstituted cyclopropene was less stable. Chandross[52] added benzyl-carbene to diphenylacetylene, then added HBr, and isolated 1,2,3-triphenyl-cyclopropenyl bromide. Perchlorate salts have also been proposed as three-carbon ring analogs of triphenylmethyl dyes.[52a]

Diazoketones decompose to give α-ketocarbenes, which add to acetylenes to form cyclopropenes. Reaction goes at 100° in the presence of copper metal. Cyclopropanes are sometimes major by-products.[53]

TABLE 2-21
Addition of Acylcarbenes to Acetylenes

$$\underset{R'-C-C-R^2}{\overset{O\quad N_2}{\overset{||\quad ||}{}}} + RC\equiv CR \xrightarrow[Cu^o]{100°} \underset{R^2\quad \overset{O}{C^{\diagdown}R'}}{\overset{R-C=C-R}{\diagdown C \diagup}} \qquad \underset{\underset{O}{\overset{||}{C-R'}}}{\overset{O\qquad\quad O}{\overset{R'-C}{\diagdown \diagup}\overset{C-R'}{}}}$$

(1)   (2)

Substituent				% Yield of	
R^1	R^2	Acetylene	Solvent	(1)	(2)
C_6H_5	C_6H_5	Diphenyl	Diglyme	—	—
C_6H_5	H	Diphenyl	Diglyme	8.1	30.8
C_6H_5	H	Di-n-butyl	Diglyme	21.1	1.8
p-$CH_3OC_6H_4$	H	Diphenyl	Diglyme	3.2	—
CH_3	H	Diphenyl	Neat	3.3	—
CH_3	H	Di-n-butyl	Neat	2.5	—

11.2. Dihalocarbenes

Vol'pin first added a dihalocarbene to an acetylene in 1959.[54] Dibromocarbene and diphenylacetylene gave diphenylcyclopropenone after hydrolysis. Similar reactions have given 85–90% yields.[55] Parham[57] in 1963, and Kirmse[56] in 1964, wrote reviews on the addition of halocarbenes to acetylenes and to olefins. Allenes add one dihalocarbene to give methylenecyclopropanes. Internal acetylenes add carbenes easily, but terminal acetylenes do not. As with carbenes, dihalocarbenes add only to double bonds of enynes. This allows additional flexibility in the preparation of cyclopropylacetylenes. Some reactions are complex. Dichlorocarbene from phenyl(trichloromethyl)mercury adds to diiodoacetylene to give an assortment of chloroiodocyclopropenes, plus 13% yield of $CCl_3C{\equiv}Cl$.[58]

12. ADDITION OF NITRENES AND SILENES

Nitrenes are similar to carbenes. Meinwald[59] generated nitrenes from azidoformates and added them to olefins to form azirane carboxylates. Acetylenes do not react the same way. Products are isoxazoles and a bicyclic product:

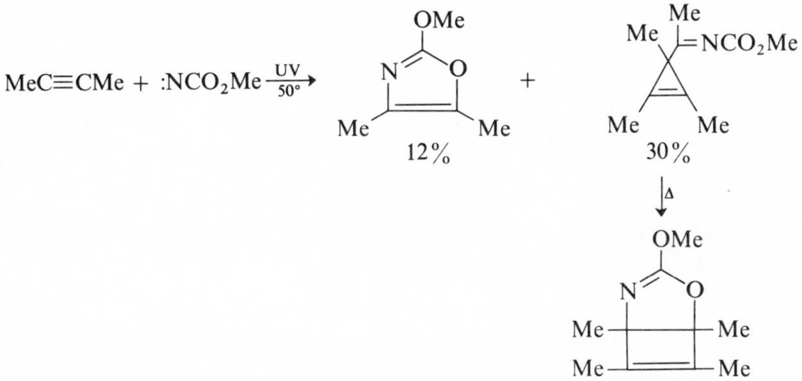

Silenes, generated by thermolysis of polysilanes, methoxypolysiloxanes or 7-silanobornadienes, can be "trapped" by acetylenes or by dienes. Acetylenes add silene to form a silacyclopropene intermediate, which dimerizes to form a disilacyclohexadiene:[59a]

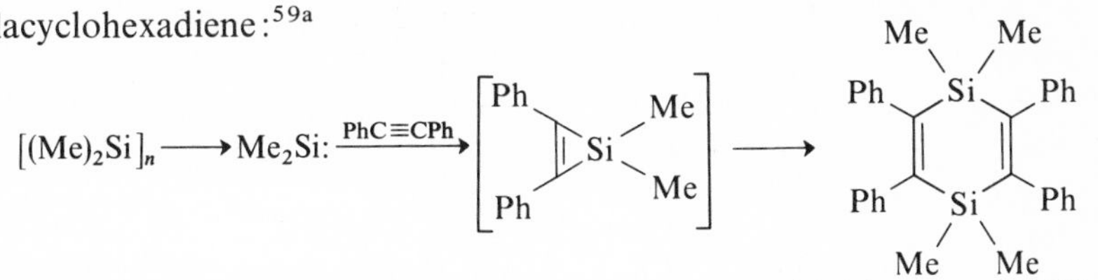

1,1,4,4-tetramethyl-2,3,5,6-tetraphenyl-1,4-disilacyclohexadiene

13. THERMAL ADDITION OF TRIALKYL PHOSPHITES

Electrophilic olefins undergo nucleophilic attack by trialkyl phosphites. This reaction has been studied extensively,[60] but the reaction with acetylenes has received little attention. Pudovik in 1950 reported that an Arbuzov reaction occurred when tertiary acetylenic chlorides underwent S_N2' attack by trialkyl phosphites, to form dialkyl allenylphosphonates.[61] The allenylphosphonates were attacked further by phosphite to form bis(phosphonates). The proposed mechanism shows the dipolar intermediate collapsing by an oxygen-to-carbon migration:

$$(RO)_3P: + \overset{\overset{\displaystyle Cl}{|}}{HC \equiv CCR'_2} \longrightarrow [(RO)_3\overset{+}{P}-CH=C=CR'_2]\xrightarrow{Cl^-}$$

$$\overset{\overset{\displaystyle O}{\uparrow}}{(RO)_2P}-CH=C=CR'_2 \xrightarrow{P(OR)_3} \left[\overset{\overset{\displaystyle O}{\uparrow}}{(RO)_2P}-\bar{C}H-\underset{\underset{\displaystyle {}^+P(OR)_3}{|}}{C}=CR'_2 \right] \longrightarrow$$

$$\overset{\overset{\displaystyle O \ \ R'}{\uparrow \ \ |}}{(RO)_2P}-CH-\underset{\underset{\displaystyle O \leftarrow P(OR)_2}{|}}{C}=CR'_2$$

Dialkyl propargyl phosphites undergo a fast S_Ni' rearrangement to form dialkyl allenylphosphonates.[62-64] Trialkyl phosphites add 1,4 to acetylenic acids to form dialkyl β-carboalkoxyvinylphosphonates.[65] The first adduct is probably enolic allene, formed by an internal oxygen-to-oxygen transalkylation. The enolic allene tautomerizes to the final product:

$$(RO)_3P: + R'C \equiv CCO_2H \longrightarrow (RO)_3\overset{+}{P}-\underset{\underset{\displaystyle R'}{|}}{C}=C=C\overset{\diagup O^-}{\diagdown OH} \longrightarrow$$

$$\overset{\overset{\displaystyle O}{\uparrow}}{(RO)_2P}\underset{\underset{\displaystyle R'}{|}}{C}=C=C\overset{\diagup OR}{\diagdown OH} \longrightarrow (RO)_3\overset{+}{P}\underset{\underset{\displaystyle R'}{|}}{C}=CH\overset{\overset{\displaystyle O}{||}}{C}OR$$

Taking all the facts into account, Griffin and Mitchell[66] concluded that an intermolecular attack of trialkyl phosphites on unactivated acetylenes incapable of isomerizing to allenes might be a reasonable reaction. The initial product (1) might undergo oxygen-to-carbon alkyl transfer to form a vinylphosphonate (2):

$$(RO)_3\overset{+}{P}-\underset{\underset{\displaystyle R'}{|}}{C}=\bar{C}R' \longrightarrow \overset{\overset{\displaystyle O}{\uparrow}}{(RO)_2P}-\underset{\underset{\displaystyle R'}{|}}{C}=C\overset{\diagup R}{\diagdown R'}$$
$$\qquad\qquad (1) \qquad\qquad\qquad\qquad (2)$$

This reaction would be a new synthetic method for vinylphosphonates, and would allow study of the mode of collapse of quasi-phosphonium salts (1) without additional nucleophiles. When Griffin and Mitchell heated a 2:1 molar mixture of triethyl phosphite and phenylacetylene at 150°, and observed changes in infrared spectra, they noted that after 36 hours the terminal acetylenic band at 3300 cm^{-1} disappeared and bands at 1610 (C=C) and 1242 cm^{-1} (P→O) appeared. The product, diethyl β-styrylphosphonate (5), was isolated in 6% yield, along with much intractable tar. A reaction in diglyme increased the yield to 15% but did not eliminate the tar. They could not isolate any products which resulted from the postulated oxygen-to-carbon alkyl transfer.

$$(EtO)_3P: + \; HC{\equiv}CPh \longrightarrow (EtO)_2\overset{+}{P}{-}CH{=}\overset{-}{C}Ph$$

(3)   (4)

(6)

(b) + PhC≡CH

(a) cyclic *cis* β-elimination of ethylene (68 % yield C_2H_4)

$$PhC{\equiv}C^- + (EtO)_3\overset{+}{P}CH{=}CHPh \longrightarrow (EtO)_2\overset{O}{\overset{\uparrow}{P}}{-}CH{=}CHPh \; (\textit{trans} \text{ by PMR})$$

(7)   (5)

Two mechanisms are reasonable.[66] Nucleophilic attack by triethyl phosphite on the terminal carbon of phenylacetylene can produce dipolar intermediate (6) which can undergo (a) internal proton transfer by cyclic *cis* β-elimination to give (5) and ethylene, or (b) protonation by phenylacetylene to give quasi-phosphonium cation (7) and phenylacetylide anion. Nucleophilic dealkylation of (7) by attack of either triethyl phosphite or phenylacetylide anion would form (5). Evidence supporting the *cis* β-elimination process (a) is the fact that 68% yield of ethylene was isolated from the reaction and that *trans*-styryl-phosphonate required by path (a), is the product found.

The *cis* β-elimination reaction is general in the reactions of trialkyl phosphites with acetylenes, but it has synthetic usefulness only for monoaryl-acetylenes. In these acetylenes, stabilization of the negative charge in the initial adducts is the driving force for reaction, which is not provided by alkynes.

14. ADDITION OF TETRACYANOETHYLENE OXIDE (TCNEO)

Tetracyanoethylene oxide is quite different from other epoxides in nucleophilic reactions. Abnormal reactions occur with olefins, acetylenes and aromatic hydrocarbons.[67] The usual epoxide ring-opening reactions do not take place, because the four electronegative cyano groups decrease the electron density on the oxirane ring. Attack by nucleophilic reagents is extremely fast, and cleavage of the C—C bond usually occurs. Ethylene and TCNEO at 130–150° give 80% yield of 1:1 adduct, shown to be 2,2,5,5-tetracyanotetrahydrofuran. Allene and TCNEO react to form 3-methylene-2,2,5,5-tetracyanotetrahydrofuran (1):

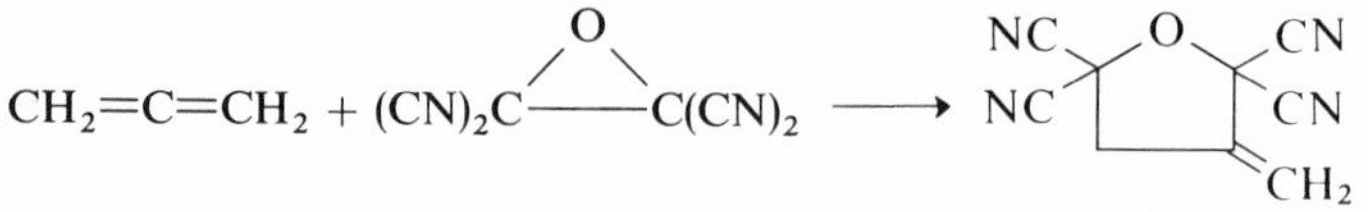

$$CH_2=C=CH_2 + (CN)_2C\overset{O}{\triangle}C(CN)_2 \longrightarrow$$

(1)

Acetylene and TCNEO react under similar conditions to give 70% yield of 2,2,5,5-tetracyanodihydrofuran.

Mono- and disubstituted acetylenes add TCNEO equally well to form the tetracyanodihydrofurans. 2,4-Hexadiyne reacts to form dihydrofuran (2), but even excess TCNEO and forcing conditions fail to give a diadduct. Insertion of methylene groups between the acetylene groups, as in 1,6-heptadiyne, allows synthesis of either monoadduct (3) or diadduct (4):

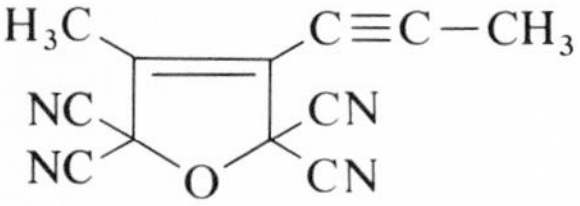

(2) 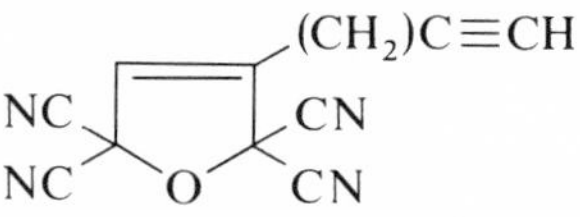   (3)

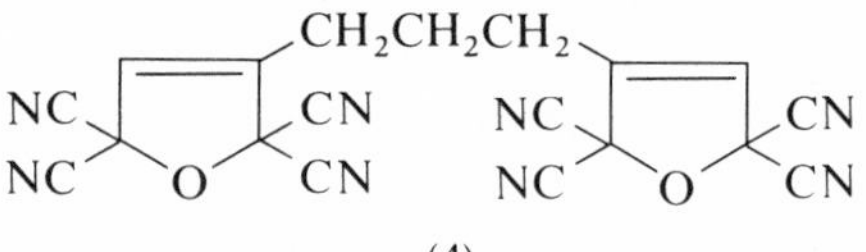

(4)

Competitive reactions show that ethylene reacts about ten times faster than acetylene.

15. ADDITION OF ISOCYANATES

Phenylacetylene and phenylisocyanate react to form cyclic products:[68,69]

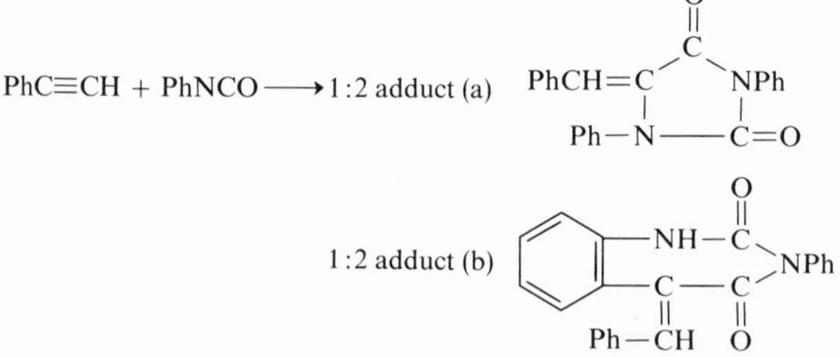

$$PhC\equiv CH + PhNCO \longrightarrow 1:2\ adduct\ (a)$$

1:2 adduct (b)

16. ADDITION OF CYANATES

These are probably 1,3-dipolar addition reactions and are a general route to isoxazoles.

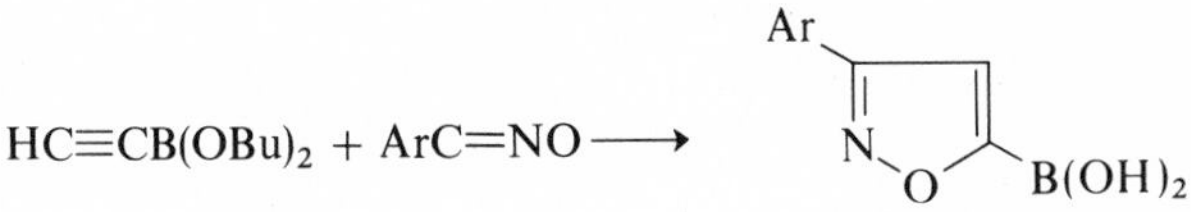

$$HC \equiv CB(OBu)_2 + ArC \equiv NO \longrightarrow$$

(60–74% yield)

The reaction of phenyl cyanate (a nitrile oxide) with phenylacetylene gives 3,5-diphenylisoxazole:

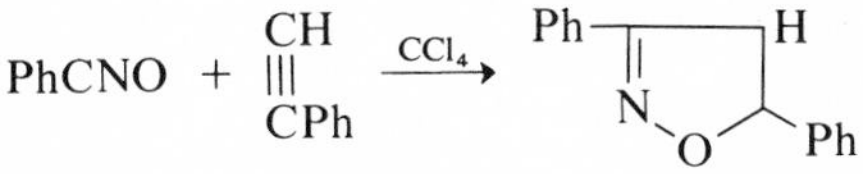

The reaction is second order in CCl_4 solution. The effect on rate of substituents in both the nitrile oxide and the phenyl group of phenylacetylene tends to confirm the 1,3-dipolar nature of the reaction.[70a]

17. ADDITION OF ORGANOSILANES

Olefins tend to isomerize during the platinum-catalyzed addition of silicon hydrides. 2-Butyne does not isomerize. Methyldichlorosilane adds *cis* to the triple bond to form *cis*-2-methyldichlorosilylbutene-2:[71]

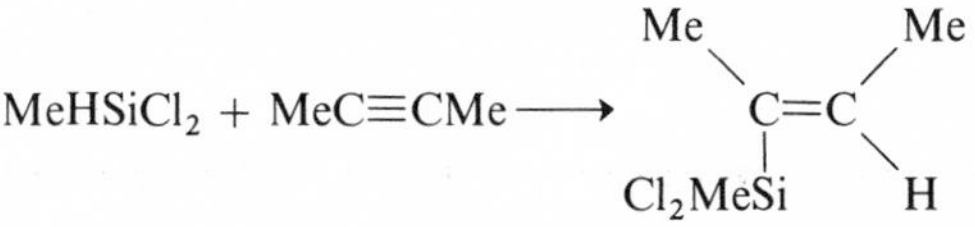

In 1958, Benkeser[72] reviewed the literature on addition of silicochloroform to acetylenes. Peroxides give radical chain reactions, but the reaction catalyzed by Pt-charcoal is probably ionic. Peroxide catalysts give *cis* products by *trans* addition to alkynes and phenylacetylene:

$$RC \equiv CH + HSiCl_3 \xrightarrow[\text{addition}]{\text{radical}} cis\text{-}RCH{=}CHSiCl_3$$

Platinum-catalyzed addition gives *trans* products by *cis* addition. Benkeser obtained 73–93% yields in platinum-catalyzed reactions, but the peroxide reaction gives only 36–47% yields. Formation of *trans* products over Pt catalyst indicates that the Si—H bond is flat on the catalyst surface and is polarized. The polarized bond can add to acetylene from one side only.

When the ratio of trichlorosilane to acetylene is 2, long reaction times give diaddition products. 1,6-Bis(trichlorosilyl)hexane can be obtained as 82% of the total product, along with 15% of the 1,2-isomer. By using more catalyst and shorter reflux time in isopropanol, the 1,2-isomer can be obtained as 40% of the total product. Total yields are 80–90%.[72a]

Trialkylsilanes add exothermically to acetylenic carbinols in the presence of chloroplatinic acid to form trialkylsilyl ethylenic alcohols.[73] Silicon adds to terminal carbon of 1-alkynes and R adds to carbon 2 of the triple bond. Methyldichlorosilane adds only to the triple bond of enynes.[74] Silanes add to alkynylsilanes to form 1,1-disilylalkenes if peroxide is the catalyst, but they form 1,2-disilylalkenes if platinum is the catalyst.[75] Acetylene and dichloromethylsilane react over Pd-alumina at 150° to give silylethylene, butadiene and other products of dimerization.[77]

18. ADDITION OF ORGANOGERMANIUM HYDRIDES

Propargyl chloride and organogermanium hydrides react by three different routes: (a) addition of R_3GeH to form linear and branched olefinic products; (b) reduction through substitution of chlorine by hydrogen from the hydride; (c) substitution of the chlorine by germanium.[78] Which reaction predominates depends on the organogermanium hydride and on the conditions. Addition (a) across the triple bond is usually the major reaction (Table 2-22).

Platinum-catalyzed addition of triethylgermanium hydride to acetylenic glycols gives triethylgermanium-substituted ethylenic glycols in 15–40% yields.[80] Chloroplatinic acid-catalyzed addition of tributylgermanium hydride

TABLE 2-22
Reaction of Organogermanium Hydrides with Propargyl Chloride[78]

Hydride	Temp. (°C)	Conditions Catalyst	Solvent	% Yield from Reaction (a)	(b)	(c)
Et_3GeH	25	H_2PtCl_6	—	65	35	0
	80	H_2PtCl_6	CH_3CN	20	80	0
$Et_2(Cl)GeH$	150	—	—	80	20	0
	80	AIBN · or UV	—	85	15	0
	80	—	CH_3CN	45	55	0
$Et(Cl_2)GeH$	25	—	—	90	~ 0	10
	80	—	CH_3CN	75	0	25
Cl_3GeH^a	25	—	Ether	0	0	~ 90

[a] In refluxing ether, both substitution and diaddition occur to form $(Cl_3Ge)_2CHCH_2CH_2GeGl_3$.[79]

to terminal acetylenes gives anti-Markownikoff addition. The triple bond in isopropenylacetylene reacts preferentially.[81]

19. ADDITION OF ORGANOLEAD HYDRIDES

Tri-n-butyllead hydride is more stable than lower alkylleads and is best for synthetic work. It adds to acrylonitrile, acrylates and phenylacetylene. The exothermic reaction with phenylacetylene gives $PhCH{=}CHPb(Bu)_3$.[82]

20. ADDITION OF ORGANOTIN HYDRIDES AND ORGANOARSENIDES

Acetylenic bonds add organotin hydrides faster than olefinic bonds do, and activation of the triple bond by negative groups is unnecessary.[83] Diaddition gives symmetrical bis(trisubstitutedstannyl)ethanes. Triphenyltin hydride adds to acetylenes with small substituted groups to give both *cis* and *trans* products, but it adds to acetylenes with large substituent groups to give *trans* products only. Propargyl alcohol gives both isomers, but phenylacetylene gives only the *trans* product. Yields are usually 40–80%. The catalysts used for organosilanes are ineffective, as are Lewis acids and alkalies.

Lithium diphenylarsenide adds to diphenylacetylene to form the vinylarsine. If a secondary amine is present in the tetrahydrofuran solution, the product is *cis*. If a primary amine is present, the product is *trans*:[83a]

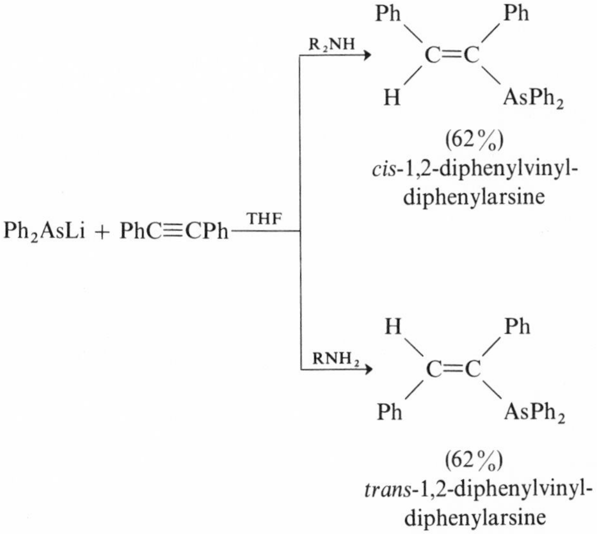

$$Ph_2AsLi + PhC{\equiv}CPh \xrightarrow{THF}$$

R$_2$NH →

Ph   Ph
  \   /
   C=C
  /   \
H   AsPh$_2$

(62%)
cis-1,2-diphenylvinyl-
diphenylarsine

RNH$_2$ →

H   Ph
  \   /
   C=C
  /   \
Ph   AsPh$_2$

(62%)
trans-1,2-diphenylvinyl-
diphenylarsine

However, phenylacetylene gives *cis* products in the presence of either primary or secondary amines:

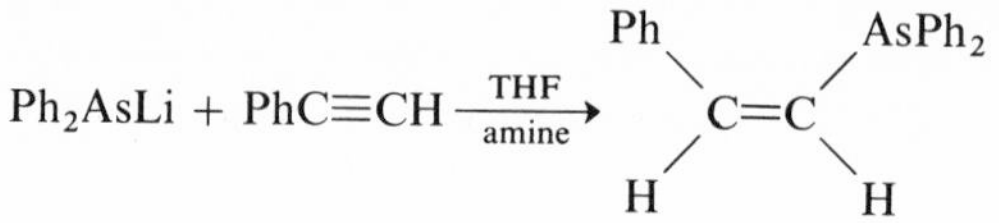

$$Ph_2AsLi + PhC{\equiv}CH \xrightarrow[\text{amine}]{\text{THF}}$$

$$\begin{array}{c} Ph \\ \diagdown \\ \end{array} C{=}C \begin{array}{c} AsPh_2 \\ \diagup \\ \end{array}$$

(67%)

cis-β-styryldiphenyl-
arsine

21. ADDITION OF METAL HALIDES

Disubstituted boron halides, R_2BX, add across triple bonds to give $R_2BCH{=}CHX$. Diaddition gives symmetrical products.[84-88] Yields are usually high. Selenium tetrachloride[89,90] and tellerium tetrachloride also add.[89] Germanium iodide at 220° with tolan gives hexaphenylbenzene and

$$PhC\overset{\displaystyle GeI_2}{=\!\!=\!\!=}CPh.$$

[91] Antimony chloride adds to give *cis*- and *trans*-tris-β-chlorovinylstibines.[92] Lewisite is made by adding arsenic trichloride across acetylene in the presence of aluminum chloride or mercuric chloride. The product is $ClCH{=}CHAsCl_2$.[93]

22. ADDITION OF METAL SALTS

Tolan reacts with mercuric acetate in acetic acid at 110° to form

$$PhC\overset{\displaystyle OAc \quad HgOAc}{=\!\!=\!\!=}CPh.$$

[94] Mercuric chloride in methanolic lithium chloride adds as a nucleophile to triple bonds, via a π complex.[95] Dimsylsodium (from sodium and dimethyl sulfoxide) adds across triple bonds to form $RCH{=}C(CH_2\overset{\displaystyle O}{\overset{\|}{S}}Me)R'$.[96] The product is mostly *trans*. Nickel sulfide and tolan at 160° in toluene give tetraphenylthiophene and an unusual nickel complex:[97]

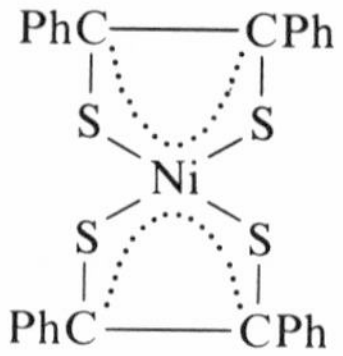

Lithium aroyltricarbonylnickelates add to acetylene and to acetylenic hydrocarbons to form 1,4-diketones.[97a]

$$2Li[Ar\overset{\overset{\displaystyle O}{\|}}{C}-Ni(CO)_3] + RC\equiv CH \xrightarrow[\text{(2) H}^+]{\text{(1) } -70°, \text{ ether}} Ar\overset{\overset{\displaystyle O}{\|}}{C}CHRCH_2\overset{\overset{\displaystyle O}{\|}}{C}Ar$$
$$25–75\% \text{ yield}$$

This is a reasonable synthesis for a variety of 1,4-diketones.

23. ADDITION OF LITHIUM ALKYLS

23.1. Intermolecular Addition

In hydrocarbon solution, lithium alkyls have the unusual ability to polymerize isoprene to all-*cis*-1,4-polymer. The covalent nature of lithium is obviously important, and polymer growth must proceed through a series of concerted *cis* additions. In basic solvents, the polymer is random. In 1966, Mulvaney[98,99] noted that the literature contains little about addition of alkali metal alkyls to "unactivated" acetylenes. The magnesium in Grignard reagents may coordinate with triple bonds. Tsutsui[100] refluxed phenylmagnesium bromide with tolan in xylene-tetrahydrofuran. Hexaphenylbenzene and octaphenyl-cyclooctatetraene were the products. Butyllithium does not react with tolan in pentane, but in ethyl ether solution, addition does occur. Butyllithium also metalates the *ortho* position of one of the benzene rings:[99,100]

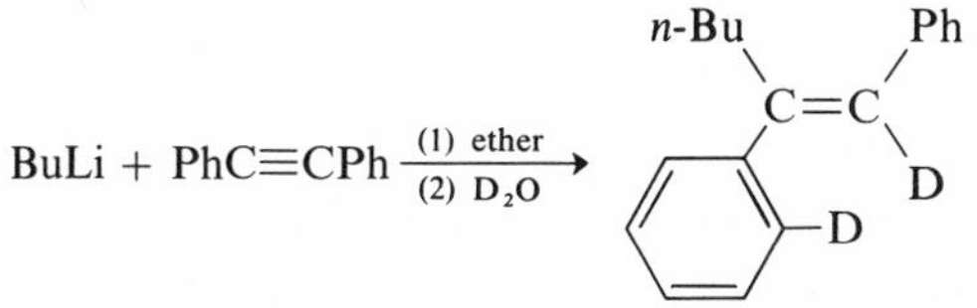

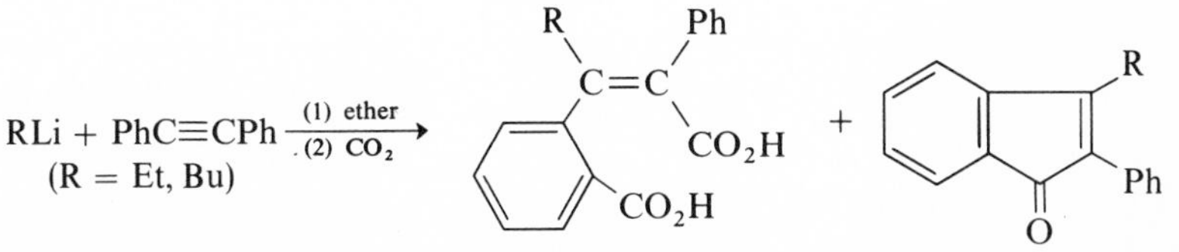

Methyllithium does not react. Phenyllithium adds but does not metalate the benzene ring of tolan; the product is triphenyl acrylic acid (11% yield), as Eisch reported earlier.[101]

23.2. Intramolecular Addition

The 1- and 8-positions in naphthalene are well suited for the study of inter-actions in a system in which a σ-bonded metal atom of an organometallic is in position to interact with a triple bond in the same molecule. Dessy and Kandil[102] prepared 8-chloromercuri-9-ethynylnaphthalene, but the ultraviolet spectra did not indicate whether Hg-triple bond interaction occurred. Polarographic half-wave potentials indicated weak interactions. Since the mercury should be deep inside the π system of the triple bond, strong interaction was expected.

Lithium and the triple bond do interact in this system. The reaction (below) is the first example of intramolecular addition of lithium to a triple bond:

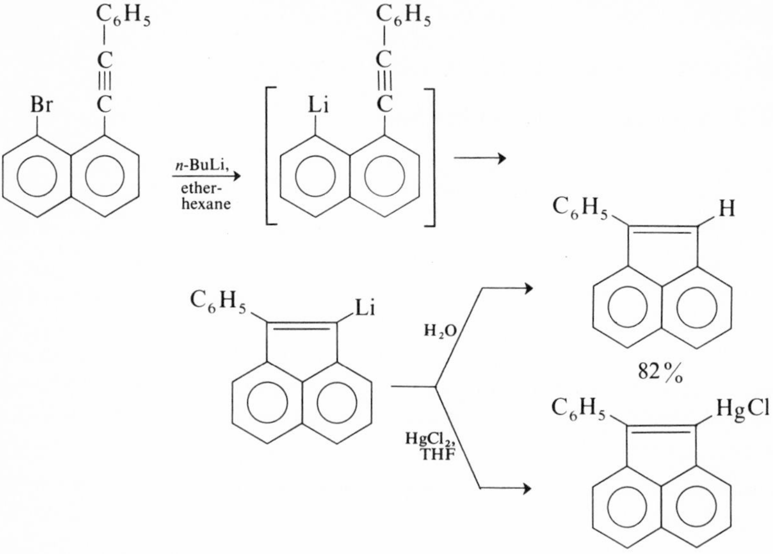

The phenylacenaphenyl anion is apparently more stable than the 1-phenylethynylnaphthyl anion, and this is the driving force for the reaction.

Kandil and Dessy[103] worked with three spatial arrangements of triple bonds and metals in the same molecule:

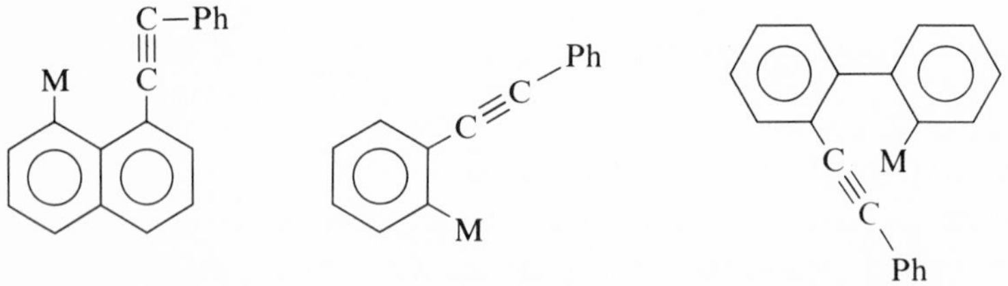

(1) 0° between groups   (2) Diverging 60°   (3) Converging 60°

Arrangement (1) gives acenaphthenes when lithium is the metal[102] (above). The chloromercuri derivative gives the same acenaphthene on electrolysis (ring closure via carbanion). Radical ring closure and dimerization occur when the magnesium derivative is coupled via the copper derivative (radical interaction).

Arrangement (2), diverging 60° substituents, gives no evidence of interaction, but arrangement (3), with converging 60° groups, does. Both lithium and magnesium derivatives give alkylidenefluoroenes, in a new synthesis of these compounds:

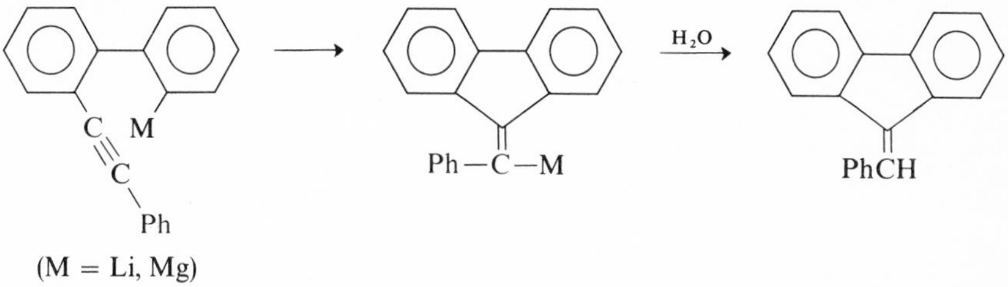

(M = Li, Mg)

The converging 60° arrangement with two phenylethynyl groups is also possible in 2,2′-bis(phenylethynyl)biphenyl, and this compound shows some triple bond-triple bond interaction:

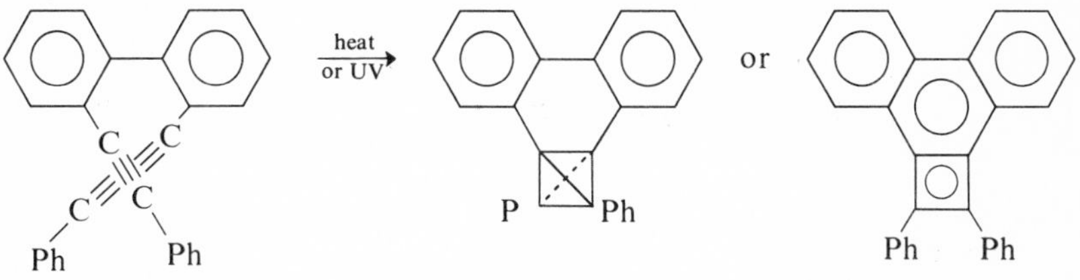

24. ADDITION OF ALUMINUM ALKYLS

Wilke[104] first reported the very easy reaction of $AlEt_3$ with acetylene to form 1-butenyldiethylaluminum by insertion of acetylene. The Al—H bond of dialkylaluminum hydrides adds across triple bonds. 1-Hexyne and diethylaluminum hydride give hexenyldiethylaluminum, for example. Dialkylaluminum hydrides can add *cis* or *trans* to unsymmetrically disubstituted acetylenes to give four products. Sometimes addition is all *cis*.[105,106] Excess R_2AlH and long times can cause *trans* addition. *Cis-trans* isomerization goes via diaddition-elimination, as it does in hydroboration reactions.[107] Al—C bonds can add to acetylenes, so the product vinylaluminum dialkyl can add to another acetylene to give two different butadienylaluminum dialkyls.[101] The direction of addition depends more on electronic factors than on steric factors. Eisch[108]reported a detailed study of the isobutylaluminum hydride-phenylpropyne reaction to determine the direction of addition. Aluminum adds mostly to the acetylenic carbon alpha to phenyl. The butadienylaluminum alkyls are intermediates to the unsymmetrical benzenes formed by cyclization:

$$R-C{\equiv}C-Ph + (i\text{-}Bu)_2AlH \longrightarrow \left[\begin{array}{c} R-C{\equiv}C-Ph \\ \downarrow \\ H\dot{A}l-R' \\ | \\ R' \end{array}\right] \longrightarrow \left[\begin{array}{c} R-\overset{+}{C}{=}C-Ph \\ \vdots \\ H\cdots\overset{-}{Al}-R' \\ | \\ R' \end{array}\right] \underset{100°}{\overset{50°}{\rightleftarrows}}$$

$$\begin{array}{c} R \quad\quad Ph \\ \diagdown \ \diagup \\ C{=}C \\ \diagup \ \diagdown \\ H \quad Al(i\text{-}Bu)_2 \\ (1) \end{array} \quad \xrightarrow[\text{slow}]{R-C{\equiv}C-H} \quad \begin{array}{c} Ph \diagdown \quad \diagup R \\ C-C \\ R-C \quad\quad C-Ph \\ \diagup \quad\quad \diagdown \\ H \quad Al(i\text{-}Bu)_2 \end{array}$$

$$\begin{array}{c} R \quad\quad Ph \\ \diagdown \ \diagup \\ C{=}C \\ \diagup \ \diagdown \\ Al(i\text{-}Bu)_2 \quad H \\ (1a) \end{array} \quad \xrightarrow{D_2O} \quad \begin{array}{c} R \quad\quad Ph \\ \diagdown \ \diagup \\ C{=}C \\ \diagup \ \diagdown \\ H \quad\quad D \\ \text{(high yield)} \end{array}$$

+

$$\begin{array}{c} R \quad\quad Ph \\ \diagdown \ \diagup \\ C{=}C \\ \diagup \ \diagdown \\ D \quad\quad H \\ \text{(low yield)} \end{array}$$

$$\begin{array}{c} R \\ Ph \diagdown \quad\quad \diagup R \\ \diagdown \quad\quad \diagup \\ \diagup \quad\quad \diagdown \\ Ph \diagup \quad\quad \diagdown Ph \\ R \end{array} \quad \xleftarrow[R-C{\equiv}C-Ph]{150°} \quad \begin{array}{c} R \diagdown \quad \diagup R \\ C-C \\ Ph-C \quad\quad C-Ph \\ \diagup \quad\quad \diagdown \\ H \quad Al(i\text{-}Bu)_2 \end{array} \quad \xleftarrow[\text{fast}]{R-C{\equiv}C-Ph}$$

(high yield if small
amount of R_2AlH
is used)

When diisobutylaluminum hydride adds to terminal alkynes, the product is usually *trans*-vinylalane, while addition to nonterminal alkynes usually gives *cis*-vinylalanes.[103a] The reaction of halogens with *trans*-vinylalanes at $-50°$ in tetrahydrofuran is an excellent synthesis of *trans*-1-halo-1-alkenes. The same reaction applied to the *cis*-vinylalanes (from nonterminal acetylenes) gives *cis*-n-halo-n-alkenes ($n = 3$ if 3-hexyne is the starting alkyne). The *cis*- and *trans*-vinylalanes in the presence of alkyllithium may form the lithium vinylate, which reacts well with CO_2 to form olefinic acids, and with aldehydes to form carbinols. Thus, the alanates can react much like Grignard reagents.

25. ADDITION OF BORON HYDRIDES

25.1. Monohydroboration

Addition of 1 mole of boron hydride to an acetylenic bond is a method of homogeneous hydrogenation, and is discussed in Chapter 2, Part 1.

25.2. Dihydroboration

It is generally agreed that 2 moles of boron hydride will add to triple bonds, but here clear-cut agreement stops. The unresolved question is: Where do the two borons add? Glycol products are more easily explained by symmetrical addition (one B to each carbon of the acetylenic bond), while aldehydes and alcohols are more easily explained by unsymmetrical addition. The glycols, aldehydes and alcohols are formed by alkaline hydrogen peroxide oxidation of the boron compounds. Brown's book on hydroboration discusses the possibilities as of 1962.[109] In 1964, Pasto[107] reported some careful studies of hydroborations undertaken to try to clarify the confusion over where the boron atoms add. He dihydroborated phenylacetylene and tolan in diglyme, and oxidized the adducts with alkaline hydrogen peroxide. Phenylacetylene gives benzaldehyde, benzyl alcohol, acetophenone, phenylacetaldehyde, 1-phenylethanol, 2-phenylethanol, phenylethanediol and styrene. Pasto's conclusions are: (1) Olefin is not formed by elimination of some B-B compound from a *vic*-diboro intermediate, as Hassner[110] suggested. Olefin is not present before hydrolysis or deuterolysis. *cis*-Olefin forms, not *trans* as expected from B-B elimination. (2) Olefin does not form by hydrolysis of a vinyl carbon-boron bond, since no deuterium is incorporated in the olefin. (3) Monoalcohol is not formed exclusively by hydrolysis of *gem*-diboro compound, followed by oxidation, or by hydroboration of an intermediate with a carbon-boron double bond, or by reduction of some intermediate carbonyl compound. (4) A common intermediate goes to olefin and to alcohol.

From these results, Pasto suggested the *vic*-diboro compound formed during dihydroboration can undergo anion formation through attack by base on boron, from the front or from the back. The boron anion gives olefin plus D_2 on deuterolysis:

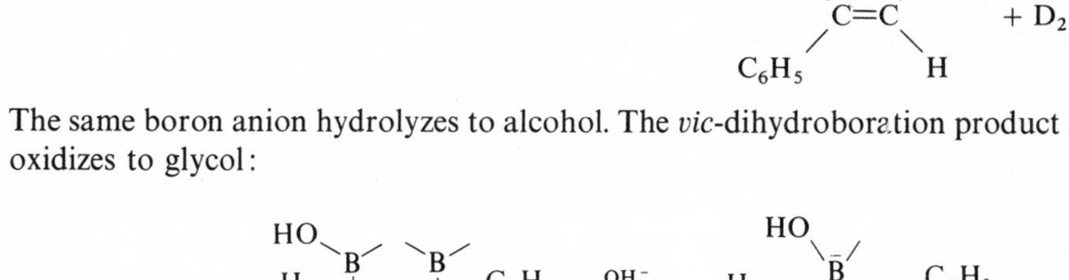

The same boron anion hydrolyzes to alcohol. The *vic*-dihydroboration product oxidizes to glycol:

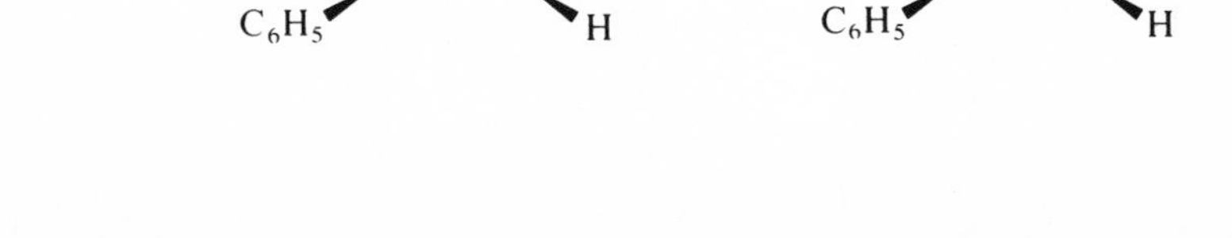

Thus, *vic*-diboro compound is the precursor to all products. Base hydrolyzes the *vic*-diboro compound very rapidly, indicating that the neighboring borons participate during hydrolysis. The *vic*-diboro compound forms a bridged monoboro compound, which deuterolyzes to another monoboro derivative, which in turn oxidizes to alcohol. It can also eliminate to form olefin.

The disagreements are not yet fully resolved. In 1967, Zweifel and Arzoumanian[111] published results which indicate that 1-alkynes hydroborate on the terminal carbon almost exclusively. When they hydroborated 1-hexyne with excess diborane and oxidized the polymeric dihydroborated product, they obtained 80% 1-hexanol and 10–12% 1,2-hexanediol. Deuteroboration of 1-hexyne and alkaline hydrogen peroxide oxidation of the product confirms that 1-hexanol is formed from the 1,1-diboro compound:

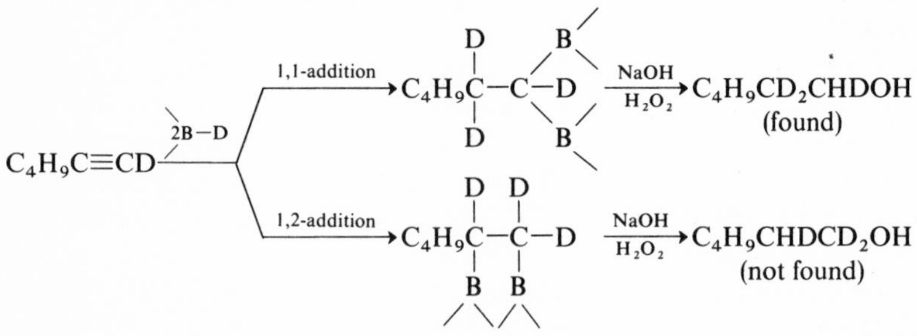

The 1,1-diboro compound is converted to carboxylic acid by oxidation with a peracid under anhydrous conditions. 2,3-Dimethyl-2-butylborane and dicyclohexylborane add to the terminal carbon of 1-alkynes to give the 1,1-diboro compounds in 90–96% yield.

With this report, the variety of products which can be made from acetylenes by hydroboration is even more impressive. These are valuable synthetic methods, and the yields are good:

(1) Internal acetylenes: *cis*-Olefins or ketones.

(2) Terminal acetylenes: Olefins, alcohols, aldehydes, carboxylic acids, and to a lesser extent, glycols.

(3) Terminal acetylenes to higher olefins: The product from dicyclohexylborane and terminal alkyne is reacted with BuLi, and aldehyde or ketone is added. This is a general reaction:[112]

$$RC \equiv CH \xrightarrow[\text{(2) R'CHO}]{\text{(1) } (C_6H_{12})_2BH,\ BuLi} RCH_2-CH=CH-R'$$

1-Alkynes and alkenynes react well with decaborane to give a carborane:

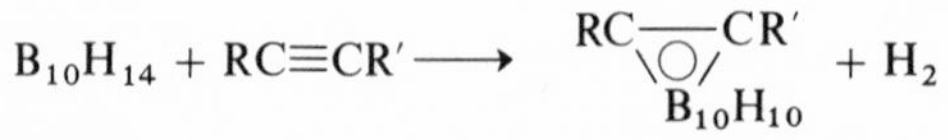

Yields are 23–77%.[113] Alcohol and acid groups in the acetylene destroy the borane, but halogen, ester and alkylamino groups do not interfere. Solutions of decarborane in dioxane are shock sensitive.[113]

26. DIELS-ALDER ADDITIONS OF ACETYLENIC HYDROCARBONS

Acetylenic hydrocarbons are not as strong dienophiles as negatively substituted acetylenes are. Acetylene reacts with cyclopentadiene at 150–400° to give 60% yield of bicyclo[2.2.1]-hepta-2,5-diene.[114,115] Acetylene also reacts with hexachlorocyclopentadiene to give 50% yield of the hexachloro derivative.[116]

Phenylacetylene reacts with tetraphenylcyclopentadienone to liberate CO and form pentaphenylbenzene derivatives.[117] Tolan gives 84% yield of hexaphenylbenzene.[120]

Acetylene and anthracene at 250° and 52 atmospheres (in toluene) give 9,10-dihydro-9,10-ethenoanthracene in 67% yield:[118]

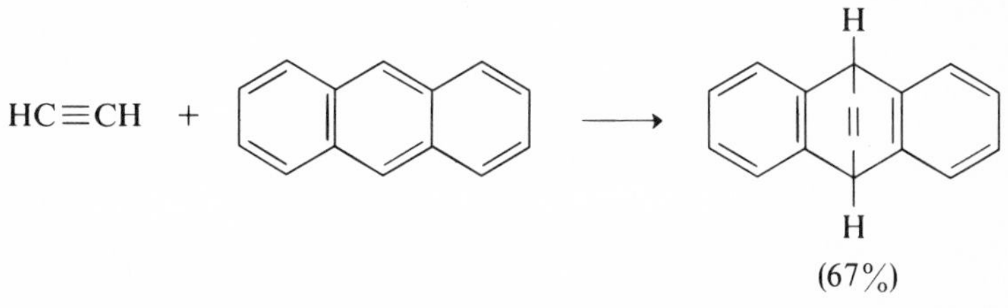

(67%)

Phenylacetylene, propyne and ethyl propiolate react with ethyl 1,3-cyclohexadiene-1-carboxylate to form bicyclic intermediates which aromatize to form substituted benzenes:[119]

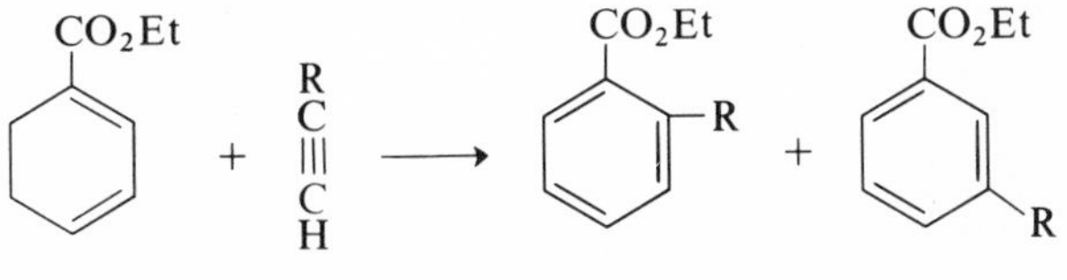

27. INTRAMOLECULAR ADDITION-CYCLIZATION REACTIONS OF PROPARGYL COMPOUNDS

Propynyl (propargyl) derivatives are widely used because the starting alcohols and halides are easy to make or can be purchased commercially. The halides react with amines to form propargylamines, and substituted propargylamines undergo many cyclization reactions which are useful synthetically.

Dillard and Easton[121] condensed 2-propynylamines with malonic esters. The products hydrolyze and dehydrate with acid or base in a general method for making 3-substituted 3-pyrrolin-2-ones:

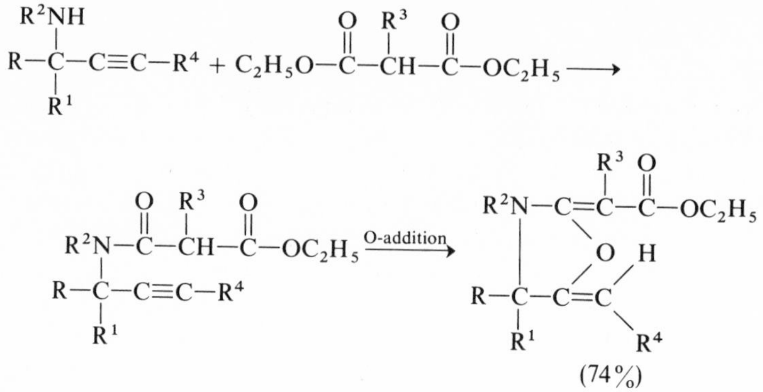

$$R-\underset{\underset{R^1}{|}}{\overset{\overset{R^2NH}{|}}{C}}-C{\equiv}C-R^4 + C_2H_5O-\overset{\overset{O}{\|}}{C}-\underset{\underset{}{\overset{R^3}{|}}}{CH}-\overset{\overset{O}{\|}}{C}-OC_2H_5 \longrightarrow$$

$$\underset{R-\underset{\underset{R^1}{|}}{\overset{}{C}}-C{\equiv}C-R^4}{R^2N-\overset{\overset{O}{\|}}{C}-\underset{\overset{R^3}{|}}{CH}-\overset{\overset{O}{\|}}{C}-OC_2H_5} \xrightarrow{\text{O-addition}}$$

(74%)

(a) $R = R^1 = R^2 = CH_3$; $R^3 = R^4 = H$
(b) $R = R^1 = R^2 = CH_3$; $R^3 = C_2H_5$; $R^4 = H$
(c) $R = R^1 = CH_3$; $R^2 = R^3 = R^4 = H$

3-Substituted N-(β-hydroxyethyl)-1,1-dialkyl-2-propynylamines cyclize in the presence of KOH to form either 1,4-oxazepines or morpholines, depending on the 3-substituent:[122]

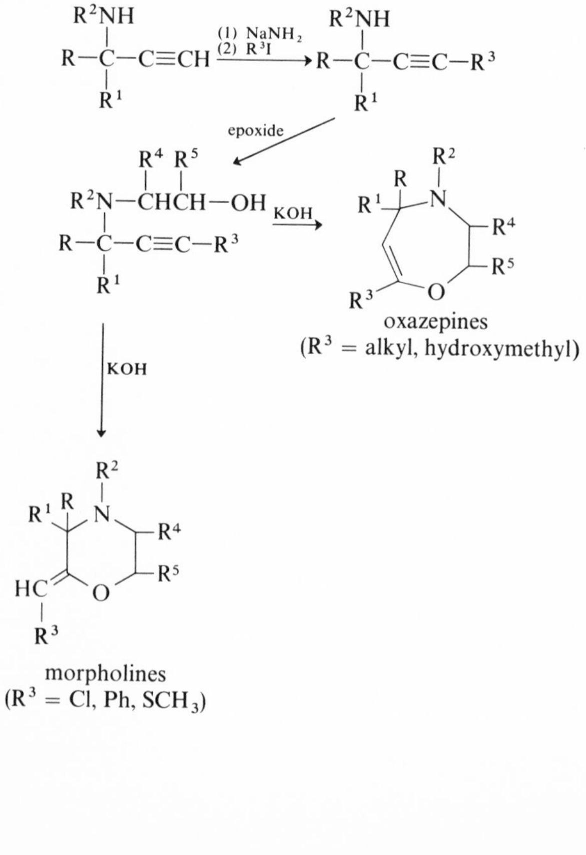

$$R-\underset{\underset{R^1}{|}}{\overset{\overset{R^2NH}{|}}{C}}-C{\equiv}CH \xrightarrow[\text{(2) } R^3I]{\text{(1) NaNH}_2} R-\underset{\underset{R^1}{|}}{\overset{\overset{R^2NH}{|}}{C}}-C{\equiv}C-R^3$$

oxazepines
(R^3 = alkyl, hydroxymethyl)

morpholines
(R^3 = Cl, Ph, SCH$_3$)

In an earlier study of the cyclization of N-alkyl-N-propargylethanolamine [(1) below] in the presence of sodium hydroxide or potassium hydroxide, results showed that the direction of reaction is dependent on solvent.[123] In water, the only cyclic product is 2-methylenemorpholine (2). In aprotic toluene, dimethyl sulfoxide or ether, the major or sole product is the 2-vinyloxazolidine (3). 2-Methylenemorpholine forms by nucleophilic addition to acetylenic carbon, and 2-vinyloxazolidine forms by prototropic rearrangement to the allenic amino alcohol followed by nucleophilic addition at carbon 1 of the allene group. The prototropic rearrangement apparently goes faster in aprotic solvents than in water:

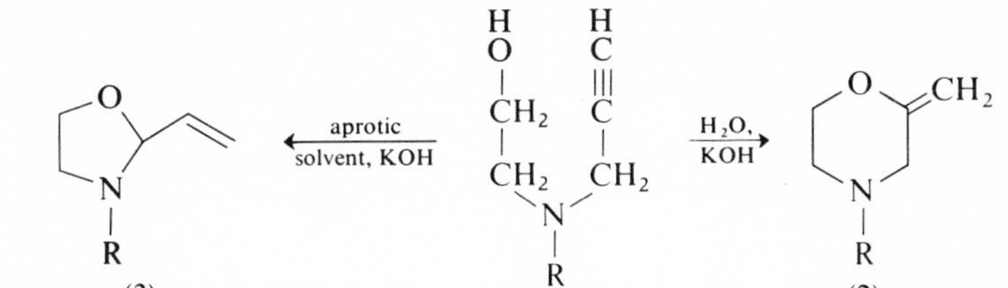
O
N
R
aprotic
solvent, KOH
H
O
CH2
CH2
H
C
C
CH2
N
R
H2O,
KOH
O
CH2
N
R

(3) 2-vinyloxazolidine (1) (2) morpholine

Two other cyclizations of N-alkyl-N-propargylethanolamines and N-alkyl-N-(2-haloallyl)ethanolamines are known. N-(2-hydroxyethyl)-N-4-dimethyl-4-amino-2-pentyne gives the seven-membered ring product 4,5,5,7-tetramethyl-2,3,4,5-tetrahydro-1,4-oxazepine when treated with KOH in boiling toluene or xylene.[124,125] The alkoxide of N-*t*-butyl-N-(2-chloroallyl)-1-amino-2-methyl-2-propanol in ether gives the corresponding 2-vinyloxazolidine and a product tentatively identified as the isomeric 5,6-dihydro-1,4-oxazine.[123]

Propargyloxyethanol (4) reacts with base to give different products:

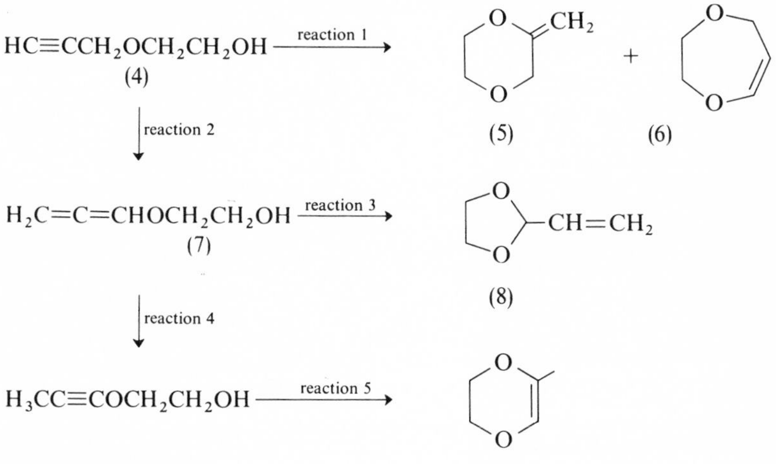
HC≡CCH2OCH2CH2OH
reaction 1
O
CH2
O
+
O
O
reaction 2
H2C=C=CHOCH2CH2OH
reaction 3
O
CH=CH2
O
reaction 4
H3CC≡COCH2CH2OH
reaction 5
O
O

(4) (5) (6) (7) (8) (9)

In water, the main reaction is 1, to form (5) and (6) in nearly equal amounts. Reaction 2 competes with reaction 1. Prototropic rearrangement of (4) forms allenyloxyethanol (7) which in water reacts (by reaction 3) to form 2-vinyl-1,3-dioxolane (8). In aprotic solvents, the rates of the prototropic rearrangements (reactions 2 and 4) are increased relative to the rates of the ring closing reactions 1 and 3, and the main products become (8) from reaction 3 and 2-methyl-1,4-dioxane (9) from reaction 5.

By analogy with the alkyl-N-propargylethanolamine reaction, formation of (5) is expected. The formation of the seven-numbered ring (6) is not expected, since no cyclization of a propargylaminoethanol to a seven-membered ring has ever been noted. The only reasonable explanation for the formation of (8) is intramolecular nucleophilic addition of alkoxide to the internal allenic carbon of allenyloxyethanol, $H_2C{=}C{=}CHOCH_2OH(7)$, formed by the base-induced prototropic rearrangement of the haloallenyl ether.

The decreased importance of the five-membered ring product when $-O-$

$$\overset{R}{\underset{|}{}}$$

is substituted for $-N-$ is caused by an increased rate of prototropic rearrangement and a decreased rate of cyclization to the intermediate allene. Most or all of (9) is formed by cyclization of 1-propynyl ether, probably via a cyclic transition state.[126]

The study of base-induced cyclization reactions of the hydroxyethyl ethers was more informative regarding the scope and mechanisms than the earlier study of the amino compounds.[123] The direction of ring closure is determined not only by the nature of the solvent, but also by a delicate balance of electronic and steric factors.

Ring closure occurs when certain propynylureas are treated with phosphorus pentachloride.[127] The reaction in refluxing benzene for 4 hours gives 6–70% yields of imidazolones:

t-Acetylenic carbinols react with isocyanates to give urethanes which cyclize to oxazolidinones when treated with sodium ethoxide.[128] The reaction is an intramolecular addition of N—H across the acetylenic bond. *t*-Acetylenic ureas also cyclize by N—H addition, to give imidazolidinones:[128a]

but HCl causes O-addition to form 2-iminooxazolidines:

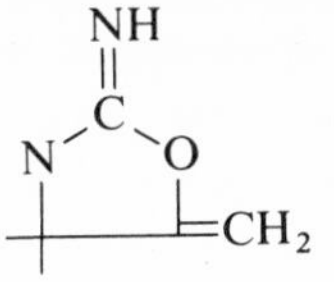

t-Acetylenic urethanes with HCl also give O-closure to form dioxolanes:

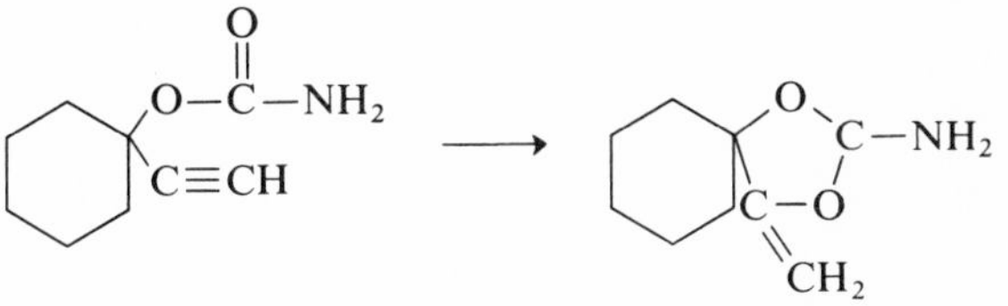

Hennion and diGiovanni[129] found that a new intramolecular coupling reaction occurred during reduction of crowded bis(propynyl)amines with sodium in ammonia:

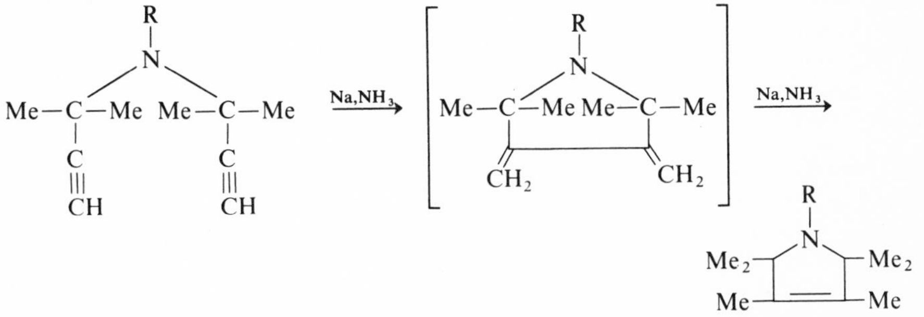

With bis(propargyl)amines, $RN(CH_2C\equiv CH)_2$, two products form: acyclic $RN(CH_2CH=CH_2)_2$, and cyclic RN⟨ring⟩. If R is *t*-butyl, the cyclic product is major, but in most other cases the acyclic product is major.[130]

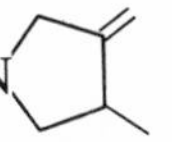

28. CYCLIZATION TO SMALL RINGS DURING SOLVOLYSIS
of β-ACETYLENIC ALCOHOL DERIVATIVES

Tosylates of β-acetylenic alcohols undergo fairly rapid solvolysis in formic acid to give small yields of alkylcyclobutanones and acylcyclopropanes:[131]

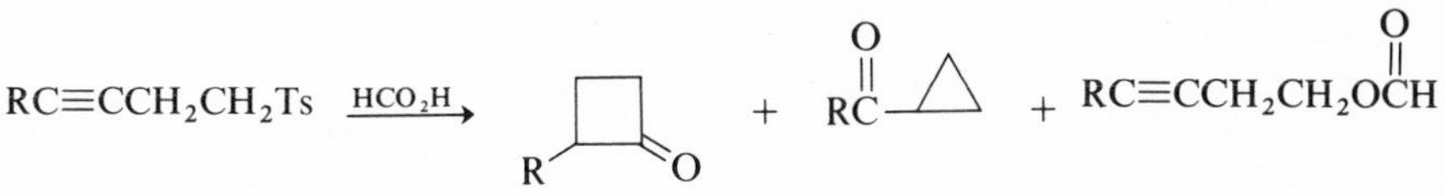

   m-Nitrobenzenesulfonates and 3,5-dinitrobenzenesulfonates are solvolyzed
slowly in trifluoroacetic acid to give nearly quantitative yields of alkylcyclobutanones. In the presence of mercuric acetate catalyst, the course of the
reaction is altered to give mainly the acylcyclopropanes. The reaction may be
represented in two ways: (1) solvolysis followed by cyclization and addition of
the acid anion, and (2) addition of the acid anion first to form a vinyl ester:

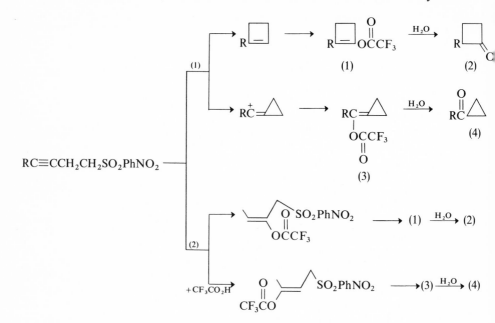

$RC{\equiv}CCH_2CH_2SO_2PhNO_2$ ———

References

1. Petrov, A. A., and Porfir'eva, Yu. I., *Dokl. Akad. Nauk. SSR*, **111**, 839 (1956); *Chem. Abstr.*, **51**, 9469 (1957).
2. Poutsma, M. L., and Kartch, J. L., *Tetrahedron*, **22**, 2167 (1966).
2a. Merritt, R. F., *J. Org. Chem.*, **32**, 4124 (1967).
3. Castro, C. E., Gaughan, E. J., and Owsley, D. C., *J. Org. Chem.*, **30**, 587 (1965).
4. Whitlock, H. W., Jr., and Sandvick, P. E., *J. Am. Chem. Soc.*, **88**, 4525 (1966).
5. Verbanc, J. J., and Hennion, G. F., *J. Am. Chem. Soc.*, **60**, 1711 (1938).
6. Baum, A. A., Vogt, R. R., and Hennion, G. F., *J. Am. Chem. Soc.*, **61**, 1458 (1939).
7. Reed, S. F., Jr., *J. Org. Chem.*, **30**, 2195 (1965).
8. Goldschmidt, S., Endres, R., and Dirsch, R., *Chem. Ber.*, **58**, 572 (1925).
9. Walling, C., Heaton, La. D., and Tanner, D. D., *J. Am. Chem. Soc.*, **87**, 1715 (1965).
10. Ried, W., and Saurez-Rivero, E., *Chem. Ber.*, **96**, 1475 (1963).
11. Neale, R. S., *J. Am. Chem. Soc.*, **86**, 5340 (1964).
11a. Grimwood, B. E., and Swern, D., *J. Org. Chem.*, **32**, 3665 (1967).
12. Herbertz, T., *Chem. Ber.*, **92**, 541 (1959).

13. Drehfahl, G., and Zimmer, C., *Chem. Ber.*, **93**, 505 (1960).
13a. Fahey, R. C., and Lee, D.-J., *J. Am. Chem. Soc.*, **89**, 2780 (1967); **90**, 2124 (1968).
14. Owai, I., Tomita, K., and Ide, J., *Chem. Pharm. Bull. (Tokyo)* (English), **13**, 118 (1965); *Chem. Abstr.*, **62**, 14541 (1965).
15. Dombrovskii, A. V., and Prilutskii, G. M., *Zh. Obshch. Khim.*, **25**, 1943 (1955).
16. Gilbert, E. E., McGough, C. J., and Otto, J. A., *Ind. Eng. Chem.*, **51**, 925 (1959).
17. VEB Farbenfabrik Wolfen, German Patent 1,024,498 (Feb. 20, 1958).
18. Bindacz, L., and Balog, A., *Chem. Ber.*, **63**, 1716, 1722 (1960).
19. Chini, P., *et al.*, *Chim. Ind. (Milan)*, **45**, 701 (1963); *Chem. Abstr.*, **60**, 10523 (1964).
20. Pohland, A. E., and Benson, W. R., *Chem. Rev.*, **66**, 161 (1966).
21. Truce, W. E., and Simms, J. A., *J. Am. Chem. Soc.*, **78**, 2756 (1956).
22. Shostakovski, M. F., *et al.*, *Zh. Obshch. Khim.*, **32**, 709 (1962); *Chem. Abstr.*, **58**, 5498 (1963).
23. Maioli, L., and Modena, G., *Bull. Sci. Fac. Chim. Ind. Bologna*, **16**, 86 (1958); *Chem. Abstr.*, **53**, 7079 (1959).
24. Kateeva, L. M., *et al.*, *Zh. Obshch. Khim.*, **32**, 3965, 3699 (1962); *Chem. Abstr.*, **59**, 4071 (1963).
25. Montanari, F., *Gazz. Chim. Ital.*, **86**, 406 (1956); *Chem. Abstr.*, **52**, 8999 (1958).
26. Calo, V., *et al.*, *Tetrahedron Letters*, 4394, 4405 (1965).
27. Chierici, L., and Montanari, F., *Bull. Sci. Fac. Chim. Ind. Bologna*, **14**, 78 (1956); *Chem. Abstr.*, **51**, 5721 (1957).
28. Schulte, K. E., Reisch, J., and Walker, H., *Chem. Ber.*, **98**, 98 (1965).
29. Vasil'eva, E. D., Kotlyarevskii, I. L., and Faiershtein, Yu. M., *Izv. Akad. Nauk SSSR, Ser. Khim.*, 322 (1965); *Chem. Abstr.*, **62**, 14619 (1965).
30. Horie, S., *Nippon Kagaku Zasshi*, **78**, 1171 (1957); *Chem. Abstr.*, **54**, 5613 (1960).
31. Horie, S., *Nippon Kagaku Zasshi*, **78**, 1795 (1957); *Chem. Abstr.*, **53**, 21868 (1959).
32. Kozlov, N. S., and Koz'minzkh, O. K., *Zh. Obshch. Khim.*, **27**, 1628 (1957); *Chem. Abstr.*, **52**, 3814 (1958).
33. Kozlov, N. S., and Pinegina, L. Yu., *Zh. Obshch. Khim.*, **27**, 1965 (1957); *Chem. Abstr.*, **52**, 5413 (1958).
34. Chekulaeva, I. A., and Kondrat'eva, L. V., *Russ. Chem. Rev.*, **34**, 669 (1965) (English translation).
35. Reimlinger, H., *Chem. Ber.*, **92**, 970 (1959).
36. European Res. Assoc., *Chem. Ber.*, **92**, 970 (1959).
37. Matteson, D. R., *J. Org. Chem.*, **27**, 4293 (1962).
38. Komendantov, M. I., *Zh. Organ. Khim.*, **1**, 209 (1965); *Chem. Abstr.*, **62**, 16168 (1965).
38a. Reimlinger, H., *Chem. Ber.*, **100**, 3097 (1967).
39. Hüttel, R., *et al.*, *Chem. Ber.*, **93**, 1425 (1960).
40. Durden, J. A., Jr., Stansbury, H. A., and Catlette, W. H., *J. Chem. Eng. Data*, **9**, 228 (1964).
41. Robson, E., Tedder, J. M., and Webster, B., *J. Chem. Soc.*, 1862 (1963).
42. Huisgen, R., *et al.*, *Tetrahedron*, **17**, 3 (1962).
43. Hartzell, L. W., and Benson, F. R., *J. Am. Chem. Soc.*, **76**, 667 (1954).
44. Dimroth, O., and Fester, G., *Chem. Ber.*, **43**, 2219 (1910).
45. Birkofer, L., Ritter, A., and Uhlenbrauck, H., *Chem. Ber.*, **96**, 2750 (1963).
46. Akimova, G. S., Chistokletov, V. N., and Petrov, A. A., *Zh. Organ. Khim.*, **1**, 2077 (1965); *Chem. Abstr.*, **64**, 9713 (1966).

46a. Hermes, M. E., and Marsh, F. D., *J. Am. Chem. Soc.*, **89**, 4760 (1967).
47. Huisgen, R., and Blasche, H., *Tetrahedron Letters*, 1409 (1964).
48. D'yakonov, I. A., and Komendantov, M. I., *Zh. Obshch. Khim.*, **29**, 1749 (1959); *Chem. Abstr.*, **54**, 8723 (1960).
49. Rozantsev, G. G., Fainzil'berg, A. A., and Novikov, S. S., *Russ. Chem. Rev.*, **34**, 69 (1965) (English translation).
50. Frey, H. M., *Chem. Ind. (London)*, 1266 (1960).
51. Doering, W. von E., and Mole, T., *Tetrahedron*, **10**, 65 (1960).
52. Chandross, E. A., and Smolinsky, G., *Tetrahedron Letters*, 19 (1960).
52a. Broser, W., and Brockt, M., *Tetrahedron Letters*, 3117 (1967).
53. Obata, N., and Moritani, I., *Bull. Chem. Soc. Japan*, **39**, 1975 (1966).
54. Vol'pin, M. E., Koreshkov, Yu. D., and Kursanov, O. I., *Izv. Akad. Nauk SSSR, Otd. Khim. Nauk*, 560 (1959).
55. Kursanov, D. N., Vol'pin, M. E., and Koreshkov, Yu. D., *Zh. Obshch. Khim.*, **30**, 2877 (1960); *Chem. Abstr.*, **55**, 16473 (1961).
56. Kirmse, W., "Progress in Organic Chemistry," Vol. 6, p. 164, Washington, D.C., Butterworth's Publishing Co., 1964.
57. Parham, W. E., and Schweizer, E. E., *Org. Reactions*, **13**, 55 (1963).
58. Cohen, H. M., and Keough, A. H., *J. Org. Chem.*, **31**, 3428 (1966).
59. Meinwald, J., and Aue, D. H., *J. Am. Chem. Soc.*, **88**, 2849 (1966).
59a. Weyenberg, D. R., and Atwell, W. H., *Chem. Eng. News*, 30 (Sept. 4, 1967).
60. Harvey, R. G., and DeSombre, E. R., in "Topics in Phosphorus Chemistry," Vol. 1, p. 92, New York, Interscience Publishers, 1964.
61. Pudovik, A. N., *Zh. Obshch. Khim.*, **20**, 92 (1950).
62. Pudovik, A. N., and Aladzheva, I. M., *J. Gen. Chem. USSR*, **33**, 700, 702, 3372 (1963).
63. Boisselle, A. P., and Meinhardt, N. A., *J. Org. Chem.*, **27**, 1828 (1962).
64. Mark, V., *Tetrahedron Letters*, 281 (1962).
65. Kirillova, K. M., Kukhtin, V. A., and Sudakova, T. M., *Proc. Acad. Sci. USSR, Chem. Sect., English Transl.*, **149**, 209 (1963).
66. Griffin, C. E., and Mitchell, T. D., *J. Org. Chem.*, **30**, 1935 (1965).
67. Linn, W. J., and Benson, R. E., *J. Am. Chem. Soc.*, **87**, 3657 (1965).
68. Tyabji, A., *J. Univ. Bombay*, **10**, 110 (1942); *Chem. Abstr.*, **37**, 6494 (1943).
69. Bird, C. W., *J. Chem. Soc.*, 5762 (1965).
70. Bianchi, G., Gogoli, A., and Gruenanger, P., *Ric. Sci.*, **36**, 132 (1966); *Chem. Abstr.*, **64**, 19650 (1966).
70a. Dondoni, A., *Tetrahedron Letters*, 2397 (1967).
71. Ryan, J. W., and Speier, J. L., *J. Org. Chem.*, **31**, 2698 (1966).
72. Benkeser, R. A., *et al.*, *J. Am. Chem. Soc.*, **83**, 4385 (1961).
72a. Benkeser, R. A., *et al.*, *J. Org. Chem.*, **32**, 2634 (1967).
73. Shchukovskaya, L. L., Pal'chuk, R. I., and Petrov, A. D., *Dokl. Akad. Nauk SSSR*, **160**, 621 (1965); *Chem. Abstr.*, **62**, 14717 (1965).
74. Stadnichuk, M. D., and Petrov, A. A., *Zh. Obshch. Khim.*, **32**, 3514 (1962); *Chem. Abstr.*, **58**, 12592 (1963).
75. Shchukovskaya, L. L., Petrov, A. D., and Egorov, Yu. P., *Zh. Obshch. Khim.*, **26**, 3338 (1956); *Chem. Abstr.*, **51**, 9474 (1957).
76. Mironov, V. F., and Nepomnina, V. V., *Izv. Akad. Nauk SSSR, Otd. Khim. Nauk*, 1419 (1960); *Chem. Abstr.*, **55**, 358 (1961).

77. Shostakovskii, M. F., *et al.*, *Izv. Akad. Nauk SSSR, Otd. Khim. Nauk*, 1452 (1957); *Chem. Abstr.*, **52**, 7134 (1958).
78. Massol, M., Satge', J., and Lesbre, M., *Compt. Rend.* (*C*), **262**, 1806 (1966).
79. Mironov, V. F., and Gar, T. K., *Izv. Akad. Nauk SSSR, Ser. Khim.*, 291 (1965); *Chem. Abstr.*, **62**, 14715 (1965).
80. Gverdtsitch, I. M., and Buachidze, M. A., *Soobshch. Akad. Nauk Gruz.*, **37**, 59 (1965); *Chem. Abstr.*, **62**, 14716 (1965).
81. Lesbre, M., and Satge', J., *Compt. Rend.*, **250**, 2220 (1960).
82. Neumann, W. P., and Kühlein, K., *Chem. Eng. News*, 49 (1965).
83. van der Kerk, G. J. M., and Noltes, J. C., *J. Appl. Chem.* (*London*), **9**, 106 (1959).
84. Gipstein, E., *et al.*, *J. Org. Chem.*, **26**, 943, 2947 (1961).
85. Arnold, H. R., U.S. Patent 2,402,589 (June 5, 1946); *Chem. Abstr.*, **40**, 5769 (1946).
86. Chambers, C., and Holliday, A. K., *J. Chem. Soc.*, 3459 (1965).
87. Ceron, P., *et al.*, *J. Am. Chem. Soc.*, **81**, 6368 (1959).
87a. Aguiar, A. M., Archibald, T. G., and Kapicak, L. A., *Tetrahedron Letters*, 4447 (1967).
88. Lynds, L., and Stern, D. R., *J. Am. Chem. Soc.*, **81**, 5006 (1959).
89. Campos, M. deM., and Petragnani, N., *Tetrahedron*, **18**, 521 (1962).
90. Riley, R. F., Flato, J., and McIntyre, P., *J. Org. Chem.*, **28**, 1138 (1963).
91. Vol'pin, M. E., and Kursanov, D. N., *Zh. Obshch. Khim.*, **32**, 1455 (1962); *Chem. Abstr.*, **58**, 9111 (1963).
92. Nesmeyanov, A. N., and Borisov, A. E., *Akad. Nauk SSSR Inst. Organ. Khim. Sintezy Organ. Soedin. Sb.*, **1**, 128, 150 (1950); *Chem. Abstr.*, **47**, 8001, 8004 (1953).
93. Jarman, G. N., "Metal-Organic Compounds," *Advan. Chem. Ser.*, **23**, 328 (1959).
94. Drehfahl, G., Hueblein, G., and Wintzer, A., *Angew. Chem.*, **70**, 166 (1958).
95. Dvorko, G. F., and Shilov, E. A., *Dopovidi Akad. Nauk Ukr. RSR*, 636 (1959); *Chem. Abstr.*, **54**, 21955 (1960).
96. Iwai, I., and Ide, J., *Chem. Pharm. Bull.* (*Tokyo*), **13**, 663 (1965); *Chem. Abstr.*, **63**, 6903 (1965).
97. Schrauzer, G. N., and Mayweg, V., *J. Am. Chem. Soc.*, **84**, 3321 (1962).
97a. Sawa, Y., *et al.*, *J. Org Chem.*, **33**, 2159 (1968).
98. Mulvaney, J. E., *et al.*, *J. Am. Chem. Soc.*, **88**, 476 (1966).
99. *Ibid.*, **85**, 3897 (1963).
100. Tsutsui, M., *Chem. Ind.* (*London*), 780 (1962).
101. Easton, N. R., *et al.*, *J. Org. Chem.*, **26**, 3772 (1961).
102. Dessy, R. E., and Kandil, S. A., *J. Org. Chem.*, **30**, 3875 (1965).
103. Kandil, S. A., and Dessy, R. E., *J. Am. Chem. Soc.*, **88**, 3027 (1966).
103a. Zweifel, G., *et al.*, *J. Am. Chem. Soc.*, **89**, 2753, 2752 (1967).
104. Wilke, G., and Müller, H., *Ann. Chem.*, **629**, 222 (1960).
105. Wilke, G., and Müller, H., *Chem. Ber.*, **89**, 444 (1956).
106. Wilke, G., and Müller, H., *Ann. Chem.*, **618**, 267 (1959).
107. Pasto, D. J., *J. Am. Chem. Soc.*, **86**, 3039 (1964).
108. Eisch, J. J., and Kaska, W. C., *J. Am. Chem. Soc.*, **88**, 2213 (1966).
109. Brown, H. C., "Hydroboration," p. 227, New York, W. A. Benjamin, Inc., 1962.
110. Hassner, A., and Braun, B. H., *J. Org. Chem.*, **28**, 261 (1963).
111. Zweifel, G., and Arzoumanian, H., *J. Am. Chem. Soc.*, **89**, 291 (1967).
112. Cainelli, G., Dal Bello, G., and Zubiani, G., *Tetrahedron Letters*, 4315 (1966).

113. Heying, T. L., *et al.*, *Inorg. Chem.*, **2**, 1089 (1963).
114. Hyman, J., Freirich, E., and Lidov, R. E. (to Shell Dev. Co.), U.S. Patent 2,875,256 (Feb. 24, 1959); *Chem. Abstr.*, **53**, 13082 (1959).
115. Plate, A. F., and Pryanshnikova, M. A., *Izv. Akad. Nauk SSSR, Otd. Khim. Nauk*, 741 (1956); *Chem. Abstr.*, **51**, 1863 (1957).
116. Anderson, J., *et al.* (to Bataafse Petrol. Maatschappij, N.V.), British Patent 841,674 (July 20, 1960); *Chem. Abstr.*, **55**, 2805 (1961).
117. Ogliaruso, M. A., and Becker, E. I., *J. Org. Chem.*, **30**, 3354 (1965).
118. Muller, J.-C., and Vergne, J., *Compt. Rend.* (*C*), **263**, 1452 (1966).
119. Wynn, C. M., and Klein, P. S., *J. Org. Chem.*, **31**, 4521 (1966).
120. Fieser, L. F., *Org. Syn.*, **46**, 44 (1966).
121. Dillard, R. D., and Easton, N. R., *J. Org. Chem.*, **31**, 2580 (1966).
122. *Ibid.*, **31**, 122 (1966).
123. Bottini, A. T., Mullikan, J. A., and Morris, C. J., *J. Org. Chem.*, **29**, 373 (1964).
124. Easton, N. R., Cassady, D. R., and Dillard, R. D., *J. Org. Chem.*, **28**, 448 (1963).
125. Easton, N. R., and Dillard, R. D., *Tetrahedron Letters*, 1807 (1963).
126. Bottini, A. T., Corson, F. P., and Böttner, E. F., *J. Org. Chem.*, **30**, 2988 (1965).
127. Sinnema, Y. A., and Arens, J. F., *Rec. Trav. Chim.*, **75**, 1423 (1956).
128. Easton, N. R., Cassady, D. R., and Dillard, R. D., *J. Org. Chem.*, **27**, 2927 (1962).
128a. Easton, N. R., Cassady, D. R., and Dillard, R. D., *J. Org. Chem.*, **29**, 1851 (1964).
129. Hennion, G. F., and DiGiovanna, C. V., *J. Org. Chem.*, **30**, 2645 (1965).
130. Hennion, G. F., and Ode, R. H., *J. Org. Chem.*, **31**, 1975 (1966).
131. Hanack, M., Herterich, I., and Vött, V., *Tetrahedron Letters*, 3871 (1967).

PART SIX

Addition to Enynes and Polyynes

1. ADDITION TO ENYNES

Enynes, with acetylenic and olefinic bonds in the same molecule, are suitable for studying relative rates and direction of addition to the unsaturated bonds. Ionic reagents can add to the double bond, to the triple bond, or 1,4 across conjugated enynes. The only known commercial use of enynes is the addition of HCl to vinylacetylene to produce chloroprene. HCl adds much faster to vinylacetylene than to acetylene or diacetylene. This is the basis for a procedure for removing vinylacetylene from arc acetylene.[1]

Vartanyan[2] reviewed the chemistry of divinylacetylene in 1964.

1.1. Addition to the Double Bond(s) of Enynes

1.1.1. ADDITION OF ArN_2Cl

Benzenediazonium chloride adds as Ph and Cl across double bonds in enynes.[3] Vinylacetylene gives $PhCH_2CHCl-C\equiv CH$. Isopropenylacetylene reacts the same way. The products can be dehydrochlorinated to form phenylvinylacetylenes. Vinylacetylene also reacts to form 1-chloroallenes by addition-rearrangement.[4]

$$CH_2=CHC\equiv CH + ArN_2Cl \longrightarrow ArCH_2CHClC\equiv CH + ArCH_2CH=C=CHCl$$
$$(Ar = Ph, 33\%) \qquad\qquad (37\%)$$

If Ar is *para* substituted with Cl or MeO, the proportion of the acetylenic product increases.

1.1.2. ADDITION TO TETRAFLUOROETHYLENE TO VINYLACETYLENE

Tetrafluoroethylene adds to ethylene to form tetrafluorocyclobutane.[5] The reaction with vinylacetylene is more complex. Products form by cyclic addition (1) to the double bond, (2) to the triple bond, and (3) to both bonds:

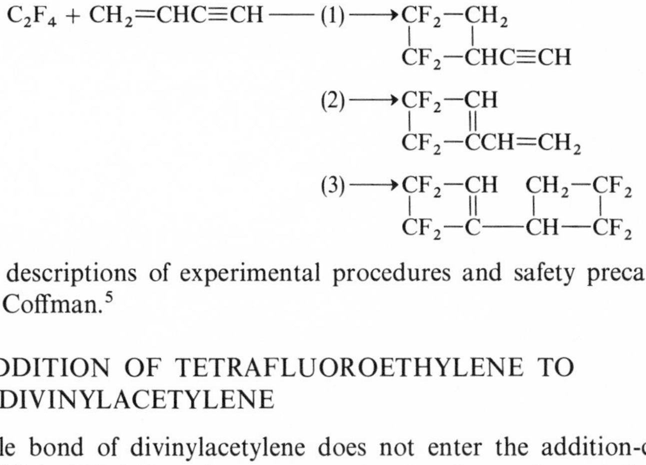

$$C_2F_4 + CH_2=CHC\equiv CH \;\longrightarrow\; (1) \longrightarrow \begin{array}{c} CF_2-CH_2 \\ | \quad\quad | \\ CF_2-CHC\equiv CH \end{array}$$

$$(2) \longrightarrow \begin{array}{c} CF_2-CH \\ | \quad\quad \| \\ CF_2-CCH=CH_2 \end{array}$$

$$(3) \longrightarrow \begin{array}{cc} CF_2-CH & CH_2-CF_2 \\ | \quad\quad \| & | \quad\quad | \\ CF_2-C\!\!-\!\!-\!\!-CH-CF_2 \end{array}$$

Detailed descriptions of experimental procedures and safety precautions are given by Coffman.[5]

1.1.3. ADDITION OF TETRAFLUOROETHYLENE TO DIVINYLACETYLENE

The triple bond of divinylacetylene does not enter the addition-cyclization reaction.[6,7] At 125–140° and autogenous pressure the products are from mono-addition to one double bond and from further addition at the second double bond:

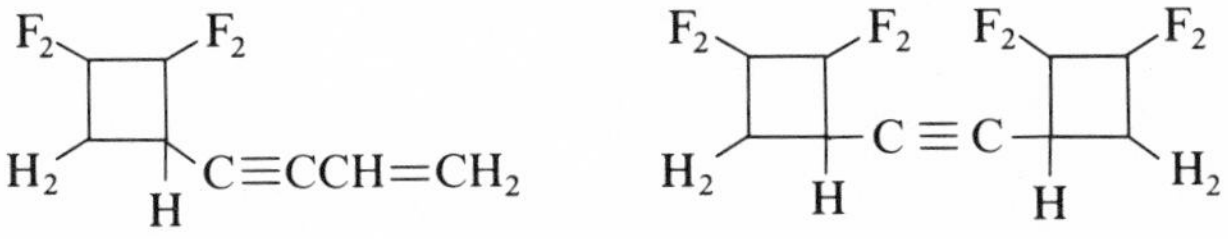

Chlorotrifluoroethylene gives similar products, and the diadducts are mixtures of stereoisomers with the chlorine at different places on the cyclobutane ring.

1.1.4. ADDITION OF β,β-DIFLUOROMETHACRYLONITRILE

Hanby reported the first example of a cycloalkylation of β,β-difluoromethacrylonitrile with a nonfluorinated olefin, divinylacetylene.[6,7] The products are the isomers:

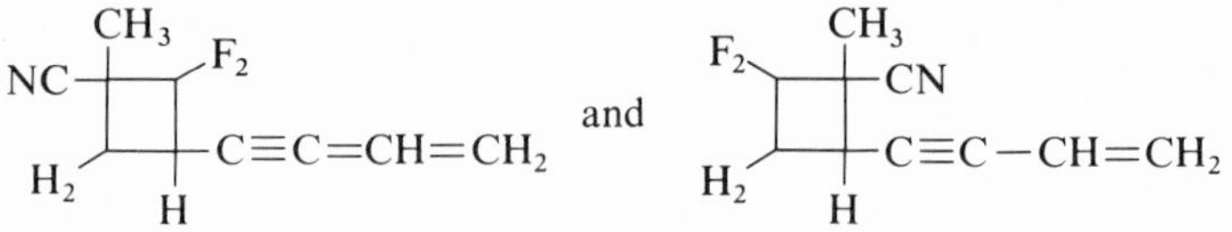

1.1.5. OTHER EXAMPLES OF SELECTIVE ADDITION TO DOUBLE BONDS

Examples of selective addition to the double bonds in enynes are listed in Table 2.23 (page 202).

1.2. Selective Addition to the Acetylenic Bond

The zinc chloride-catalyzed addition of t-alkyl halides to enyne hydrocarbons occurs at the triple bond. Electronic effects in the enynes have little effect.[18]

$$MeCH=CH-C\equiv CH + t\text{-}BuCl \longrightarrow C_9H_{15}Cl, \text{ mostly } MeCH=CH-CCl=CH-t\text{-}Bu$$
$$(30\%)$$

$$\overset{\overset{\displaystyle Me}{\displaystyle |}}{CH_2=CC\equiv CH} \text{ and } CH_2=CHC\equiv CMe \text{ give similar products.}$$

Other examples of selective addition to the acetylenic bond in enynes are listed in Table 2.24 (page 204).

1.3. 1,4-Addition to Conjugated Enynes

This is a general synthesis of allenic compounds (see Chapter 1).

α-Chloroalkyl ethers add 1,4 to vinylacetylene.[29] The product can be an allenic chloride, which isomerizes to the conjugated diene:

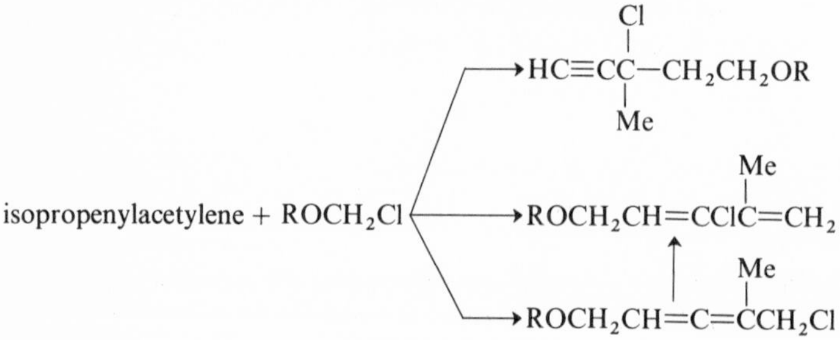

Hept-1-en-3-yne gives only the unrearranged 1,4-addition product, the allene:

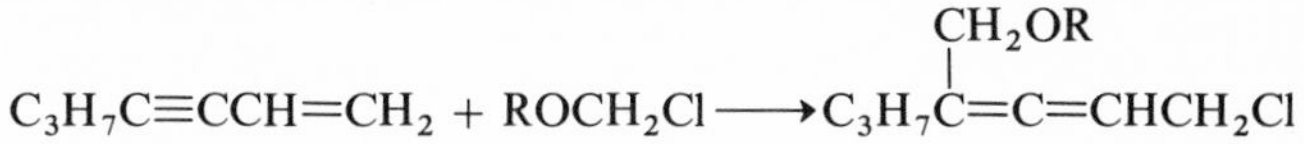

$$C_3H_7C\equiv CCH=CH_2 + ROCH_2Cl \longrightarrow C_3H_7\overset{\overset{\displaystyle CH_2OR}{|}}{C}=C=CHCH_2Cl$$

t-Vinylacetylenic carbinols and their ethers add α-chloroalkyl ethers to form an allenic chloride which isomerizes to other products:

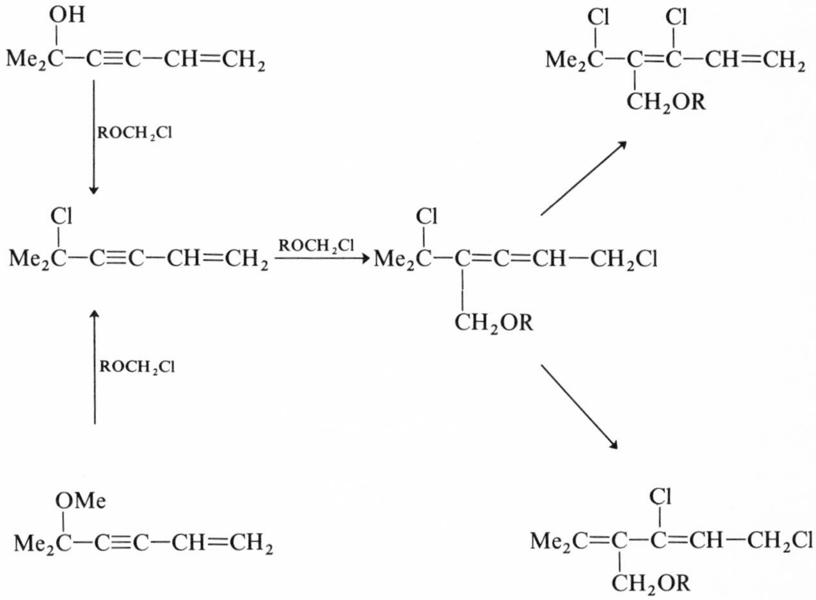

Chloromethyl ethers add to the carbon richest in electrons in homologs of divinylacetylene. This is probably an electrophilic reaction:

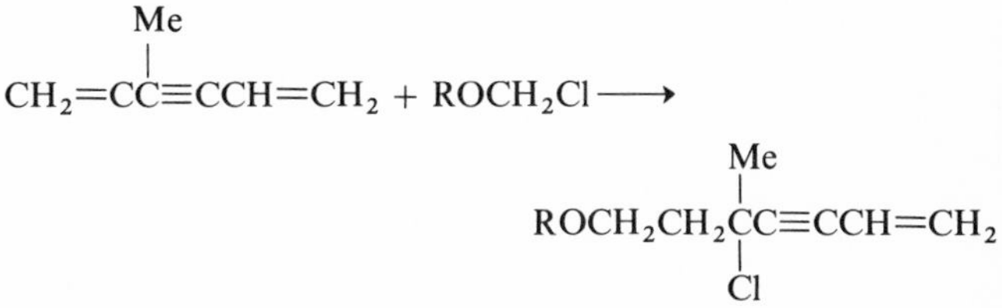

$$CH_2=C\overset{\overset{\displaystyle Me}{|}}{C}\equiv CCH=CH_2 + ROCH_2Cl \longrightarrow$$

$$ROCH_2CH_2\overset{\overset{\displaystyle Me}{|}}{\underset{\underset{\displaystyle Cl}{|}}{C}}C\equiv CCH=CH_2$$

The addition of chloromethyl ethers and other α-halo-ethers to unsaturated compounds has great synthetic possibilities. Products may be used as insecticides, fungicides, solvents and plasticizers. Some products have already found use as insecticides and as water-repellent agents for textiles.[29]

Other examples of 1,4-addition to enynes are listed in Table 2.25 (page 206).

TABLE 2-23
Selective Addition to the Olefinic Bond in Enynes

Enyne	Addend	Conditions	Product	% Yield	Reference
$RC{\equiv}CCH{=}CH_2$	Diphenylnitrone	Benzene, reflux	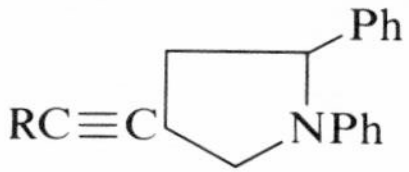 $RC{\equiv}C$ ring with Ph, NPh	21–66	8, 9
$Me_3SiC{\equiv}CCH{=}CH_2$	Et_2NLi	Ether	$Me_3SiC{\equiv}CCH_2CH_2NEt_2$ + $Me_3SiCH{=}C{=}CHCH_2NEt_2$	62	10
$HC{\equiv}CC(Me){=}CHMe$	$:CCl_2$	Hexane, 0°	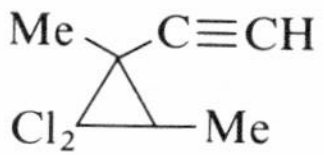 cyclopropane: Me, $C{\equiv}CH$, Cl_2, Me	25	11
$RC{\equiv}CC(Me){=}CH_2$	$:CBr_2$		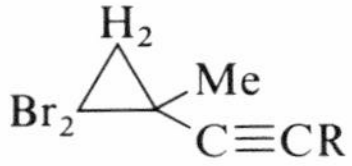 cyclopropane: H_2, Me, Br_2, $C{\equiv}CR$	56	12
$CH_2{=}CHC{\equiv}CEt$	Br_2, MeOH	−5°	$BrCH_2CH(OMe)C{\equiv}CEt$	75	13
$CH_2{=}CHC{\equiv}CEt$	PhCl		$PhCH_2CHClC{\equiv}CEt$	42	14
$CH_2{=}CHC{\equiv}CH$	RSCl	Ether, −20°	$ClCH_2CH(SR)C{\equiv}CH$		15
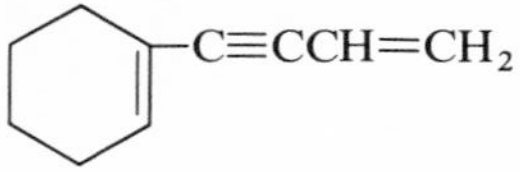 cyclohexenyl–$C{\equiv}CCH{=}CH_2$	Me_2NH		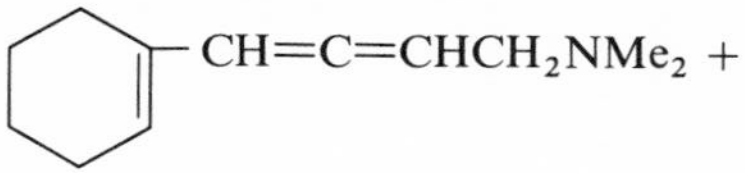 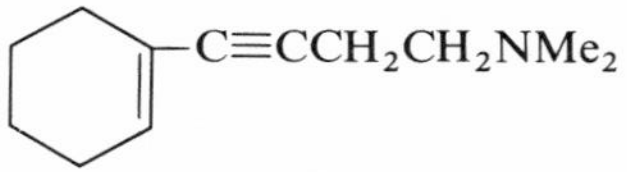 cyclohexenyl–$CH{=}C{=}CHCH_2NMe_2$ + cyclohexenyl–$C{\equiv}CCH_2CH_2NMe_2$		2

Substrate	Reagent	Conditions	Product	Ref.
$\overset{\text{Me}}{|}$ $CH_2=CC\equiv CCH=CH_2$	MeOBr		$\overset{\text{OMe}}{|}$ $BrCH_2CC\equiv CCH=CH_2$ $\underset{\text{Me}}{|}$	2
$\overset{\text{Me}}{|}$ $CH_2=CC\equiv CCH=CH_2$	ROCH$_2$Cl	Ether, ZnCl$_2$	$\overset{\text{Me}}{|}$ $ROCH_2CH_2CC\equiv CCH=CH_2$ $\underset{\text{Cl}}{|}$	16
$\overset{\text{H}}{|}$ $CH_2=CHC\equiv CCH=CMe$	MeOBr—MeOH	$-5°$	$CH_2BrCHC\equiv CCH=CHMe$ $\underset{\text{OMe}}{|}$ (only product)	15
$MeC\equiv CCH_2CH=CH_2$	:CCO$_2$Et		$MeC\equiv CCH_2\triangleright CO_2Et$	17

TABLE 2-24
Selective Addition to Acetylenic Bonds of Enynes

Enyne	Addend	Conditions	Product	% Yield	Reference
CH_2=CHC≡CH	$HGeCl_3$	Ether, reflux	CH_2=CHCH=$CHGeCl_3$		19
CH_2=$\overset{\text{Me}}{C}$C≡CH	$i\text{-}Bu_2AlH$	20°	Isoprene	72	20
CH_2=CHC≡CH	$i\text{-}Bu_2AlH$	20°, 2 hr	Butadiene	11.8	20
CH_2=$\overset{\text{Me}}{C}$C≡CH	Borane	Diglyme, 10°	Isoprene	61	21
CH_2=CHC≡CH	Borane	Diglyme, 10°	Butadiene	31	21
CH_2=CHC≡CH	R_3SiH	UV light	R_3SiCH=CHCH=CH_2		22
CH_2=CHC≡CH	HCl	CuCl solution	Chloroprene		1
CH_2=CHC≡CH	HCl	Hg vapor, C, 90°	Chloroprene	50	1
CH_2=$\overset{\text{Me}}{C}$C≡CH	HBr		HC≡CC($\overset{\text{Br}}{\vert}$)—$CH_3$ (3,4-addition)	10	23, 24
			+ H_2C=C(Br)C(Me)=CH_2 (1,2-addition)	35	
			+ CH_2=C=C($\overset{\text{Me}}{\vert}$)$CH_2$Br (1,4-addition) Mostly 1,4-addition to allene		
CH_2=CHC≡CH	RSH	KOH		55	25
CH_2=$\overset{\text{Me}}{C}$C≡CH	RSH	KOH	CH_2=C—C(SR)=CH_2 (3,4-addition)	25	25

Substrate	Reagent	Product	Yield	Ref.
Me │ $CH_2=CC≡CH$	RSH, KOH	cis-$CH_2=\overset{\underset{\displaystyle Me}{│}}{C}-CH=CHSR$	56	25a
$CH_2=CHC≡CMe$	RSH, KOH	$CH_2=CHCH=C(SR)Me$ (4,3-addition)		25
$RC≡CCH=CH_2$	I_2, EtCl, light	$RCI=CICH=CH_2$		13
cis-$HC≡CCH=CHCH_2Cl$	Sodium malonic ester	cis-$HC≡CCH=CHCH_2CH(CO_2Et)_2$ 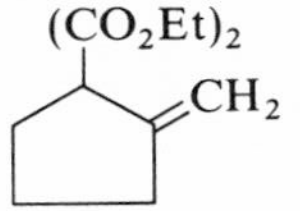 + $(CO_2Et)_2$ (cyclopentane$=CH_2$)		26
R │ $RCH=C-C≡CH$	Hg^{++}, Ethylene glycol	$RCH=\overset{\underset{}{}}{C}-\overset{}{C}-CH_3$ (dioxolane $O-CH_2$, H_2C-CH_2)	15 conversion	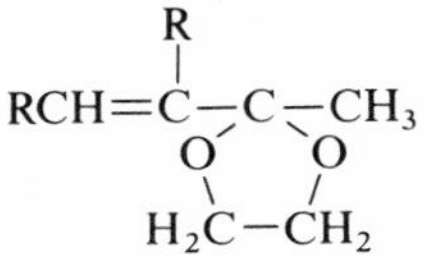 26a
$PhC≡CCH=CHPh$	$:CX_2$	$PhC=C-CH=CHPh$ (с $C=O$)		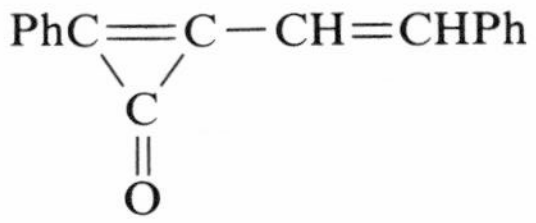 27
$HC≡CCH=CHMe$	Br_2, CuBr, HBr, ether, $-10°$	$CHBr=CBrCH=CHMe$	60	28
Me │ $MeCH=CC≡CH$	HBr	$MeCH=\overset{\underset{\displaystyle Me}{│}}{C}CBr=CH_2$	100	24
cyclohexenyl–$C≡CH$	HBr	cyclohexenyl–$CBr=CH_2$	100	23

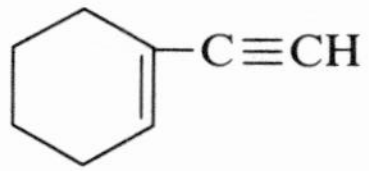

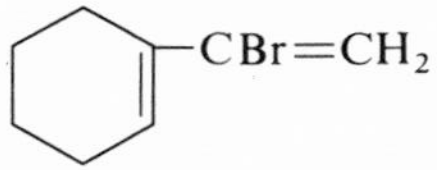

TABLE 2-25
1,4-Addition to Conjugated Enyne Systems

Enyne	Addend	Conditions	Product	% Yield	Reference
$CH_2{=}CHC{\equiv}CH$	$(i\text{-}Bu)_2AlH$	20°	Butadiene + 1-butyne		30
$CH_2{=}CHC{\equiv}CH$	$(i\text{-}Bu)_3Al$	25°	$i\text{-}Bu_2AlCH{=}CHCH{-}i\text{-}Bu$ (with CH_3 branch)		30
$R_2CC{\equiv}CCH{=}CH_2$ (with OH)	R'Li	−40°, Et_2O	$R_2CCH{=}CHCH{=}CHR'$ (with OH)		31
$t\text{-}BuC{\equiv}CCH{=}CH_2$	BuLi	−15°, $CHCl_3$	$t\text{-}BuCH{=}C{=}CHCH_2Bu$	47	32
$R_3SiC{\equiv}CCH{=}CH_2$	Br_2	−10°, $CHCl_3$	$R_3SiCBr{=}C{=}CHCH_2Br$	47	33
$Et_3GeC{\equiv}CCH{=}CH_2$	Br_2		Allenic bromide and addition to triple bond		34

2. ADDITION TO POLYYNES

2.1. Background

(1) Electrophilic additions: These are very slow with conjugated polyacetylenes. The triple bonds in non-conjugated polyynes react independently.

(2) Nucleophilic additions: Some nucleophilic additions which fail with simple acetylenes do occur with polyynes having more than three conjugated triple bonds.[35] For example:

$$Me(C\equiv C)_4Me + HCN + NaCN \xrightarrow[95°]{THF} Me(C\equiv C)_3CH=\overset{\overset{\displaystyle Me}{|}}{C}CN$$

Diacetylenes do not react without Cu_2Cl_2 catalyst. Also, ozonolysis is much faster and more complete for polyynes. One explanation is that polyynes meso-merize[35] to form positive centers: $R\overset{+}{C}=(C\overset{-}{=}C)_nR$. Nucleophilic Michael addition to olefins occurs only if the double bond is polarized by an electrophilic group. Dimethyltetraacetylene reacts with sodium malonic ester and with lithium alkyls, presumably because of such a polarization of the acetylene groups. Addition of lithium aluminum hydride gives *cis*-olefins after hydrolysis, and no solvent is required. In solvents, diacetylenes react slowly, but tetraacetylenes react very rapidly.

Bromine adds to form *trans*-dibromoethylene groups:

$$-(C\equiv C)_n\overset{\delta^+}{\overset{\frown}{C}}\equiv CH \longrightarrow R\overset{\overset{\displaystyle |\bar{Br}|^+}{|}}{C}=CH \longrightarrow R\overset{\overset{\displaystyle Br}{|}}{C}=CHBr$$
$$|\bar{Br}\smile\bar{Br}| \qquad\qquad \underset{\underset{\displaystyle Br}{|}}{}$$

Note that additions usually start at one end of the conjugated acetylene chain.

2.2. Addition of Triethylgermanium Hydride

Triethylgermanium hydride, with chloroplatinic acid catalyst, adds to one triple bond of symmetrically substituted conjugated diacetylenic glycols.[36] Larger amounts of catalyst cause violent reaction to form uncertain products. Products are

$$\underset{\underset{\displaystyle HO-R}{|}}{\overset{\overset{\displaystyle GeEt_3}{|}}{C}}=CHC\equiv C-ROH$$

yields are 10–70%. Triethylgermanium hydride adds to unsymmetrical diacetylenic glycols in which one of the hydroxyls is replaced by $-OSiEt_3$. Germanium always adds to the triple bond alpha to the free hydroxyl group.

2.3. Addition of Carbenes

One of the triple bonds in diacetylenes can add a carbene, but tri- and tetra-acetylenes do not react. Diphenyldiacetylene and dichlorocarbene give, after hydrolysis,

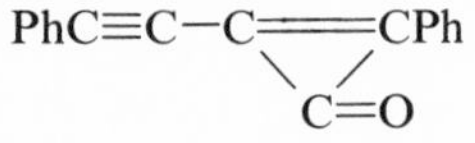

$$PhC\equiv C-C \underset{\underset{\displaystyle C=O}{\diagdown \diagup}}{=\!=\!=} CPh$$

in 15% conversion. Dichlorocarbene adds to dipropyldiacetylene. One triple bond adds carbene, the other adds HCl.[27]

2.4. Addition of Thiols

Thiols add 1,2 to diacetylene by a *trans* approach to give *cis*-substituted products:[37]

$$BuSH + HC\equiv CC\equiv CH \longrightarrow BuSCH=CHC\equiv CH \xrightarrow[\text{addition}]{\text{further}} BuSCH=CHCH=CHSBu$$

2.5. Addition of H₂S—Cyclization to Thiophenes

Thiophenes are obtained by a simple reaction of substituted diacetylenes with H₂S in ethanol or acetone solution at pH 8–10 (Table 2.26).[38]

TABLE 2-26
Addition of H₂S to Conjugated Diacetylenes to form Thiophenes[38]

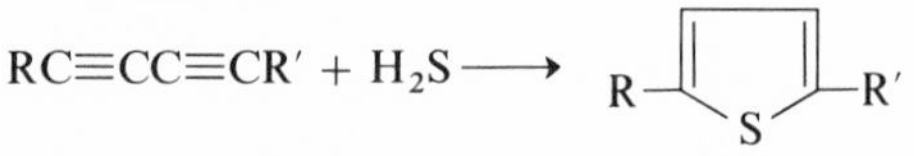

$$RC\equiv CC\equiv CR' + H_2S \longrightarrow R-\underset{S}{\boxed{}}-R'$$

| Alkyne | | % |
R	R'	Yield
H	H	20
H	Me	51
Me	Me	70
Et	Et	65
Ph	Ph	85
CO₂H	Ph	52
PhC≡C	Ph	83
PhC≡CC≡C	Ph	75

The diacetylenes and polyacetylenes are easily made by coupling as described by Armitage.[39]

α-Propynyl ketones undergo similar reactions[38a] in ethanol in the presence of H_2S and HCl. For example:

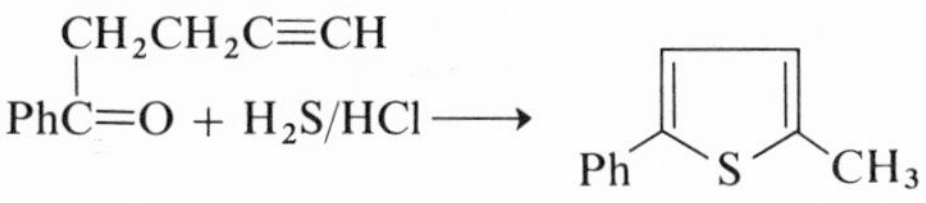

71% yield

Products are 1-methylthiophenes in all cases.

Diarylbutadiynes add SCl_2 to form dichlorothiophenes.[39a] Yields are higher than with dialkylbutadiynes.

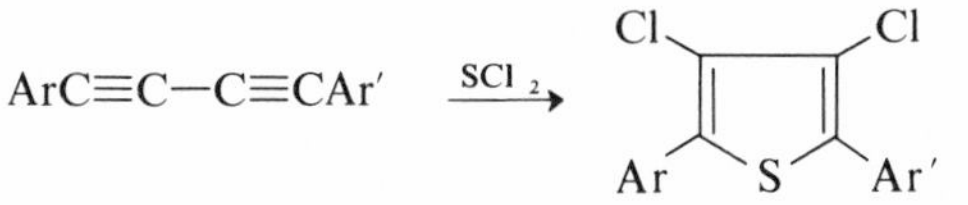

$ArC≡C-C≡CAr'$ $\xrightarrow{SCl_2}$

(16–80% yields)

Triacetylenes and tetraacetylenes usually react with H_2S at only two triple bonds, but 1,4-bis(thienyl)diacetylene reacts with H_2S to form the α-terthienyl:

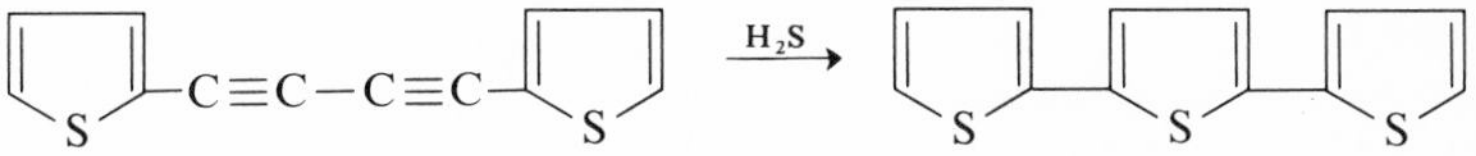

1,4-Bis(phenylbutadiynyl)benzene reacts with SCl_2 to form the bisthiophene in good yields:[39a]

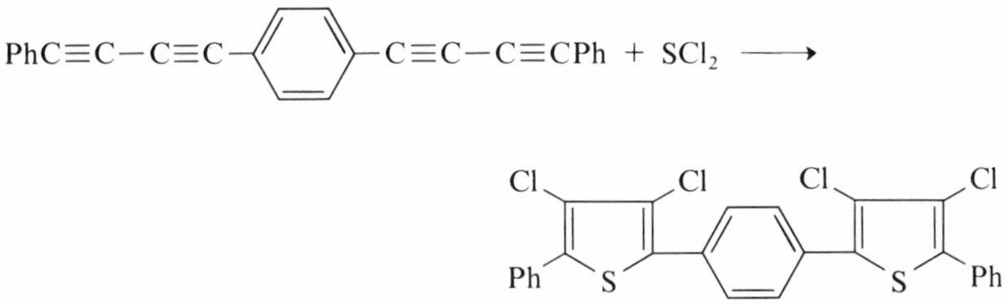

2.6. Addition of Sodium Telluride

Sodium telluride in methanol at 20° reacts with conjugated diacetylenes to form tellurophenes

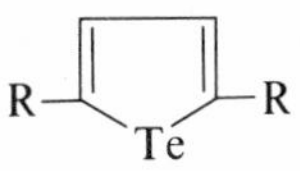

in good yield [R = H, 69%; CH_2OH, 25%; $C(Me)_2OH$, 100%].[40]

2.7. Addition of Phosphines to Form Phospholes

Phosphines add to conjugated diynes to form phospholes (analogous to pyrroles). Reaction in boiling pyridine (A) or in benzene in the presence of phenyllithium as catalyst (B) gives good yields via the phosphine anion (Table 2.27).[41]

TABLE 2-27
Addition of Phosphines to Conjugated Diynes to form Phospholes[41]

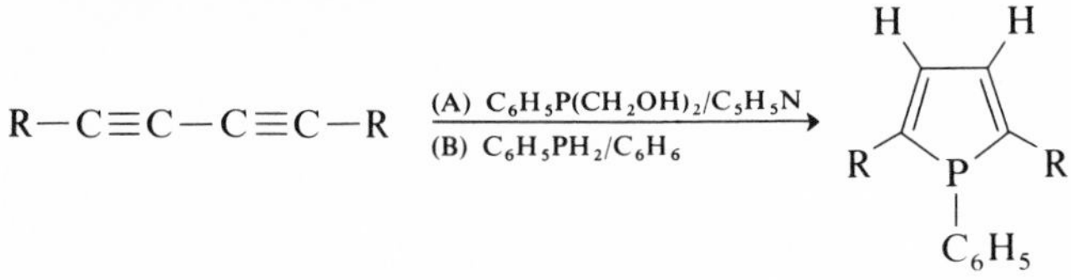

R	b.p. (°C), @ mm Hg or m.p. (°C)	% Yield A	B
Phenyl	184–186	20.5	60
2-Naphthyl	230–231	17.5	89.5
p-Tolyl	194–196	29	59
p-Bromophenyl	200–202	16.5	
CH_3	66–69, @ 0.2	0	51

2.8. Diels-Alder Reactions

Diynes have been used in three Diels-Alder reaction systems: (1) mix the reactants and heat slowly to reflux; (2) reflux in β-decalin; (3) add the diyne to the diene in hot decalin, and boil for 20 minutes.[42] Diphenyldiacetylene reacts with cyclopentadienones to form 1-aryl-2-phenylacetylenes in these systems:

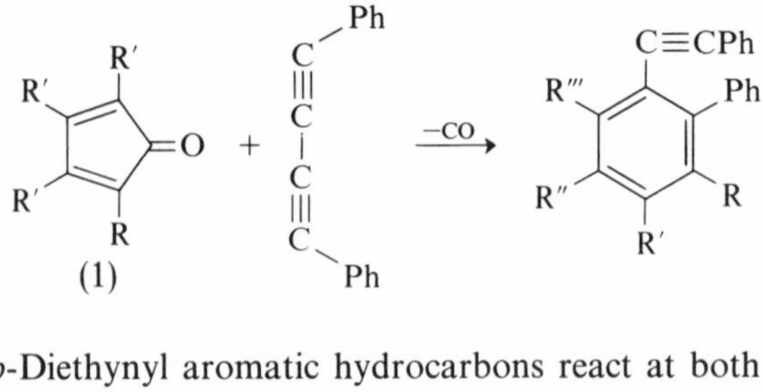

o-, m- and p-Diethynyl aromatic hydrocarbons react at both triple bonds, to form terphenyls and quinquiphenyls:

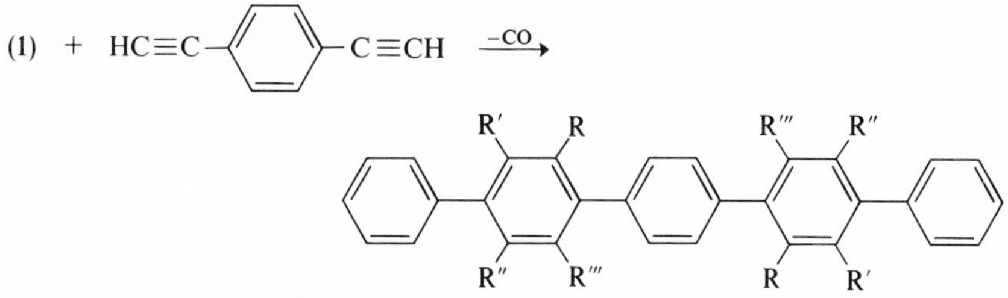

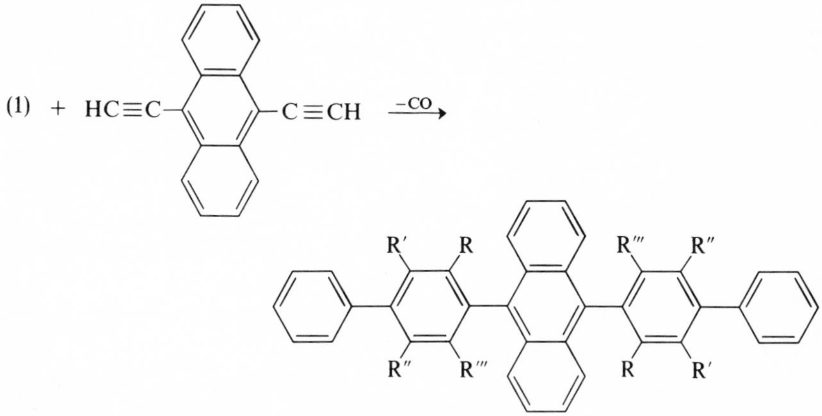

(1) $+$ HC≡C—⟨C₆H₄⟩—C≡CH $\xrightarrow{-CO}$

R' R   R''' R''

R'' R'''   R   R'

(1) $+$ HC≡C—⟨anthracene⟩—C≡CH $\xrightarrow{-CO}$

R' R   R''' R''

R'' R'''   R   R'

Diphenylacetylene reacts in similar systems:[43]

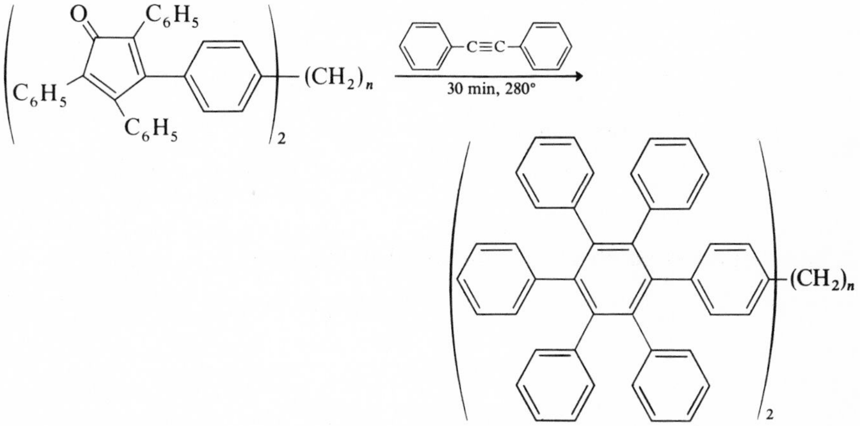

$$\left(\underset{C_6H_5}{\overset{O \quad C_6H_5}{\cdots}} \text{—}⟨C_6H_4⟩\text{—}\right)_2 \text{(CH}_2)_n \xrightarrow[\text{30 min, 280°}]{C_6H_5-C≡C-C_6H_5}$$

$$\left(\cdots\text{—}⟨C_6H_4⟩\text{—}\right)_2 \text{(CH}_2)_n$$

n	% Yield
2	58
3	65
4	59
5	12
6	11

Both triple bonds in diphenyldiacetylene can react to form decaphenylbi-phenyl:[43]

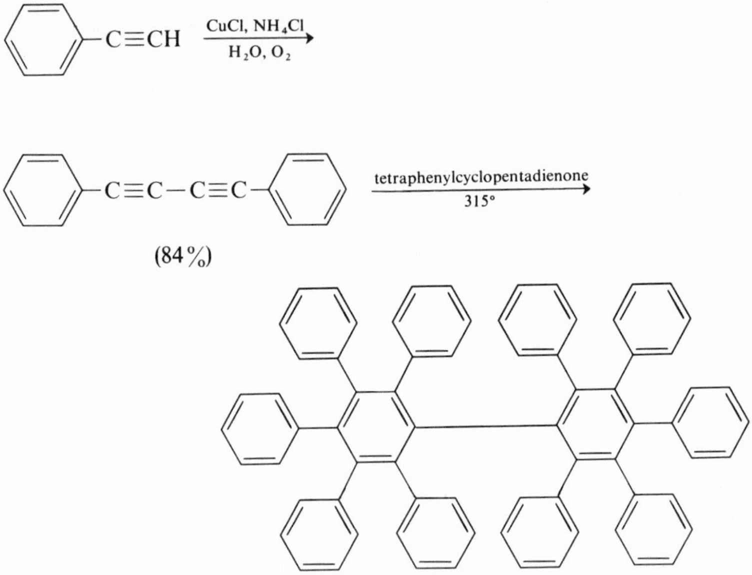

(84%)

(60% yield, highly
fluorescent)

2.9. Other Addition Reactions

Some other addition reactions are summarized in Table 2.28.

2.10. Addition to Nonconjugated Diacetylenes

Examples of additions to nonconjugated diacetylenes are given in Table 2.29.

TABLE 2-28
Addition Reactions of Conjugated Diacetylenes

Diyne	Addend	Conditions	Product	% Yield	Reference
Diacetylene	HCl	Cu_2Cl_2 solution	$CH_2=CClCCl=CH_2$	10	1
Diacetylene	HCl	$CuCl_2$, Cu_2Cl_2, HCl, water, 85°	$CHCl=CClCCl=CHCl$	88	44
Diacetylene	$EtNH_2$		$-(CH=CHNHEt)_2$		49
$Me_3SiC{\equiv}CC{\equiv}CEt$	EtLi	Ether, $-20°$, then hydrolysis	$Me_3SiC{\equiv}CCH=C(Et)_2$	37	45
$RC{\equiv}CC{\equiv}CH$	Et_3SnH	60–80°	$Et_3SnCH=CHC{\equiv}CR$		46
$RC{\equiv}CC{\equiv}CH$	Br_2	$-40°$, $CHCl_3$	To terminal carbon		47
$Me_3SiC{\equiv}CC{\equiv}CH$	Br_2		To silyl-substituted acetylenic group		48
$HC{\equiv}CC{\equiv}CC(Me)_2OR$	Me_2NH	Water, 5 min.	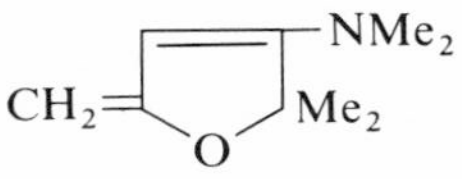 (structure with NMe_2, Me_2, CH_2, O)	90	50

TABLE 2-29
Additions to Nonconjugated Diacetylenes

Diacetylene	Addend	Conditions	Product	% Yield	Reference
$RC{\equiv}C(CH_2)C{\equiv}CH$	Br_2	$-50°$, $CHCl_3$	$HC{\equiv}CCH_2CBr=CHBr$	60	51
$RC{\equiv}CCH_2C{\equiv}CR'$	Br_2	$-50°$, $CHCl_3$	To $C{\equiv}C$ with least branched R		51
$R_3SiC{\equiv}CCH_2C{\equiv}CR'$	Br_2	$-50°$, $CHCl_3$	To $C{\equiv}C$ attached to R'		51
$MeSC{\equiv}C(CH_2)_2C{\equiv}CH$	Br_2	$-20°$, $CHCl_3$	$MeSCCl=CCl(CH_2)_2-C{\equiv}CH$	46	52
$HC{\equiv}C(CH_2)_2C{\equiv}CMe$	Br_2	$-20°$, $CHCl_3$	Mostly to Me-substituted $C{\equiv}C$		53

References

1. Herbertz, T., *Chem. Ber.*, **92**, 541 (1959).
2. Vartanyan, S. A., *Russ. Chem. Rev. (English Transl.)*, 243 (1964).
3. Dombrovskii, A. V., *Zh. Obshch. Khim.*, **27**, 3050 (1957); *Chem. Abstr.*, **52**, 8087 (1958).
4. Kheruze, Yu. I., and Petrov, A. A., *ibid.*, **31**, 428 (1961); *Chem. Abstr.*, **55**, 2330 (1961).
5. Coffman, D. D., *et al., J. Am. Chem. Soc.*, **71**, 490 (1949).
6. Handy, C. T., and Benson, R. E., *J. Org. Chem.*, **27**, 39 (1962).
7. Hartmann, H., Beerman, C., and Czempik, H., *Z. Anorg. Allgem. Chem.*, **287**, 261 (1956); *Chem. Abstr.*, **51**, 8030 (1957).
8. Chistokletov, V. N., and Petrov, A. A., *Zh. Obshch. Khim.*, **32**, 2385 (1962); *Chem. Abstr.*, **58**, 9040 (1963).
9. Chistokletov, V. N., Vagina, L. K., and Petrov, A. A., *Zh. Organ. Khim.*, **1**, 369 (1965); *Chem. Abstr.*, **62**, 16092 (1965).
10. Petrov, A. A., Kormer, V. A., and Stadnichuk, M. D., *Zh. Obshch. Khim.*, **31**, 1135 (1961); *Chem. Abstr.*, **55**, 23330 (1961).
11. Danilkina, L. P., D'yakonov, I. A., and Roslovtseva, G. I., *Zh. Organ. Khim.*, **1**, 465 (1965); *Chem. Abstr.*, **63**, 1688 (1965).
12. Vo-Quang, L., and Vo-Quang, Y., *Compt. Rend.*, **263**(C), 640 (1966).
13. Petrov, A. A., *et al., Zh. Obshch. Khim.*, **28**, 2320, 2325 (1958); *Chem. Abstr.*, **53**, 5030 (1959).
14. Kheruze, Yu. L., and Petrov, A. A., *Zh. Obshch. Khim.*, **30**, 2528 (1960); *Chem. Abstr.*, **55**, 21002 (1961).
15. Radchenko, S. I., and Petrov, A. A., *Zh. Organ. Khim.*, **1**, 47 (1965); *Chem. Abstr.*, **62**, 14484 (1965).
16. Petrov, A. A., *et al., Zh. Obshch. Khim.*, **31**, 1518 (1961); *Chem. Abstr.*, **55**, 23330 (1961).
17. Danilkina, L. P., and D'yakonov, I. A., *Zh. Obshch. Khim.*, **34**, 3129 (1964); *Chem. Abstr.*, **61**, 14541 (1964).
18. Maretina, I. A., and Petrov, A. A., *Zh. Obshch. Khim.*, **31**, 419 (1961); *Chem. Abstr.*, **55**, 23329 (1961).
19. Mikolajczak, K. L., *et al., J. Org. Chem.*, **29**, 318 (1964).
20. Mironov, V. F., and Gar, T. K., *Izv. Akad. Nauk SSSR Ser. Khim.*, 291 (1965); *Chem. Abstr.*, **62**, 14715 (1965).
21. Markova, V. V., Kormer, V. A., and Petrov, A. A., *Zh. Obshch. Khim.*, **35**, 1669 (1965); *Chem. Abstr.*, **63**, 17870 (1965).
22. Montclair Res. Corp., British Patent 684,597 (Dec. 24, 1952); *Chem. Abstr.*, **48**, 2761 (1954).
23. Traynard, J. C., *Bull. Soc. Chim. France*, 19 (1962).
24. Cocordano, M., *Bull. Soc. Chim. France*, 738 (1962).
25. Kupin, B. S., and Petrov, A. A., *Zh. Organ. Khim.*, **1**, 244 (1965); *Chem. Abstr.*, **62**, 14442 (1965).
25a. Jacobs, T. L., and Mikailovski, A., *Tetrahedron Letters*, 2607 (1967).
26. Mavrov, M. V., Derzhinskii, A. R., and Kucherov, V. F., *Izv. Akad. Nauk SSSR, Ser. Khim.*, 1460 (1965); *Chem. Abstr.*, **63**, 16223 (1965).
26a. Normant, H., *Compt. Rend. (C)*, **265**, 522 (1967).
27. Dehmlow, E. V., *Tetrahedron Letters*, 2317 (1965).

28. Petrov, A. A., and Porfir'eva, Yu. I., *Dokl. Akad. Nauk SSSR*, **111**, 839 (1956); *Chem. Abstr.*, **51**, 9469 (1957).
29. Vartanyan, S. A., and Tosunyan, A. O., *Russ. Chem. Rev. (English Transl.)*, **34**, 267 (1965).
30. Kormer, V. A., and Petrov, A. A., *Dokl. Akad. Nauk. SSSR*, **146**, 1343 (1962); *Chem. Abstr.*, **58**, 8883 (1963).
31. Kormer, V. A., and Petrov, A. A., *Zh. Obshch. Khim.* **30**, 3890 (1960); *Chem. Abstr.*, **55**, 23328 (1961).
31b. Kooyman, J. G. A., *et al., Rec. Trav. Chim.*, **87**, 69 (1968).
32. Petrov, A. A., Maretina, I. A., and Kormer, V. A., *Zh. Obshch. Khim.*, **33**, 413, 416, 419 (1963); *Chem. Abstr.*, **59**, 1462 (1963).
33. Stadnichuk, M. D., and Petrov, A. A., *Zh. Obshch. Khim.*, **30**, 3890 (1960); *Chem. Abstr.*, **55**, 23328 (1961).
34. Stadnichuk, M. D., and Petrov, A. A., *Zh. Obshch. Khim.*, **35**, 700 (1965); *Chem. Abstr.*, **63**, 4322 (1965).
35. Bohlmann, F., *Ann. Chem.*, **604**, 207 (1957).
36. Gverdtsitch, I. M., and Buachidze, M. A., *Soobshch. Akad. Nauk Gruz.*, **37**, 323 (1965); *Chem. Abstr.*, **62**, 14719 (1965).
37. Prilezhaeva, E. N., Tsymbal, L. V., and Shostakovskii, M. F., *Zh. Obshch. Khim.*, **31**, 2487 (1961); *Chem. Abstr.*, **56**, 9944 (1962).
38. Schulte, K. E., Reisch, J., and Hoerner, L., *Chem. Ber.*, **95**, 1943 (1962).
38a. Schulte, K. E., Reisch, J., and Bergenthal, D., *Chem. Ber.* **101**, 1540 (1968).
39. Armitage, I. B., *et al., J. Chem. Soc.*, 147 (1954).
39a. Schulte, K. E., Walker, H., and Rolf, L., *Tetrahedron Letters*, 4819 (1967).
40. Mack, W., *Angew. Chem. Intern. Ed.*, **5**, 896 (1966).
41. Märkel, G., and Potthast, R., *Angew. Chem. Intern. Ed.*, **6**, 86 (1967).
42. Ried, W., and Bönnighausen, K. H., *Chem. Ber.*, **93**, 1769 (1960).
43. Ogliaruso, M. A., and Becker, E. I., *J. Org. Chem.*, **30**, 3354 (1965).
44. Mkryan, G. M., *Armyansk. Khim. Zh.*, **19**, 192 (1966); *Chem. Abstr.*, **65**, 10479 (1966). (1966).
45. Shokhovskoi, B. G., Stadnichuk, M. D., and Petrov, A. A., *Zh. Obshch. Khim.*, **35**, 1031 (1965); *Chem. Abstr.*, **63**, 9978 (1965).
46. Zavgorodnii, V. S., and Petrov, A. A., *Zh. Obshch. Khim.*, **35**, 1313 (1965); *Chem. Abstr.*, **63**, 11601 (1965).
47. Porfir'eva, Yu. I., *et al., Zh. Obshch. Khim.*, **34**, 1873, 1881 (1964); *Chem. Abstr.*, **61**, 8151 (1964).
48. Shokhovskoi, B. G., Stadnichuk, M. D., and Petrov, A. A., *Zh. Obshch. Khim.*, **35**, 1714 (1965); *Chem. Abstr.*, **64**, 2119 (1966).
49. Shostakovskii, M. F., *et al., Izv. Akad. Nauk SSSR Otd. Khim. Nauk.*, 2217 (1962); *Chem. Abstr.*, **58**, 12401 (1963); Chekulaeva, Y. A., *et al., Izv. Akad. Nauk SSSR* **8**, 1829 (1967) (English Translation).
50. Nazarova, I. I., Gusev, B. P., and Kucherov, V. F., *Izv. Akad. Nauk SSSR Ser. Khim.*, 729 (1965); *Chem. Abstr.*, **63**, 2890 (1965).
51. Porfir'eva, Yu. I., Turbanova, E. S., and Petrov, A. A., *Zh. Organ. Khim.*, **2**, 777 (1966); *Chem. Abstr.*, **65**, 10480 (1966).
52. Petrov, A. A., and Forost, M. P., *Zh. Organ. Khim.*, **1**, 1550 (1965).
53. Petrov, A. A., and Forost, M. P., *Zh. Obshch. Khim.*, **34**, 3292 (1964); *Chem. Abstr.*, **62**, 3915 (1965).

PART SEVEN

Addition to 1-Acetylenic Ethers and Thioethers

1. BACKGROUND

In 1940, Jacobs[1] reported continuation of some research on phenoxyacetylene started by Slimmer in 1903.[2] According to Slimmer, phenoxyacetylene is an unstable oil which polymerizes to a dark viscous polymer in a few hours. Jacobs was interested in acetylenic ethers because they are members of an "yne-ol" system related to aldoketenes: $-C{\equiv}C-OH \rightleftarrows -CH{=}C{=}O$. He expected acetylenic ethers to be very reactive, by analogy with ketenes. Jacobs first prepared phenoxyacetylene by Slimmer's method. Phenoxyacetylene solidified in dry ice, and could be stored without difficulty at $-78°$. It polymerized at room temperature; at $100°$ in a sealed tube, it exploded violently and left a carbon residue.

Jacobs reports marked a resurgence of interest in acetylenic ethers. The most prolific workers in this field are J. F. Arens and his associates. By 1966, this group had published about eighty papers. Professor Arens published a comprehensive review of his work and of pertinent literature in 1960.[3]

Volger and Arens,[4] and Arens,[3] reviewed addition reactions of acetylenic ethers and thioethers. Using ethoxyethyne as the model, the most likely polarization in 1-acetylenic ethers is

$$H \overset{\frown}{-}C{\equiv}\overset{\frown}{C}-OEt$$
$$\quad\quad \beta \quad\; \alpha$$

Electrophilic addends should attach at the β-carbon, and nucleophilic addends at the α-carbon. Ethylthioethyne can polarize the same way, and electrophilic reagents should add to the β-carbon. But nucleophilic reagents also attach to the β-carbon of thioethers. The explanation is that sulfur can accommodate a decet of electrons in its valence shell and can therefore act as an electron acceptor as well as donor:

$$H-C\overset{\frown}{\equiv}C\overset{\frown}{-}SEt$$
$$\quad\; \beta \quad\; \alpha$$

Approach of a nucleophilic fragment induces this polarization. Most known addition reactions confirm the predictions, but there are some exceptions. Some of the additions can best be represented as concerted cyclic attacks (formaldehyde and formaldehyde + amine; water is essential):

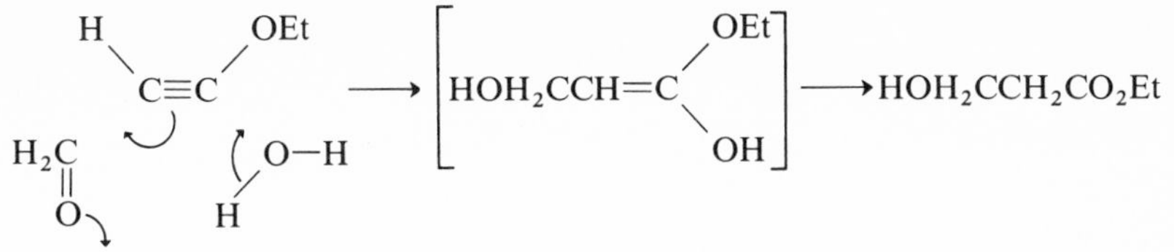

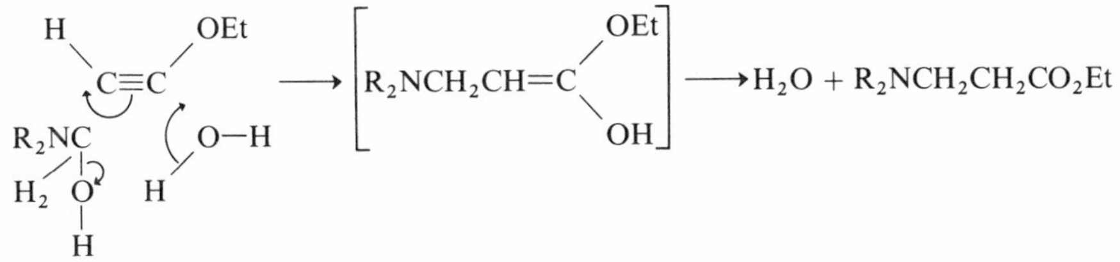

Since $RC{\equiv}COEt$ reacts the same way, these are not Mannich attacks on the acetylenic hydrogen.

2. ADDITION OF CARBOXYLIC ACIDS

Zwanenburg[5] made 1-ethoxyvinyl ethers by adding carboxylic acids to ethoxyethyne, using a procedure similar to Wasserman's[6] (Table 2.30).

TABLE 2-30[5]

$$HC{\equiv}COEt + RCOOH \xrightarrow{Hg^{++}} H_2C{=}C(OEt)OCOR$$

R	% Yield
Phenyl	84
Methyl	64
Chloromethyl	78
Trichloromethyl	88 (no catalyst needed)
Benzyl	68

o-Benzoylbenzoic acid adds across ethoxyacetylene, and the resulting product cyclizes via the new [3.2.1] bicyclic reaction path.[7] One or two moles of o-benzoylbenzoic acid can react with ethoxyacetylene:

1:1

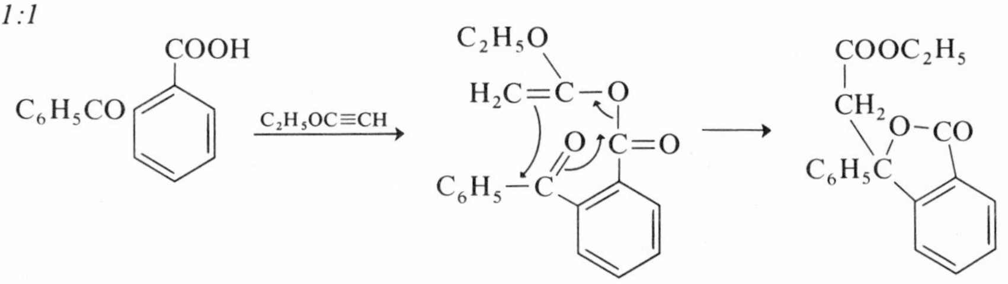

2:1

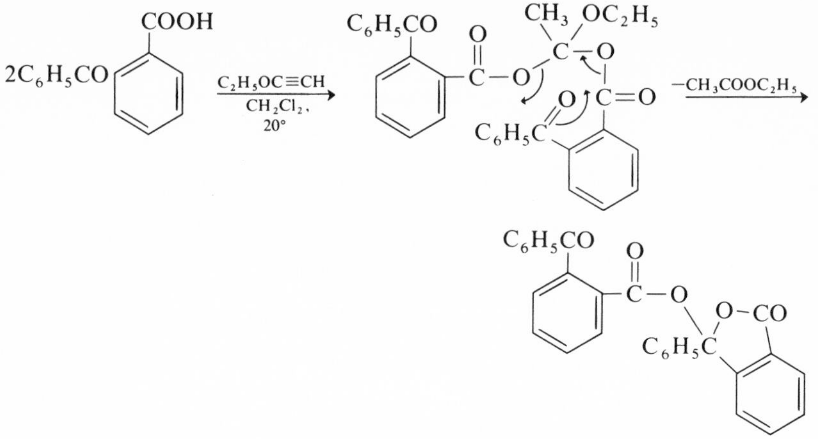

COOH

$2C_6H_5CO$ [benzene ring] $\xrightarrow[\substack{CH_2Cl_2, \\ 20°}]{C_2H_5OC\equiv CH}$ C_6H_5CO [structures]

CH_3 OC_2H_5
C_6H_5CO ... $\xrightarrow{-CH_3COOC_2H_5}$

C_6H_5CO [structure]

The reaction of N-acylamino acids with ethoxyethyne to form the acid anhydrides is a good preparative method if nonpolar solvents and low temperatures are used. The reaction is vigorous in the presence of a drop or two of HCl. Crude anhydride precipitates as a crystalline mass. Peptide synthesis via acid anhydrides[12] is more likely than alternative mechanisms.[8]

acylNHCHRCO$_2$H + HC≡COEt ⟶ acylNHCHRCO$_2$C(OEt)=CH$_2$ $\xrightarrow{+\text{acylamino acid}}$

(acylNHCHRCO)$_2$O + EtOAc

anhydride + amino acid ester ⟶ acylNHCHRCO—NHCHR'CO$_2$R″ + acylamino acid
(regenerated)

3. ADDITION OF CARBONYL COMPOUNDS

Of the known methods for preparing α,β-unsaturated thiolesters, two did not require preparation of the acid first. One of the better methods is rearrangement of carbinols from carbonyls and lithium alkylthioethynylides, but the thiolesters are sometimes contaminated with the corresponding β-hydroxythiolesters.[13] Arens[3] predicted that the BF$_3$-catalyzed reaction between acetylenic ethers and carbonyls might become the best and most convenient way to make α,β-unsaturated esters.

Later, Bos and Arens[14] found that this is an excellent way to make several types of α,β-unsaturated thiolesters. Acetylenic thioethers are easy to prepare.[15] They react with saturated or unsaturated aliphatic and aromatic aldehydes and ketones in the presence of BF$_3$, and the yields of α,β-unsaturated thiolesters are usually 70–80%. Ethers are the best solvents, presumably because they complex the BF$_3$ and release it slowly as it is needed in the reaction. Aldol condensation is the major side reaction. The reaction is usually stereoselective, giving mostly *trans* products, particularly from aldehydes. Unsymmetrical ketones give *cis-trans* mixtures. Cyclohexanone gives a product with the double bond in the ring or in the side chain, but the products from cyclopentanone and cycloheptanone are all exo (double bond in side chain).

The most likely mechanism involves a four-membered ring:

$$\underset{R'}{\overset{R}{>}}\overset{+}{C}-O\cdots\overset{-}{BF_3} + R''C\equiv CSEt \longrightarrow \left[\begin{array}{c} R'' \quad SEt \\ \boxed{\begin{array}{cc} 3 & 4 \\ 2 & 1 \end{array}} O \\ R \qquad \qquad \\ R' \qquad BF_3 \end{array} \right] \longrightarrow$$

$$\underset{R'}{\overset{R}{>}}C=C\underset{\underset{\displaystyle O}{\overset{\|}{C}-SEt}}{\overset{R''}{<}} \quad \text{and/or} \quad \underset{R}{\overset{R'}{>}}C=C\underset{\underset{\displaystyle O}{\overset{\|}{C}-SEt}}{\overset{R''}{<}}$$

The four-membered ring is flat, and the groups on positions 3 and 4 are in the plane of the ring. The groups on the 2-position are perpendicular to this plane. BF_3 is probably outside the plane of the ring, and is *trans* to the larger group on position 2. The smaller group on position 2 will turn toward BF_3 and give a *trans*-α,β-unsaturated thiolester.

When 1 equivalent of 1-alkynyl ether is added to an ether solution of equivalent amounts of carbonyl compound and $BF_3 \cdot Et_2O$ at 0–15°, and the finished reaction mix is hydrolyzed by aqueous potassium carbonate, substituted acrylate esters are formed.[16] Aliphatic aldehydes give low yields, but aromatic and heterocyclic aldehydes give 50–70% yields. Dialkyl ketones and alkyl aryl ketones give good yields, but diaryl ketones are poor. Esters give low yields, as does dimethylformamide. The reaction mechanism is similar to the one proposed for the analogous reaction with acetylenic thioethers.

4. EXAMPLES OF OTHER ADDITIONS

Some examples of electrophilic, nucleophilic and cycloaddition reactions of ethoxyethyne and thioethoxyethyne are listed in Table 2–31. The examples are taken from Aren's review[3] unless otherwise noted.

TABLE 2-31
Addition to Ethoxyethyne and Thioethoxyethyne[3]

Products from Addition to

Reagent	Adds First	HC≡C—O—Et	HC≡C—S—Et
		Electrophilic Addition	
(1) HCl	H^+	H_3CCCl_2—OEt	H_2C=C—SEt $\xrightarrow[H^+]{EtOH}$ CH_3CO_2Et + EtSH, with Cl substituent
(2) RCO_2H	H^+	$\left[H_3C-\overset{OCOR}{\underset{OCOR}{C}}-OEt \right] \longrightarrow (RCO_2)O + CH_3CO_2Et$	
(3) $H_2O + H^+$	H^+	$[H_2C{=}\overset{+}{C}-OEt] \xrightarrow{+H_2O} CH_3CO_2Et + H^+$	
(4) ROH or ArOH (BF_3)	H^+	$[CH_3C(OEt)_3] \longrightarrow CH_3CO_2Et + R_2O$ (or Ar_2O)	
(5) $(Cl_3CCO)_2O$	Cl_3CCO^+	$\left[\overset{O}{\overset{\|}{Cl_3CCCH}}{=}C\overset{OEt}{\underset{OCCCl_3}{\overset{\|}{\underset{O}{}}}} \right] \xrightarrow{ROH}$ $\overset{O}{\overset{\|}{Cl_3CCCH_2C}}-OEt + Cl_3CCO_2R$	

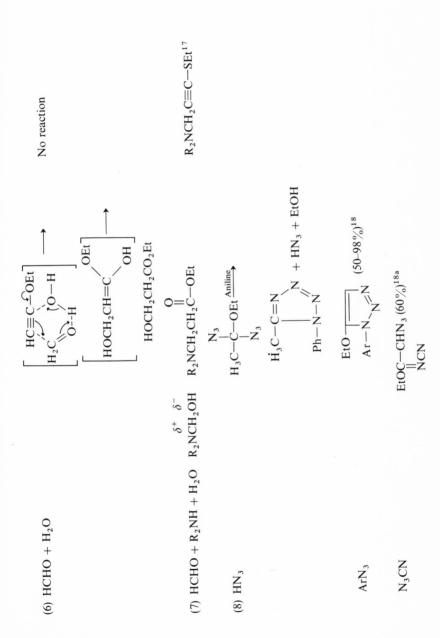

ACETYLENES AND ALLENES

TABLE 2-31—(continued)

Reagent	Adds First	Products from Addition to	
		HC≡C—O—Et	HC≡C—S—Et
		Electrophilic Addition—(continued)	

(9) $HO_2CC\equiv CCO_2H$

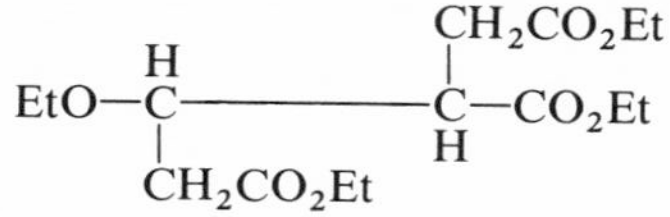

(10) $\underset{/}{\overset{\backslash}{C}}=O + BF_3$ — Adds first: $\underset{(+)}{\overset{\backslash}{C}}-\overset{(-)}{O}BF_3$

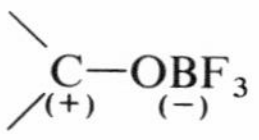

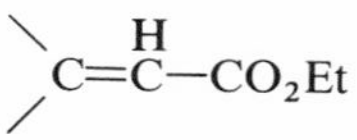

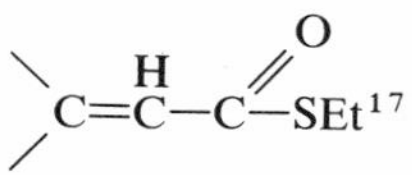

(11) $COCl_2$ — Adds first: $\overset{(+)}{Cl_2}\overset{(-)}{C}=O$

[19]

(12) SCl_2 — Adds first: $\overset{(+)}{S}\overset{(-)}{Cl_2}$

[19]

(13) $RCO_2Et(BF_3\cdot Et_2O)$ — Adds first: R^+

$EtO_2—C\overset{H\ R}{=}C—C=C—OEt$ [20—22]

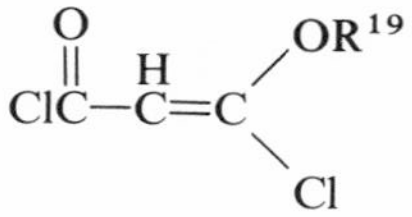

HC≡C—S—Et product (row 10):

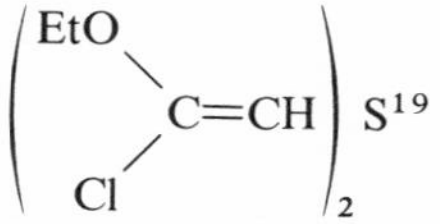

$\underset{/}{\overset{\backslash}{C}}=C\overset{H}{-}\overset{O}{\overset{\|}{C}}-SEt$ [17]

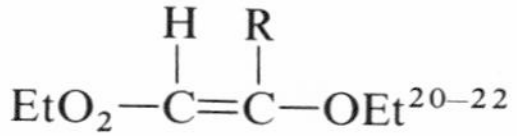

Nucleophilic Addition

(1) RNH_2 $H_3C-\overset{OR}{\underset{}{C}}=NR$ $RNHCH=CH-SEt$

     $2BuNH_2$ $H_3C-\overset{NHBu}{\underset{}{C}}=NBu$

(2) R_2NH $H_2C=\overset{OEt}{\underset{}{C}}\!-NR_2$ $R_2NCH=CH-SEt$

(3) $RNH_2 + H_2O$ $H_3C-\overset{O}{\overset{\|}{C}}-NHR + EtOH$

     $R_2NH + H_2O$ $H_3C-\overset{O}{\overset{\|}{C}}-NR_2 + EtOH$

     $R_3N + H_2O$ $\left[H_2C=\overset{OEt}{\underset{}{C}}\!-\overset{+}{N}R_3 \right] OH^-$

(4) $RO^- + ROH$ RO^- $H_3C-\overset{OEt}{\underset{}{C}}(OR)_2$ $ROCH=CH-SEt$

TABLE 2-31—(continued)

Products from Addition to

Reagent	Adds First	HC≡C—O—Et	HC≡C—S—Et
		Nucleophilic Addition—(continued)	
(5) OH⁻ + H₂O	OH⁻	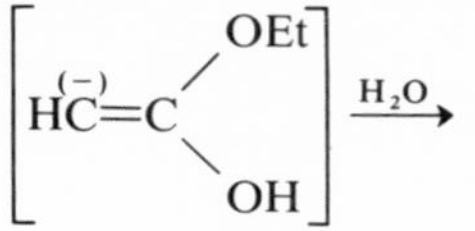 $\left[HC\overset{(-)}{=}C\overset{OEt}{\underset{OH}{}} \right] \xrightarrow{H_2O}$ $CH_3CO_2^- + EtOH$	
(6) Ethylenediamine		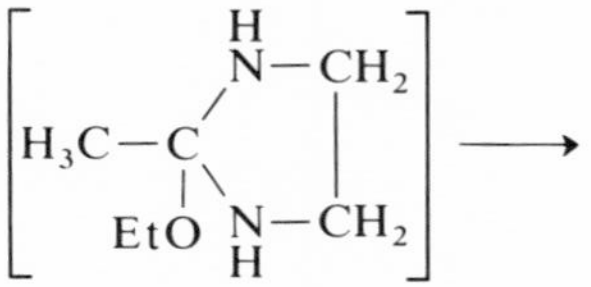 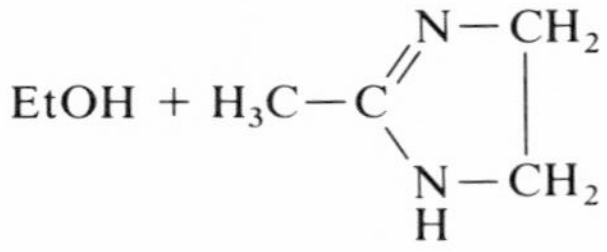 $\left[H_3C-C\overset{\overset{H}{N-CH_2}}{\underset{\underset{H}{N-CH_2}}{\underset{EtO}{}}} \right] \longrightarrow$ $EtOH + H_3C-C\overset{N-CH_2}{\underset{\underset{H}{N-CH_2}}{}}$	
(7) RS⁻ + R'OH	1 mole RS⁻	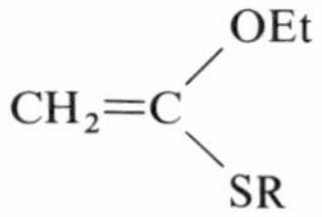 $CH_2=C\overset{OEt}{\underset{SR}{}}$	RSCH=CHSEt *cis*
	2 moles RS⁻		(RS)₂CHCH₂SR

(8) Ethanolamine

$$H_3C-C \overset{O}{\underset{N}{\diagdown}} \overset{CH_2}{\underset{C}{\diagup}} H_2 \quad + \quad EtOH$$

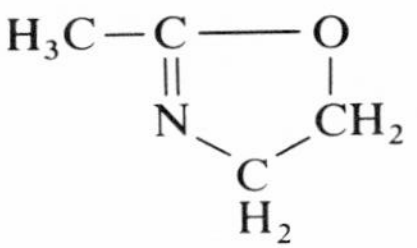

(9) CH_2N_2

$\overset{(-)}{C}H_2\overset{(+)}{N_2}$

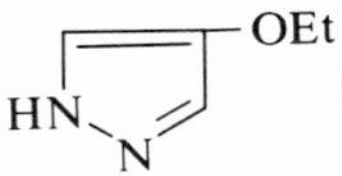

(pyrazole) OEt, HN—N (very slow)

SEt, N=, N—H (faster)[23]

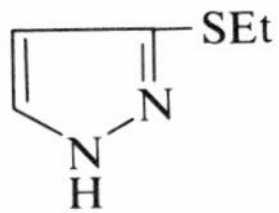

(10) :CCl_2

:CCl_2 → $Cl_2C=CHCHO$

(11) $R_2C=NOH$

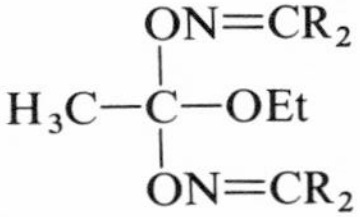

$$H_3C-\overset{ON=CR_2}{\underset{ON=CR_2}{C}}-OEt$$

(12) EtSH

$$H_2C=C\overset{OEt}{\underset{SEt}{}}{}^{24}$$

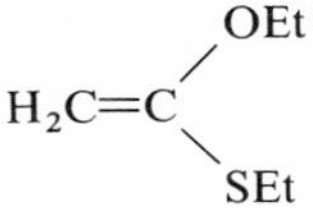

[made directly from $ClCH_2CH(OEt)_2$, $NaNH_2$, and EtSH in NH_3]

TABLE 2-31—(continued)

Products from Addition to

Reagent	Adds First	HC≡C—O—Et	HC≡C—S—Et
		Addition to Both Carbons Followed by Ring Closure	
(1) Ketenes	$\overset{\delta^+ \;\; \delta^-}{R_2C=C=O}$	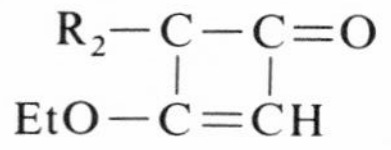 $\begin{array}{c} R_2-C-C=O \\ \;\;\;\; \vert \;\;\;\; \vert \\ EtO-C=CH \end{array}$	
(2) Diphenylketene (+ RC≡C—OEt)	$\overset{\delta^+ \;\; \delta^-}{Ph_2C=C=O}$	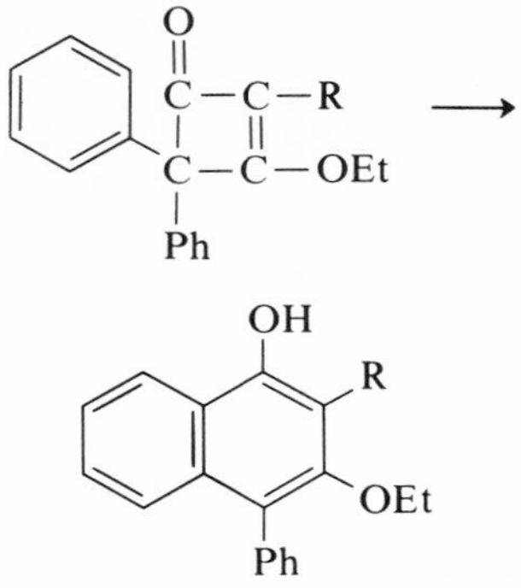 (see structure below)	
(3) PhNCO		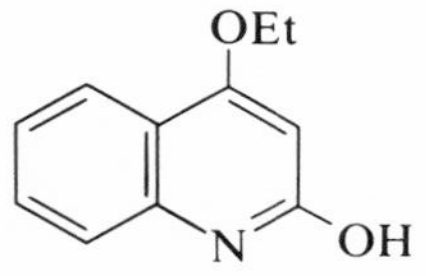 (see structure below)	

Structure for (2):

$$O=\overset{}{C}-\overset{}{C}=\overset{}{C}-R \quad\quad \longrightarrow$$
$$\;\;\;\; \vert \;\;\;\;\;\; \vert \;\;\;\; \vert$$
$$\;\;\;\; EtO-C \;\; C-OEt$$
$$\;\;\;\;\;\;\;\;\;\; \vert$$
$$\;\;\;\;\;\;\;\;\;\; Ph$$

(naphthalene ring product with OH, R, OEt, Ph substituents)

(quinoline ring product with OEt, OH, N substituents)

TABLE 2-31A
Addition to Substituted Acetylenic Ethers and Thioethers

Ether	Addend	Conditions	Product	% Yield	Reference
$MeSC \equiv C(CH_2)_2C \equiv CH$	HCl	$CHCl_3$, $-6°$	$MeSCCl = CH(CH_2)C \equiv CH$	77	25
$Me_2C(OH)C \equiv COEt$	$3PhCH_2SH$		$Me_2C = CCH(SCH_2Ph)_2$ (with SCH_2Ph)	75	26
$MeCC \equiv CSEt$ (with $=O$)	Et_2NH		$MeCC = CHSEt$ (with NEt_2, $=O$)		27
$EtOCC \equiv CSEt$ (with $=O$)	R_2NH		$EtOCC = CHSEt$ (with ONR_2)		27
$CH_2 = CHC \equiv C-SR$	$HX (HCl, H_2O, Cl_2)$		$CH_2 = CHC = C-SR$ (with X, H)		28
$CH_2 = CHC \equiv C-SR$	PrLi		$BuCH = C = CHSR$ (60%)		28

TABLE 2-31A—(continued)

Ether	Addend	Conditions	Product	% Yield	Reference
$\overset{O}{\underset{\|\|}{}}$ EtOC—C≡C—SEt	PhNHNH₂	Reflux	EtO₂C—C—CH=NNHPh + (=NNHPh) 3-(ethylthio)-indole-2-carboxylate		27
RC≡CCO₂Et	PhNHNH₂		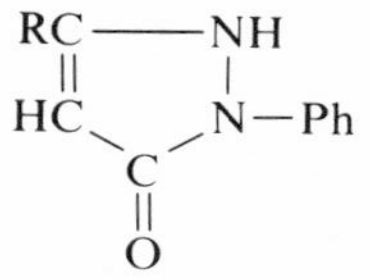 RC—NH, HC...N—Ph, C=O 3-alkyl-1-phenyl-5-pyrazolones		24
EtSC≡CCO₂Et	Butadiene (Diels-Alder addition)	130°, 5 hr	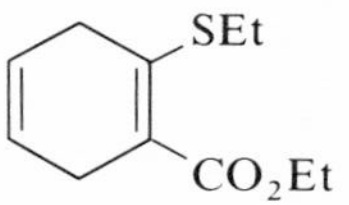 (SEt, CO₂Et cyclohexene)	32	27

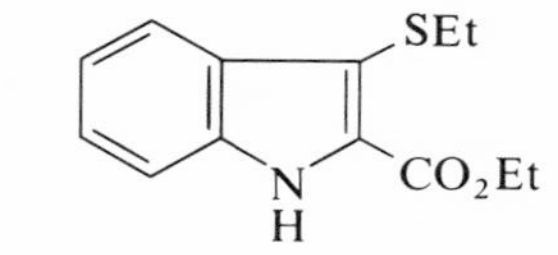

5. THERMAL REARRANGEMENT-CYCLIZATION OF ALKYNYL ETHERS

Aliphatic acetylenic ethers decompose thermally via aldoketene to eliminate olefin and form cyclobutenone ethers. Loss of olefin requires at least one β-hydrogen. This is a *cis* elimination and probably involves a cyclic transition state. Acetylenic methyl or neopentyl ethers have no β-hydrogen and polymerize instead of losing alkene when heated.[30,31] Acetylenic *t*-butyl ethers are not thermally stable. At 80° they evolve isobutylene briskly and form cyclobutenone ethers. 1-Ethoxy-1-alkynes react slowly even at 100°:

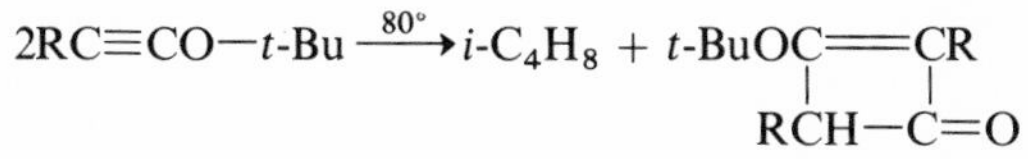

$$2RC\equiv CO-t\text{-Bu} \xrightarrow{80°} i\text{-}C_4H_8 + t\text{-BuOC}\!\!=\!\!\!\!\underset{\substack{| \\ RCH-C=O}}{}\!\!CR$$

In refluxing CCl_4, the vigorous thermal rearrangement of acetylenic benzyl ethers can be controlled to give high yields of 1,3-dialkyl-2-indanones which are difficult to make by other methods:[31a]

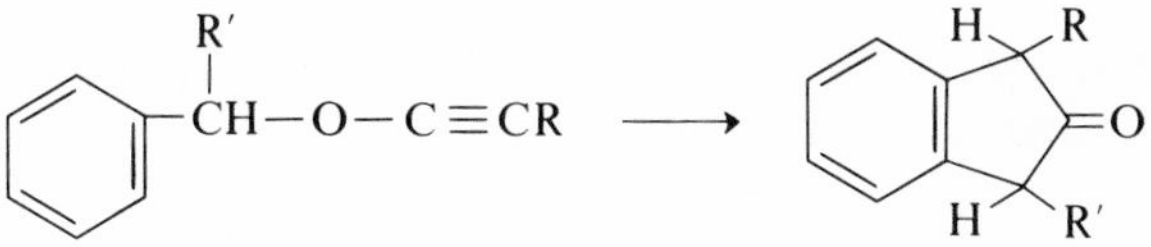

In the decomposition of 1-ethoxy- and 1-isopropoxy-1-heptyne, the rate-determining steps are monomolecular. Intermediate aldoketenes apparently form in a concerted process, followed by slower addition of aldoketene to ether:

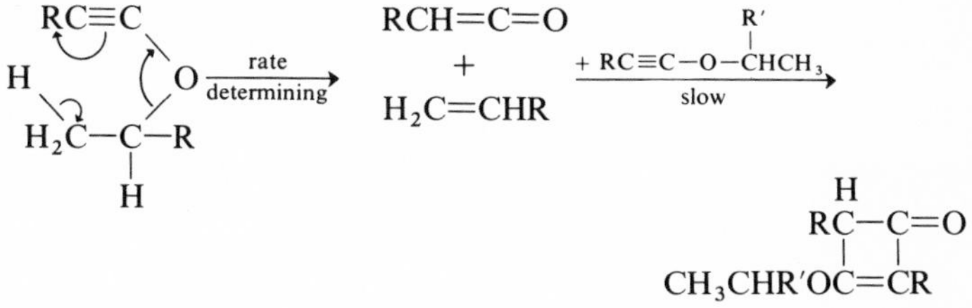

Ketene adds to acetylenic ethers at 0° to form cyclobutenone ethers.[31b]

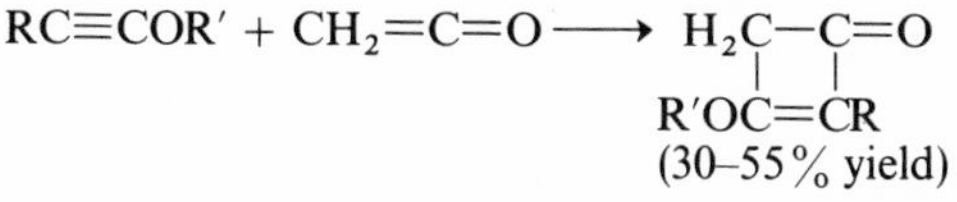

$$RC\equiv COR' + CH_2\!\!=\!\!C\!\!=\!\!O \longrightarrow \underset{\substack{| \quad | \\ R'OC=CR}}{H_2C-C=O}$$
(30–55% yield)

This substantiates the proposed mechanism. Ethoxyacetylene and diphenyl-ketene do not give cyclobutenone[3] but instead give 3-ethoxy-4-phenyl-α-naphthol.

The intermediate is probably 1-ethoxy-3-exo-3a-phenyl-3,3a-dihydroazulene,[32,33] from ring enlargement of one of the phenyl groups:

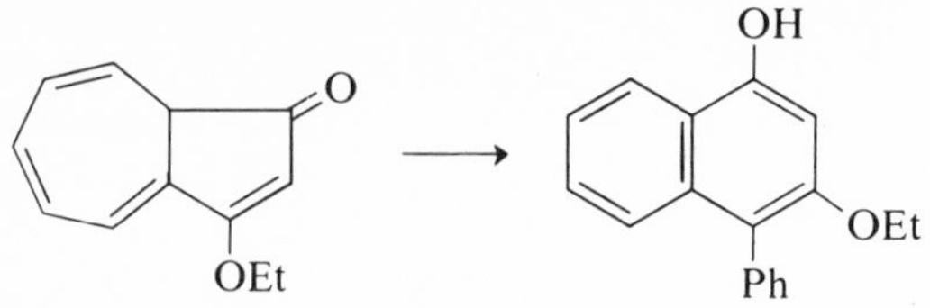

Thioketene intermediates are presumed to exist in some addition reactions. Lithium alkynylides react with olefins to form RC≡CSLi, which adds RSH to form dithioesters.[34]

$$R'C{\equiv}CSLi + RSH \rightleftarrows R'C{\equiv}CSH + RSLi$$

$$\updownarrow$$

$$R'CH{=}C{=}S \xrightarrow{+RSH} R'CH_2{-}C(SR){=}S \quad (15\text{–}50\%)$$

6. ADDITION OF ORGANOLITHIUM COMPOUNDS TO 1-ALKYNYL ETHERS

Lithium dialkylamides react with 1-alkynyl ethers to form 1-alkynylamines in a simple synthesis of the ynamines:

$$R'C{\equiv}C{-}OEt + LiNR_2 \xrightarrow{\text{ether}} \left[R'C(Li){=}C(NR_2)(OEt) \right] \longrightarrow R'C{\equiv}C{-}NR_2 + LiOEt$$

Here, the intermediate is probably the addition product shown. Other organolithium compounds react with 1-alkynyl ethers in a similar way:[31b]

$$C_4H_9C{\equiv}C{-}OEt + RLi \xrightarrow{\text{ether}} C_4H_9C{\equiv}C{-}R + LiOEt$$

R	% yield
n-Bu	70
t-Bu	50
Me	34
Ph	60
1-Cyclohexenyl	77
1-Thienyl	33

Ethoxyhexyne and hexynyllithium in dioxane form two products, 30% total yield:

$$C_4H_9C{\equiv}COC_2H_5 + C_4H_9C{\equiv}CLi \xrightarrow[100°]{dioxane}$$

$$\xrightarrow[-LiOC_2H_5]{heat} C_4H_9C{\equiv}C{-}C{\equiv}CC_4H_9$$

References

1. Jacobs, T. L., Cramer, R., and Weiss, F. T., *J. Am. Chem. Soc.*, **62**, 1849 (1940).
2. Slimmer, M., *Chem. Ber.*, **36**, 289 (1903).
3. Arens, J. F., in "Advances in Organic Chemistry: Methods and Results," Vol. 2, p. 117, New York, Interscience Publishers, 1960.
4. Volger, H. C., and Arens, J. F., *Rec. Trav. Chim.*, **77**, 1170 (1958).
5. Zwanenburg, B., *Rec. Trav. Chim.*, **82**, 593 (1963).
6. Wasserman, H. H., and Wharton, P. S., *J. Am. Chem. Soc.*, **82**, 661 (1960).
7. Newman, M. S., and Courduvelis, C., *J. Am. Chem. Soc.*, **88**, 781 (1966).
8. Arens, J. F., *Rec. Trav. Chim.*, **74**, 769 (1955).
9. Arens, J. F., *Festschr. Arthur Stoll*, 468 (1957).
10. Panneman, H. J., Marx, A. F., and Arens, J. F., *Rec. Trav. Chim.*, **78**, 487 (1959).
11. Sheehan, J. C., and Hlovka, J., *J. Org. Chem.*, **23**, 635 (1958).
12. Tadema, G., *et al.*, *Rec. Trav. Chim.*, **83**, 345 (1964).
13. Arens, J. F., *et al.*, *Rec. Trav. Chim.*, **75**, 1459 (1956).
14. Bos, L. B., and Arens, J. F., *Rec. Trav. Chim.*, **82**, 157 (1963).
15. Brandsma, L., Wijers, H. A., and Arens, J. F., *Rec. Trav. Chim.*, **81**, 583 (1962).
16. Vieregge, H., *et al.*, *Rec. Trav. Chim.*, **85**, 929 (1966).
17. Boonstra, H. J., and Arens, J. F., *Rec. Trav. Chim.*, **79**, 866 (1960).
18. Grunanger, P., Finzi, P. V., and Fabri, E., *Gazz. Chim. Ital.*, **90**, 413 (1960); *Chem. Abstr.*, **55**, 11397 (1961).
18a. Hermes, M. E., and Marsh, F. D., *J. Am. Chem. Soc.*, **89**, 4760 (1967).
19. van den Bosch, G., Bos, H. J. T., and Arens, J. F., *Rec. Trav. Chim.*, **85**, 567 (1966).
20. Krasnaya, Zh. A., and Kucherov, V. F., *Izv. Akad. Nauk SSSR Ser. Khim.*, **110**, (1965); *Chem. Abstr.*, **62**, 11682 (1965).
21. Krasnaya, Zh. A., *Izv. Akad. Nauk SSSR Ser. Khim.*, 313 (1965); *Chem. Abstr.*, **62**, 14483 (1965).
22. Krasnaya, Zh. A., and Kucherov, V. F., *Izv. Akad. Nauk SSSR Otd. Khim. Nauk*, 484 (1962); *Chem. Abstr.*, **57**, 16383 (1962).
23. Groen, S. H., and Arens, J. F., *Rec. Trav. Chim.*, **80**, 879 (1961).
24. Alkema, H. J., and Arens, J. F., *Rec. Trav. Chim.*, **79**, 1257 (1960).
25. Petrov, A. A., and Forost, M. P., *Zh. Organ. Khim.*, **1**, 1550 (1965); *Chem. Abstr.*, **64**, 577 (1966).
26. Wieland, J. H. S., and Arens, J. F., *Rec. Trav. Chim.*, **75**, 1358 (1956).
27. Bonnema, J., and Arens, J. F., *Rec. Trav. Chim.*, **79**, 1137 (1960).

28. Radchenko, S. I., and Petrov, A. A., *Zh. Anorg. Khim.,* **1**, 987 (1965); *Chem. Abstr.,* **63**, 11340 (1965).
29. Zoss, A. O., and Hennion, G. F., *J. Am. Chem. Soc.,* **63**, 1151 (1941).
30. Nazarov, I. N., and Semenovskii, A. V., *Izv. Akad. Nauk. SSSR Otd. Khim. Nauk.,* 1772 (1959); *Chem. Abstr.,* **54**, 10827 (1960).
31. van Daalen, J. J., Kraak, A., and Arens, J. F., *Rec. Trav. Chim.,* **80**, 810 (1961).
31a. Olsman, H., Graveland, A., and Arens, J. F., *Rec. Trav. Chim.,* **83**, 301 (1964).
31b. Rosebeek, B., and Arens, J. F., *Rec. Trav. Chim.,* **81**, 549 (1962).
32. Barton, D. H. R., *et al., Proc. Chem. Soc.,* 21 (1962); *J. Chem. Soc.,* 2708 (1962).
33. Druey, J., *et al., Helv. Chim. Acta.,* **45**, 600 (1962).
34. Schuijl, P. J. W., Brandsma, L., and Arens, J. F., *Rec. Trav. Chim.,* **85**, 889 (1966).

PART EIGHT

Addition to Negatively Substituted "Activated" Acetylenes: Acids, Esters, Ketones, Nitriles, Sulfones and Sulfoxides

INTRODUCTION

Winterfeldt published a review of nucleophilic additions to "activated" triple bonds in 1967.[1a]

1. SIMPLE ADDITION REACTIONS

1.1. Halogen Acids

Halogen acids add anion first to DMAD (dimethyl acetylenedicarboxylate) in nucleophilic addition, followed by addition of a proton.[1] Nucleophilic chloride adds in the rate-determining step, but the stronger nucleophiles bromide, iodide, and thiocyanate add very rapidly, and the rate-determining step is subsequent addition of a proton. HI adds 192 times as fast to DMAD as to methyl propiolate.[2]

1.2. Hydrazoic Acid

HN_3 adds to DMAD in 90% acetic acid to form *trans*-MeO_2CC=CCO_2Me.[3]
(with N_3 and H substituents shown on the double bond carbons)

HN_3 adds to aryl ethynyl ketones to form 45–75% yield of isoxazoles and some arylvinylazides.[4]

1.3. Methanol

Methanol and sulfuric acid esterify monopotassium acetylenedicarboxylate, but some alcohol adds across the triple bond to form methoxyfumarate, methoxymaleate and α,α-dimethoxysuccinate.[5] DMAD and ammonia gave mostly diamide, but some aminofumaramide and α-aminofumaraminate form by addition of ammonia.

1.4. Thiols

Truce and Heine[6] tested Truce's general rule[7] that nucleophilic additions to acetylenes usually go *trans* to form *cis* products. In addition of *p*-toluenethiol, ethyl propiolate obeys the rule, and gives *cis* product. Sodium propiolate gives mostly *trans* product. Phenyl ethynyl ketone gives *cis* product. The usual *trans* addition is caused by coulombic repulsion between the attacking negative thiolate ion and the pair of electrons being displaced from the triple bond. This forces the displaced electron pair to the *trans* position where it abstracts a proton from solvent. With propiolate anion, competing repulsion between carboxyl and thiolate anions forces these groups into *trans* positions, and overall addition goes *cis*.

Electronegative substituents have a powerful effect on the reactivity of the triple bond toward nucleophilic attack. The three compounds (above) react very rapidly and exothermically, while 1-alkynes require drastic conditions for thiolate addition.[8] The reactivity of substituted acetylenes in the addition of thiolate varies with the substituent:

$$ArC{=}O > CO_2R > CO_2^- > Ph > Cl > ArS > alkyl$$

Propiolic esters add thioglycolic acid esters, and the products cyclize to form thiophene esters[9]

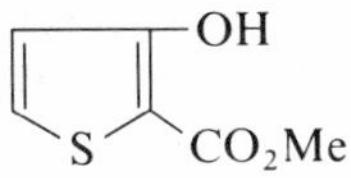

Bohlmann[10] applied such reactions to the synthesis of naturally occurring acetylenic thiophenes. *Cis* products cyclized to one product, while *trans* products cyclized to another:

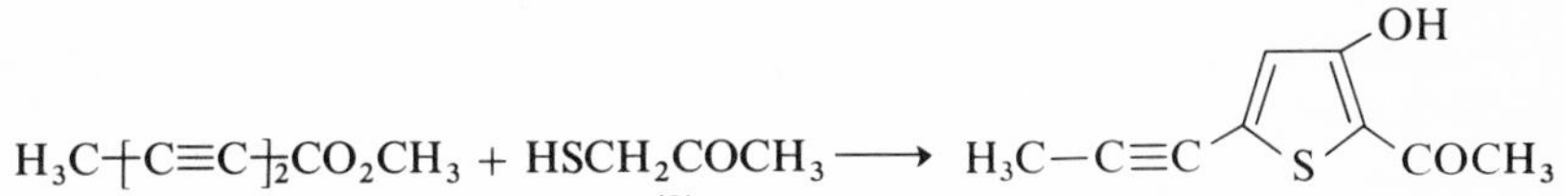

$$H_3C{+}C{\equiv}C{\}_2CO_2CH_3 + HSCH_2COCH_3 \longrightarrow H_3C{-}C{\equiv}C$$
(1)   (2)

(3)

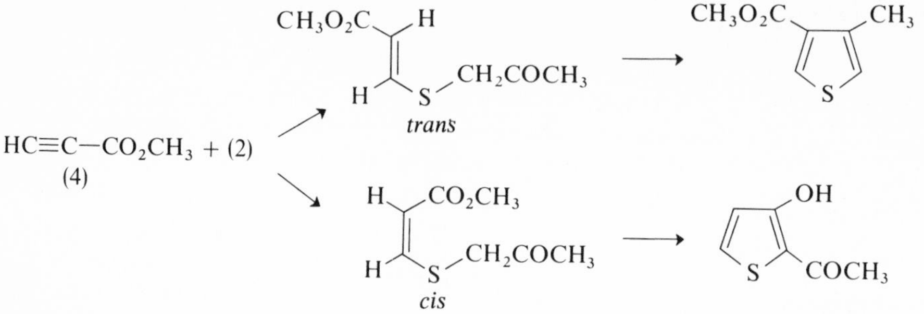

Bohlmann[11] had found earlier that mercaptans add mostly *cis* to benzoyldiacetylene if resonance stabilization of the primary carbanion is possible. Thiolacetone is similar to thiolglycolate esters in its reaction with methyl propiolate. Both *cis* and *trans* products form. Tetrolic and phenylpropiolic esters give *cis* additions with thiolglycolate esters. The *cis* addition is not changed by β-substituents in the acetylenic esters but depends on resonance stabilization of the initial anions. Where this stabilization is not possible, as in simple acetylenics, addition goes directly *trans*.

1.5. Dithiocarbamic and Dithioformic Acids

Addition of dithiocarbamic acid and its N-substituted derivatives to olefins is well known. Garraway[12] in 1962 first applied these additions to acetylenic bonds:

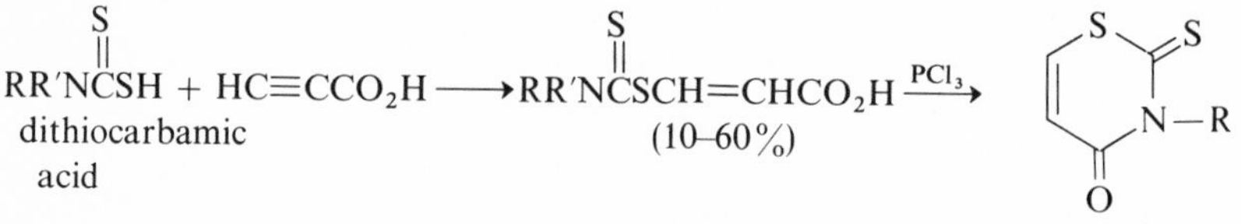

thiothiazines

Garraway also was first to add alkoxydithioformic acids to α,β-unsaturates. The *cis*-3-substituted acrylic acids form by *trans* addition. Yields from alkoxydithioformic acids are 5–20%.

$$\underset{\text{alkoxydithioformic acid}}{\mathrm{RO\overset{S}{\overset{\|}{C}}SH}} + \mathrm{HC{\equiv}CCO_2H} \longrightarrow \mathrm{RO\overset{S}{\overset{\|}{C}}SCH{=}CHCO_2H}$$

1.6. Amines

Secondary amines add exclusively *cis* to DMAD to give *trans* products. Alcohols also add *cis* in the presence of tertiary amine catalysts.[13] Both *cis*[6,14–16] and *trans*[16] addition of amines to various acetylenics have been reported. A given amine addition reaction can be *cis* or *trans* depending on the solvent.[17] Aziridine adds to DMAD in methanol to give 67% *trans*-ester (fumarate) and 33% *cis*-ester (maleate). In dimethyl sulfoxide, fumarate is 5%; maleate, 95%. Reaction with methyl propiolate is similar. The zwitterion

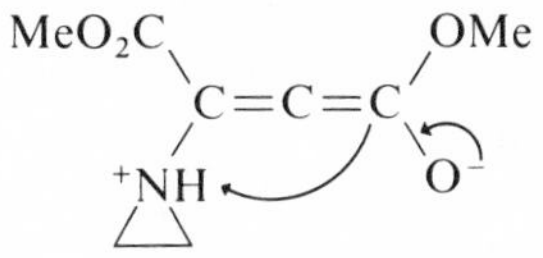

may be the intermediate.[17] In the absence of external protons, as in DMSO, the zwitterion collapses stereospecifically via intramolecular protonation to form *cis*-ester groups. In protic methanol, solvation of the incipient Michael anion causes protonation by methanol. The reaction goes as well in cyclohexane-benzene as in DMSO.

Hendrickson[16] has written "whenever mobile protons are available, the first product of additions to acetylenedicarboxylic esters is the simple *trans* adduct." Dolfini[17] feels that this is an over-simplification of the reaction mode and may be erroneous, except where the "mobile protons are derived from solvent." Although Hendrickson reported *trans* addition in aprotic benzene, he did not show that the product was not formed at least partly from isomerization. Dolfini proved that his products did not isomerize to each other.

When 1,2,3-triphenylaziridine reacts with diethyl acetylenedicarboxylate, the C—C bond cleaves as it does in tetracyanoethylene oxide.[8]

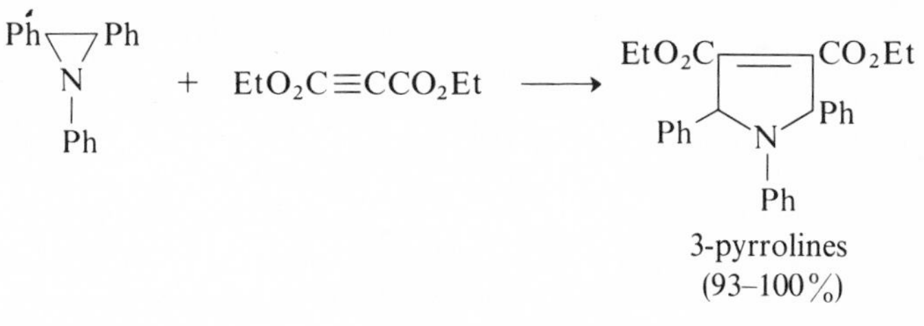

3-pyrrolines
(93–100%)

Azetidines add to DMAD to give maleates if aprotic solvents are used.[8a]
Primary amines and diacetylenic ketones give γ-pyrilones[19]

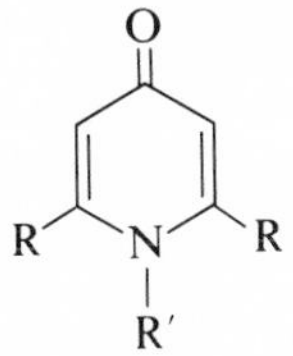

NH adds across both triple bonds, in steps. Aniline gives mostly pyrrolinone with bis(phenylethynyl) ketone, but if one of the phenyls of the ketone is substituted *para* with NO_2, only the pyrilone forms.

Addition of secondary amines to activated acetylenes at 20° in methanol gives only *trans*-enamines, except in one case: 1-propynyl benzyl sulfoxide gives a little *cis* product.[20] Ethyleneimine is an exceptional amine. With tolyl-*p*-sulfonylacetylene it gives only the *cis* product, and in other cases mixtures of *cis* and *trans* products, as it does with ethyl propiolate. Yields are 93–99% with ketones, esters, sulfones and sulfoxides.

Addition of primary amines is not as clear-cut. The ratio of *cis* to *trans* product depends on the solvent, the amine and the structure of the acetylenic compound. Deuterochloroform favors the *cis* isomer, dimethyl sulfoxide favors the *trans*. In deuterochloroform, little hydrogen bonding with solvent is possible, while in dimethyl sulfoxide, hydrogen bonding with solvent can occur. If hydrogen bonding with solvent is impossible, intramolecular hydrogen bonding occurs, and this favors the *cis* isomer.[20] Truce and Brady[21] reached the same conclusion from their recent study.

TABLE 2-32
Adducts from Activated Acetylenes and Primary Amines[20]

Acetylene	Amine	% Yield	Total Product	
			% *cis*	% *trans*
$PhCOC\equiv CH$	$PhCH_2NH_2$	98	100	0
$PhCOC\equiv CH$	Me_3CNH_2	94	100	0
$PhCOC\equiv CH$	$PhNH_2$	88	75	25
$MeOCOC\equiv CH$	$PhCH_2NH_2$	99	70	30
$MeOCOC\equiv CH$	Me_3CNH_2	64	100	0
$PhN(Me)COC\equiv CH$	$PhCH_2NH_2$	86	85	15
$PhCH_2SOCH_2C\equiv CH$	$PhCH_2NH_2$	99	80	20
$PhCH_2SO_2CH_2C\equiv CH$	$PhCH_2NH_2$	99	35	65

Isatoic anhydrides in refluxing methanolic sodium methylate add to dimethyl acetylenedicarboxylate. The product *trans*-enamines probably result from Michael addition of the amine group of the methyl anthranilates generated *in situ*. The enamines cyclize to 2,8-dicarboalkoxy-4(1H)-quinolines when heated:[20a]

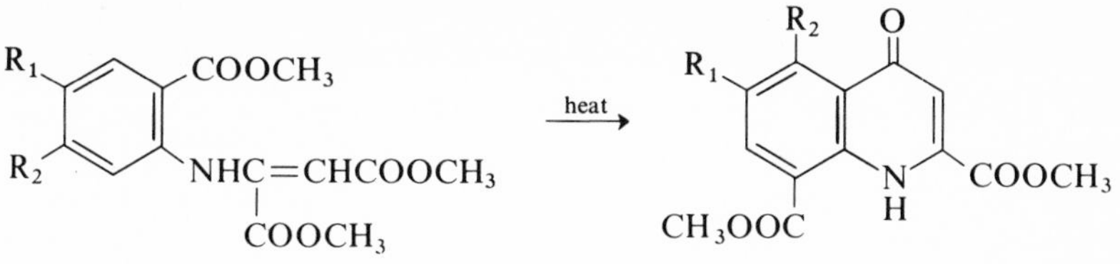

(60–80%)

p-Toluoylacetylene and *p*-toluoylacetaldehyde react with α-phenylethylamine to give the same product:[22]

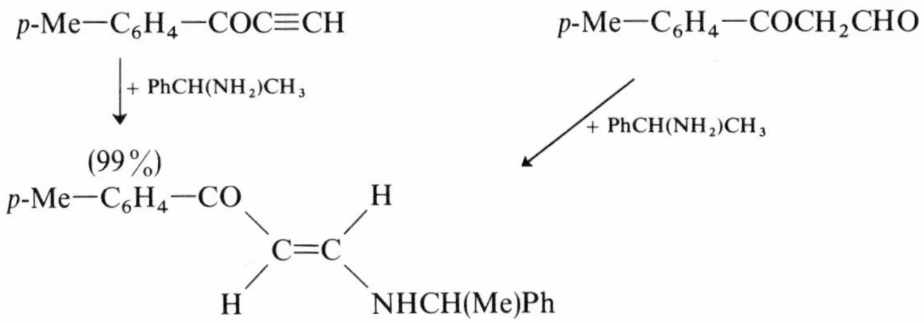

Tertiary amine salts also add.[21] This is the basis of a new synthesis of 1,5-diphenylpyrazolone-3.[24]

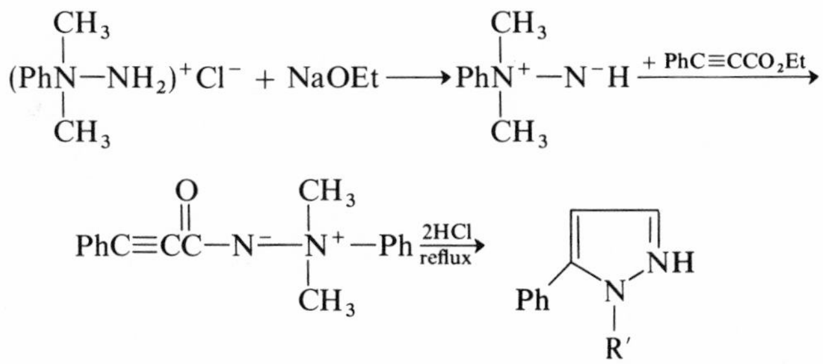

When methyl propiolate is added to primary or secondary amines in the cold, 1:1 adducts form. However, when primary amines and methyl propiolate are heated to 100°, diadducts form.[24a] If monoadducts (1:1) are heated with methyl propiolate, the diadducts form. Secondary amines do not give diadducts. A possible mechanism for formation of the diadducts is:

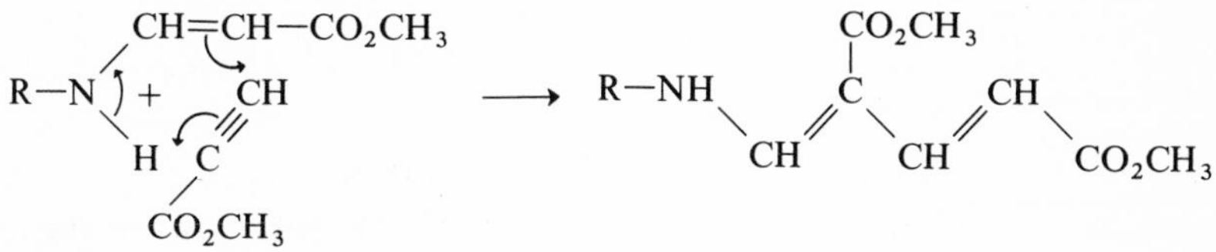

monoadduct + methyl propiolate

Secondary amines add to bis(2-propynyl)sulfones to form divinylsulfones or cyclic dienamines.[23b] The reaction with piperidine is typical:

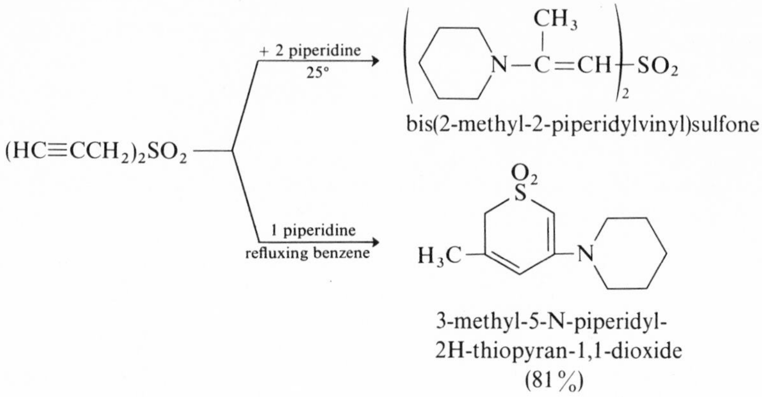

bis(2-methyl-2-piperidylvinyl)sulfone

$(HC{\equiv}CCH_2)_2SO_2$

3-methyl-5-N-piperidyl-
2H-thiopyran-1,1-dioxide
(81%)

1.7. Hydroxylamines

DMDA reacts with N-phenylhydroxylamine to form an unstable 2:2 and a stable 2:1 (ester:hydroxylamine) product:[23]

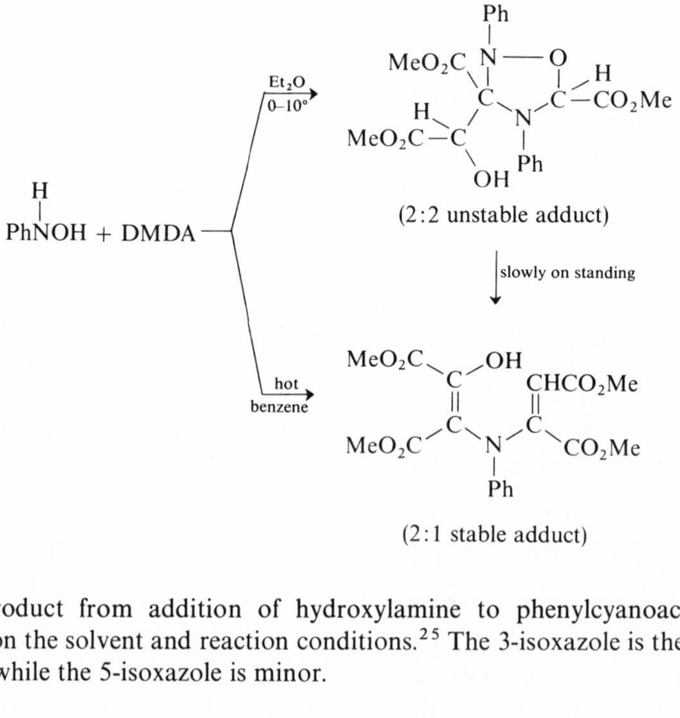

(2:2 unstable adduct)

slowly on standing

(2:1 stable adduct)

The product from addition of hydroxylamine to phenylcyanoacetylene depends on the solvent and reaction conditions.[25] The 3-isoxazole is the major product, while the 5-isoxazole is minor.

1.8. Phosphines and Phosphites

Trialkyl or triarylphosphines add to negatively substituted ethylenes to form ethylphosphonium salts.[26] Phosphines add *cis* to negatively substituted acetylenes:[27]

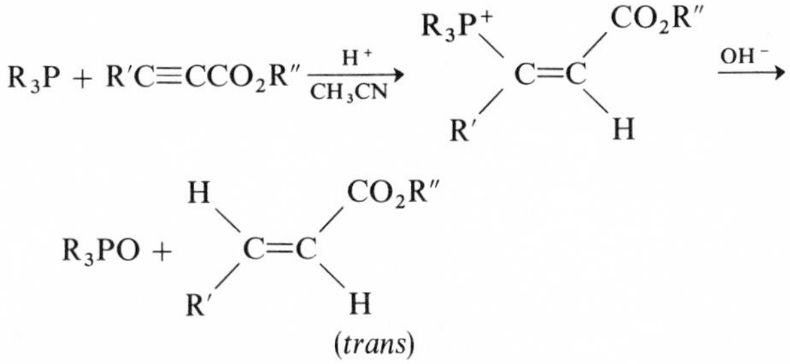

$$R_3P + R'C{\equiv}CCO_2R'' \xrightarrow[CH_3CN]{H^+} \underset{R'}{\overset{R_3P^+}{\diagdown}}C{=}C\underset{H}{\overset{CO_2R''}{\diagup}} \xrightarrow{OH^-}$$

$$R_3PO + \underset{R'}{\overset{H}{\diagdown}}C{=}C\underset{H}{\overset{CO_2R''}{\diagup}}$$

(*trans*)

Phenylacetylene does not react, but diphenylacetylene does. Diphenyldiacetylene adds phosphine very slowly. Acids, esters, amides and aldehydes react well (25–90% yields).

Two moles of dicyanoacetylene react with triphenylphosphine in acetonitrile at room temperature to give a different kind of product, a percyanophosphole[28] (13.5%)

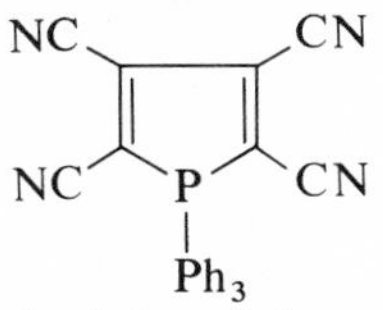

At −50° in ether, DMAD and triphenylphosphine give different products in nitrogen and in CO_2.[29]

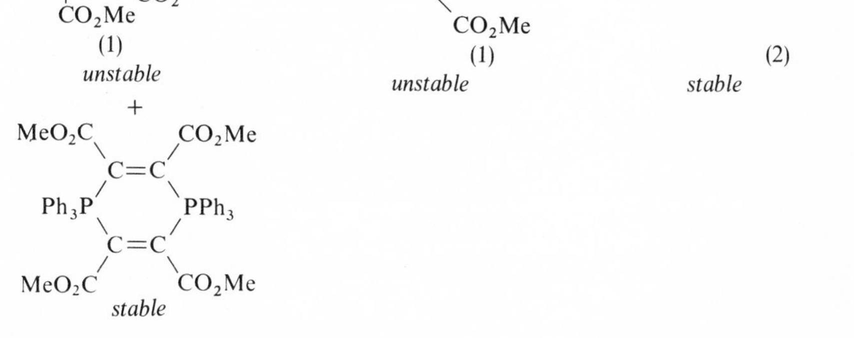

$$DMAD + (Ph)_3P \xrightarrow{Et_2O, N_2, -50°}$$

(1) unstable

(1) unstable

(2) stable

+

stable

Dialkylphosphites add to the carbonyl group of propargyl aldehyde to form compounds such as[29a]

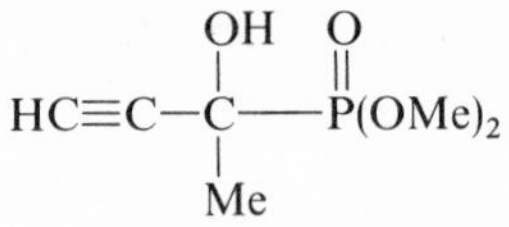

$$
\begin{array}{cc}
& \text{OH} \quad \text{O} \\
& | \qquad \parallel \\
\text{HC} \equiv \text{C} - \text{C} - \text{P(OMe)}_2 \\
& | \\
& \text{Me}
\end{array}
$$

These products react with $(RO)_2PCl$ via an allenic intermediate to give

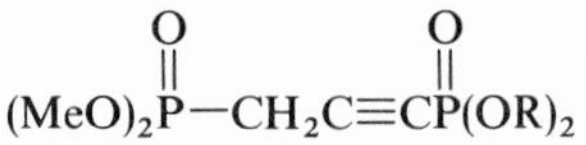

$$
\begin{array}{cc}
\text{O} & \text{O} \\
\parallel & \parallel \\
(\text{MeO})_2\text{P} - \text{CH}_2\text{C} \equiv \text{CP(OR)}_2
\end{array}
$$

Dialkyl phosphorodithioates add across the acetylenic bond of propargyl aldehyde, not to the carbonyl group, to give products such as

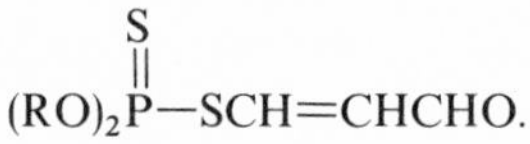

$$
\begin{array}{c}
\text{S} \\
\parallel \\
(\text{RO})_2\text{P} - \text{SCH} = \text{CHCHO}.
\end{array}
$$

1.9. Other Additions

Other examples of simple addition reactions of negatively substituted acetylenes are given in Table 2-33.

2. MICHAEL ADDITION

Michael addition to acetylenic bonds is a special case of nucleophilic addition: A carbanion adds across the triple bond of a negatively substituted acetylene, such as an ester, ketone or nitrile (the acceptor). The number of addends, or donors, is large, and includes malonic and cyanoacetic esters, which usually require a basic catalyst to ionize the acidic α-C—H bond to form carbanion. Some Michael additions go without catalysts, especially if the addend is already partly ionized. Bergmann[47] reviewed Michael addition reactions in 1959. His review shows that much more work has been done with ethylenic acceptors than with acetylenic acceptors.

Acetylenic acceptors sometimes add two molecules of addend, or the initial products often cyclize. The tendency to cyclize has been utilized to make numerous heterocyclic compounds. Only a few recent examples of Michael addition are reviewed here.

2.1. Malonate Esters as Addends

Dimethyl acetylenedicarboxylate reacts with dimethyl malonate to give (1) and (2), which react further in the presence of potassium acetate to form (3).[48]

TABLE 2-33
Addition Reactions of Negatively Substituted Acetylenes

(A) Addition to Monoacids, Nitriles and Ketones

Acetylenic Compound	Addend	Conditions	Product	% yield	Reference
Ethyl propiolate	MeNHNHMe	Ligroin (60°)	$(EtO_2CCH{=}CHN{-})_2$ (Me on N)		30
Ethyl propiolate	MeNHNHMe	Et_2O, HCl—HOAc	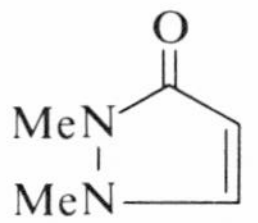 1,2-Dimethyl-5-pyrazolone (MeN–MeN, C=O ring)		30
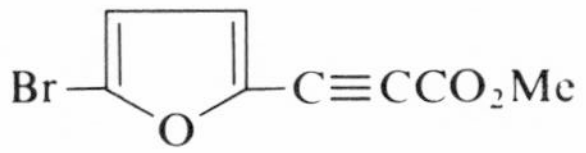 Br–furyl–$C{\equiv}CCO_2Me$	N_2H_4	MeOH, 1 hr	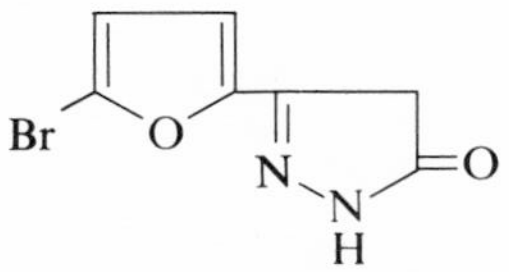 3-(5-bromo-2-furyl)-pyrazolone-5	90	31
Methyl propiolate	2-Amino-pyridine	10–20°	2H-Pyrido[1,2-a]-pyrimidine-2-one + cyclic monoadduct + acyclic monoadduct		32
$PhC{\equiv}CCOR$ (OO)	$ArNH_2$	Ether, 20°	$ArNHC{=}CHCCOR$ (Ph, OO)	42–85	33
Propiolic acid	HCl	Cu_2Cl_2, cold	$cis\text{-}ClCH{=}CHCO_2H$	82	34

TABLE 2-33—(continued)

(A) Addition to Monoacids, Nitriles and Ketones—continued

Acetylenic Compound	Addend	Conditions	Product	% Yield	Reference
MeC≡CCN	HBr	Ether, 0°	15% cis-, 85% trans-MeCBr=CHCN	55	35
MeC≡CCN	HBr	Nonpolar solvent, 0°	MeCBr₂CH₂CN		35
MeC≡CCO₂Et	HBr	Ether, 0°	10% cis-, 85% trans-MeCBr=CHCO₂Et	45	35
$\overset{O}{\overset{\|}{Me}C}$≡CEt	HBr		$\overset{O}{\overset{\|}{Me}C}$CH=CBrEt	20	36
Propiolic acid	ArSeH	NaOEt catalyst, −10°	trans-ArSeCH=CHCO₂H		37
HC≡CCO₂R (R = H or Me)	Cyclohexanone 2 moles	KOH or KO—t-Bu	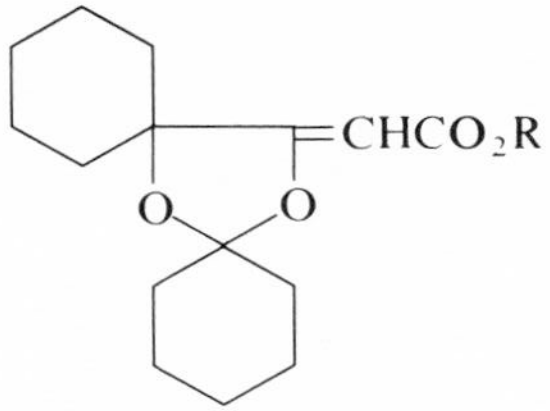 cis and trans	85	37a
DMAD (Dimethyl acetylenedicarboxylate)	o-Phenylene-diamine		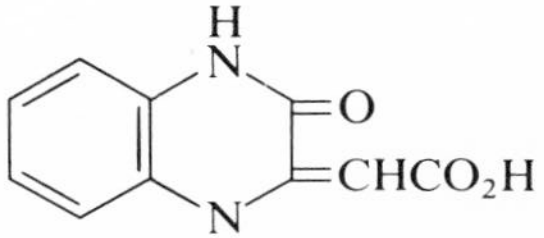 quinoxaline		38

DMAD	Et₃N, Et₃NHBr	40°	Diethylaminomaleate	100	39
DMAD	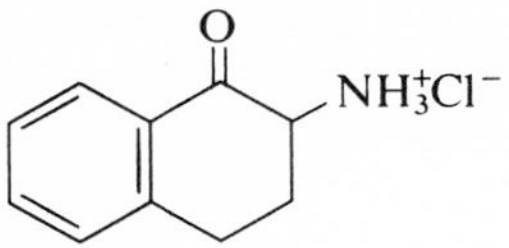 [structure] NH₃⁺Cl⁻	MeOH, 0°	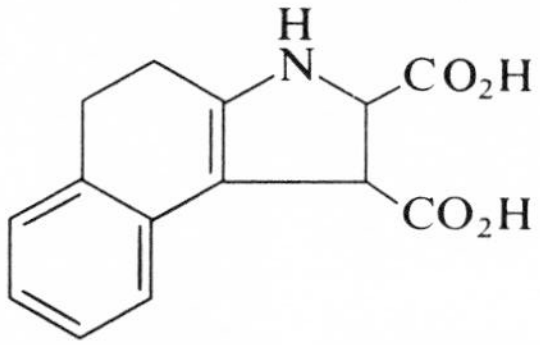 H N—CO₂H, CO₂H (new route to 4,5-dihydro-3H-benz(e)indoles)	65	41
DMAD	cis-2,3-Diphenyl-aziridine	Benzene, reflux 12 hr	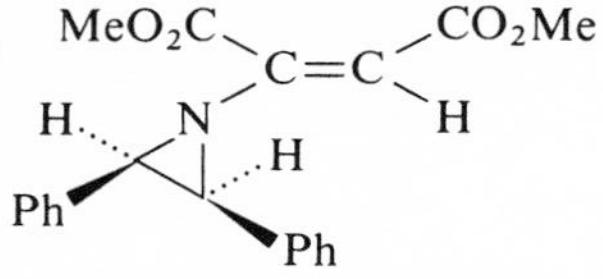 MeO₂C—C=C—CO₂Me H N Ph, Ph H	85	42
DMAD	trans-2-Phenyl-3-benzoyl-aziridine		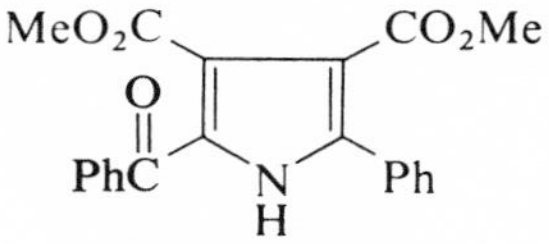 MeO₂C, CO₂Me PhC N Ph H O	80	42
DEAD (Diethyl acetylenedicarboxylate)	Guanidine		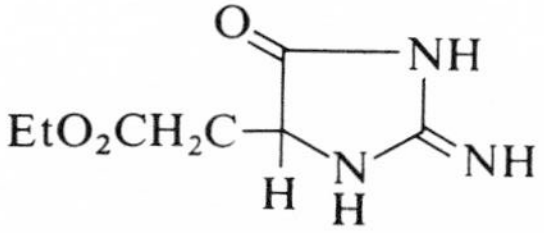 EtO₂CH₂C NH NH N H H O, 2-imino-4-oxo-5-ethoxycarbonyl-methyleneimidazoline	low	40
DEAD	Thiourea		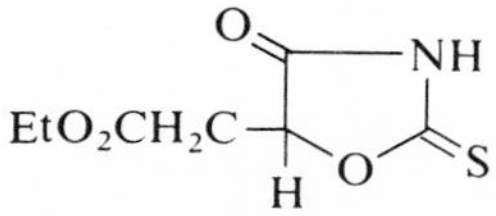 EtO₂CH₂C NH O S H O	82	40

TABLE 2-33—(continued)

(A) Addition to Monoacids, Nitriles and Ketones—continued

Acetylenic Compound	Addend	Conditions	Product	% yield	Reference
DMAD	1,3-Dimethyl-thiourea	MeOH, 20°	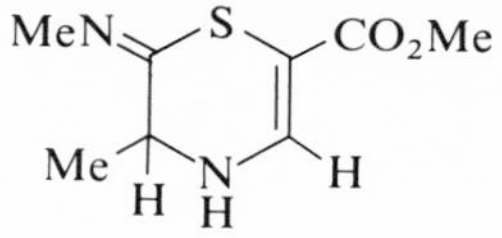 (thiazine structure: MeN–S–CO₂Me ring)	75	40a
DMAD	Thiourea	Dioxane, 20°	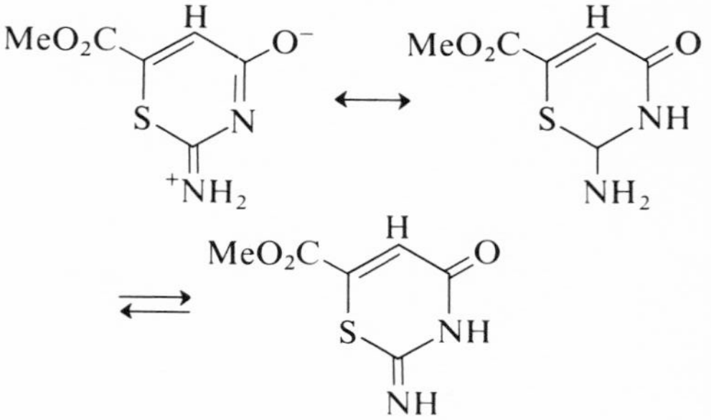 (three resonance/tautomeric structures of thiazinone, MeO₂C, S, NH, NH₂ ⟷ O⁻ ... ⇌ NH)		40b
DMAD	Tetramethyl-thiourea	Dioxane, 20°	(1) 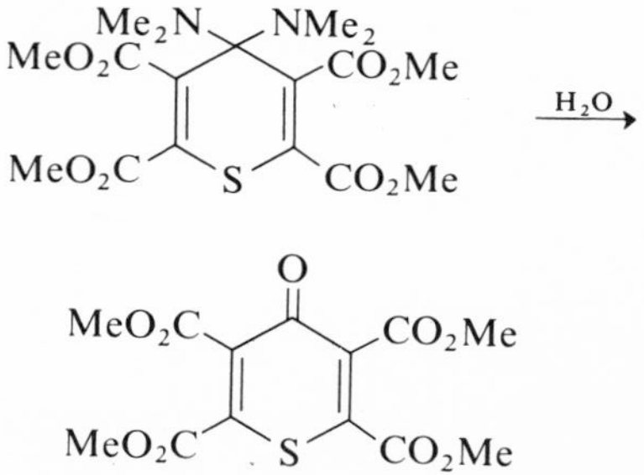 (Me₂N, MeO₂C, CO₂Me dihydrothiine structure) →H₂O→ (thiopyranone: MeO₂C, CO₂Me, S, O)		41a

(2)

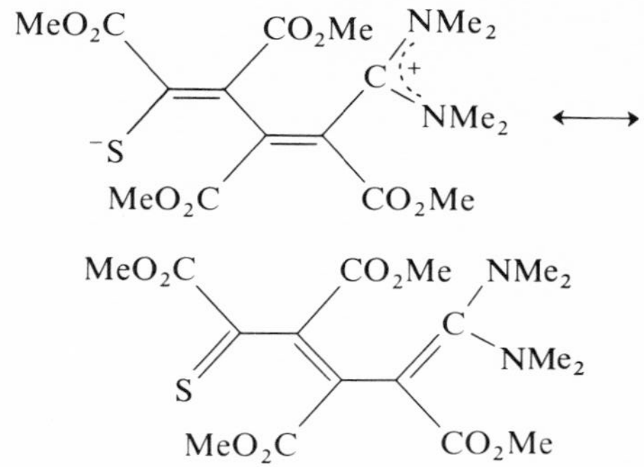

MeO_2C CO_2Me NMe_2 C^+ NMe_2 CO_2Me MeO_2C $-S$ ⟷ MeO_2C CO_2Me NMe_2 C NMe_2 CO_2Me MeO_2C S

DMAD	$RNHNH_2$	MeOH	$RNHN{=}C{-}CH_2CO_2Me$, CO_2Me	40a
DMAD	4-Methyl thiosemi-carbazide	MeOH	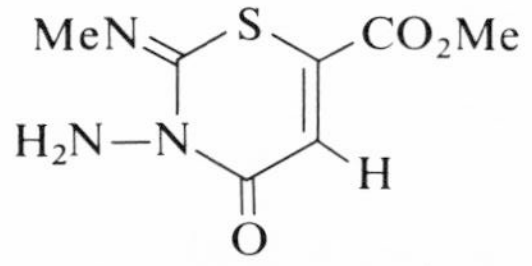 $MeN{\bigvee}S{-}CO_2Me$ ... $H_2N{-}N{-}O$ ... H (3-amino-2,3-dihydrothiazin-4-ones)	40a
DMAD	1,2,3-triphenyl-guanidine	MeOH	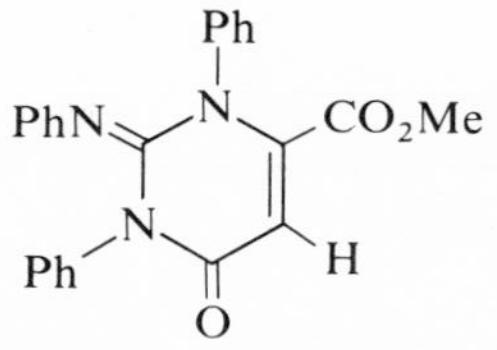 Ph, PhN, CO_2Me, N, O, Ph, H (pyrimidone)	40a

TABLE 2-33—(continued)

(A) Addition to Monoacids, Nitriles and Ketones—continued

Acetylenic Compound	Addened	Conditions	Product	% yield	Reference
DMAD	R̄CR'—P⁺Ph₃		RR'C=C—CO₂Me, Ph₃P=C—CO₂Me (not the Michael product)	50–77	43
ADA (acetylene-dicarboxylic acid)	ArSO₂H	Hot HOAc	—[CH(SO₂Ph)CO₂H]₂	31–80	44, 45
DMAD	Dimethyl sulfoxide	Heat	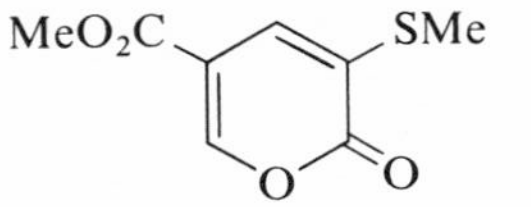 SMe pyrone (MeO_2C substituted)	Good	46
DMAD	Dimethyl sulfoxide	$SOCl_2$	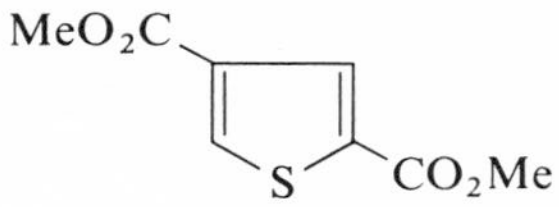 thiophene (MeO_2C, CO_2Me, S)	100	46
DMAD	N₂CHCOEt (with NCN)	65°	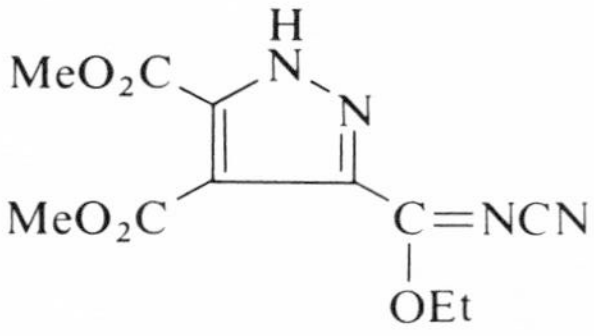 pyrazole (MeO_2C, MeO_2C, C=NCN, OEt, N–N, H)	75	46a

The acid corresponding to (3) is as strong as HCl:

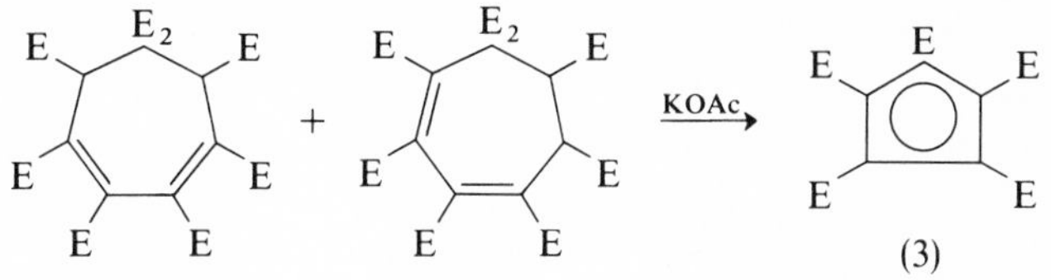

$$\text{(1)} \quad + \quad \text{(2)} \quad \xrightarrow{\text{KOAc}} \quad \text{(3)}$$

(1) (2) (3)
$(E = CO_2Me)$

LeGoff[50] proposed this mechanism:

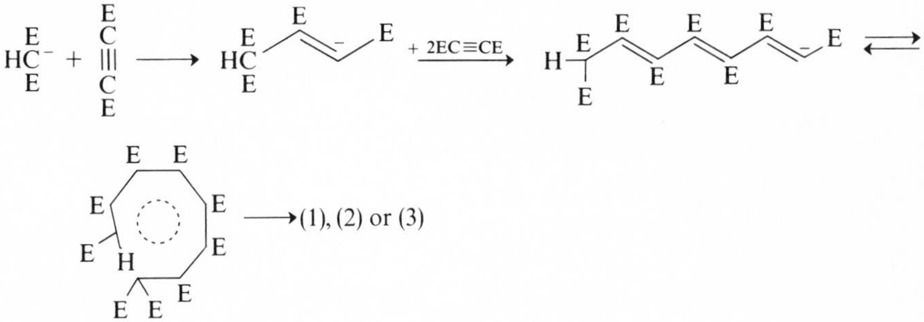

$$HC^- + \text{(EC≡CE)} \longrightarrow HC \xrightarrow{+ 2EC≡CE} H \rightleftharpoons$$

$\longrightarrow$ (1), (2) or (3)

2.2. Cyanoacetic Ester as Addend

Dimethyl acetylenedicarboxylate and cyanoacetic ester in pyridine form products which contain groups from one cyanoacetic ester, one pyridine, and two dimethyl acetylenedicarboxylates. These products are (1) and (2); (2) reacts in acid to form (3) and (4) $(E = CO_2Me)$:[48]

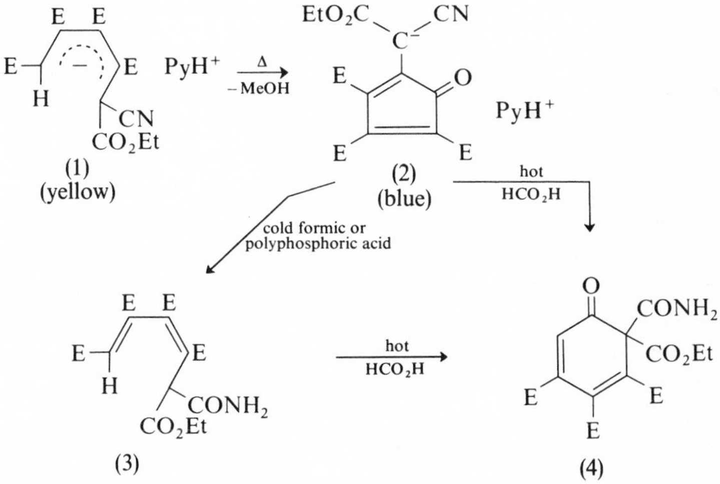

(1)
(yellow)

$PyH^+ \xrightarrow[-MeOH]{\Delta}$

(2)
(blue)

PyH^+

$\xrightarrow[\text{HCO}_2\text{H}]{\text{hot}}$

cold formic or polyphosphoric acid

(3)

$\xrightarrow[\text{HCO}_2\text{H}]{\text{hot}}$

(4)

Phenylpropiolic ester gives a different product:

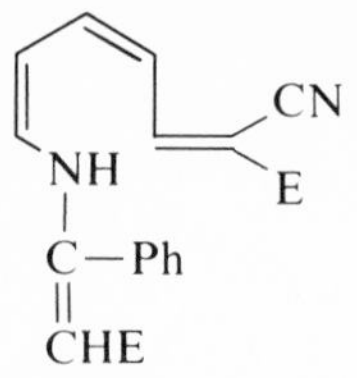

This is one of the many examples of cleavage of a pyridinium ring by a nucleophilic reagent. The first intermediate is probably a 1,3-dipole:

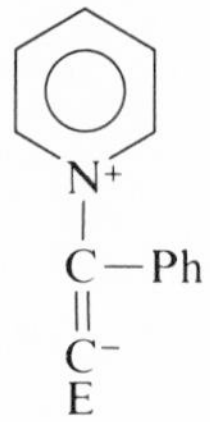

2.3. α-Amino Ketones and α-Hydroxy Ketones as Addends

Michael addition of a nucleophile B: to an acetylenic carbonyl compound, followed by cyclization, is a general synthesis of heterocycles:[16]

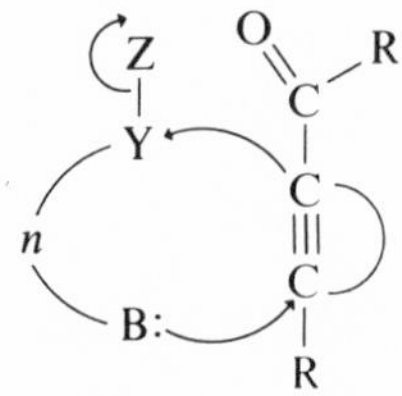

In the best nucleophiles, YZ is ketone. In the reaction of B: with acetylenedicarboxylate esters, the intermediate enolate is formed. In at least one case, the enolate has been isolated and identified. The π electrons on the central carbon can overlap on either side with an electron-deficient material to give *cis* or *trans* products, but the preferred movement of a coordinated proton from the ester carbonyl, and the development of sterically favored fumarate, work in the same direction to give *trans* addition in simple cases.

   In the general reaction, α-amino ketones and dimethyl acetylenedicarboxylate give pyrroles, and α-hydroxy ketones give dihydrofurans:

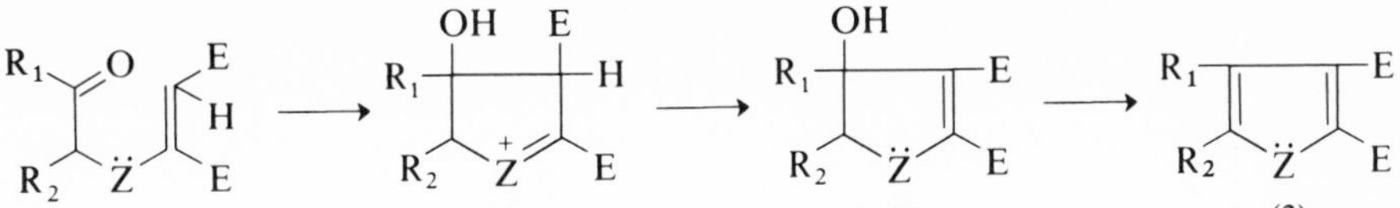

$$(2)\ Z = O, R_1 = R_2 = H;\ 44\%\ \text{yield}$$
$$(3)\ Z = N, R_1 = R_2 = Me;\ 20\%\ \text{yield}$$
$$R_1 = R_2 = Ph;\ 44\%\ \text{yield}$$
$$R_1 = Ph, R_2 = Me;\ 80\%\ \text{yield}$$

Furans, thiophenes, quinolines, pyrrolinones and thiazolinones can also be made.[16]

2.4. Phosphoranes as Addends

When triphenylphosphorylideneacetophenone reacts with dimethyl acetylenedicarboxylate in protic methanol, the product is a substituted fumarate (1) from *trans*-Michael addition:[51]

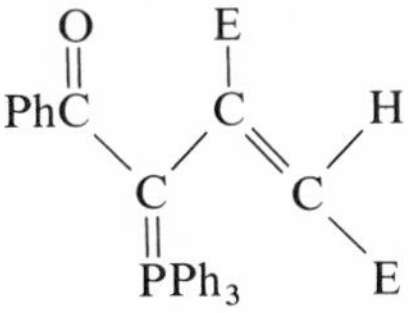

In aprotic dry ether, the product is an isomer, more stable than the fumarate because of more conjugation. This isomer probably forms by way of a four-membered phosphorane, which is less strained than a cyclobutene ($E = CO_2Me$):

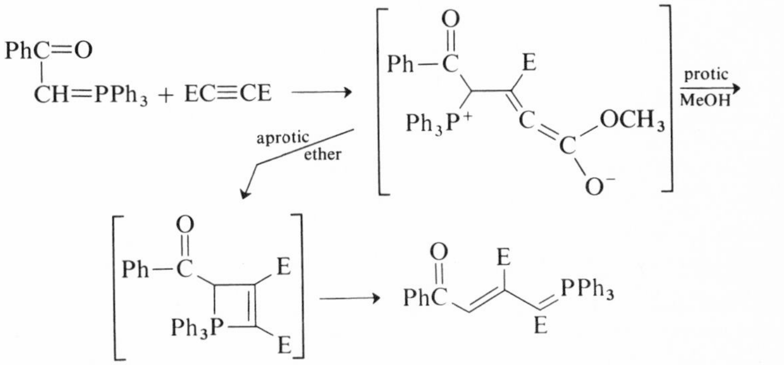

Earlier, Bestmann had proposed a similar sequence to account for the products from ylides and dimethyl acetylenedicarboxylate.[43]

Iminophosphoranes add to DMAD provided there are no electron-with-drawing groups on nitrogen or phosphorus which prevent formation of the intermediate, such as

$$Ph_3P-NR$$
$$\overset{|\underline{\quad}|}{E\qquad E}$$

(E = CO_2Me). Final products are formed according to this reaction:

$$Ph_3P=N-C_6H_4-p\text{-}Br + DMAD \longrightarrow Ph_3P=\overset{\overset{\displaystyle E}{|}}{C}-\overset{\overset{\displaystyle E}{|}}{C}=NC_6H_4-p\text{-}Br$$

Acetylenes which are less electron deficient than dimethyl acetylenedicarboxy-late will not react with ylides.[43a]

2.5. Double Michael addition to Vinyl Ethynyl Compounds— Synthesis of Griseofulvin

Stork proposed that it might be possible to prepare cyclohexenones by double Michael reactions (adding a suitable methylene compound to cross-conjugated vinyl ethynyl ketones).[52] Griseofulvin might be made by reacting the correct coumaranone with methoxyethynyl propenyl ketone. This would give the enol ether and construct the griseofulvin rings in essentially one reaction. Griseoful-vin is an important orally active antifungal antibiotic.

Alkoxyethynyl ketones were unknown in 1964, at the time of this work. Further, double Michael addition to doubly unsaturated ketones was not a known reaction. The great lability of the alkoxyethynyl propenyl ketones explains why they had not been made before, although the alkylthioethynyl ketones were reported.[53] Stork made the ketone (1) as follows:

$$LiC\equiv COEt + \text{crotonaldehyde} \xrightarrow[-15^\circ]{\text{ether}} EtOC\equiv C\overset{\overset{\displaystyle OH}{|}}{C}HCH=CHCH_3$$
$$(43\text{–}63\%)$$

$$\xrightarrow[]{MnO_2,\ CH_2Cl_2} EtOC\equiv C\overset{\overset{\displaystyle O}{\|}}{C}CH=CHCH_3$$
$$(1)\ (\text{stable at } -15^\circ)$$

Hydration experiments showed that either OH^- or water adds selectively to the acetylenic bond.

Successful double Michael reaction must first involve an addition to the triple bond, since addition to the double bond will not give the correct spatial arrangement. The triple bond is more reactive in some compounds in which carbonyl is cross-conjugated with triple and double bonds. Even unactivated acetylenes can add amines, so this selectivity is expected. The ketone Stork[52] prepared is different, since the triple bond is flanked on the other side by alkoxy, which might decrease the reactivity by increasing the electron density at the triple bond. Since conjugation is not expected to be very effective through the triple bond, the inductive effect of the alkoxy group should favor addition. Stork was able to prepare *dl*-griseofulvin by the predicted double Michael addition:

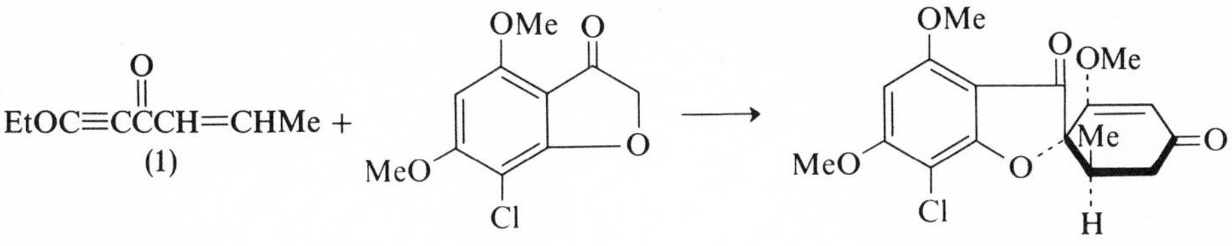

griseofulvin

2.6. Heterocyclics as Addends

2.6.1. SIMPLE MICHAEL ADDITIONS

Pyrroles react if they have a free α- or β-position, and provide protons for the reaction:[54]

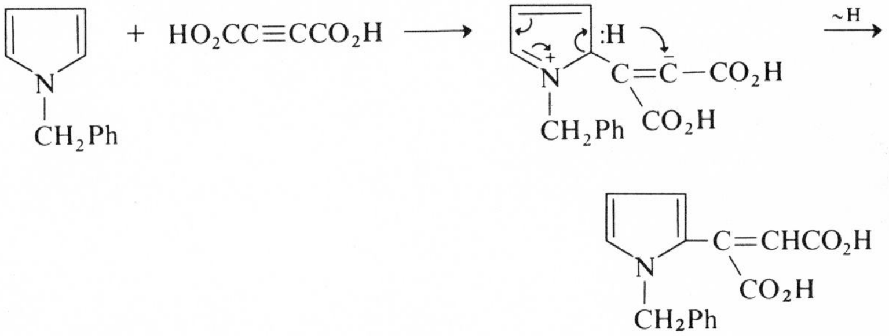

Imidazoles add to dimethyl acetylenedicarboxylate to form

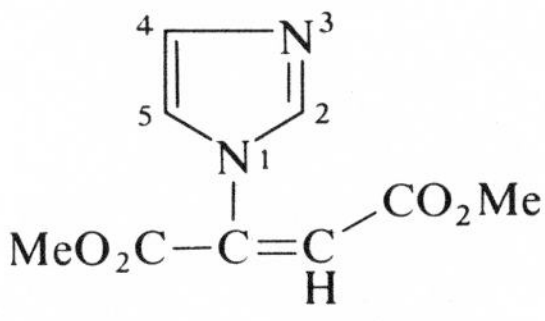

(4-methyl and 2-methyl)[54a]

The protons required by the zwitterion can be furnished by solvent. When acridine reacts with dimethyl acetylenedicarboxylate in methanol, the zwitterion abstracts a proton from methanol, and the intermediate adds methoxy[55] (E = CO₂Me):

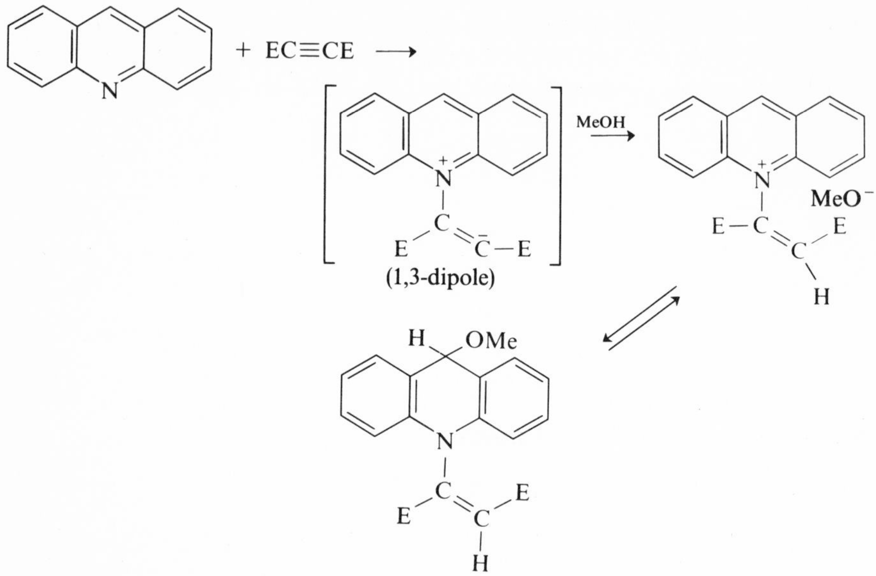

(1,3-dipole)

Other additions of pyridine bases are discussed in connection with 1,3- and 1,4-dipolar addition, Sections 3 and 4.

2.6.2. CYCLIZATION OF MICHAEL ZWITTERIONS

Diels and co-workers did extensive work on the reaction of heterocyclic amines with dimethyl acetylenedicarboxylate.[56,57] For the six-membered heterocycles pyridine and derivatives, they thought the products were zwitterions; the products are now known to be quinolizine derivatives. For five-membered heterocycles such as pyrrole, the main reaction is substitution to form the α,β-dicarbomethoxyvinyl residue.[54]

Pyrrocoline, which has a nitrogen common to a six- and a five-membered ring, reacts well. This is a convenient route to cycl[3.4.5]azine derivatives. Pd/C catalyst is necessary for good yields[56,57] (E = CO₁Me):

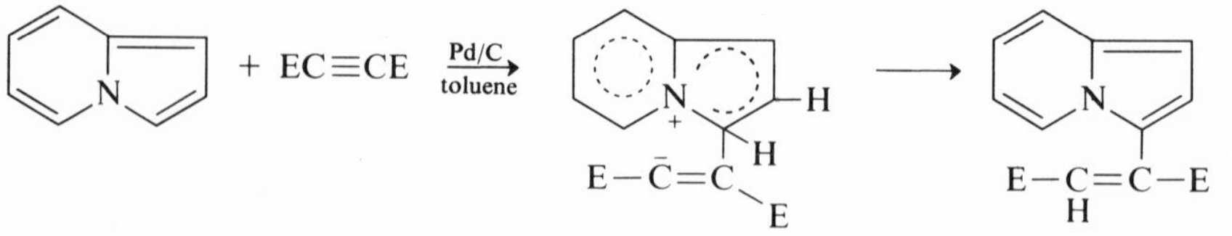

(by-product, formed only if Pd/C not used)

Boekelheide[58] extended this work, and reported in 1961 that the reaction between heterocyclic zwitterions and dimethyl acetylenedicarboxylate is a general method for making the pyrrocolines. Generally yields are fairly low, but the products in many cases are unavailable by other reactions.

The Michael zwitterion from 1-methylbenzimidazole and dimethyl acetylenedicarboxylate cyclizes[54a] to form: ($E = CO_2Me$)

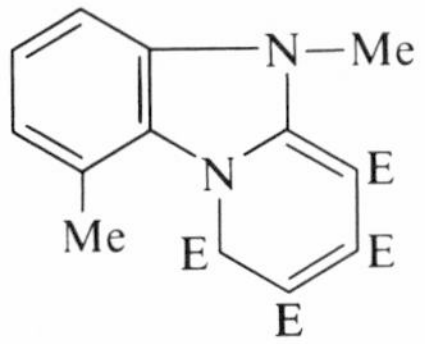

3. 1,3-DIPOLAR ADDITIONS

3.1. Background

Molecules with free or induced 1,3-dipoles add stereoselectively to acetylenes and olefins, and usually give cyclic products. 1,3-Dipolar addition is represented as:[59]

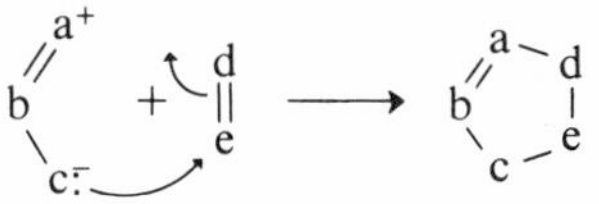

In 1960, Huisgen[59] reviewed the work his group had done on 1,3-dipolar cycloaddition. At that time, he had been working only three years, and had made and characterized over 500 heterocyclics, mostly from 1,3-dipoles and olefins, with a few acetylenes. Huisgen published another review in 1963.[60] Some 1,3-dipolar systems have double bonds:

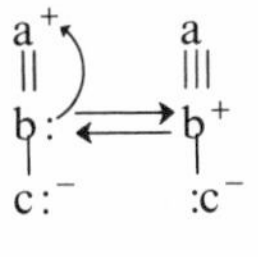

Examples are: nitrile ylides, nitrilimines, nitrile oxides, diazoalkanes, azides and nitrous oxide. Some 1,3-dipolar systems have no double bonds (b = N or O):

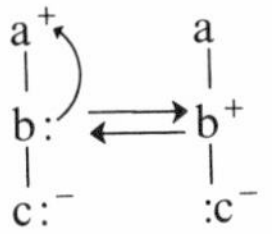

Some examples are: azomethine ylides, azomethine imines, nitrones, azonium imines, azoxy compounds and nitro compounds. These all have nitrogen at the center of the dipole. Some have oxygen at the center: carbonyl ylides, carbonyl imines, carbonyl oxides, nitroso oxides, nitroso imines, and ozone. Some 1,3-dipoles have no octet stabilization: vinylcarbenes, iminocarbenes, ketocarbenes, vinylazenes, iminoazenes and ketoazines. Some of these have no double bonds.

3.2. 3,1′-Dimethylaminovinylindolizines as 1,3-Dipoles

These add to dimethyl acetylenedicarboxylate in aprotic solvents to form cyclopenta[c]quinolizines:[61]

3,1′-dimethylaminovinylindolizines + EC ≡ CE →

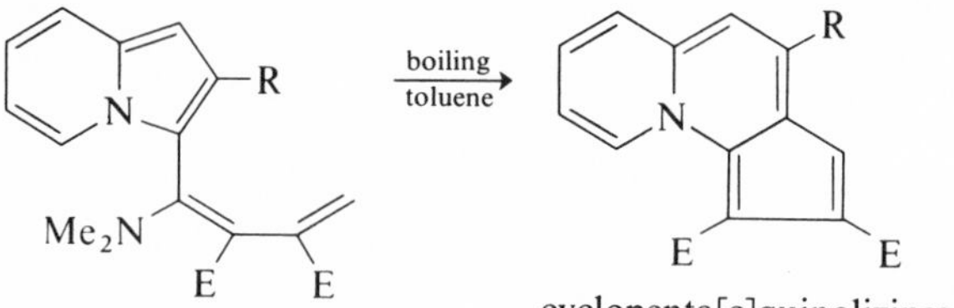

cyclopenta[c]quinolizines

3.3. Diphenyldiazomethane as a 1,3-Dipole

Usually, acetylenes react faster than the corresponding olefins. Kinetics of addition of diphenyldiazomethane in dimethyl formamide at 40° illustrate this[62] (Table 2-34). Acetylenes react as follows:

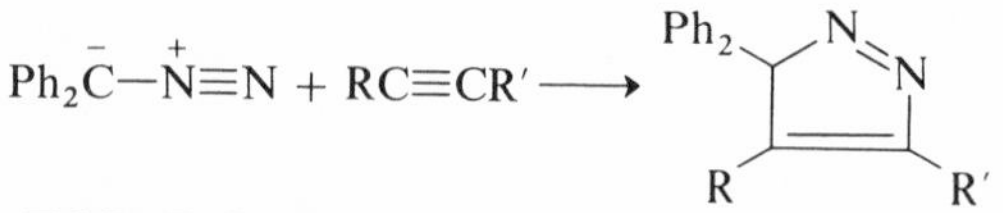

TABLE 2-34
Rate of Reaction of Unsaturates with Diphenyldiazomethane[62]

Unsaturate	$10^5 k_2$ (liters/mole-sec)
Phenylacetylene	1.18
Styrene	1.10
HC≡C—CH(OPr)(OH)	2.22
PhC≡CCO$_2$Et	3.33
Methyl Propiolate	1065
Methyl acrylate	707
Diethyl acetylenedicarboxylate	9680
Diethyl fumarate	2450
Dimethyl maleate	68.5

Diethyl acetylenedicarboxylate reacts 4 times as fast as fumarate and 15 times as fast as maleate.

Diazopropyne adds to acetylenes to form ethynylpyrazoles.[62a]

$$RC\equiv CR' + N_2CHC\equiv CH \longrightarrow$$

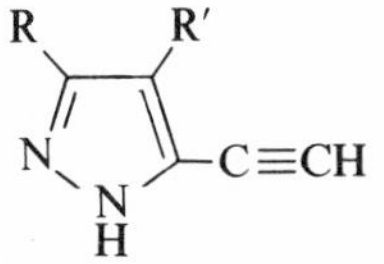

R	R'	% yield
H	H	17
CO$_2$Me	H	62

As expected, "activated" acetylenic bonds are more reactive than "unactivated" acetylenic bonds or "activated" olefinic bonds (as in dimethyl maleate).

3.4. 3,4,5,6-Tetrachlorobenzene-2-diazo-1-oxide as a 1,3-Dipole

When 3,4,5,6-tetrachlorobenzene-2-diazo-1-oxide is heated with acetylenes at 130°, benzo[b]furans are formed.[63] Thermolysis of the diazo oxide gives mesomeric non-octet stabilized ketocarbenes, which add to the triple bond. The intermediate cyclizes (Table 2-35):

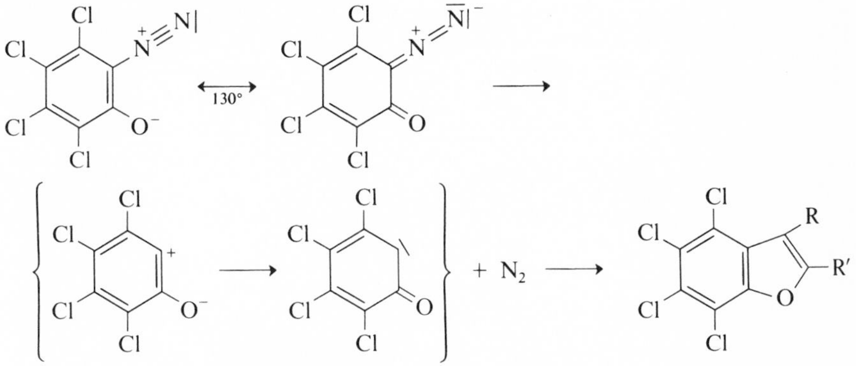

TABLE 2-35
1,3-Dipolar Addition of 3,4,5,6-Tetrachloro-benzene-2-diazo-1-oxide to Acetylenes[63]

Acetylenic Dipolarophile	% Yield
Phenylacetylene	39
Diphenylacetylene	28
Dimethyl acetylenedicarboxylate	48
Ethyl phenylpropiolate	38
1-Phenyl-2-benzoylacetylene	21

3.5. Sydnones as 1,3-Dipoles

1,3-Dipolar addition of sydnones is practically quantitative. Heating N-phenyl-C-methylsydnone with dimethyl acetylenedicarboxylate at 120° in xylene for 1 hour causes evolution of 1 equivalent of CO_2.[64] Product is 1-phenyl-5-methylpyrazole-3,4-dicarboxylic acid dimethyl ester, 99% yield. The order of reactivity of acetylenes is: acetylene < alkylacetylenes < diphenylacetylene < phenylacetylene < propargylaldehyde acetal < propargyl alcohol < phenylpropiolic acid ester < 1-phenyl-2-acylacetylenes < propiolic acid esters < acetylenedicarboxylic acid esters.

Sydnones are nearly aromatic, and can be described only by zwitterion formulas. Form (c) can be regarded as an aromatic azomethinimine, and enters a 1,3-dipolar addition with the acetylene:

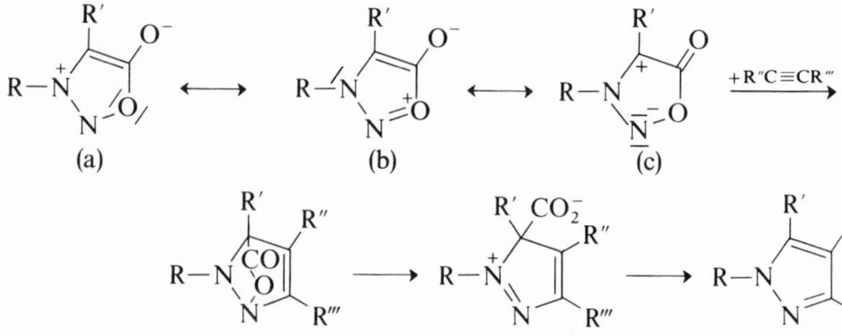

Stille used the 1,3-dipolar addition of sydnones as a polymer forming reaction (Chapter 4).

3.6. Trithiones as 1,3-Dipoles

The exocyclic S in trithiones is nucleophilic and is part of a 1,3-dipole which adds to acetylenes:[65]

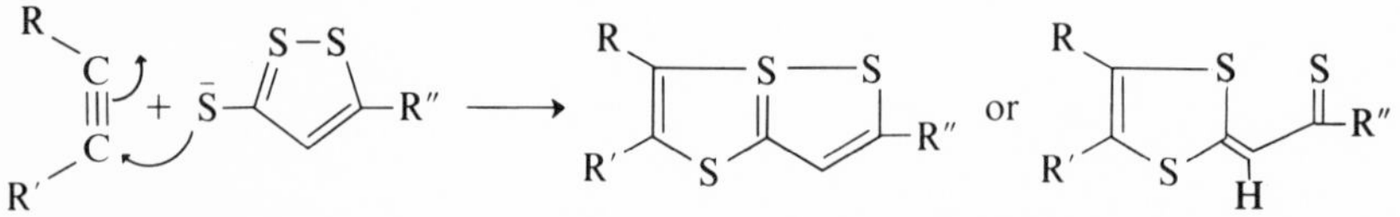

If R and/or R' is electronegative, like carboxyl or aldehyde, the yield of product is 70–90%. Phenylacetylenes give 40–60% yields, but diphenylacetylene does not react.

Formamidinium carbodithioates are similar to trithiones. Negatively substituted acetylenes react to form 53–95% yields of the spiroheterocycles:[65a]

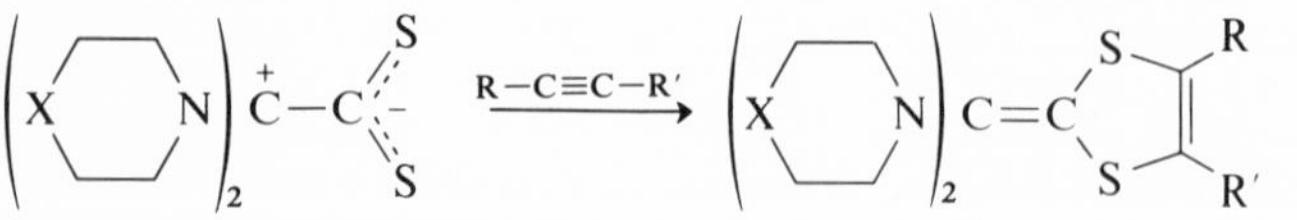

X = O or $-CH_2-$

3.7. Nitrogen Heterocycles as 1,3-Dipoles

In 1965 Acheson[66,67] reported on his continuing studies of addition of nitrogen heterocycles to dimethyl acetylenedicarboxylate. Thiazoles and oxazoles add, probably via 1,3-dipoles ("zwitterions"), to form 1:2 adducts ($E = CO_2Me$):

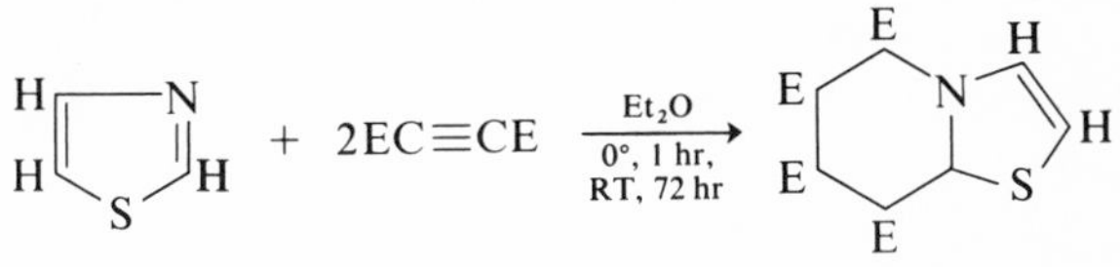

tetramethyl-8aH-benzo[b]-
thiazole-5,6,7,8-tetra-
carboxylate

2,4-Dimethylthiazole, 2-ethylbenzothiazole, 2-methylbenzothiazole, and 2-methylbenzoselenazole give azepines in which two of the original active hydrogens are in different positions. The azepines may form via the 1,3-dipole produced in the first addition ($E = CO_2Me$):

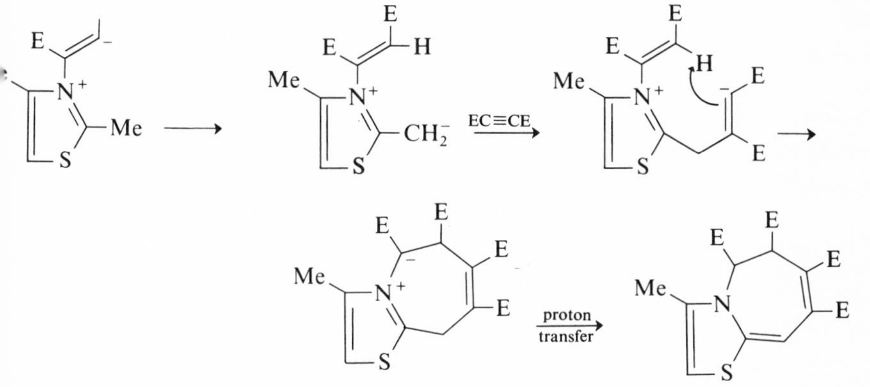

Carbazoles form when dimethyl acetylenedicarboxylate reacts with indole[68] ($E = CO_2Me$):

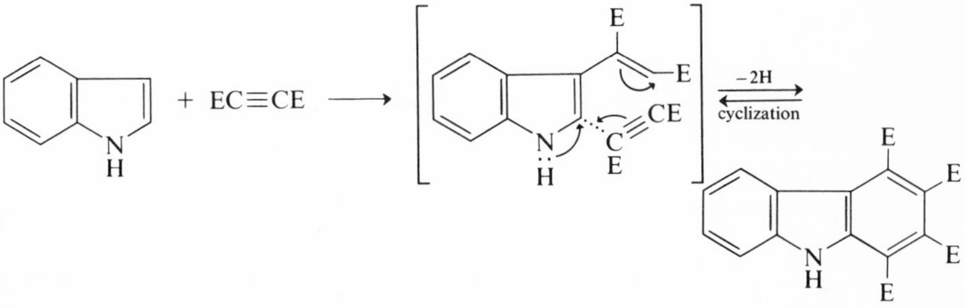

3.8. *o*-Substituted Anilines as 1,3-Dipoles

o-Hydroxyaniline and *o*-thiolaniline give 1:1 products with dimethyl acetylene-dicarboxylate ($E = CO_2Me$):

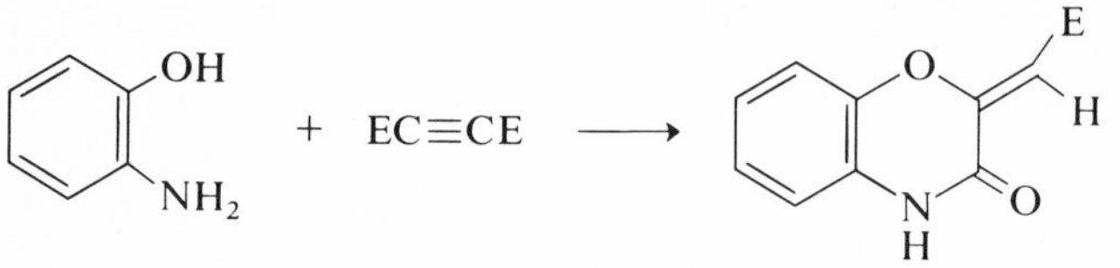

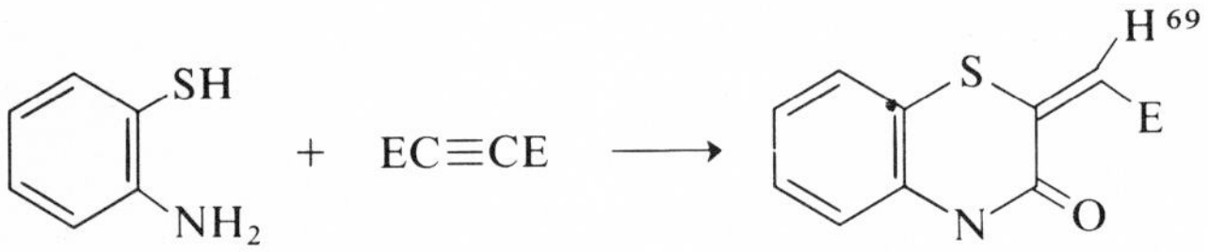

3.9. Organic Azides as 1,3-Dipoles

Reaction of azides with acetylenes to form 1,2,3-triazoles is the oldest known 1,3-dipolar addition reaction, but is not the best understood (see reference 70). Generally, activated acetylenes react faster than their olefinic counterparts.[70a] Electron-rich triple bonds react especially well with electron-poor azides, and vice versa. Huisgen[71] posed the question: Will electron-poor azides add to electron-poor triple bonds? *p*-Tosylazide has an electron-poor azide group, and acetylenic acid esters have electron-poor triple bonds. *p*-Tosylazide adds to acetylenic acid esters to give good yields of 1,2,3-triazoles after 6–13 days at 70–80° (Table 2-36).

TABLE 2-36

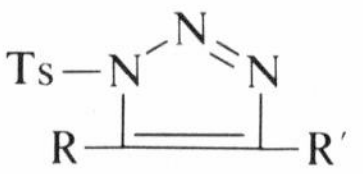

R	R'	% Yield
CO$_2$Me	CO$_2$Me	75
Ph	H	49
Ph	CO$_2$Me	Trace
H	CO$_2$Me	63
H	CH(OEt)$_2$	0

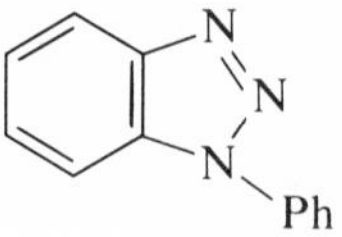

Note that phenyl azides and methyl propiolate or ethoxyethyne give good yields. Benzyne reacts with phenylazide to form 50% yield of

o-Diazidobenzene reacts with two molecules of dimethyl acetylenedicarboxylate in ether at room temperature to give 65% yield of 1,2-bis(4,5-dicarbomethoxy-1,2,3-triazol-1-yl)benzene:[71a]

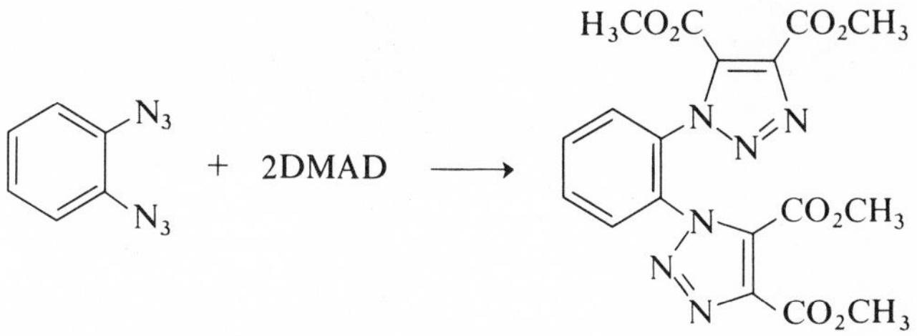

o-Azidoaniline has two different adding groups. The product, dimethyl 2-(4,5-dicarbomethoxy-1,2,3-triazol-1-yl)anilinofumarate, results from one 1,3-dipolar addition and one Michael addition.

3.10 Cyclic α-Amino Ketones as 1,3-Dipoles

Pandit and Huisman[72] were the first to react cyclic α-amino ketones with an acetylenic compound in a reaction formally similar to 1,3-dipolar addition (E = CO$_2$Me):

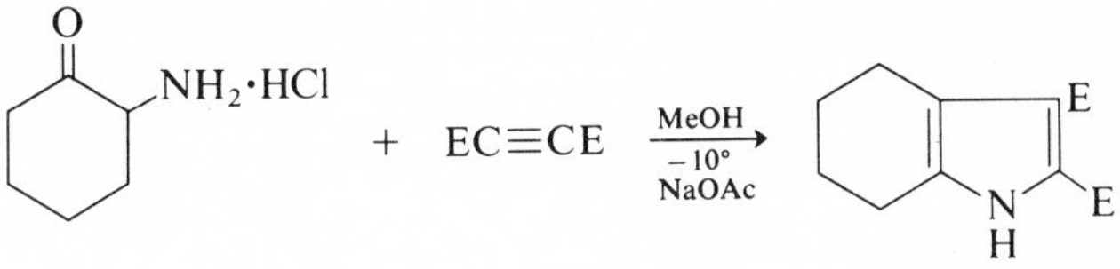

(25%)

Some α-amino ketones add to the triple bond, but the product does not cyclize:

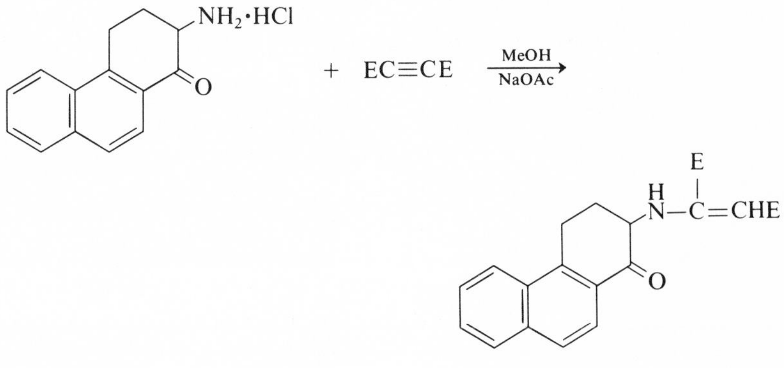

(75%)

3.11. Nitrilamines as 1,3-Dipoles

The most highly substituted ethylenic or acetylenic carbon atom goes into the 5-position of the Δ^2-pyrazoline or pyrazole products. Acetylenes give pyrazoles:

$$
\left[
\begin{array}{c}
C_6H_5-C\equiv\overset{(+)}{N}-\overset{(-)}{N}-C_6H_5 \\
\updownarrow \\
C_6H_5-\overset{(-)}{\underset{}{C}}=\overset{(+)}{N}=N-C_6H_5
\end{array}
\right]
\underset{\text{diphenylnitrilamine}}{} + C_2H_2 \xrightarrow[\text{Et}_3\text{N}]{15°,\ 10\ \text{days}}
$$

$$
\underset{\substack{\text{1,3-diphenylpyrazole}\\(81\%)}}{}
$$

1-Hexyne gives only 24% reaction at 75° (8 hours). 4-Octyne does not react even at 100°.[72a] A negatively substituted acetylene is not necessary.

4. 1,4-DIPOLAR ADDITIONS

4.1. Background

A molecule a—b—c—d is a 1,4-dipole if a has an electron deficiency (formal positive charge), and d has at least one free electron pair (formal negative charge).[73] The 1,4-dipole can react with a multiple bond system (dipolarophile) to form a six-membered ring:

1,4-Dipoles with an electron sextet on a can undergo internal or external octet stabilization; internal stabilization occurs if b has a free electron pair:

$$
\overset{+}{a}-\overset{\frown}{b}-c-\overset{..}{d}{}^- \rightleftharpoons a=\overset{+}{b}-c-\overset{..}{d}{}^-
$$

Huisgen[73] feels that the analogy with the 1,3-dipole and its cycloadditions is not as close as it seems: The characteristic partial charge compensation in 1,3-dipoles in octet forms and the interchangeability of the charge centers in the sextet form do *not* occur in the normal way in the 1,4-dipoles; 1,3-dipoles have various forms:

$$
\underset{\text{octet forms}}{a=\overset{+}{b}-\overset{-}{c}: \rightleftharpoons :\overset{-}{a}-\overset{+}{b}=c} \rightleftharpoons \underset{\text{sextet forms}}{\overset{+}{a}-b-\overset{-}{c}: \rightleftharpoons :\overset{-}{a}-b-\overset{+}{c}}
$$

In contrast to the 1,3-dipoles, electrophilic and nucleophilic centers in the 1,4-dipoles are fixed. This causes definite mechanistic differences.

Since the 1,4-dipole cannot rearrange into mesomeric sextet forms, no double bond can form between b and c. A double bond here would make the 1,4-dipole like a 1,3-diene, which could undergo Diels-Alder reactions. 1,4-Dipolar addition is different from the cyclic electron displacement of the Diels-Alder reaction. Electrons cannot migrate in a cyclic intermediate if c is a tetrahedral center.

Most known examples which can be classified as 1,4-dipolar additions form the 1,4-dipole *in situ*, and it cannot be isolated. It forms from nucleophile a=b and electrophile c=d:

$$a=b + c=d \rightleftarrows a=\overset{+}{b}-\overset{-}{c}-d: \rightleftarrows \overset{+}{a}-b-c-\overset{..}{\overset{-}{d}}:$$

4.2. Addition of Pyridines

The most studied reaction of this type is the addition of two molecules of acetylenedicarboxylic acid esters to pyridines, which has been used to make many six-membered heterocyclics. The electrophilic acetylenic ester (c=d in the above scheme) attaches itself to the pyridine N, in which the aromatic C=N bond functions as a=b. The 1,4-dipole combines with a second molecule of acetylenic ester to form 9aH-quinazolines, which easily isomerize to derivatives of 4H-quinazolines[74] (E = CO$_2$Me):

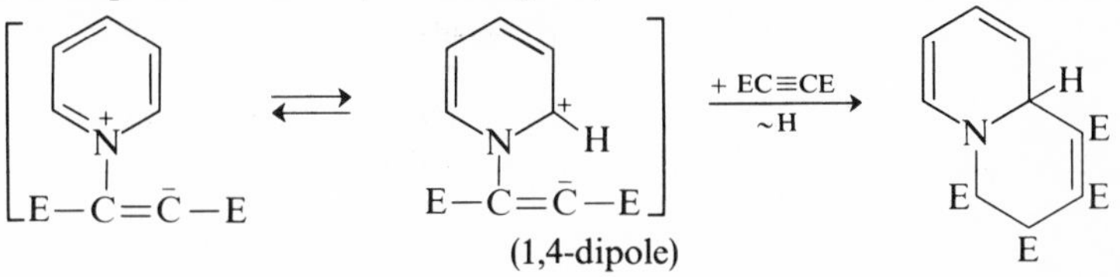

(1,4-dipole)

Note that this picture is slightly different from Acheson's,[67] who uses only the 1,3-zwitterion in his representations (see Michael additions).

Quinoline reacts the same way.[75] Huisgen[76] has shown that the 1-4 dipole can be "trapped" by various dipolarophiles, including phenylisocyanate. Phenylisocyanate does not react with quinoline, but it does react with the 1,4-dipole formed by adding dimethyl acetylenedicarboxylate to the mixture. Either quinoline or pyridine can be used:

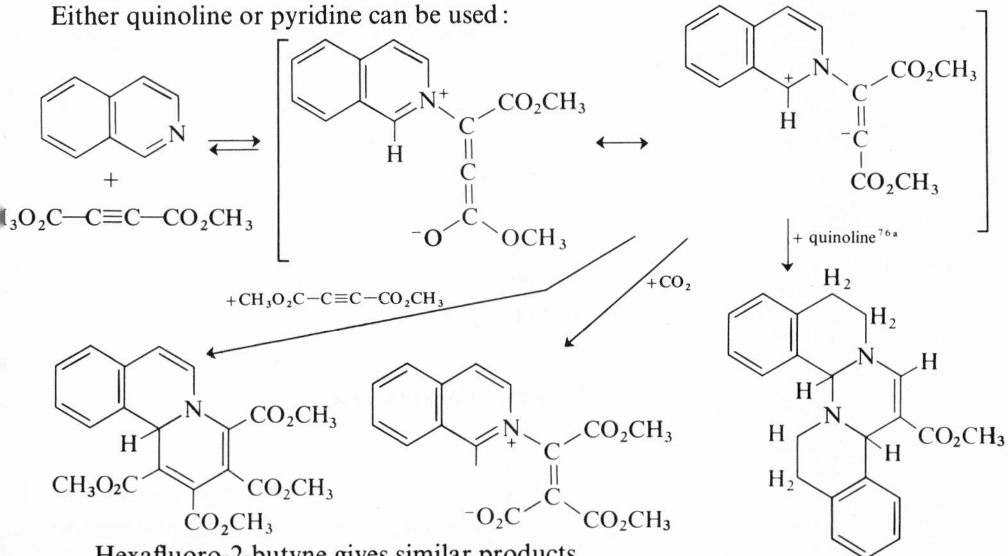

Hexafluoro-2-butyne gives similar products.

4.3. Addition of Azomethines

3,4-Dihydroisoquinoline reacts with two molecules of dimethyl acetylenedi-carboxylate to form a product in 34% yield, but the product can also be formed from two molecules of 1,4-dihydroisoquinoline and one molecule of diester :[73]

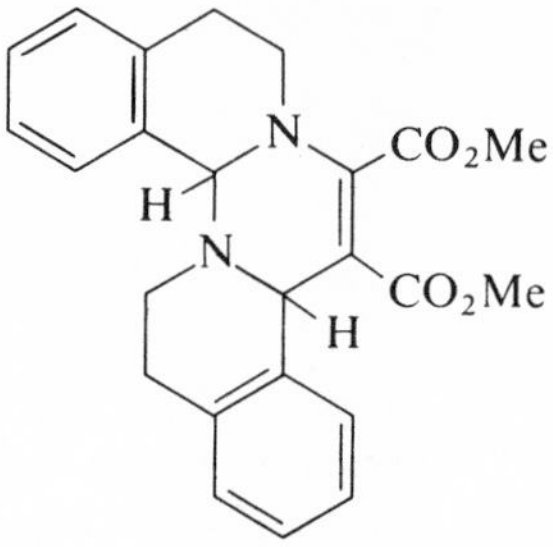

(64% yield)

Benzalaniline does not react with dimethyl acetylenedicarboxylate in boiling toluene, but other benzalamines do react (see reference 77).
($E = CO_2Me$):

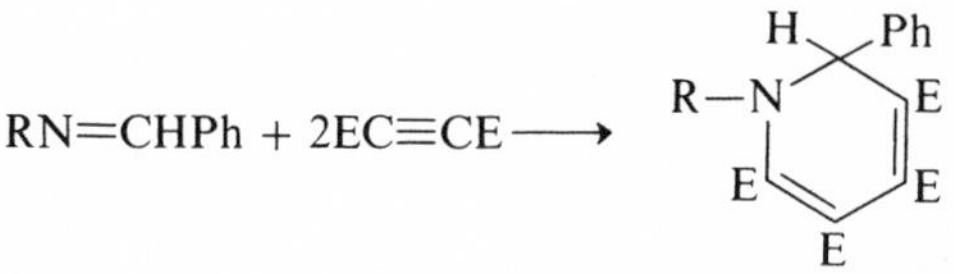

$$RN{=}CHPh + 2EC{\equiv}CE \longrightarrow$$

R = Me, 55% yield;
benzyl, 47% yield
(via 1,4-dipole)

5. "ENE" SYNTHESIS

Olefins add to acetylenes or to activated olefins in the "ene" synthesis. Olefins add to maleic anhydride to form alkenylsuccinates, and to acetylenedicarboxylic acid to form maleic acid derivatives by *cis* addition.[73] At 180°, olefins add to dimethyl acetylenedicarboxylate to form $RCH{=}CHCH_2\overset{\overset{\displaystyle CO_2Me}{|}}{C}{=}CHCO_2Me$.
Methyl propiolate gives mostly *trans* product by *cis* addition (80% *trans-β*-acrylic acid derivative), but it also gives 20% α-acrylate derivative.[78]

6. DIELS-ALDER OR DIENE SYNTHESIS

6.1. Background

Alder[78] observed that the diene synthesis gives only one configuration in the product, indicating that both bonds form simultaneously. Diels-Alder additions always give *cis* products (Alder's "*cis* principle"). Activated acetylenic bonds are good dienophiles. Butadiene, cyclopentadiene, furan and other dienes react well with α-acetylenic acids. The general diene synthesis is represented as:[73]

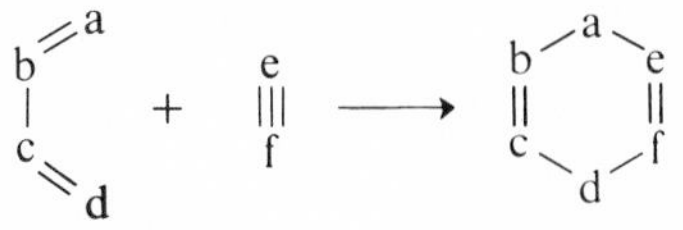

Acheson[67] in 1963 reviewed the literature and his own work on the addition of heterocyclic nitrogen compounds to acetylene-carboxylic acids by Diels-Alder, Michael and 1,3-dipolar additions. In the 1930's the reaction of nitrogen heterocyclics with acetylenic acids received considerable attention, but interest died out until the middle 1950's, when new analytical methods were used to reexamine some of the earlier products. Some of the structures of early products were found to be wrong and required changes.

6.2. Furans

Dimethyl acetylenedicarboxylate reacts with 2 moles of furan in ether to give 65% yield of Diels-Alder adduct after 2 weeks. The product is *exo-exo*:[79]

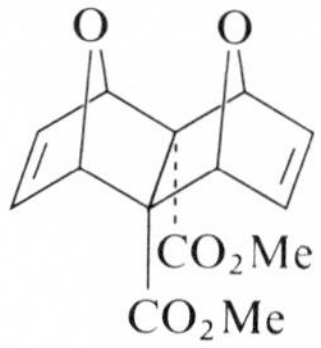

6.3. Pyrroles

A few pyrroles react with acetylenedicarboxylic acid by Diels-Alder addition across the 2,5-positions:[81,82]

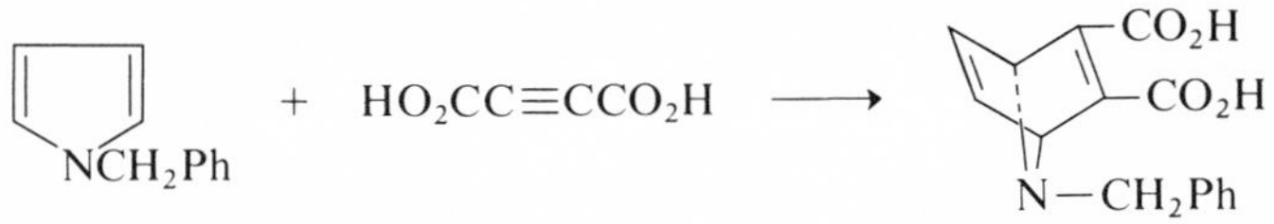

The Diels-Alder products can react further (E = CO_2Me):

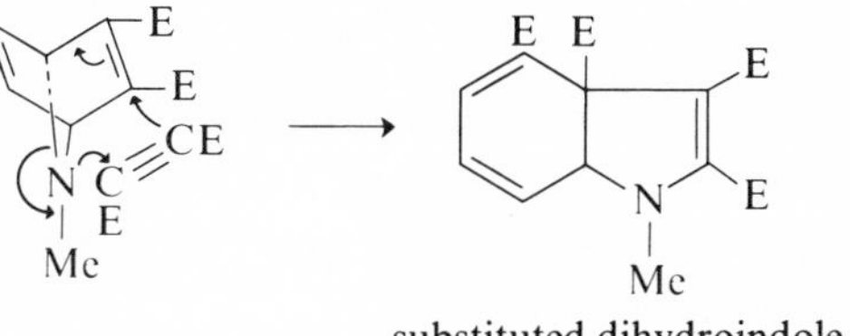

substituted dihydroindole

6.4. Tetrachlorocyclopentadienone Dimethyl Ketal

This ketal reacts with dimethyl acetylenedicarboxylate to form different products at different temperatures:[84]

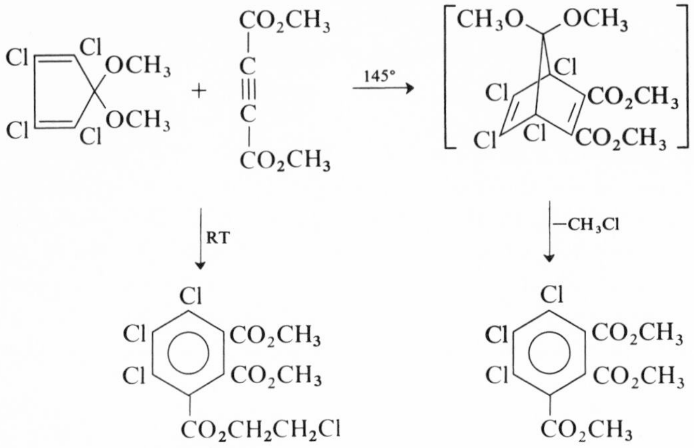

Phenylacetylene, which is a much weaker electron acceptor than the diester, gives norbornadiene ketals:

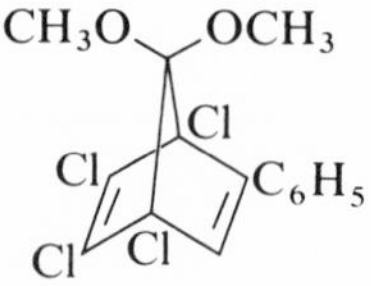

Hexachlorocyclopentadiene and bis(trimethyltin)acetylene undergo Diels-Alder addition in refluxing butyl ether to give 50–70 % yields of 2,3-bis(trimethyltin)-1,4,5,6,7,7-hexachlorobicyclo[2.2.1]hepta-2,5-diene:[85]

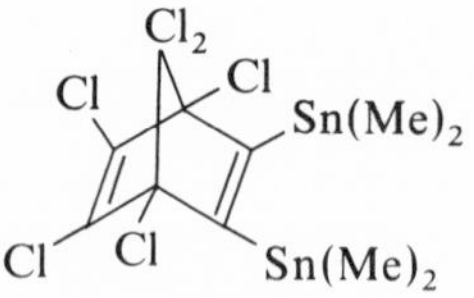

Other tin-acetylenic compounds react similarly.

6.5. Cyclopentadiene, Cyclohexadiene, Aromatic Hydrocarbons and Bicyclopentane

Dicyanoacetylene is a more active dienophile than acetylenedicarboxylic acid esters. Cyclopentadiene and cyclohexadiene add at $-10°$ to form 60 and 54 %

yields, respectively, of Diels-Alder adducts.[86a] Aromatic hydrocarbons such as naphthalene and anthracene require 3–24 hours at 100° to form 30% yields of adducts. The product from cyclopentadiene is typical:

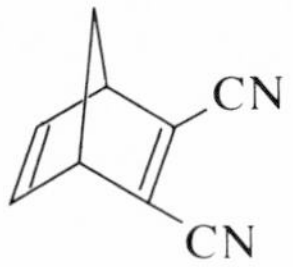

Alkynylsilanes undergo similar Diels-Alder additions with cyclopentadiene, but are much less reactive than negatively substituted acetylenes. This is true even when silanes such as $Me_3SiC{\equiv}CCO_2Et$ are used.[86c] Dimethyldiethynyl-silane reacts with cyclopentadiene in benzene at 250° to give 48% yield of dimethylbis-(2-bicyclo[2.2.1]heptadienyl)silane after 48 hours:

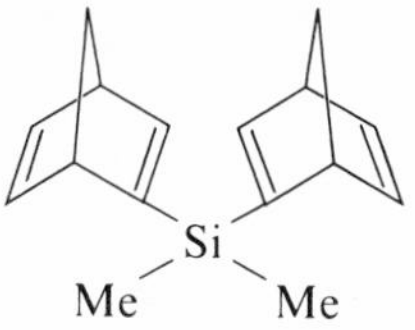

The first reported reactions of bicyclopentane with electron-deficient acety-lenes indicated that the reaction was analogous to the Diels-Alder addition.[86] Later work, however, shows that the reaction is more complex:[86b]

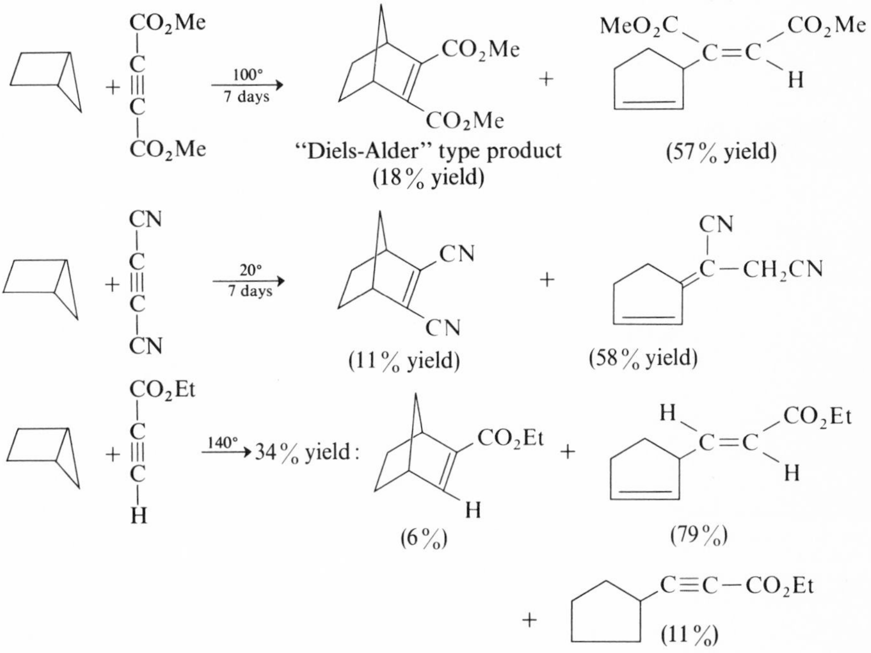

"Diels-Alder" type product
(18% yield)

(57% yield)

(11% yield)

(58% yield)

(6%)

(79%)

(11%)

Hexafluoro-2-butyne, a very powerful dienophile, does not react with bicyclopentane. Thus, the reaction is not a normal Diels-Alder addition. An intermediate diradical accounts for the products, as shown for the ethyl propiolate reaction:[86b]

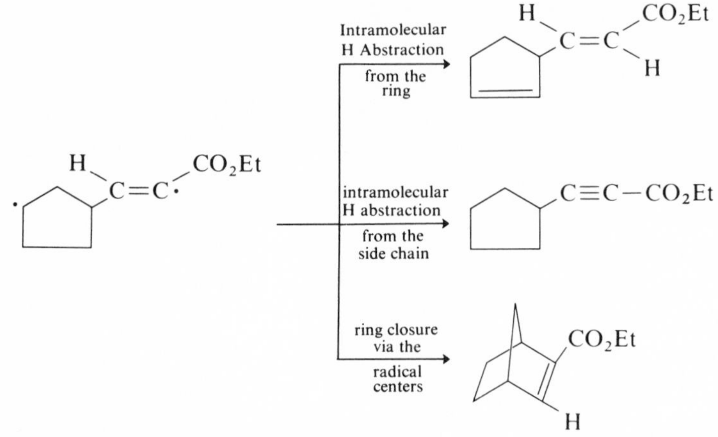

6.6. Rates of Diels-Alder Reactions

The Diels-Alder reaction of phenylpropiolic acids with tetraphenylcyclopentadienone (tetracyclone) is an exceptionally clean, irreversible reaction, and no polymer is formed.[87] Progress is easily followed by the rate of evolution of CO. This is a good reaction for studying substituent effects. The product of the reaction is:

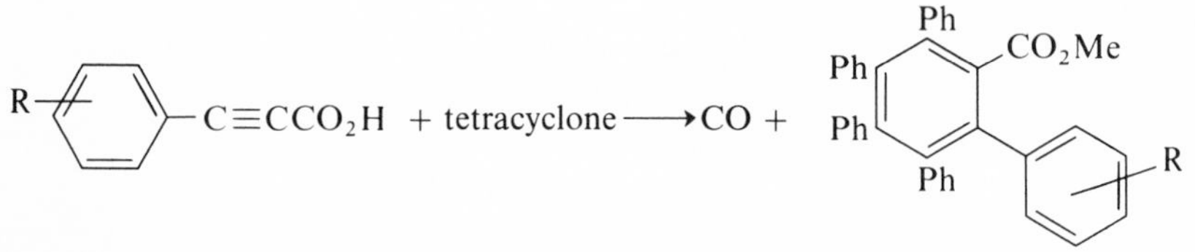

The reaction is second order, and the rate-determining step is probably formation of the adduct. Evolution of CO is rapid. Stronger acids react more quickly (Table 2-37).

The effect of some substituents in the *para* positions of the phenyl groups in tetraphenylcyclopentadienone

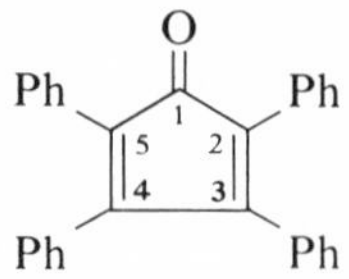

TABLE 2-37
pK$_a$ versus Rate of Reaction of Phenylpropiolic Acids with Tetracyclone[87]

RPhC≡CCO$_2$H R	Apparent pK$_a$ (by titration)	Specific Rate Constants, ($\times 1000$, molal^{-1} sec^{-1})
H	3.40	1.48
m-Me	3.44	1.61
p-Me	3.53	1.25
o-Cl	3.26	2.97
m-Cl	3.15	2.87
p-Cl	3.20	2.25
m-NO$_2$	2.96	5.73
p-NO$_2$	2.87	7.75
m-OMe	3.36	1.73
p-OMe	3.63	1.19

on the rate of the Diels-Alder reaction with methyl phenylpropiolate was determined.[88] Chlorine substitution in tetracyclone increases the rate constant. Methoxy group accelerates when in the *para* positions of the 2 and 5 phenyls, and in all four *para* positions, but decelerates when in the *para* positions of the 3 and 4 phenyls. Methoxy decelerates when it is in the phenylpropiolic acid in the *para* position. This is the first time chlorine has been observed to accelerate the Diels-Alder reaction whether in diene or dienophile.

6.7. Intramolecular Diels-Alder Reaction

Klemm[89] reported an intramolecular Diels-Alder reaction:

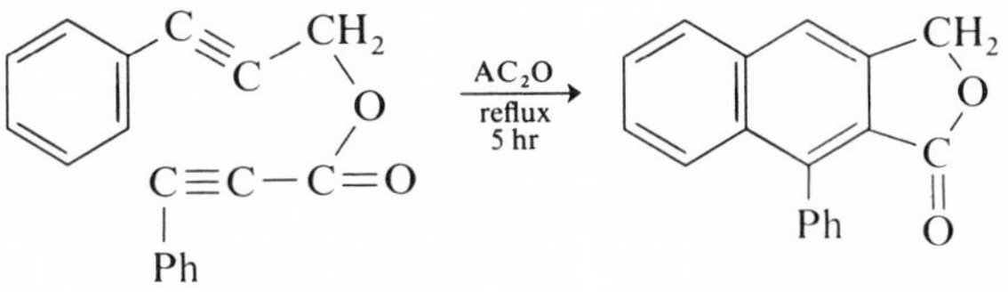

(36%)

6.8. Other Diels-Alder Reactions

Table 2-38 lists other examples of Diels-Alder reactions of "activated" acetylenes.

TABLE 2-38
Diels-Alder Reactions of "Activated" Acetylenes

Acetylene	Addend	Product	% Yield	Reference
$MeO_2CC≡CCO_2Me$	Butadiene	Dimethyl 3,6-dihydrophthalate	75	90, 91
$MeO_2CC≡CCO_2Me$	Piperylene	6-Methyl derivative	85	91
$MeO_2CC≡CCO_2Me$	Isoprene	4-Methyl derivative	87	91
$MeO_2CC≡CCO_2Me$	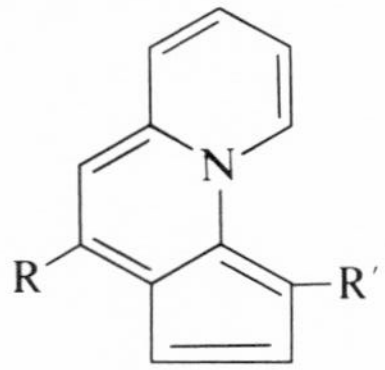	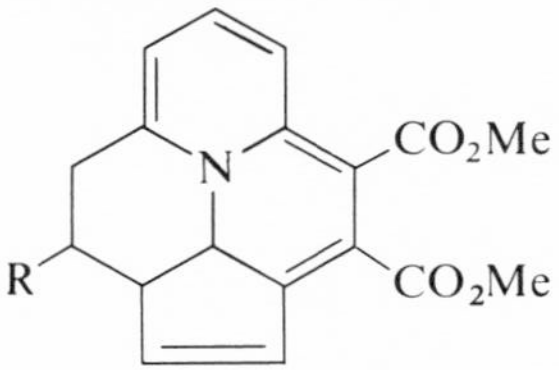		92
$MeO_2CC≡CCO_2Me$	AcOCH=CHCH=CHOAc	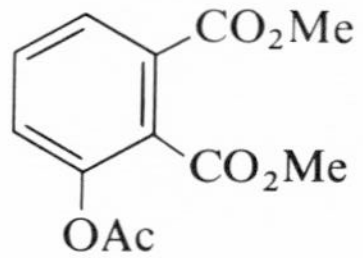	49	93
$PhC≡CCO_2H$	AcOCH=CHCH=CHOAc	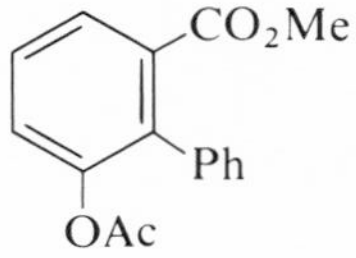	28	93
$PhC≡CCO_2Me$	AcOCH=CHCH=CH$_2$	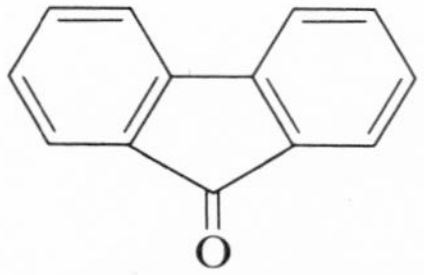	5.6	93

Acetylenic compound	Substrate	Product	Yield (%)	Ref.
$MeO_2CC{\equiv}CCO_2Me$	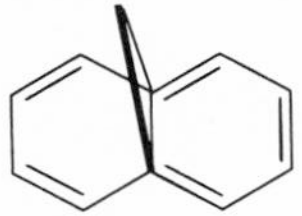 (second stage) (400°) [structure]	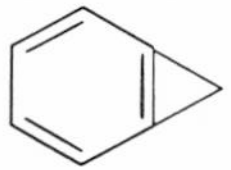 [cyclohexadiene/benzene structure]	45	94
$EtO_2CC{\equiv}CCO_2Et$	2-Pyrones	[structure]	75–90	95
$MeO_2CC{\equiv}CH$		3,4-Dialkylphthalic acids		96
$MeO_2CC{\equiv}CCO_2Me$	4-Oxo-2a,3,4,5-tetrahydro-naphthostyril	[structure: OH, CO_2Me, HN, O]	63	97
$MeC_2C{\equiv}CCO_2Me$	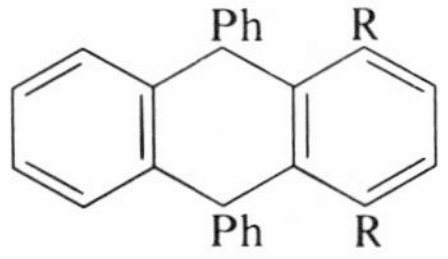 Chloroprene [structure: R, Ph, Ph, R]	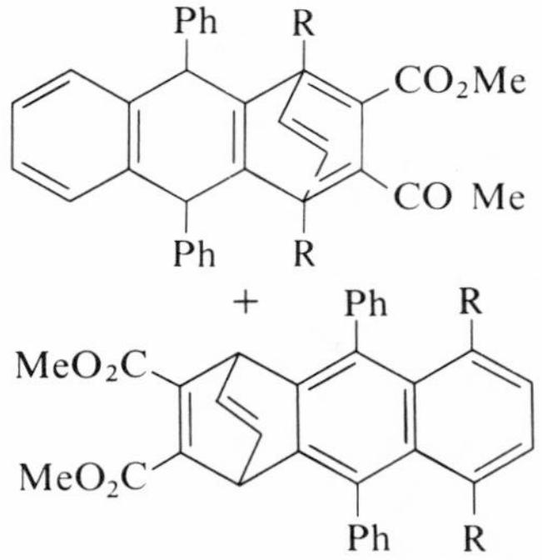 Dimethyl 4-chloro-$\Delta^{1,4}$-dihydrophthalate [structure: CO_2Me, R, Ph, Ph] $\left(\begin{array}{l}R = Me, \\ R = MeO \\ \text{(sole product)}\end{array}\right)$ [$CO\ Me$] [structure + : MeO_2C, MeO_2C, R, Ph, Ph, R] $\begin{array}{l}R = Me \\ R = Ph\end{array}$		98
$MeO_2CC{\equiv}CCO_2Me$	9,10-Dimethyl-anthracene	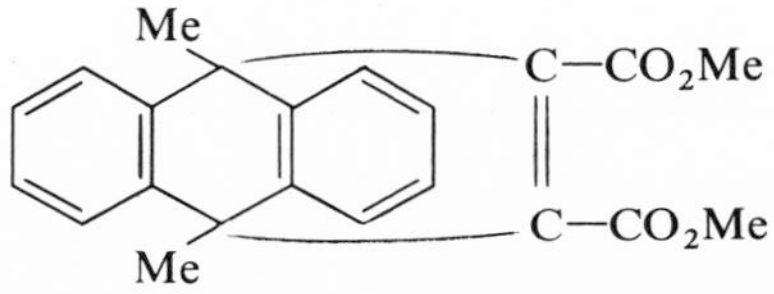 [structure: $C{-}CO_2Me$, $C{-}CO_2Me$, Me, Me]		99

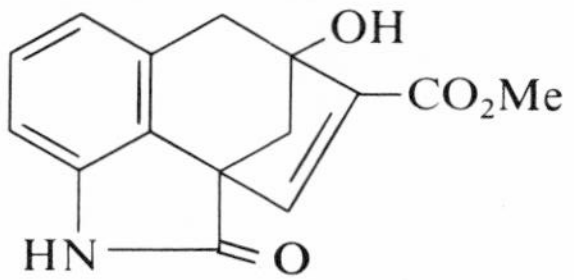

TABLE 2-38—(continued)

Acetylene	Addend	Product	% Yield	Reference
$MeO_2CC{\equiv}CCO_2Me$	1-Carbomethoxypyrrole	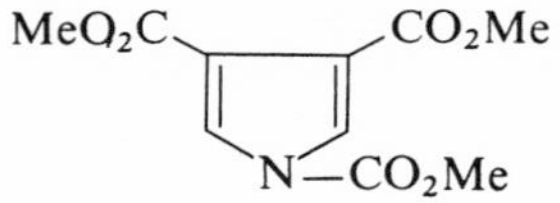		100, 82
$MeO_2CC{\equiv}CCO_2Me$	Cyclooctatetraene	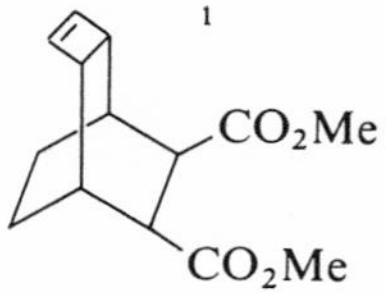		101
$MeO_2CC{\equiv}CCO_2Me$	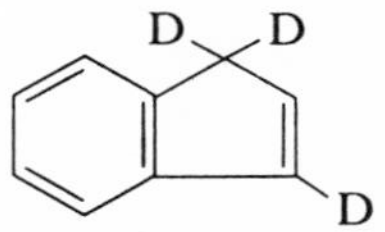	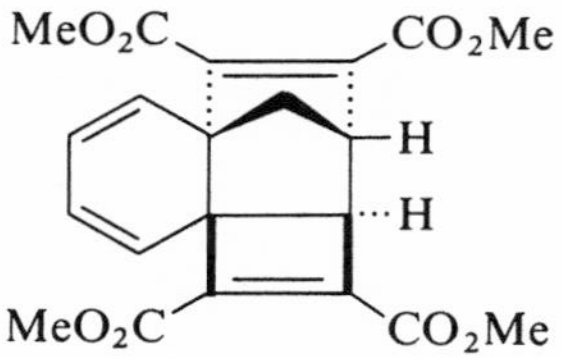		102
$HC{\equiv}CCHO$	Butadiene (and other diolefins)	2,5-Dihydrobenzaldehyde	82	103

$PhC{\equiv}CCCH_3$ (O=)	Butadiene	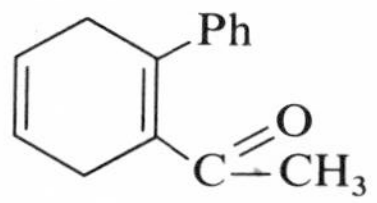		105
$HC{\equiv}CCCH_3$ (O=)	Butadiene	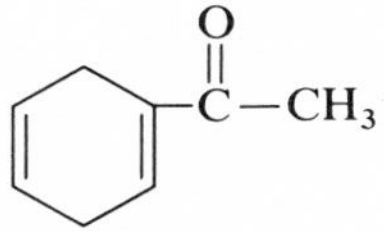		104
$HC{\equiv}CCCH{=}CHMe$ (O=)	Butadiene		78	106
$NCC{\equiv}CCN$	Benzene, $AlCl_3$ catalyst	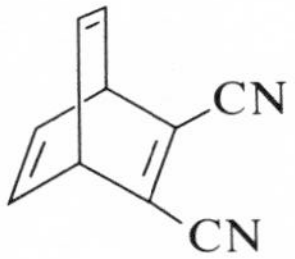	63	106a

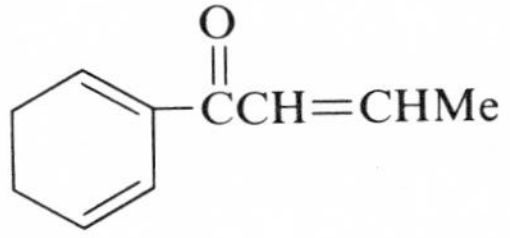

7. ADDITION OF TRIALKYLTIN HYDRIDES (HYDROSTAN-NATION)

Trialkyltin hydrides add to mono- and disubstituted acetylenes to give both terminal and nonterminal 1,2-adducts.[107] Examples of exclusive terminal 1,2-addition of trialkyltin hydrides to phenylacetylene and to p-diethynylbenzene are not typical. Trimethyltin hydride adds to 1-hexyne and to phenylacetylene to give mostly the terminal cis and trans adducts.

Electron-withdrawing substituents on the acetylenic carbon strongly favor nonterminal adducts. It is hard to explain nonterminal addition as a free radical mechanism (as proposed for the addition of organotin hydrides to olefins). Some additions to negatively substituted acetylenes are tabulated (Table 2-39).

TABLE 2-39
Hydrostannation of Negatively Substituted Acetylenes

Acetylene	R_3Sn adds to Carbon Number:	Product Is	% of Total	Reference
$H\diagdown \ \diagup CO_2R'$ $\underset{1}{C}\!\!\equiv\!\!\underset{2}{C}$	2		Most	
	1	cis	Next	108
	1	trans	Least	
$\underset{1}{H}\overset{2}{C}\!\!\equiv\!\!\underset{}{C}CH_2OH$	2			107
	1	cis		
	1	trans		
$Me\underset{1}{C}\!\!\equiv\!\!\underset{2}{C}CO_2Et$	2	cis	50	107
	2	trans	13	
	1	cis	37	
$\underset{1}{H}\overset{2}{C}\!\!\equiv\!\!\underset{}{C}CN$	2		100	107
$MeO_2C\underset{1}{C}\!\!\equiv\!\!\underset{2}{C}CO_2Me$	2	trans		107
	2	cis		

Some hydrostannations are free radical, accelerated by radical generators and retarded by inhibitors.[109,110] Hydrostannation of strongly electrophilic acetylenes is a polar reaction. Some acetylenes, such as methyl propiolate, give products by both ionic and radical addition. Generally, these conclusions apply to electronegatively substituted acetylenes:

(1) They hydrostannate faster than ordinary acetylenes.

(2) As the size of substituents on tin increases, more terminal product forms.

(3) The rate of addition of tin hydrides decreases as R changes: ethyl, butyl > methyl ≫ phenyl.

(4) The reaction is first order on hydride and on acetylene, and shows small isotope effect with trialkyltin deuterides.

(5) Nucleophilic attack of organotin hydride hydrogen on acetylenic carbon is the rate-determining step.

(6) The mechanism is probably:

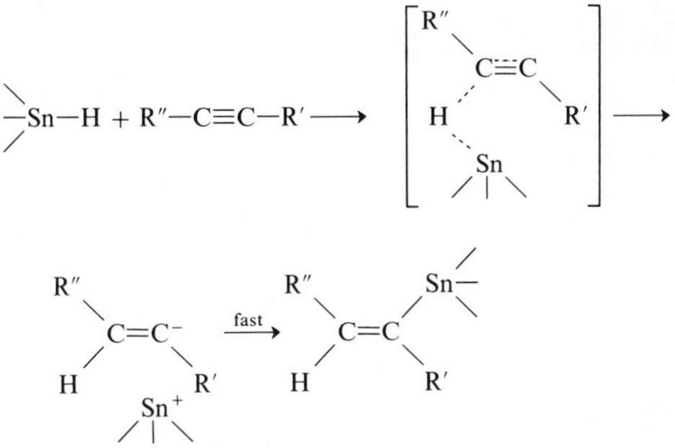

$$-\overset{|}{\underset{|}{Sn}}-H + R''-C\equiv C-R' \longrightarrow \left[\begin{array}{c} R'' \\ \diagdown \\ \quad C\equiv C \\ H \diagup \qquad \diagdown R' \\ \vdots \\ Sn \\ \diagup | \diagdown \end{array} \right] \longrightarrow$$

$$\underset{H}{\overset{R''}{\diagdown}} C=C \underset{\underset{\diagup | \diagdown}{Sn^+}}{\overset{-}{\diagdown}} R' \quad \xrightarrow{\text{fast}} \quad \underset{H}{\overset{R''}{\diagdown}} C=C \underset{R'}{\overset{Sn-}{\diagdown}}$$

8. HYDROPLUMBATION

The lead-hydrogen bond is relatively unstable, and little is known about it. It adds to olefinic bonds and to acetylenic bonds. Trimethyllead hydride and cyanoacetylene at −70° give three products, in a ratio which depends on the solvent:[111]

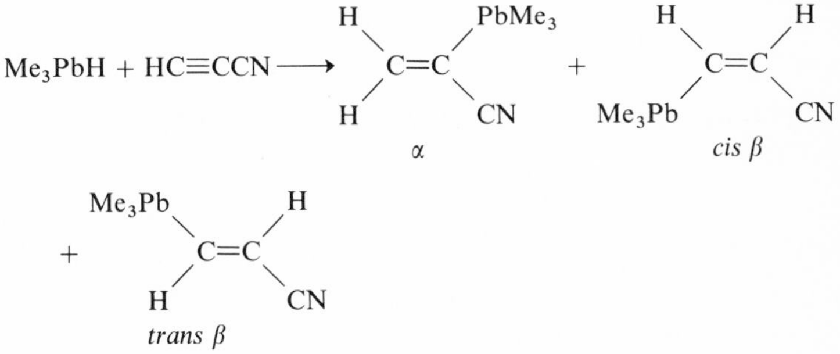

$$Me_3PbH + HC\equiv CCN \longrightarrow \underset{H}{\overset{H}{\diagdown}} C=C \underset{CN}{\overset{PbMe_3}{\diagup}} + \underset{Me_3Pb}{\overset{H}{\diagdown}} C=C \underset{CN}{\overset{H}{\diagup}}$$

$$\alpha \qquad\qquad cis\ \beta$$

$$+ \quad \underset{H}{\overset{Me_3Pb}{\diagdown}} C=C \underset{CN}{\overset{H}{\diagup}}$$

$$trans\ \beta$$

Solvent	% of Each Product		
	α	cis β	trans β
Neat	50	45	5
Butyronitrile	76	24	
Diethyl ether	2	92	6

Since formation of the α-product is not a radical reaction, but formation of the β-products is radical, fortuitous initiators or inhibitors could influence these results.

The α-adduct forms by a *trans* mechanism, as shown by additions to deutero-cyanoacetylene:

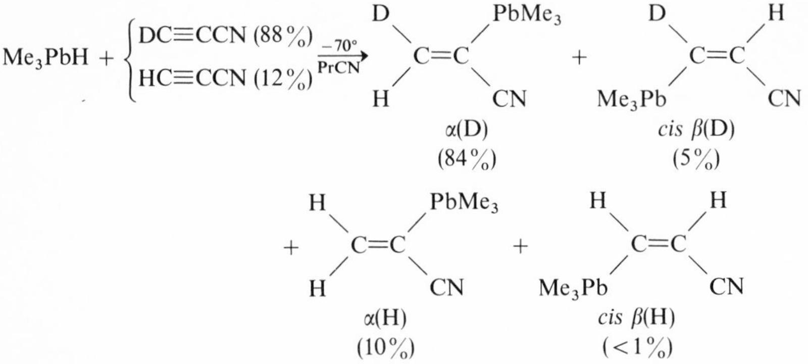

$$\text{Me}_3\text{PbH} + \begin{cases} \text{DC}{\equiv}\text{CCN } (88\%) \\ \text{HC}{\equiv}\text{CCN } (12\%) \end{cases} \xrightarrow[\text{PrCN}]{-70°}$$

$$\alpha(\text{D})\ (84\%) \qquad \textit{cis}\ \beta(\text{D})\ (5\%)$$

$$\alpha(\text{H})\ (10\%) \qquad \textit{cis}\ \beta(\text{H})\ (<1\%)$$

Thus, the *trans* β-isomer probably forms by rearrangement of the *cis* β-isomer.

Addition of trimethyllead hydride to diethyl acetylenedicarboxylate is mostly *trans* regardless of solvent:

Products are

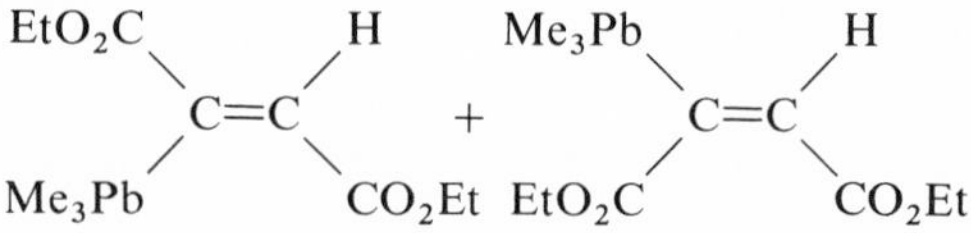

$$(16:1\ \text{ratio})$$

Methyl propiolate gives mostly the *cis* β-adduct, and in butyronitrile gives about 15% of the α-adduct. The mechanism is probably the same as for hydrostannation.

9. ADDITION OF DICYCLOPENTADIENYLRHENIUM HYDRIDE

Dicyclopentadienylrhenium hydride adds very rapidly to dimethyl acetylene-dicarboxylate in tetrahydrofuran at 20°.[112] The product is the maleate:

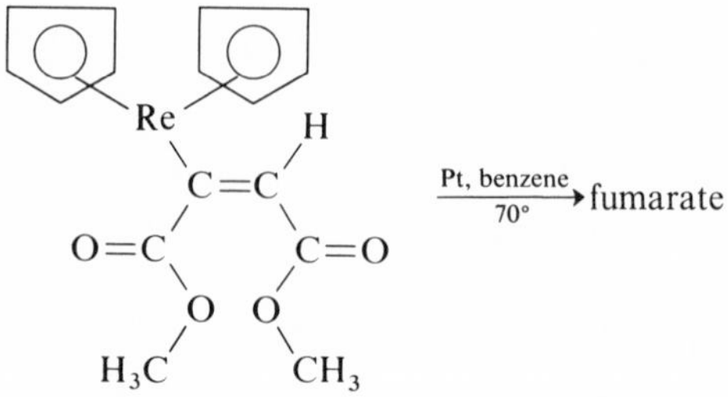

$$\xrightarrow[70°]{\text{Pt, benzene}} \text{fumarate}$$

maleate

The product is unstable in air, but inert to water. Methyl propiolate forms a β-substituted acrylic ester. Thus, addition is *cis*. The reaction is not a simple radical or ionic addition. The intermediate is probably a four-center transition complex.

References

1. Dvorko, G. F., and Mironova, D. F., *Ukr. Khim. Zh.*, **31**, 195 (1965); *Chem. Abstr.*, **63**, 2865 (1965).
1a. Winterfeldt, E., *Angew. Chem., Intern. Ed.*, **6**, 423 (1967).
2. Dvorko, G. F., and Karpenko, T. F., *Ukr. Khim. Zh.*, **31**, 75 (1965); *Chem. Abstr.*, **62**, 14465 (1965).
3. Otsuka, S., and Murahashi, S., *Kogyo Kagaku Zasshi*, **59**, 511 (1956); *Chem. Abstr.*, **52**, 3818 (1958).
4. Turck, U., and Behringer, H., *Chem. Ber.*, **98**, 3020 (1965).
5. Arai, A., Kado, M., and Chiyomaru, I., *Yuki Gosei Kagaku Kyokai Shi*, **23**, 435 (1965); *Chem. Abstr.*, **63**, 6855 (1965).
6. Truce, W. E., and Heine, R. F., *J. Am. Chem. Soc.*, **79**, 5311 (1957).
7. Truce, W. E., Klein, H. G., and Kruse, R. B., *J. Am. Chem. Soc.*, **83**, 4636 (1961).
8. Truce, W. E., and Simms, J. A., *ibid.*, **78**, 2756 (1956).
8a. Chen, T.-Y., Kato, H., and Ohta, M., *Bull. Chem. Soc. Japan*, **40**, 1964 (1967).
9. Fiesselmann, H., Schipprak, P., and Zeither, L., *Chem. Ber.*, **87**, 841 (1954).
10. Bohlmann, F., and Bresinsky, E., *Chem. Ber.*, **97**, 2109 (1964).
11. Bohlmann, F., Bornowski, H., and Kramer, D., *Chem. Ber.*, **96**, 584 (1963).
12. Garraway, J. L., *J. Chem. Soc.*, 4077 (1962).
13. Winterfeldt, E., and Preuss, H., *Angew. Chem.*, **77**, 679 (1965).
14. Wieland, J. H. S., and Arens, J. F., *Rec. Trav. Chim.*, **79**, 1293 (1960).
15. Jones, E. R. H., and Whiting, M. C., *J. Chem. Soc.*, 1423 (1949).
16. Hendrickson, J. B., Rees, R., and Templeton, J. F., *J. Am. Chem. Soc.*, **86**, 107 (1964).
17. Dolfini, J. E., *J. Org. Chem.*, **30**, 1298 (1965).
18. Heine, H. W., and Peavy, R., *Tetrahedron Letters*, 1359 (1965); *J. Org. Chem.*, **31**, 3924 (1966).
19. Brand, J. C. D., Eglinton, G., and Tyrrell, J., *J. Chem. Soc.*, 5914 (1965).
20. McMullen, C. H., and Stirling, C. J. M., *J. Chem. Soc. (B)*, 1217 (1966).
20a. Taylor, E. C., and Heindel, N. D., *J. Org. Chem.*, **32**, 3339 (1967).
21. Truce, W. E., and Brady, D. G., *J. Org. Chem.*, **31**, 3543 (1966).
22. McMullen, C. H., and Stirling, C. J. M., *J. Chem. Soc. (B)*, 1221 (1966).
23. Agosta, W., *J. Org. Chem.*, **26**, 1724 (1961).
23a. Skatebøl, L., Boulette, B., and Soloman, S., *J. Org. Chem.*, **33**, 548 (1968).
24. Schiessl, H. W., and Appel, R., *J. Org. Chem.*, **31**, 3851 (1966).
24a. Bottomley, W., *Tetrahedron Letters*, 1997 (1967).
25. Lopez, L., and Barrans, J., *Compt. Rend.*, **263**(C), 557 (1966).
26. Hoffmann, H., *Chem. Ber.*, **94**, 1331 (1961).
27. Hoffmann, H., and Diehr, H. J., *Chem. Ber.*, **98**, 363 (1965).
28. Reddy, G. S., and Weis, C. D., *J. Org. Chem.*, **28**, 1822 (1963).
29. Johnson, A. W., and Tebby, J. C., *J. Chem. Soc.*, 2126 (1961).

29a. Pudovik, A. N., and Durova, O. S., *J. Gen. Chem. USSR (Engl. Transl.)*, **36**, 1465 (1966).

30. Lingers, F., and Schneider-Bernlohr, H., *Ann Chem.*, **686**, 134 (1965).

31. Vereshchagin, L. I., and Korshunov, S. P., *Zh. Organ. Khim.*, **1**, 960, 955 (1965); *Chem. Abstr.* **63**, 6943 (1965).

32. Lappin, G. R., *J. Org. Chem.*, **26**, 2350 (1961).

33. Lapkin, I. I., and Andreichikov, Yu. S., *Zh. Organ. Khim.*, **1**, 1212 (1965); *Chem. Abstr.*, **63**, 13111 (1965).

34. Kurtz, A. N., *et al.*, *J. Org. Chem.*, **30**, 3141 (1965).

35. Vessiere, R., and Theron, F., *Compt. Rend.*, **260**, 597 (1965).

36. Chelpanova, L. F., *et al.*, *Izv. Vysshikh Uchebn. Zavedenii Khim. i. Khim. Tekhnol.*, **7**, 945 (1964); *Chem. Abstr.*, **62**, 14487 (1965).

37. Chierici, L., and Montanari, F., *Boll. Sci. Fac. Chim. Ind. Bologna*, **14**, 78 (1956); *Chem. Abstr.*, **51**, 5721 (1957).

37a. Grob, C. A., and Kaiser, A., *Helv. Chim. Acta.*, **50**, 1599 (1967).

38. Iwanami, Y., *Nippon Kagaku Zasshi*, **83**, 593 (1962); *Chem. Abstr.*, **59**, 5153 (1963).

39. Alaimo, R. J., and Farnum, D. G., *Can. J. Chem.*, **43**, 200 (1965).

40. Sasaki, H., Sakata, H., and Iwanami, Y., *Nippon Kagaku Zasshi*, **85**, 704 (1964); *Chem. Abstr.*, **62**, 14678 (1965).

40a. Lown, J. W., and Ma, J. C. N., *Can. J. Chem.*, **45**, 939, 953 (1967).

40b. Winterfeldt, E., and Nelke, J. M., *Chem. Ber.*, **100**, 3671 (1967).

41. Pandit, U. K., and Huisman, H. O., *Rec. Trav. Chim.*, **84**, 50 (1964).

41a. Winterfeldt, E., *Chem. Ber.*, **100**, 3679 (1967).

42. Padwa, A., and Hamilton, L., *Tetrahedron Letters*, 4363 (1965).

43. Cymerman-Craig, J., and Moyle, M., *Proc. Chem. Soc.*, 149 (1962).

43a. Brown, G. W., *J. Chem. Soc. (C)*, 2018 (1967).

44. Garraway, J. L., *J. Chem. Soc.*, 4077 (1962).

45. Grinblatt, E. I., and Postovskii, I. Ya., *Zh. Obshch. Khim.*, **31**, 389 (1961); *Chem. Abstr.*, **55**, 22203 (1961).

46. Winterfeldt, E., *Angew. Chem. Intern. Ed.*, **5**, 741 (1966).

46a. Hermes, M. E., and Marsh, F. D., *J. Am. Chem. Soc.*, **89**, 4760 (1967).

47. Bergmann, E. D., Ginsburg, D., and Pappo, R., *Org. Reactions*, **10**, 179 (1959).

47a. Heindel, N. D., *et al.*, *J. Org. Chem.*, **32**, 2678 (1967).

48. Bamfield, P., Crabtree, A., and Johnson, A. W., *J. Chem. Soc.*, 4355 (1965). Cookson, R. C., *et al.*, *J. Chem. Soc. (C)*, 1986 (1967).

49. Bamfield, P., *et al.*, *Chem. Ind. (London)*, 1313 (1964).

50. LeGoff, E., and LaCount, R. B., *J. Org. Chem.*, **29**, 423 (1964).

51. Hendrickson, J. B., *et al.*, *J. Org. Chem.*, **30**, 3312 (1965).

52. Stork, G., and Tomasz, M., *J. Am. Chem. Soc.*, **86**, 471 (1964).

53. Bonnema, J., and Arens, J. F., *Rec. Trav. Chim.*, **79**, 1137 (1960).

54. Mandell, L., and Blanchard, W. A., *J. Am. Chem. Soc.*, **79**, 6198 (1957).

54a. Acheson, R. M., *et al.*, *J. Chem. Soc. (C)*, 882 (1967).

55. Acheson, R. M., and Burstall, M. L., *J. Chem. Soc.*, 3240 (1954).

56. Galbraith, A., Small, T., and Boekelheide, V., *J. Org. Chem.*, **24**, 582 (1959).

57. Galbraith, A., *et al.*, *J. Am. Chem. Soc.*, **83**, 453 (1961).

58. Boekelheide, V., and Fahrenholz, *J. Am. Chem. Soc.*, **83**, 458 (1961); *Tetrahedron Letters*, 651 (1963).

59. Huisgen, R., *Proc. Chem. Soc.*, 357 (1961).
60. Huisgen, R., *Angew Chem. Intern. Ed.*, **2**, 565 (1963).
61. Gibson, W. K., and Leauer, D., *Proc. Chem. Soc.*, 330 (1964).
62. Huisgen, R., *et al.*, *Angew. Chem.*, **73**, 170 (1961).
62a. Reimlinger, H., *Ann. Chem.*, **713**, 113 (1968).
63. Huisgen, R., Binsch, G., and Koenig, H., *Chem. Ber.*, **97**, 2884 (1964).
64. Huisgen, R., *Angew. Chem.*, **74**, 29 (1962).
65. Behringer, H., and Wiedenmann, R., *Tetrahedron Letters*, 3705 (1965).
65a. Behringer, H., and Falkenberg, J., *Tetrahedron Letters*, 1895 (1967).
66. Acheson, R. M., Foxton, M. W., and Miller, G. R., *J. Chem. Soc.*, 3200 (1965).
67. Acheson, R. M., in "Advances in Heterocyclic Chemistry," Vol. 1, p. 125, New York, Academic Press, 1963. See also Acheson, R. M., *et al.*, *J. Chem. Soc. (C)*, 348–389 (1968).
68. Noland, W. E., Kuryla, W. C., and Lange, R. F., *J. Am. Chem. Soc.*, **81**, 6010 (1959).
69. Kalbag, S. M., *et al.*, *Tetrahedron*, **23**, 1911 (1967).
70. Benson, F. R., and Savell, W. L., *Chem. Rev.*, **46**, 1 (1950).
70a. Huisgen, R., Szeimies, G., and Möbius, L., *Chem. Ber.*, **100**, 2494 (1967).
71. Huisgen, R., *et al.*, *Chem. Ber.*, **98**, 4014 (1965).
71a. Khetan, S. K., and George, M. V., *Can. J. Chem.*, **45**, 1993 (1967).
72. Pandit, U. K., and Huisman, H. O., *Rec. Trav. Chim.*, **85**, 311 (1966).
72a. Huisgen, R., *et al.*, *Chem. Ber.*, **100**, 1580 (1967).
73. Huisgen, R., and Herbig, K., *Ann. Chem.*, **688**, 98 (1965).
74. Crabtree, A., Johnson, A. W., and Tebby, J. C., *J. Chem. Soc.*, 3497 (1961).
75. Acheson, R. M., *et al.*, *Proc. Chem. Sov.*, 281 (1960).
76. Huisgen, R., *et al.*, *Chem. Ber.*, **100**, 1094 (1967).
76a. Morikawa, M., and Huisgen, R., *Chem. Ber.*, **100**, 1616 (1967).
77. Gagan, J. M. F., *J. Chem. Soc. (C)*, 1121 (1966).
78. Alder, K., and Brachel, H. V., *Ann. Chem.*, **651**, 141 (1962).
79. Kallos, J., and Deslongchamps, P., *Can. J. Chem.*, **44**, 1239 (1966).
80. Mandell, L., and Blanchard, W. A., *J. Am. Chem. Soc.*, **79**, 2343 (1957).
81. *Ibid.*, 6198 (1957).
82. Acheson, R. M., and Vernon, J. M., *J. Chem. Soc.*, 457 (1961).
83. *Ibid.*, 1148 (1962).
84. Lemal, D. M., Gosselink, E. P., and McGregor, S. D., *J. Am. Chem. Soc.*, **88**, 582 (1966).
85. Seyferth, D., and Evnin, A. B., *J. Am. Chem. Soc.*, **89**, 1468 (1967).
86. Gassmann, P. G., and Mansfield, K., *Chem. Commun.*, 391 (1965).
86a. Cookson, R. C., Dance, J., and Godfrey, M., *Tetrahedron*, **24**, 1529 (1968).
86b. Gassmann, P. G., and Mansfield, K. T., *J. Am. Chem. Soc.*, **90**, 1517, 1524 (1968).
86c. Kraihanzel, C. S., and Losee, M. L., *J. Org. Chem.*, **33**, 1983 (1968).
87. Benghiat, I., and Becker, E. I., *J. Org. Chem.*, **23**, 885 (1958).
88. Romanelli, M. G., and Becker, E. I., *J. Org. Chem.*, **27**, 662 (1962).
89. Klemm, L. H., *et al.*, *J. Org. Chem.*, **31**, 2376 (1966).
90. Diels, O., Alder, K., and Nienburg, H., *Ann. Chem.*, **490**, 236 (1931).
91. Sapov, N. P., and Miklashevskaya, V. S., *Zh. Obshch. Khim.*, **26**, 1914 (1956); *Chem. Abstr.*, **51**, 4968 (1957).
92. Gibson, W. K., and Leaver, D., *Chem. Commun.*, 11 (1965).

93. Hill, R. K., and Carlson, R. M., *J. Org. Chem.*, **30**, 2414 (1965).
94. Vogel, E., Grimme, W., and Korte, S., *Tetrahedron Letters*, 3625 (1965).
95. Shusherina, N. P., Levina, R. Ya., and Shostakovskii, V. M., *Zh. Obshch. Khim.*, **29**, 3237 (1959); *Chem. Abstr.*, **54**, 13057 (1960).
96. Kappeler, H., and Renk, E., *Helv. Chim. Acta.*, **44**, 1541 (1961).
97. Miklashevskaya, V. S., and Petrov, A. A., *Zh. Obshch. Khim.*, **28**, 1125 (1958); *Chem. Abstr.*, **52**, 20027 (1958).
98. Rigaudy, J., Guillaume, J., and Cuong, N. K., *Compt. Rend.*, **259**, 4729 (1964).
99. Rigaudy, J., and Cuong, N. K., *Compt. Rend.*, **253**, 1705 (1961).
100. Gabel, N. W., *J. Org. Chem.*, **27**, 301 (1962).
101. Avran, M., Mateescu, G., and Neninetzescu, C. D., *Ann. Chem.*, **636**, 174 (1960).
102. Muir, K. W., *et al.*, *Chem. Ind.* (*London*), 1581 (1964).
103. Petrov, A. A., *Zh. Obshch. Khim.*, **24**, 2136 (1954); *Chem. Abstr.*, **50**, 233 (1956).
104. Petrov, A. A., *Zh. Obshch. Khim.*, **17**, 497 (1954); *Chem. Abstr.*, **42**, 1881 (1948).
105. Nightingale, D., and Wadsworth, F., *J. Am. Chem. Soc.*, **67**, 416 (1945).
106. Bondarev, G. N., and Petrov, A. A., *Zh. Organ. Khim.*, **2**, 1005 (1966); *Chem. Abstr.*, **65**, 16874 (1966).
106a. Ciganek, E., *Tetrahedron Letters*, 3321 (1967).
107. Leusink, A. J., Marsman, J. W., and Budding, H. A., *Rec. Trav. Chim.*, **84**, 689 (1965).
108. Leusink, A. J., *et al.*, *Rec. Trav. Chim.*, **84**, 567 (1965).
109. Leusink, A. J., and Marsman, J. W., *Rec. Trav. Chim.*, **84**, 1123 (1965).
110. Neumann, W. P., and Sommer, R., *Ann. Chem.*, **675**, 10 (1964).
111. Leusink, A. J., and van der Kerk, G. J. M., *Rec. Trav. Chim.*, **84**, 1617 (1965).
112. Dubeck, M., and Schell, R. A., *Inorg. Chem.*, **3**, 1757 (1964).

PART NINE

Addition to Polyfluoroacetylenes

1. SIMPLE NUCLEOPHILIC ADDITION

Polyfluoroacetylenes behave as negatively substituted acetylenes in their addition reactions. Polyfluoroacetylenes add nucleophiles in the "abnormal" fashion:[1]

$$R_FC\equiv CH + XH \xrightarrow{\text{catalyst}} R_FCH=CHX \quad (\text{XH is HF, HCN, } RNH_2, \text{ or } R_2NH)$$

HF adds to 1,1,1-trifluoro-2-propyne to form 92% of 1,3,3,3-tetrafluoropropene after 2 days at room temperature in the presence of BF_3 catalyst. HCN in the presence of $KCN\text{-}Cu_2Cl_2$ gives 52% yield of γ,γ,γ-trifluorocrotonitrile after 1 hour at 110°. Diethylamine adds to trifluoropropyne in the presence of cuprous chloride to give 28% yield of 1-diethylamino-3,3,3-trifluoropropene after 4 hours at room temperature, followed by heating at 100°.

Arsenic compounds which have As—As, As—H, and As—Cl bonds add across the triple bond of hexafluoro-2-butyne. Phosphorus and nitrogen compounds also add.[1a] Diethylphosphine adds vigorously at $-78°$ to give 45% yield of trans-2-diethylphosphino-3H-hexafluoro-2-butene. Ultraviolet irradiation is required to induce reaction with bis(trifluoromethyl)phosphine. The main product is bis[bis(trifluoromethyl)phosphino]-1,1,1,4,4,4-hexafluoro-butane (21% yield). This is probably a free radical reaction.

As expected, these nucleophilic attacks give mostly trans-olefinic products. Triphenylphosphine polymerizes hexafluoro-2-butyne.[1a] Water adds vigorously at 20° in the presence of triethylamine. The major product is the bis(butenyl) ether, 31% yield. Simple hydration to the butanone is a minor reaction here. Chlorodimethylphosphine adds violently to form the 1:1 product

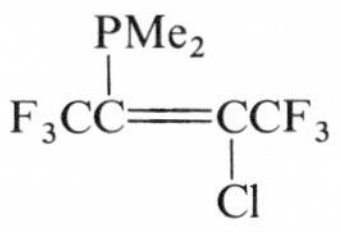

$$\underset{\underset{Cl}{|}}{\overset{\overset{PMe_2}{|}}{F_3CC}}=CCF_3$$

Chlorodimethylamine adds slowly at 85° to form the corresponding 1:1 adduct.

2. DOUBLE ADDITION OF A CARBENE

Bicyclobutane hydrocarbons have been made by double carbene additions.[2] Difluorocarbene adds analogously to hexafluoro-2-butyne to form 1,2-bis(trifluoromethyl)-3,3-difluorocyclopropene (1) which adds another mole of difluorocarbene to form 25% yield of 1,3-bis(trifluoromethyl)-2,2,4,4-tetrafluorobicyclobutane (2):

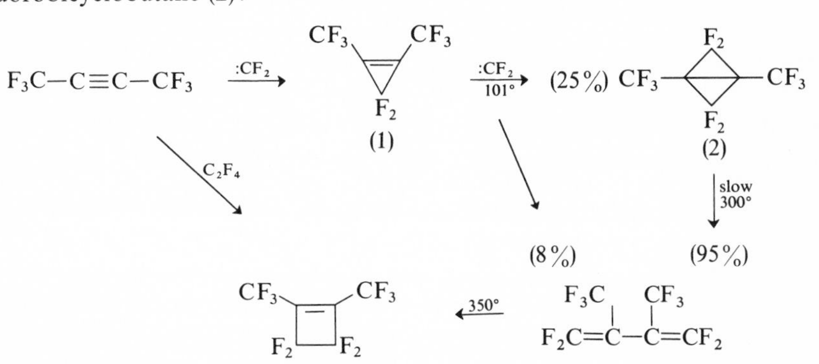

3. ADDITION OF (CF₃P)₄ AND (CF₃P)₅

Reaction of $(CF_3P)_4$ and $(CF_3P)_5$ with excess hexafluoro-2-butyne at 170° for 70 hours gives 55% yield of 1,2,3,4-tetrakis-(trifluoromethyl)-3,4-diphosphacyclobutene (1) and 31% yield of 1,2,3,4,5-pentakis(trifluoromethyl)-3,4,5-triphosphacyclopentene (2), respectively:[3]

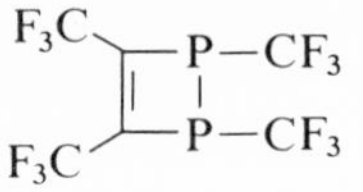

(1) 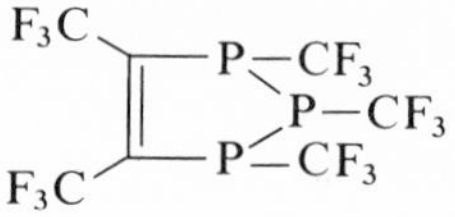   (2)

4. ADDITION OF PHOSPHORUS AND OF ARSENIC

Krespan[4] reported another remarkable reaction of hexafluoro-2-butyne. He prepared two new heterocyclic compounds which have the isolated bicycloocta-triene skeleton. When he heated hexafluoro-2-butyne with red phosphorus and iodine catalyst at 200°, he obtained 43% yield of hexakis(trifluoromethyl)-1,4-diphosphabicyclo-[2.3.3]octa-2,5,7-triene, a representative of a new class of heterocycles:

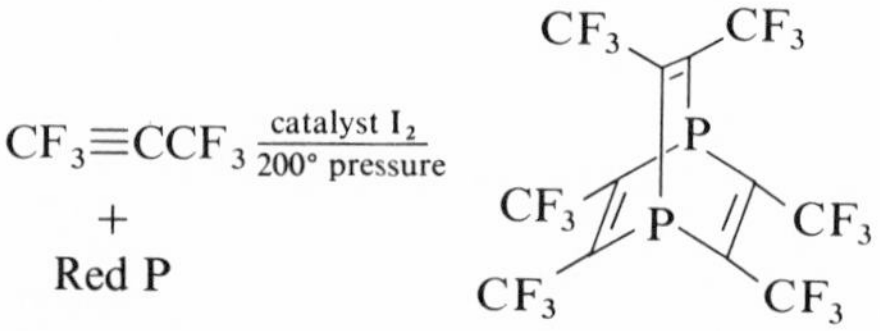

(43%)

The phosphorus compound is a colorless, very volatile solid. Arsenic gives analogous products.

5. DIELS-ALDER ADDITIONS

5.1. Addition of Dithietenes

The S—S linkage in bis(trifluoromethyl)-1,2-diethietene is weak.[5] It can be broken by weak base to give dimer, or by heat to cause addition to hexafluoro-2-butyne. This dithietene adds smoothly to olefins and to other acetylenes. Especially reactive are olefins and acetylenes which have a high density of loosely bound electrons. The reaction is general.[6] 3-Hexyne has two electron-donating alkyl groups, and the triple bond is more susceptible to attack than acetylene, so reactions go at only 25° to give a *p*-dithiin. Dimethyl acetylenedicarboxylate has a less susceptible triple bond and reacts at 70° to give a *p*-dithiin. The most likely mechanism is illustrated by the reaction with acetylene itself:

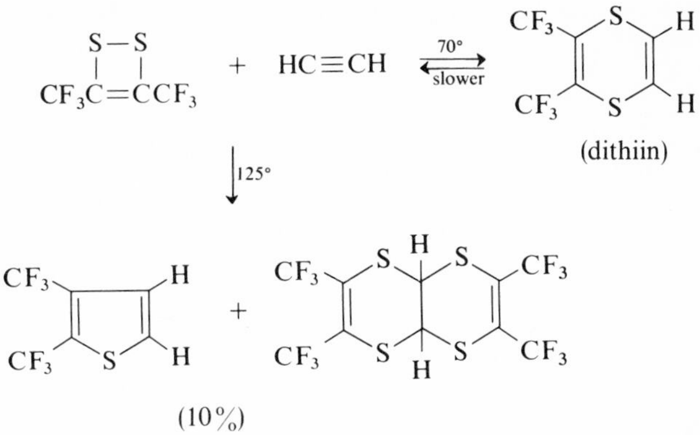

5.2. Addition of Aromatic Hydrocarbons

Only very reactive aromatics such as anthracene and furan undergo Diels-Alder reactions with acetylenes.[7] Benzene or its simple derivatives do not react with acetylenes or olefins by 1,4-addition.[8] However, hexafluoro-2-butyne adds 1,4 to durene to give 40% yield of 2,3,5,6-tetramethyl-7,8-bis(trifluoromethyl)-bicyclo[2.2.2]octa-2,5,7-triene. This is an excellent way to prepare the "elusive" bicyclooctatriene skeleton:[8]

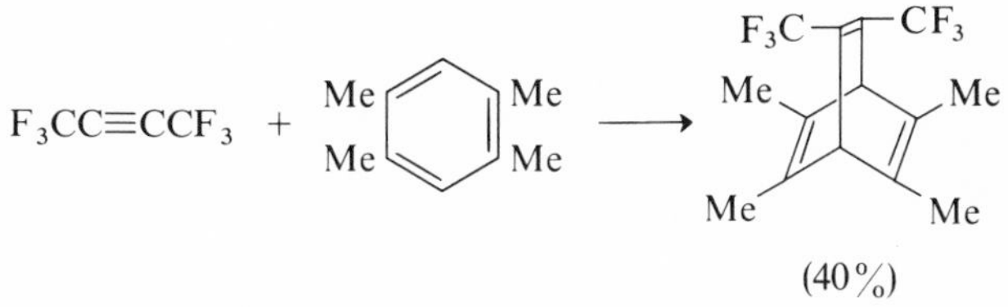

(40%)

Other bis(polyfluoroalkyl)acetylenes also add to durene, in an apparently general reaction. Neither acetylene nor 1,1,1-trifluoropropyne reacts. This shows that the bis(polyfluoroalkyl)-acetylenes are remarkably reactive dienophiles, probably because the electrophilic character of the triple bond is greatly increased by the flanking perfluoroalkyl groups.

Both hexafluoro-2-butyne and 1,6-dichlorooctafluoro-3-hexyne react easily with anthracene at the 9- and 10-positions. This is the first example of reaction of a simple naphthalene nucleus with an acetylene.[8]

References

1. Haszeldine, R. N., British Patent 772,109, 772,110 (April 10, 1957); *Chem. Abstr.*, **51**, 16518 (1957).
1a. Cullen, W. R., and Dawson, D. S., *Can. J. Chem.*, **45**, 2887 (1967).
2. Mahler, W., *J. Am. Chem. Soc.*, **84**, 4600 (1962).
3. *Ibid.*, **86**, 2306 (1964).
4. Krespan, C. G., *J. Am. Chem. Soc.*, **83**, 3432 (1961).
5. *Ibid.*, **83**, 3434 (1961).
6. Krespan, C. G., and McKusick, B. S., *J. Am. Chem. Soc.*, **83**, 3438 (1961).
7. Holmes, H. L., *Org. Reactions* **4**, 81 (1949).
8. Krespan, C. G., McKusick, B. C., and Cairns, T. L., *J. Am. Chem. Soc.*, **83**, 3428 (1961).

Chapter Three

VINYLATION

Introduction

Vinylation reactions are special cases of ionic addition across the triple bond of acetylene. Additions across triple bonds in substituted acetylenes are included where these are similar to vinylations.

Detailed descriptions of vinylations and lists of vinyl products are not given here, since these are available elsewhere.[1] Mechanisms are stressed where possible, but surprisingly little work on mechanisms has been reported. Some unusual vinylations are included, because some of them may provide leads to future developments.

This chapter is divided into five parts:
(1) Vinylation of alcohols and mercaptans.
(2) Vinylation of carboxylic acids, halogen acids, and phosphorus acids and esters.
(3) Reaction of acetylenes with carbon monoxide (addition of H and CO_2H).
(4) Vinylation of HCN and nitrogen compounds.
(5) Vinylation of "activated" C—H bonds.

PART ONE

Vinylation of Alcohols and Mercaptans

1. VINYLATION OF ALCOHOLS

1.1. Liquid Phase

1.1.1. ALCOHOLS

Base-catalyzed vinylation of alcohols is usually carried out in the presence of
KOH or NaOH. The mechanism is probably :[2,3]

$$ROH + KOH \xleftrightarrow{\text{fast}} ROK + H_2O$$

$$ROK + C_2H_2 \xrightarrow[k_2]{} ROCH{=}CHK \xrightarrow[+ROH]{\text{fast}} ROCH{=}CH_2 + ROK$$

$$\text{rate} = k_2[ROK][C_2H_2]$$

The work of Reppe and others with KOH led to the general belief that KOH
is the best and most convenient vinylation catalyst. The widespread use of
KOH probably accounts for earlier failures to vinylate tertiary alcohols.[4]
Before 1959, there was only one report of vinylation of a tertiary alcohol.[5]
When KOH is dissolved in absolute ethanol, about 94% of the KOH forms
KOEt. Secondary alcohols and KOH give little alkoxide and tertiary alcohols
form no alkoxide at all. The rate of vinylation depends on the concentration of
ROK. Thus, in the octanol series, the relative rate of vinylation of primary and
secondary octanol is 7, using 4 mole % KOH at 190°. With potassium *t*-butoxide
catalyst, tertiary butanol vinylates nearly as fast as primary or secondary
butanol. The rate of solution of acetylene is probably rate determining.[4]
Alcohols boiling above 150° vinylate easily at atmospheric pressure.[6] The
best yield is obtained by passing acetylene through a 5–10% solution of KOH
in alcohol and distilling out the vinyl ether as it forms. Products contain 90%
vinyl ether and 1–3% acetal. The reaction is usually 10–15 times slower
than in pressure reactions. However, stearyl alcohol gave 85% yield of vinyl
ether in only $1\frac{1}{2}$ hours at 180° and atmospheric pressure in the presence of 5%
KOH.[7] Monovinyl ethers of glycols can be made using similar systems.[8]
Glycols give significant amounts of cyclic acetal by-products. Chatterjee[9]
vinylated fatty alcohols at atmospheric pressure and 180°, using 5% KOH
as catalyst. Yields were 78–90% (Table 3-1).
Amino alcohols vinylate in the presence of bases to give vinyl ethers but in the
presence of zinc acetate or cadmium acetate, they give oxazolidines.[2] Amino

TABLE 3-1
Vinyl Ethers of Fatty Alcohols[9]

Alcohol	b.p. (°C, @ mm Hg)	Alcohol	b.p. (°C, @ mm Hg)
Decyl	102, @ 10	Oleyl	164, @ 1.5
Lauryl	122, @ 4	Linoleyl	160, @ 5
Myristyl	134, @ 3	Safflower	162–180, @ 6
Cetyl	156, @ 1.5	• Dehydrated castor oil	158–162, @ 4

alcohols with primary or secondary amino groups beta or gamma to the hydroxyl group give 2-methyloxazolidines or 2-methyltetrahydro-1,3-oxazines,
even in the absence of catalysts[10]

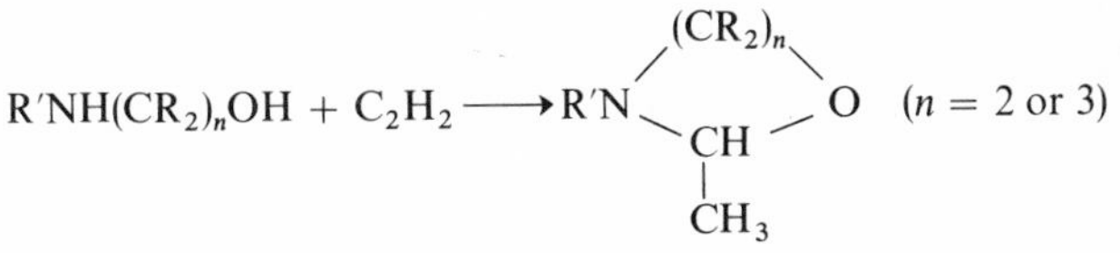

$$R'NH(CR_2)_nOH + C_2H_2 \longrightarrow R'N \underset{\underset{\underset{CH_3}{|}}{CH}}{\overset{(CR_2)_n}{\diagup}} \diagdown O \quad (n = 2 \text{ or } 3)$$

Since alcohols do not vinylate without base, the reaction must involve vinylation of the amino group, followed by addition of OH across the vinyl double
bond.

Isobutyl glycolate and acetylene at room temperature with HgO-BF$_3$
catalyst form the acetal in 70% yield. Acetals of glycolic esters with 6–12 carbon
atoms are claimed to be good plasticizers and petroleum additives.[11]

1.1.2. PHENOLS

Phenols vinylate at 160° in aqueous base.[12] The reaction is first order on KOH
and on acetylene pressure, and zero order on phenol. Phenols with electron-
donating groups vinylate faster. Thiophenols are even more reactive, vinylating
rapidly in aqueous base at 80°. Shostakovskii[13] vinylated mixed phenols from
Lurgi semi-coking, using NaOH or KOH catalyst at 14 atmospheres acetylene
and 190° (Table 3-2, page 286).

1.1.3. GLYCOLS AND POLYOLS

Monovinyl ethers of glycols must be handled in the presence of base because
they rearrange explosively to cyclic acetals in the presence of traces of acids.[14]
Nedwick[15] used his liquid-full reactor for some vinylation reactions. Glycerol
gave trivinylglycerol and monovinyl cyclic acetal. Comparison of batch and

liquid-full systems shows that the latter give higher yields of trivinylglycerol:

| | % Yield | |
Product	Batch[16]	Liquid-full
Trivinyl ether	5–10	50
Vinylacetal	50–60	20–30

TABLE 3-2
Vinyl Ethers of Phenols from Coking[13]

Vinyl Ether of	% Yield[a]	b.p. (°C @ 50 mm Hg)
Phenol	7.9	76.5– 77.5
o-Ethylphenol	6.4	83.5– 85
o-Cresol	5.5	86.5– 88.5
m-Cresol	18.2	91.5– 92.2
p-Cresol	3.7	92.2– 93.5
p-Ethylphenol	3.2	111.5–113.5
m-Ethylphenol	5.7	113.7–116
m-Xylenol	3.3	116.0–118.2

[a] from mixture

The liquid-full procedure also gives better results in vinylations of mercaptans and lactams.

Table 3-3 lists the vinyl ethers of several polyols.

TABLE 3-3
Vinyl Ethers of Polyols

Polyol	Vinyl Ether	b.p. (°C, @ mm Hg)	Reference
Ethylene glycol	Di		8
	Tri	71, @ 17; 170, @ 760	16, 18
	Monovinyl acetal		19
Isosorbide	Mono	m.p. 44–45	20
	Di	62, @ 0.08	20
Diethylene glycol	Mono	90, @ 6	21
	Di	125, @ 713	21
Triethylene glycol	Mono and di	—	21
Pentaerythritol	Di	m.p. 63–64	22
Mannitol	Hexa	80–90, @ 0.5–1	23
3,4-Isopropylidenemannitol	1,2,5,6-Tetra	97–98, @ 0.5–1	23
1,2:3,4-Diisopropylidene-mannitol	5,6-Di	m.p. 61–62	23

1.1.4. SUGARS

Hydroxyl groups in sugars vinylate in the presence of KOH at 140–160° and under 10–15 atmospheres acetylene pressure. Some products are given in Table 3-4.

TABLE 3-4
Vinyl Ethers of Sugars

Product	% Yield	b.p. (°C @ mm Hg)	Reference
1,2:3,4-Di-O-isopropylidene-6-O-vinyl-D-galactose	67	95, @ 0.5	24
2,5,6-O-Trivinyl-1,2-O-isopropylidene-D-glucose	44	129, @ 1.5	24
Divinyl-D-glucose	5.8	108, @ 2	24, 25
Vinyl methyl-2-O-α-D-glucopyranoside (also by vinyl chloride method)	—	—	26
2,5,6-O-Trivinyl-1,2-O-isopropylidene-D-glucose	44	129, @ 1.5	25
3-O-Vinyl-1,2:5,6- di-O-isopropylidene-D-glucose	54	115, @ 1.5	24, 25
1-O-Vinyl-2,3:4,5-di-O-isopropylidenefructopy-ranose	—	132, @ 3	27
6-O-Vinyl-1,2:3,4-di-O-isopropylidene-D-galacto-pyranose	58	155, @ 10	28

1.1.5. COTTON

Cotton can be vinylated with vinyl chloride, ethylene dibromide, acetylene or vinyl ethers (transvinylation).[29] The direct acetylene vinylation is best from the standpoint of amount of vinylation, retention of fiber structure and color. Vinylation decreases breaking strength to 55–85% of the original, with 20–50% elongation at break.[30]

1.1.6. SALTS OF VINYL ALCOHOL

Nesmeianov[31] reported the first metallic derivatives of vinyl alcohol. He prepared lithium and sodium salts:

$$Hg(CH_2CHO)_2 + Li(NH_3) \longrightarrow Li/Hg + 2CH_2\!=\!CHOLi$$
$$(78\% \text{ yield,}$$
$$70\% \text{ for Na})$$

The products are colorless crystals. Lithium vinylate is soluble in ether and benzene, but sodium vinylate is not. The structure was proved by chemical analysis, titration with acid and hydrolysis to acetaldehyde. The infrared spectrum showed a band for double bond absorption at 1600 cm^{-1}.

The metal salts did not alkylate to vinyl ethers or acylate to vinyl esters with reagents such as butyl bromide and benzoyl chloride.

1.2. Vapor Phase

Very little has been reported on this potentially best system for vinylating alcohols. Soda lime, sodium carbonate and KOH on carbon are used to vinylate methanol in the vapor phase.[32] The optimum temperature is 250° with a methanol:acetylene ratio 5:1. Catalyst life is 6–8 hours, and optimum conversion is 20–30%. Very little acetal, aldehyde, etc., form as by-products. With 50% KOH on carbon as catalyst at 240°, the conversion of ethanol to ethyl vinyl ether is 43%.[33]

Ethanol, butanol and isoamyl alcohol can be vinylated over commercial zinc oxide at 310–320°. Conversions to vinyl ethers are 25–28%, with no acetals. Zinc acetate on carbon, or KOH on carbon, gives only 7–9% yield. Zinc phosphate promoted with copper dehydrogenates the alcohol. Zinc acetate-cuprous acetylide catalyst is used to vinylate butanol (48% conversion).[35]

1.3. Transvinylation

1.3.1. ALCOHOLS

Mercuric acetate is the best catalyst for vinyl ethers.[36] The usual catalyst for transvinylation to form vinyl esters is mercuric sulfate-sulfuric acid. Mercuric acetate is less acidic and does not catalyze the addition of alcohol to vinyl ether to form acetal as extensively as mercuric sulfate does. Formation of acetal is an equilibrium reaction, so a good yield of vinyl ether is possible only if the ether is distilled from the mixture as it forms. This imposes severe limitations on this reaction system. Yields of vinyl ethers are: ethyl, 98%; allyl, 75%; 2-methallyl, 75%; 2-methoxyethyl, 62%; 2-chloroethyl, 52%; furfuryl, 28%.[36]

Transvinylation of a propargylic-allylic alcohol gives an aldehyde[37]

$$R_1CH{=}CHCHC{\equiv}CR_2 + CH_2{=}CHOEt \xrightarrow{\text{Hg(OAc)}} [\text{vinyl ether}]$$

$$\underset{\text{OH}}{|}$$

$R_1 = H, CH_3$
$R_2 = CH_3, Et$

$$\underset{\text{(40% yield)}}{O{=}\overset{H}{\underset{|}{C}}{-}CH_2{-}\overset{R_1}{\underset{|}{C}}HCH{=}CHC{\equiv}CR_2}$$

1.3.2. DIOLS

Jones[38] tried to use vinyl interchange with vinyl acetate in the presence of mercuric acetate and sulfuric acid catalyst to prepare divinyl ethers of glycols.

This method is not applicable because cyclic acetals form. Reaction of 1 mole of diol with 8 moles of vinyl butyl ether in the presence of mercuric acetate gives low to fair yields of divinyl ethers (Table 3-5).

TABLE 3-5
Divinyl Ethers of Diols by Transvinylation[38]

Divinyl Ether of	% Yield	b.p. (°C, @ mm Hg)
1,3-Propanediol	20	143, @ 760
1,5-Pentanediol	24	75–77, @ 11
1,11-Undecanediol	50	144, @ 4
1,4-Cyclohexanediol	13.5	81–83, @ 2
Diethylene glycol	16	85–87, @ 14
Di-1,3-propylene glycol	12	82–84, @ 7
Bis(β-hydroxyethyl) sulfide	14	112–113, @ 12
Triethylene glycol	12	136–138, @ 19

2. SUBSTITUTED ACETYLENES IN VINYLATION REACTIONS

2.1. Vinylacetylenes

2.1.1. KOH CATALYST

Alcohols at 150° in the presence of KOH add normally to the triple bond of vinylacetylenes, but in a direction opposite to the direction of HX addition.[39]

MeC≡CCH=CHMe in methanol with KOH for 12 hours at 150° reacts to form a dienyl ether. The methoxy group adds to the triple bond carbon next to methyl (the carbon atom in the 2-position). Water adds to carbon 3.[40]

2.1.2. KOR CATALYST

2-Butynyl ethers are formed by adding alcohols to the double bond of vinyl-acetylene in the presence of sodium or potassium alcoholates at 110° of atmospheric pressure.[41] This is a new reaction. Isoamyl alcohol gives 83.9% yield of 2-butynyl isoamyl ether. For lower-boiling alcohols, alkoxide and alcohol dissolved in excess isoamyl ether react with vinylacetylene as it enters the solution. Methanol reacts twice as fast in the presence of potassium alkoxide as in the presence of sodium alkoxide. At normal pressure, the polymerization of vinylacetylene is limited. The best reaction temperature is 105–110°. The reaction is reversible. At higher temperatures, vinylacetylene and the alcohol are regenerated.

2.2. Diacetylenes

Shostakovskii[42] prepared vinylethynyl ethers and thioethers from diacetylene and alcohols or thiols. He simply dissolved KOH in the alcohol, added diacetylene, heated, and distilled off the products. Some reactions went easily at room temperature (Table 3-6).

TABLE 3-6[42]

HC≡CCH=CHXR

XR	React. Temperature (°C)	% Yield	b.p. (°C, @ mm Hg)
OBu	100–110	60	144, @ 760
O—n-hexyl	95– 98	40	85, @ 8
O—benzyl	116–118	51	104–104.5, @ 3
SEt	65–70	58	65–65.5, @ 12
SPh	20–50	31	92–92.5, @ 2
OCH$_2$CH$_2$NEt$_2$	20	80–90	98, @ 10

Herbertz[43] found the conditions for reacting diacetylene with methanol in the presence of vinylacetylene, and used this as a method for separating vinylacetylene and obtaining a valuable new acetylenic intermediate at the same time. This is important because both acetylenes are components of arc acetylene:

One hundred and fifty grams of diacetylene (as a mixture with 40% vinylacetylene) is added to a liter of methanol containing sodium methylate from 12 grams of sodium. After 4 hours at 75° (very mild conditions), the vinylacetylene is unreacted. 1-Methoxybuten-1-yne-2 forms in 77% yield (b.p. 63° at 80 mm of Hg, m.p. −7°). This ether polymerizes at room temperature after several days.

$$\text{MeOH} + \text{HC}\equiv\text{C}-\text{C}\equiv\text{CH} \xrightarrow[\text{4 hr, 75°}]{\text{NaOMe}} \text{MeOCH}=\text{CH}-\text{C}\equiv\text{CH}$$
$$(77\%)$$

(very unusual odor, breathing vapors causes loss of appetite)

Nonconjugated diynes are much less reactive:[44]

$$\text{HC}\equiv\text{C}-\text{CH}_2-\text{C}\equiv\text{CH} + \text{MeOH} \xrightarrow[\substack{\text{KOH}\\ \text{3 hr, 125°}}]{\phi\text{H}} \text{MeOCH}=\text{CHCH}_2\text{C}\equiv\text{CH}$$

3. VINYLATION OF MERCAPTANS

3.1. Vinyl Thioethers

Mercaptans (RSH) vinylate in the presence of base much as alcohols do. Vinyl mercaptans have not been commercialized. Miller[45] recently reviewed the literature on vinyl mercaptans.

Schneider[46] pointed out that Reppe vinylation of mercaptans[2] is limited to the few available mercaptans. Alkyl halides react with thiourea in the presence of base to form isothiuronium salts,[1] and mercaptan vinylation is also base catalyzed. Schneider combined the two reactions to prepare high yields of vinyl sulfides:

$$RX + thiourea \xrightarrow[NaOH]{} RS\underset{\underset{(1)}{NH_2}}{\overset{\|}{C}}=NH \cdot HX \xrightarrow[\substack{aqueous\ solution \\ OH^-,\ 90-95°, \\ 450\ psig}]{+acetylene} RSCH=CH_2$$

In the mechanism, 1 mole of base is required to neutralize the isothiuronium salt, but it is regenerated by the vinylation step. The reaction works well with dialkylaminoalkyl halides and supplements the Reppe vinylation because few amino alcohols are available (Table 3-7, page 292).

α-(Alkylmercapto)styrenes are formed by base-catalyzed elimination of mercaptan from 1-phenyl-1,2,bis(alkylmercapto)ethanes in the presence of acetylene:[46]

$$styrene + RSSR \xrightarrow{I_2} PhCH\underset{\underset{(2)}{SR}}{\overset{}{C}}H_2SR \xrightarrow[\substack{ROH \\ (^-OR), \\ 90°}]{+C_2H_2} PhC\underset{\underset{(3)}{SR}}{\overset{}{=}}CH_2 + CH_2=CHSR$$

If acetylene is omitted, the yield is 31%; if acetylene is present, the yield is 89%:

$$(2) + RO^- \rightleftharpoons [PhC^-\underset{\underset{\longrightarrow(3)}{SR}}{\overset{}{-}}CH_2SR] + ROH[+ RS^-] \xrightarrow{C_2H_2} RSCH=CH_2 + RO^-$$

Elimination from α-(alkylmercapto)styrenes forms ketene mercaptals:[46]

$$RSH + RSCH_2CH(OMe)_2 \xrightarrow{H^+} RSCH_2CH(SR)_2 \xrightarrow[90-120°,\ 500\ psig]{C_2H_2,\ OH^-,\ R'OH}$$

$$CH_2=C(SR)_2 + CH_2=CHSR'$$
$$(92\%\ conversion)$$

TABLE 3-7[46]

$$RSCH{=}CH_2$$

R	% Yield	b.p. (°C, @ mm Hg)
CH_3-	89	68, @ 760
C_2H_5-	90	92, @ 760
$n\text{-}C_4H_9-$	90	70, @ 60
$sec\text{-}C_4H_9-$	75	74, @ 115
$C_4H_9CH(C_2H_5)CH_2-$	80	89, @ 10
$n\text{-}C_{14}H_{29}-$	91	105, @ 0.15
$C_6H_{11}-$	30	71, @ 9
$C_6H_5CH_2-$	91	98, @ 10
$C_6H_5CH_2CH_2-$	73	98, @ 10
$C_6H_5OCH_2CH_2-$	12	89, @ 0.65
$-(CH_2)_4SCH{=}CH-$	68	80, @ 0.6
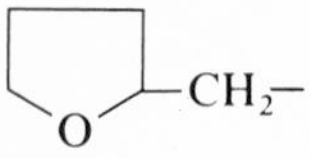 ![furan-CH2] O ring $-CH_2-$	25	84, @ 10.5
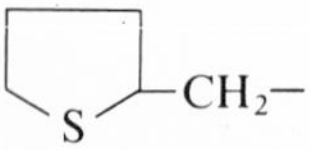 ![thiophene-CH2] S ring $-CH_2-$	55	78, @ 3
$(C_2H_5)_2NCH_2CH_2-$	78	92, @ 20
$(CH_3)_2NCH_2CH_2-$	67	68, @ 17
$CH_3NHCH_2CH_2-$	27	71, @ 24
$NH_2CH_2CH_2-$	14	67, @ 20
$NH_2CH_2CH_2CH_2-$	30	68, @ 7
$(CH_3)_2NCH(CH_3)CH_2-$	39	76, @ 16
$(CH_3)_2NCH_2CH_2CH_2-$	75	79, @ 18
$(CH_3)_2NCH_2CH_2SCH_2CH_2-$	—	83, @ 0

A better reaction is:

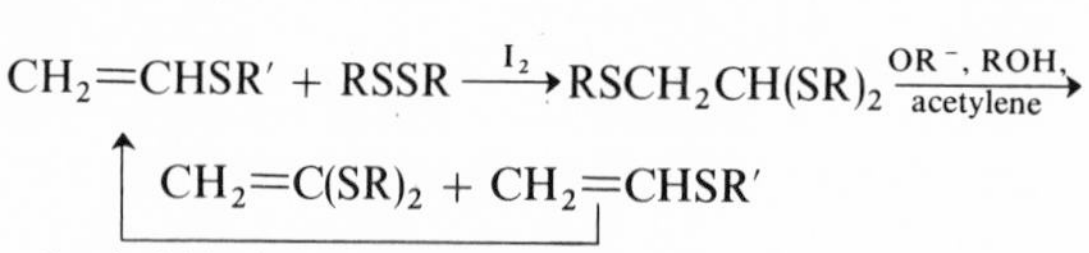

$$CH_2{=}CHSR' + RSSR \xrightarrow{I_2} RSCH_2CH(SR)_2 \xrightarrow[\text{acetylene}]{OR^-,\ ROH}$$

$$\uparrow \quad CH_2{=}C(SR)_2 + CH_2{=}CHSR'$$

(the vinyl thioether is the "acetylene" carrier)

3.2. Vinyl Thiols

Stacey[47] in 1963 reported the first preparation of vinyl thiols, $RCH{=}CHSH$. He used the x-ray induced chain addition of H_2S to the acetylenic bonds in propyne, 2-butyne, 3,3,3-trifluoropropyne and phenylacetylene.

When Strausz[48] irradiated a mixture of acetylene and hydrogen sulfide at −78° with a mercury arc lamp, he obtained vinyl thiol, $CH_2=CHSH$, in good yield. 1 : 1 mixtures exploded, but 1 : 10 ($C_2H_2 : H_2S$) mixtures were satisfactory. The reaction is a typical chain radical reaction via ·SH. Vinyl thiol polymerizes readily at room temperature. (Note: Addition of ·SR to triple bonds is discussed more completely in Chapter 2).

References

1. Miller, S. A., "Acetylene. Its Properties, Manufacture, and Uses," Vol. 2, New York, Academic Press, 1966.
2. Copenhaver, J. W., and Bigelow, M. H., "Acetylene and Carbon Monoxide Chemistry," New York, Reinhold Publishing Corp., 1949.
3. Miller, S. I., and Shkapenko, G., J. Am. Chem. Soc., 77, 5038 (1955).
4. Holly, E. D., J. Org. Chem., 24, 1752 (1959).
5. Otsuka, S., Matsui, Y., and Murahashi, S., Nippon Kagaku Zasshi, 77, 766 (1956); Chem. Abstr., 52, 8935 (1957).
6. Mikhant'ev, B. I., and Tarasova, A. V., Ref. Zh. Khim., 76196, (1956); Tr. Voronezhsk., Gos. Univ., 42, 55 (1956); Chem. Abstr., 53, 11194 (1959).
7. Teeter, H. M., J. Am. Oil Chem. Soc., 33, 399 (1956).
8. Deutsche Solvay-Werke G.m.b.h., British Patent 773,331 (April 24, 1957); Chem. Abstr., 52, 1199 (1958).
9. Chatterjee, D. C., Dakshinamurty, H., and Aggarwaal, J. S., Indian J. Technol., 2, 335 (1964); Chem. Abstr., 62, 8993 (1965).
10. Watanabe, W. H., and Conlon, L. E., J. Am. Chem. Soc., 79, 2825 (1957).
11. Coffman, D. D. (to E. I. du Pont de Nemours & Co.), U.S. Patent 2,387,495 (Oct. 23, 1945); Chem. Abstr., 40, 1870 (1948); J. Org. Chem., 13, 223 (1948).
12. Ban, M., Tamamoto, T., and Otsuka, S., Nippon Kagaku Zasshi, 77, 176 (1956); Chem. Abstr., 52, 300 (1958).
13. Shostakovskii, M. F., et al., Izv. Sibirsk. Otd. Akad. Nauk SSSR., 36 (1961); Chem. Abstr., 55, 19843 (1961).
14. Hill, H. S., and Pidgeon, L. M., J. Am. Chem. Soc., 50, 2718 (1928).
15. Nedwick, J. J., Ind. Eng. Chem. Process Design Develop., 1, 137 (1962).
16. Schachat, N., et al., J. Org. Chem., 26, 3712 (1961).
17. Walling, E., and Faerbe, G. (to Deutsche Solvay Werke), U.S. Patent 2,997,393 (Mar. 28, 1961); Chem. Abstr., 52, 423 (1958).
18. Huggett, C. M., and Nedwick, J. J. (to Rohm and Haas Co.), U.S. Patent 2,969,400 (Jan. 24, 1961).
19. Shostakovskii, M. F., and Gracheva, E. P., Zh. Obshch. Khim., 19, 1250 (1949); Chem. Abstr., 44, 3439 (1950).
20. Panzer, H. P., and Ferro, M. A., Abstracts of 145th American Chemical Society Meeting, New York, page 25d, September 1963.
21. Shostakovskii, M. F., Atavin, A. S., and Trofimov, B. A., J. Gen. Chem. USSR, (English Transl.), 34, 2125 (1964).
22. Schachat, N., and Bagnell, J. J., Jr., J. Org. Chem., 27, 471 (1962).

23. Mikhant'ev, B. I., Lapenko, V. L., and Pavlov, L. P., *Zh. Obshch. Khim.* (*English Transl.*), **32**, 2505 (1962).
24. Mikhant'ev, B. I., and Lapenko, V. L., *Zh. Obshch. Khim.*, **27**, 2840, 2972 (1957); *Chem. Abstr.*, **52**, 8054 (1958).
25. Mikhant'ev, B. I., and Lapenko, V. L., *Zh. Obshch. Khim.*, **31**, 1843 (1961); *Chem. Abstr.*, **55**, 27072 (1961).
26. Deutschman, A. J., Jr., and Kircher, H. W., *J. Am. Chem. Soc.*, **83**, 4060 (1960).
27. Reppe, W., and Hechth, O. (to I. G. Farbenind. A.-G.), U.S. Patent 2,157,347 (May 9, 1939); *Chem. Abstr.*, **33**, 6342 (1939).
28. Whistler, R. L., Panzer, H. P., and Goatley, J. L., *J. Org. Chem.*, **27**, 2961 (1962).
29. Chiddix, M. C., Glickman, S. A., and Hecht, O. F., *Textile Res. J.*, **35**, 942 (1965).
30. Chiddix, M. C., *Textile Res. J.*, **35**, 965 (1965).
31. Nesmeianov, A. N., Lutsenko, I. F., and Khomutov, R. M., *Proc. Acad. Sci. USSR, Chem. Sect.* (*English Transl.*), **120**, 473 (1959).
32. Harshman, R. C., *Univ. Microfilms*, **24109**, 166 pp.; *Chem. Abstr.*, **52**, 7116 (1958).
33. Sen, S. N., and Bhattacharyya, S. K., *J. Sci. Ind. Res.* (*India*), **18B**, 405 (1959); *Chem. Abstr.*, **54**, 14101 (1960).
34. Gorin, Yu. A., and Kalaus, A. E., *Zh. Obshch. Khim.*, **29**, 3575 (1959); *Chem. Abstr.*, **54**, 19458 (1960).
35. Kozlov, N. S., and Chumakov, S. Ya., *Zh. Prikl. Khim.*, **30**, 318 (1957); *Chem. Abstr.*, **51**, 12816 (1957).
36. Watanabe, W. H., and Conlon, L. E., *J. Am. Chem. Soc.*, **79**, 2828 (1957).
37. Cresson, P., and Atlani, M., *Compt. Rend.* (*Ser. C*), **262**, 1433 (1966).
38. Jones, D. M., and Wood, N. F., *J. Chem. Soc.*, 1560 (1965).
39. Kupin, B. S., and Petrov, A. A., *Zh. Obshch. Khim.*, **29**, 1151 (1959); *Chem. Abstr.*, **54**, 1278 (1960).
40. Kupin, B. S., and Petrov, A. A., *Zh. Obshch. Khim.*, **34**, 1897 (1964); *Chem. Abstr.*, **61**, 8145 (1964).
41. Mkryan, G. M., Mndzhoyan, Sh. L., and Gasparyan, S. M., *Izv. Akad. Nauk Arm. SSR Khim. Nauki*, **17**, 643 (1964); *Chem. Abstr.*, **63**, 8183 (1965).
42. Shostakovskii, M. F., Bogdanova, A. V., and Chekulaeva, I. A., *Izv. Vysshikh Uchebn. Zavedenii Khim. i Khim. Tekhnol.*, **2**, 769 (1959); *Chem. Abstr.*, **54**, 8598 (1960).
43. Herbertz, T., *Chem. Ber.*, **85**, 475 (1952).
44. *Ibid.*, **92**, 541 (1959).
45. Miller, S. A., "Acetylene. Its Properties, Manufacture, and Uses," Vol. 2, p. 317, New York, Academic Press, 1966.
46. Schneider, H. J., and Bagnell, J. J., *J. Org. Chem.*, **26**, 1980, 1982, 1984, 1987 (1961).
47. Stacey, F. W., and Harris, J. F., *J. Am. Chem. Soc.*, **85**, 963 (1963).
48. Strausz, O. P., Hikida, T., and Gunning, H. E., *Can. J. Chem.*, **43**, 717 (1965).

PART TWO

Vinylation of Carboxylic Acids, Halogen Acids, and Phosphorus Acids and Esters

1. CARBOXYLIC ACIDS

1.1. Vapor Phase, Catalytic

Vapor phase catalytic vinylation of carboxylic acids is the only commercially important system for making vinyl esters, the most important of which is vinyl acetate. Any carboxylic acid which has sufficient vapor pressure at 160–200° can be vinylated in the vapor phase. Vinyl acetate synthesis is a good example of a highly efficient catalytic acetylene reaction. The catalyst is easy to make, is cheap, and has very long life. The system is operated at atmospheric pressure, avoiding the hazards of acetylene explosions under pressure. Monomer grade vinyl acetate and the by-products (very small amounts) are easily separated from unreacted acetylene and acetic acid. Although vapor phase vinylation is an important industrial operation, surprisingly little work on mechanisms has been published, and this mostly by Japanese scientists. Miller[1] has reviewed vinylation of carboxylic acids in great detail.

1.1.1. THE CATALYST

Mitsutami[2] found that the best catalyst for vinyl acetate was 0.3 gram of zinc acetate per gram of carbon. Acetylene does not chemisorb on zinc acetate or on carbon alone, but does strongly adsorb on zinc acetate-carbon.[3] All of the zinc acetate can be extracted from spent catalyst. Alumina, silica and alumina gel are poor catalyst carriers (see reference 4). The catalyst is even slightly active for vinylating anthracene, anthraquinone, β-naphthol, benzil and benzophenone.

Initially, catalyst composition is $2ZnO:3Zn(OAc)_2$.[5] Activity reaches a maximum and then decreases as carbon is deposited. At the steady state, catalyst composition is $3–4ZnO:1Zn(OAc)_2$. As the amount of zinc acetate increases, the surface area decreases exponentially. The activity is proportional to surface area and to the third power of the zinc acetate content.[6] When proprionic acid is passed over zinc acetate-carbon catalyst, vinyl acetate forms first and then vinyl propionate. This shows exchange of anion in the catalyst and may indicate that the acid reacts as the anion of the zinc salt.[7]

Other catalysts have been tested,[1,8,9] including copper acetylide.[10] γ Irradiation or electron beam irradiation of the activated carbon before addition of zinc acetate increases the adsorption of zinc acetate and the rate of acetylene adsorption, and therefore increases the activity of the catalyst.[11]

1.1.2. BY-PRODUCTS

Polymer and carbon form from acetylene. Acetylene is also converted to butene-1, vinylacetylene, butadiene, 1-octene, 1-hexene, benzene, toluene and 1,7-octadiene.[2] Other by-products are acetaldehyde, acetone, crotonaldehyde, methyl ethyl ketone, isopropyl acetate and ethylidene diacetate.

1.1.3. MECHANISM

Yamada[12] studied the mechanism by determining the exchange of deuterium between acetylene and acetic acid on zinc acetate-carbon catalyst. He concluded that the mechanism is:

$$C_2H_2 + H^+ \rightleftharpoons CH_2{=}CH^+ \xrightarrow{+OAc^-} \text{vinyl acetate}$$

1.2. Liquid Phase, High Pressure

Zinc or cadmium salts are catalysts for liquid phase high-pressure vinylation of carboxylic acids. This is not an important commercial method, because of the hazards of acetylene under pressure and because it is sometimes difficult to purify product vinyl esters to polymerization grade monomers. The by-product ethylidene diesters are inhibitors for radical polymerization and usually must be removed by distillation. For high-boiling acids and their vinyl esters, this is a difficult separation.

Reaction temperature is usually 180–220°, and the pressure is 10–20 atmospheres of acetylene, diluted with nitrogen. The reaction is usually slow, and conversion only moderate. Miller[1] lists a large number of vinyl esters made by this (and other) methods.

When Otsuka[13] added pyridine to crotonic acid-zinc crotonate catalyst and then added acetylene, he found that the pyridine decreased the yield of vinyl crotonate but had no effect on the yield of ethylidene dicrotonate. Triphenylphosphine, another Lewis base, decreased the rate, but increased the yield of vinyl crotonate and decreased the yield of ethylidene diester.

Aromatic acids can vinylate without any catalyst. In toluene, the rate of vinylation of benzoic acids is first order on acetylene and independent of initial acid concentration above 0.18 mole/liter. In nitrobenzene, the rate is slower and increases with concentration of acid. Lewis bases decrease the rate of zinc salt-catalyzed vinylations. Vinylation of monoesters of dibasic acids in

Preparation of Vinyl Esters by High-pressure Liquid Phase Vinylation Reactions (in Toluene or Other Solvent)

Vinyl Ester	Catalyst	Pressure (atm)	Tempera-ture (°C)	b.p. (°C, @ mm Hg)	Reference
Hydroxyacetate	Cd, Zn salt	20–30	160–200	60–70, @ 38 (crude)	16
Propionate	Cd, Zn propionate	20–30	100–200	105–106, @ 60	16
Crotonate	Zn crotonate	—	—	47, @ 28	13
i-Butyrate	Cd, Zn i-butyrate	20–30	160–200	104–105	16
Caproate	Zn, Cd caproate	20–30	160–200	80–81, @ 13	16
Stearate	Zn, Cd stearate	20–30	160–200	174–180, @ 2	16
Laurate	Zn, Cd laurate	20–30	160–200	142, @ 10	16
Oleate	Zn, Cd oleate	20–30	160–200	162.5, @ 1	16
Linoleate	Zn, Cd linoleate	20–30	160–200	154.5, @ 2	16
p-Dimethylaminobenzoate	None	20	200	m.p. 64.5–66	13, 18
m-Chlorobenzoate	Zn salt			85, @ 0.025	18
m-Methoxybenzoate	Zn salt			105, @ 0.03	18
p-Nitrobenzoate	Zn salt			103, @ 0.5	18
o-Phthalimidobenzoate	Zn salt			m.p. 108	18
p-Phthalimidobenzoate	Zn salt			m.p. 170	18
Furan-2-carboxylate	Zn salt			72, @ 10	18
Thiophene-2-carboxylate	Zn salt			85, @ 10	18
Nicotinate	None	40	190–200	69, @ 4	13
				106, @ 12	18
Adipate (di)	HgO-chloroacetic acid	Atmospheric		118–125, @ 10	17
Isophthalate (mono)	Zn salt			m.p. 146	15, 18
Isophthalate (di)	Zn salt			m.p. 59–60	15, 18
Methyl terephthalate	Zn salt			m.p. 62	18
Terephthalate (mono)	Zn salt			m.p. 195	15, 18
Terephthalate (di)	Zn salt			m.p. 83–84	15, 18

the presence of cadmium acetate is also first order on acetylene.[14] Generally, anhydrous systems are best, and acetic anhydride can be added to maintain anhydrous conditions.[15]

Table 3-8 lists vinyl esters of a few acids made by high-pressure liquid phase vinylations.

1.3. Liquid Phase, Low-pressure, Mercury-catalyzed Vinylation

This system is attractive because it can be operated in ordinary laboratory or plant equipment. It presents the same purification problems as the zinc-catalyzed pressure vinylation, but in some cases the reaction is clean enough to be economical. The mechanism has been studied a little, and intermediate addition products of catalyst and acetylene are reasonably well established. Some of the mechanisms are similar to the mechanisms of addition of other molecules to acetylene.

1.3.1. UNCATALYZED VINYLATION

Most vinylations require a catalyst, but trichloroacetic acid adds to phenylacetylene in benzene without a catalyst.[19] 1-Phenylvinyl trichloroacetate is always at least 95% of the total product—a very clean monoaddition. A little acetophenone and low molecular weight polymer are by-products. The reaction is first order on phenylacetylene and second order on trichloracetic acid. The first step is addition of a proton to the triple bond to form an ion solvated by trichloracetic acid. Solvated anion then adds to the cation to give vinyl ester and free trichloroacetic acid.

1.3.2. MERCURY CATALYSIS

In 1936, Hennion[20] called attention to the versatility of the mercury catalyst in additions to the acetylenic bond: (1) Both mercuric salt and strong acid are usually necessary. (2) Internal acetylenic bonds are reactive, so acetylenic hydrogen is not essential. (3) Mercuric catalyst adds to the triple bond, and reaction of any available acetylenic hydrogen is not too important. (4) Vinyl ethers can add alcohols in the presence of an acid catalyst only. (5) Substituted acetylenes add only one molecule of acid, but acetylene itself can add two molecules. (6) Catalytic amounts of mercuric salt and acid are sufficient, so the mercury addition compound is very reactive and easily regenerated. (7) Essentially, catalysis is independent of the nature of the acetylenic compound and the hydroxyl compound added.

A general scheme illustrating these points is:[20]

$$
\begin{array}{c}
\text{OB HgA} \\
| \quad | \\
\xrightarrow{+\text{BOH}} \text{R}-\text{C}-\text{CHR}' \\
| \\
\text{A}
\end{array}
$$

$$
\begin{array}{c}
\text{HgA} \\
| \\
\xrightarrow{+\text{R}-\text{C}\equiv\text{C}-\text{R}'} \text{R}-\text{C}=\text{C}-\text{R}' \rightarrow \\
| \\
\text{A}
\end{array}
$$

$$\big\downarrow -\text{HA}$$

$$
\text{HgA}_2
$$

$$
\begin{array}{c}
\text{OB HgA} \\
| \quad | \\
\xrightarrow[-\text{HA}]{+\text{BOH}} \text{R}-\text{C}=\text{C}-\text{R}' \xrightarrow{\text{(if R' is H)}}
\end{array}
$$

$$
\begin{array}{c}
\text{OB HgA} \\
| \quad | \\
\text{R}-\text{C}=\text{C} \\
\diagdown \\
\text{Hg} \\
\diagup \\
\text{R}-\text{C}=\text{C} \\
| \quad | \\
\text{OB HgA} \\
?
\end{array}
$$

$$\xrightarrow[-\text{HA}]{+\text{BOH}} \text{BO}-\text{HgA} \xleftarrow{+\text{R}-\text{C}\equiv\text{C}-\text{R}'}$$

$$\big\downarrow {+\text{HA} \atop -\text{HgA}_2}$$

$$
\text{R}-\text{C(OB)}_2-\text{CH}_2\text{R}' \xleftarrow{+\text{BOH}} \begin{array}{c} \text{R}-\text{C}=\text{CHR}' \\ | \\ \text{OB} \end{array} \xrightarrow{\text{(if B is H)}} \text{R}-\text{CO}-\text{CH}_2\text{R}'
$$

B = alkyl or aryl ⟶ vinyl ethers
B = anion of acid ⟶ vinyl esters

From their study of the reaction of 3-hexyne with acetic acid using mercuric acetate-perchloric acid catalyst, Lemaire and Lucas[21] proposed a mechanism. The initial rate of addition of acetic acid to the triple bond is independent of 3-hexyne concentration and is roughly proportional to the concentration of mercuric acetate and perchloric acid when these are present in equal concentrations. The first step is fast formation of a complex, followed by slower reaction of the complex with acetic acid:

$$
\text{EtC}\equiv\text{CEt} + \text{H}^+ + \text{Hg(OAc)}_2 \longrightarrow \left[\begin{array}{c} \text{HgOAc} \\ | \\ \text{EtC}\overset{+}{=}\text{CEt} \end{array} \right] + \text{HOAc}
$$

$$\big\downarrow$$

$$\text{EtCH}=\text{C(OAc)Et} + \text{HgOAc}^+$$

The reaction of the complex may also be a rearrangement with acetate migrating from mercury to carbon, to form another mercury complex. This may explain the rapid loss of catalytic activity in this system.

The complex ions of mercury probably are:

$$
\begin{array}{ccc}
\text{HgOAc} & \text{}^+\text{HgOAc} & \text{HgOAc} \\
| & \diagup\diagdown & | \\
\text{EtC}\overset{+}{=\!=\!=}\text{CEt} \longleftrightarrow \text{EtC}=\!=\!=\text{CEt} \longleftrightarrow \text{EtC}\overset{+}{=}\text{CEt}
\end{array}
$$

Another possibility is dipositive 3-hexynemercurinium ion, formed by excess perchloric acid. It is also resonance stabilized:

$$
\begin{array}{c}
\text{Hg}^{++} \\
\diagup\diagdown \\
\text{EtC}=\!=\!=\text{CEt}
\end{array}
$$

On the basis of products and the effect of variables on the HgO-catalyzed vinylation of chloroacetic acid, Pare[22] proposed a vinylation mechanism in which a definite mercury compound (not a complex) is an intermediate:

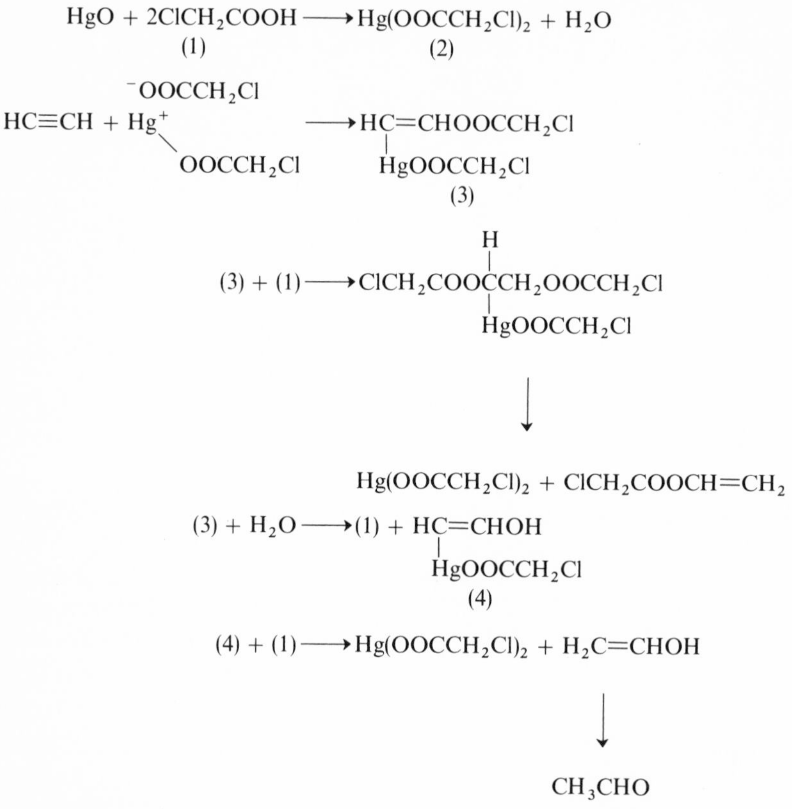

$$HgO + 2ClCH_2COOH \longrightarrow Hg(OOCCH_2Cl)_2 + H_2O$$
$$(1) \qquad\qquad\qquad (2)$$

$$HC\equiv CH + Hg^+ \overset{^-OOCCH_2Cl}{\underset{OOCCH_2Cl}{\diagdown}} \longrightarrow \underset{\underset{(3)}{HgOOCCH_2Cl}}{HC=CHOOCCH_2Cl}$$

$$(3) + (1) \longrightarrow \underset{HgOOCCH_2Cl}{ClCH_2COO\overset{\overset{H}{|}}{C}CH_2OOCCH_2Cl}$$

$$\downarrow$$

$$Hg(OOCCH_2Cl)_2 + ClCH_2COOCH=CH_2$$

$$(3) + H_2O \longrightarrow (1) + \underset{\underset{(4)}{HgOOCCH_2Cl}}{HC=CHOH}$$

$$(4) + (1) \longrightarrow Hg(OOCCH_2Cl)_2 + H_2C=CHOH$$

$$\downarrow$$

$$CH_3CHO$$

Compound (3) is analogous to an addition product of mercuric acetate and diphenylacetylene $PhC \underset{\underset{|}{HgOAc}}{\overset{\overset{OAc}{|}}{=\!=\!=\!=}} CPh$, and to the intermediates proposed by Hennion.[20]

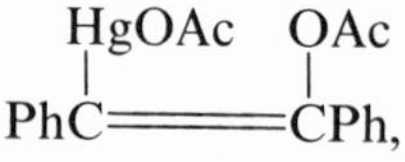

Addition of acetic acid to methoxyvinylacetylene requires a silver catalyst; mercury catalysts are inactive.[23]

1.3.3. INTRAMOLECULAR ADDITION OF CARBOXYL GROUPS

The intramolecular addition of acids across the triple bond in arylacetylenes is so closely related to vinylation that the mechanism is important. 2,2'-Tolandiboronic acid in solution at pH 9.1 cyclizes:[21]

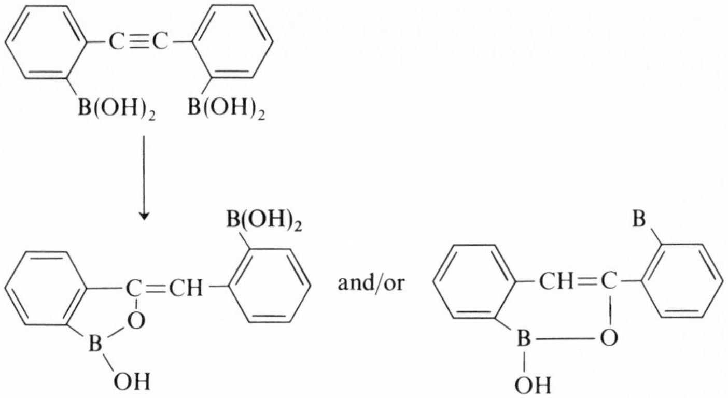

2,2-Stilbenediboronic acid, 2-vinylbenzeneboronic acid, and 2-tolanboronic acid do not undergo intramolecular addition.[24] Thus, both boronic acid groups in 2,2′–tolandiboronic acid play a role, indicating concerted donation of a proton by one boronic acid group and a hydroxide ion by the other.

2,2′-Tolandicarboxylic acid exhibits the same effects. One carboxyl group assists the addition of the other.[25] The rates of addition of carboxyl in the intramolecular cyclization are:

	k (min^{-1})	Relative Rate
2,2′-Tolandicarboxylic acid	8.44×10^{-1}	11,000
2-Tolancarboxylic acid	3.80×10^{-5}	1
2,4′-Tolandicarboxylic acid	9.80×10^{-5}	2.6

Carboxyl in the *para* position assists the addition of the *ortho* carboxyl to a small extent. The assisted addition is the first example of intramolecular catalysis of an addition reaction to an acetylenic bond. Silver ion catalyzes the cyclization in 2,4′-tolandicarboxylic acid but has no effect on 2,2′-tolandicarboxylic acid. The rates at different pH values indicate that the lactonization probably involves concerted attack. Carboxylate anion adds to one carbon, and a proton from unionized carboxyl adds to the other:

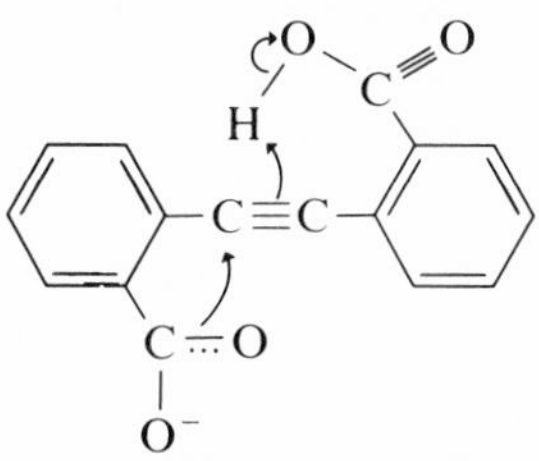

Tertiary alkyl halides dehydrohalogenate when treated with sodiomalonic ester, but tertiary acetylenic halides are zwitterions and therefore are able to alkylate sodiomalonic ester.[26] In the product diacid, one carboxyl group adds across the triple bond to form a lactone:

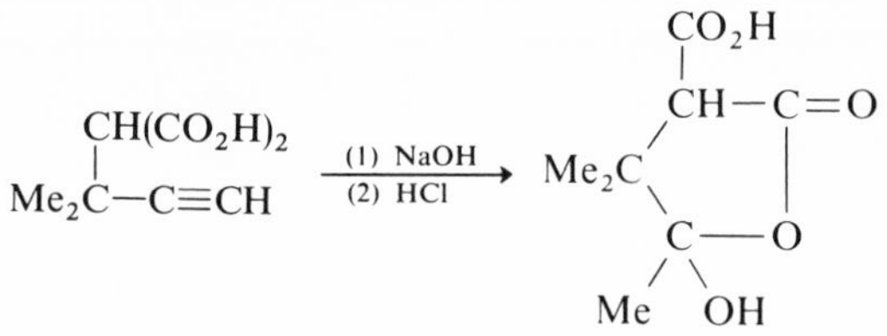

Intramolecular addition of carboxyl across the triple bond in some other molecules is catalyzed by silver ion:[27]

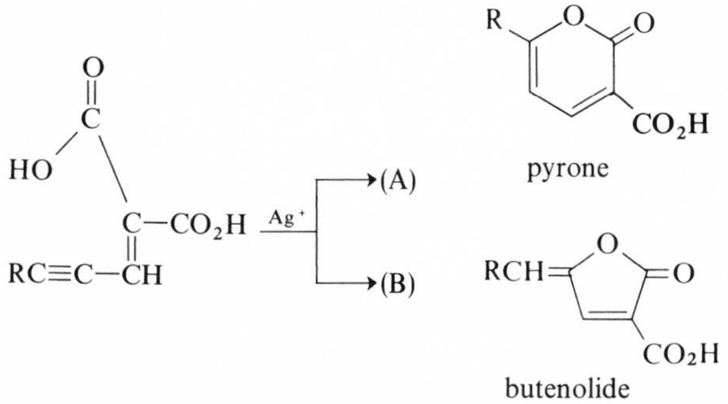

Whether (A) or (B) is the major product depends on the structure of the acid. Silver ion favors the butenolide product.

1.4. Vinylation of Carboxylic Acids by Vinyl Interchange

Vinyl acetate is cheap and available, so the interchange method is very good for vinyl esters of acids boiling above acetic acid. The reaction is done in conventional equipment, is not hazardous, and under proper conditions gives excellent yields. Both strong acid and mercuric salts are required.

Adelman[28] in 1949 published his studies on the reaction. Inactive or only slightly active catalysts were: acetates of lead, manganese(ous), cadmium and tin. Mercuric salts were the only active catalysts found. According to Adelman, vinyl acetate disproportionates to acetic acid and acetylene, and the acetylene-mercuric complex reacts with the second acid to form vinyl ester. This seems logical, because direct vinylations go under the same conditions. He correctly pictured the reaction as a series of equilibria. In developing a high-yield vinyl interchange system, Rutledge and co-workers found that acetylene poisoned the interchange reaction. Recent NMR evidence also indicates that Adelman's mechanism is incorrect.[29] Best yields of purified vinyl esters and favorable recycle of mercury catalyst require specific amounts of mercuric salt and sulfuric acid.[30] The equilibrium in the interchange can be shifted to favor the product vinyl ester if the acetic acid is removed by entrainment in excess vinyl acetate.

With mercuric acetate-boron trifluoride catalyst, the intermediate complex is probably $(AcO)_2CHCH_2HgOAc$ (by NMR), and one of the acetoxy groups is displaced by benzoyloxy when benzoic acid is added. This complex then protonates, and eliminates acetic acid and mercuric acetate to give vinyl benzoate.[29]

The carboxyl group in salicyclic acid undergoes transvinylation, and the hydroxyl group then adds across the vinyl group to give a cyclic dioxolone.[18] p-Hydroxybenzoic acid reacts at both functional groups: the carboxyl group is transvinylated, and the hydroxy group adds across another molecule of vinyl acetate. Saccharin gives 60% yield of N-vinylsaccharin. (The yield is only 10–25% in the zinc-catalyzed direct vinylation.)

Some of the unusual acids vinylated by interchange are listed in Table 3-9.

TABLE 3-9
Vinyl Esters Prepared by Interchange with Vinyl Acetate

Acid	b.p. (°C, @ mm Hg) of Vinyl Ester	Reference
m-Nitrobenzoic	117–119, @ 2	18
Coumarilic	135, @ 0.5	18
2-Thionaphthenecarboxylic	124–129, @ 0.2	18
2-Anthracenecarboxylic	82–83 (m.p.)	18
Diphenylacetic	145–150, @ 0.2	18
2-Naphthylacetic	152, @ 0.4	18
2-Naphthoxyacetic	170–175, @ 0.3	18
p-Methoxycinnamic	150–155, @ 0.1	18
Stilbene-α-carboxylic	170–173, @ 0.3	18
8-(1,3-Dioxolan-2-yl)octanoic	59.5–60.5 (m.p.)	31
9,9-Dimethoxynonanoic	—	31
9,9-Dibutoxynonanoic	—	31
9,9-Dioctadecanoxynonanoic	53–54 (m.p.)	31
12-(1,3-Dioxolan-2-yl)dodecanoic	77–78 (m.p.)	31

2. VINYLATION OF HALOGEN ACIDS

Addition of HX to substituted acetylenes is discussed in Chapter 2. Unlike substituted acetylenes, acetylene requires a catalyst for efficient addition of halogen acids. The commercial vinyl chloride synthesis uses mercuric chloride-carbon catalyst at atmospheric pressure. Thermal dehydrochlorination of dichloroethane has become competitive with direct HCl-acetylene production of vinyl chloride in recent years. Generally, the combined process is most economical. Ethylene is chlorinated, and ethylene dichloride is cracked to vinyl chloride and HCl. The HCl adds to acetylene to form another mole of vinyl chloride. Direct reaction of HF with acetylene to form vinyl fluoride is a recent development. Vinyl bromide is made by a liquid phase addition of HBr to acetylene.

2.1. Vinyl Chloride

Miller[32] has reviewed vinyl chloride preparation thoroughly. Catalysis and mechanism are emphasized in this brief discussion.

2.1.1. CATALYST

Mercuric chloride on carbon is usually used, in a vapor phase atmospheric pressure reaction. In a study of the reaction taken to only 2–3% conversion, Wesselhoft[33] deduced a rate equation which indicates acetylene adsorbs on one type of site, HCl on another, and vinyl chloride on yet a third type. Acetylene and HCl, on adjacent sites, react to form the vinyl chloride. Adsorption and desorption are fast compared to rate of reaction. Watanabe[34] also concluded that the rate-determining step is the reaction of the two adsorbed reactants. He identified 2-chlorovinylmercuric chloride $ClCH=CH-HgCl$ on the catalyst surface. Mercuric chloride is catalytically active even if not spread on a large carbon surface.[35] Unreacted acetylene and water deactivate the catalyst by reducing mercuric ion to mercurous ion and to mercury. Activated alumina alone is a catalyst at 350°.[36] This is the optimum temperature for dehydrochlorinating dichloroethane, so the two steps of the combined process can be done simultaneously.

    2.1.1.1. Molten Salt Catalyst. Sundermeyer[40] used a molten salt bath containing 5% $HgCl_2$ for both the dehydrohalogenation and the direct vinylation reaction to form vinyl chloride. At 200°, temperature control was easy, and conversion was 70% (90% yield). A mixture of 1,1-dichloroethane and acetylene gave vinyl chloride from both dehydrochlorination and direct vinylation.

2.1.2. THE SOURCE OF ACETYLENE

Lynn[37] used dilute acetylene directly from the Schach process. His catalyst was $1BaCl_2 : 1HgCl_2$ on carbon. At 150°, the conversion of acetylene was 95%, with up to 300 cubic feet of charge gas per cubic foot of catalyst per hour. The main disadvantage of dilute acetylene is the need for greater pumping capacity and a special vinyl chloride recovery system. These results were later confirmed,[38] but calculations indicated that some pressure was desirable for an efficient reaction with dilute acetylene.[39]

2.2. Vinyl Fluoride

CrO_3-carbon at 200–350° catalyzes the HF-acetylene reaction to form vinyl fluoride.[41] As temperature is increased, the conversion of acetylene to vinyl fluoride increases from 12 to 46%, and conversion to vinylidene fluoride decreases from 28 to 5.5%. Dehydrofluorination of 1,1-difluoroethane (from acetylene and HF) over aluminum fluoride pellets at 275° gives 64% yield of vinyl fluoride.[42]

2.3. Vinyl Bromide

$HgBr_3$ complex is the active catalyst in the liquid phase addition of HBr to acetylene.[43]

2.4. Vinyl Iodide

Dehydrohalogenation of ethylene diiodide by sodium ethoxide is a convenient laboratory synthesis of vinyl iodide. The yield is 35% (b.p. 56–56.5°). Vinyl iodide oxidizes in air, but it does not polymerize as easily as vinyl bromide or chloride.[44]

3. VINYLATION OF PHOSPHORUS ACIDS AND ESTERS

Direct vinylation of acids and esters of phosphorus has not been reported. Baer[45] prepared monovinyl phosphate by vinyl interchange. The most critical variable was temperature. Above 30°, the product hydrolyzed. Crystalline orthophosphoric acid and vinyl acetate in the presence of mercuric acetate catalyst gave vinyl phosphate in 1 hour. Final purification was possible only through lithium or pyridinium salts. References to preparation of vinyl phosphorus esters by dehydrohalogenation reactions are listed in Baer's report.[45]

Nesmeyanov[46] prepared various vinyl phosphite esters by reacting $Hg(CH_2CHO)_2$ with chlorophosphite esters in the presence of base:

$$Hg(CH_2CHO)_2 + ClP(OR)_2 \longrightarrow CH_2=CHOP(OR)_2$$
$$(60-70\%)$$

$$+ Cl_2POR \longrightarrow (CH_2=CH)_2POR$$
$$(50-60\%)$$

In a similar reaction, 30–35% yields of vinyl esters of various phosphorus acids can be obtained in one step:[47]

$$RR'P(O)Cl + CH_3CHO + NEt_3 \longrightarrow RR'P(O)-OCH=CH_2 + NEt_3 \cdot HCl$$

$$RP(O)Cl_2 + 2CH_3CHO + 2NEt_3 \longrightarrow RP(O)(OCH=CH_2)_2 + 2NEt_3 \cdot HCl$$

R and R' can be alkyl, vinyl, $-OCH=CH_2$, $-OCH_2CH_2O-$,

References

1. Miller, S. A., "Acetylene. Its Properties, Manufacture, and Uses," Vol. 2, p. 247, New York, Academic Press, 1966.
2. Mitsutami, A., and Kominami, T., *Nippon Kagaku Zasshi*, **80**, 886, 888, 890, 893, 895 (1959); *Chem. Abstr.*, **55**, 4348 (1961).
3. Mitsutami, A., *Nippon Kagaku Zasshi*, **81**, 298 (1960); *Chem. Abstr.*, **56**, 313 (1962).
4. Yates, D. J. C., and Lucchesi, P. J., *J. Chem. Phys.*, **35**, 243 (1961).
5. Mizazawa, S., *Kogyo Kagaku Zasshi*, **64**, 1460 (1961); *Chem. Abstr.*, **57**, 2900 (1962).
6. Yamada, N., *Kogyo Kagaku Zasshi*, **62**, 1458 (1959); *Chem. Abstr.*, **57**, 10563 (1962).
7. Mizutani, K., Nakajima, S., and Nakajima, T., *J. Chem. Soc. Japan Ind. Chem. Sect.*, **59**, 101 (1956); *Chem. Abstr.*, **51**, 1030 (1957).
8. Mizazawa, S., *Kogyo Kagaku Zasshi*, **68**, 485 (1965); *Chem. Abstr.*, **63**, 5523 (1965).
9. Schmidt, W. (to Farbwerke Hoechst, A.-G.), German Patent 1,013,278 (Aug. 8, 1957); *Chem. Abstr.*, **54**, 782 (1960).
10. Kozlov, N. S., and Chumakov, S. Ya., *Zh. Prikl. Khim.*, **31**, 143 (1958); *Chem. Abstr.*, **52**, 13619 (1958).
11. Kurashiki Rayon Co., Ltd., British Patent 902,866 (Aug. 9, 1962); *Chem. Abstr.*, **58**, 1355 (1963).
12. Yamada, Y., *Bull. Chem. Soc. Japan*, **30**, 263 (1957); *Chem. Abstr.*, **52**, 251 (1958).
13. Otsuka, S., *Nippon Kagaku Zasshi*, **75**, 884, 995, 999, 1003, 1015 (1954); *Chem. Abstr.*, **51**, 14614 (1957).
14. Friedlin, G. N., Adamov, A. A., and Zaitsev, P. M., *Zh. Organ. Khim.*, **1**, 666 (1965); *Chem. Abstr.*, **63**, 6849 (1965).
15. Hopf, H., and Lüssa, H. (to Lonza Elektrizitätwerke und Chemische Fabriken, A.-G.), German Patent 1,093,354 (Nov. 24, 1960); *Chem. Abstr.*, **55**, 26527 (1961).
16. Otsuka, S., Matsumoto, Y., and Murahashi, S., *Nippon Kagaku Zasshi* **75**, 798 (1954); *Chem. Abstr.*, **51**, 13749 (1957).
17. Schur, A. M., *Uch. Zap. Kishinevsk. Gos. Univ.*, **56**, 87 (1960); *Chem. Abstr.*, **56**, 5827 (1962).
17a. Chang, S.-P., Miwa, T. K., and Wolff, I. A., *J. Polymer Sci. (A-1)*, **5**, 2547 (1967).
18. Hopff, M., *Bull. Soc. Chim. France*, 1283 (1958).
19. Evans, A. G., Owen, E. D., and Phillips, B. S., *J. Chem. Soc.*, 5021 (1964).
20. Hennion, G. F., Vogt, R. R., and Nieuwland, J. A., *J. Org. Chem.*, **1**, 159 (1936).
21. Letsinger, R. L., and Nazy, J. R., *J. Am. Chem. Soc.*, **81**, 3013 (1959).
22. Pandit, U. K., and Huisman, H. O., *Rec. Trav. Chim.*, **84**, 50 (1964).
23. Herbertz, T., *Chem. Ber.*, **45**, 475 (1952).
24. Letsinger, R. L., et al., *J. Org. Chem.*, **26**, 1271 (1961).
25. Letsinger, R. L., Oftedahl, E. N., and Nazy, J. R., *J. Am. Chem. Soc.*, **87**, 742 (1965).
26. Easton, N. R., and Dillard, R. D., *J. Org. Chem.*, **27**, 3602 (1962).
27. Belil, C., Pascual, J., and Serratosa, F., *Tetrahedron*, **20**, 2701 (1964).

28. Adelman, R. L., *J. Org. Chem.,* **14**, 1057 (1949).
29. Slinckx, G., and Smets, G., *Tetrahedron,* **22**, 3163 (1966).
30. Buselli, A. J., and Rutledge, T. F. (to Air Red'n. Co.), U.S. Patent 2,949,480 (Aug. 16, 1960); *Chem. Abstr.,* **55**, 3438 (1961).
31. Petrov, A. A., and Kupin, B. S., *Zh. Obshch. Khim.,* **30**, 2430 (1960); *Chem. Abstr.,* **55**, 11283 (1961).
32. Miller, S. A., "Acetylene. Its Properties, Manufacture and Uses," Vol. 2, p. 62, New York, Academic Press, 1966.
33. Wesselhoft, R. D., Woods, J. M., and Smith, J. M., *A.I.Ch.E. J.,* **5**, 361 (1959).
34. Watanabe, H., and Onozuba, M., *Kogyo Kagaku Zasshi,* **62**, 125 (1959); *Chem. Abstr.,* **57**, 11900 (1962).
35. Janda, J., *Chem. Zvesti.,* **12**, 37 (1958); *Chem. Abstr.,* **52**, 13610 (1958).
36. Nagiev, M. F., *et al., Azerb. Khim. Zh.,* 37 (1964); *Chem. Abstr.,* **63**, 1688 (1965).
37. Lynn, R. E., Jr., and Kobe, K. A., *Ind. Eng. Chem.,* **46**, 633 (1954).
38. Trandofirescu, Gh., Koslovski, A., and Banck, E., *Rev. Chim. (Bucharest),* **8**, 147 (1957); *Chem. Abstr.,* **51**, 17115 (1957).
39. Skupinski, A., *Przemysl Chem.,* **43**, 484 (1964); *Chem. Abstr.,* **62**, 10325 (1965).
40. Sundermayer, W., Glemser, O., and Kleine-Weischede, K., *Chem. Ber.,* **95**, 1829 (1962).
41. Skiles, B. F. (to E. I. du Pont de Nemours & Co.), U.S. Patent 2,892,000 (June 23, 1959); *Chem. Abstr.,* **54**, 1296 (1960).
42. Diamond Alkali Co., Netherland Appl. 6,405,140 (Nov. 11, 1964); *Chem. Abstr.,* **62**, 10334 (1965).
43. Flid, R. M., and Miranov, V. A., *Dokl. Akad. Nauk SSSR,* **114**, 347 (1957); *Chem. Abstr.,* **52**, 235 (1958).
44. Spence, J., *J. Am. Chem. Soc.,* **55**, 1290 (1933).
45. Baer, E., Ciplizauskas, L. J., and Visser, T., *J. Biol. Chem.,* **234**, 1 (1959); *Chem. Abstr.,* **53**, 10009 (1959).
46. Nesmeyanov, A. N., *et al., Proc. Acad. Sci. USSR (English Transl.),* **124**, 155 (1959).
47. Gefter, E. L., and Kabachnik, M. I., *Dokl. Akad. Nauk SSSR (English Transl.),* **114**, 525 (1958).

PART THREE

Reaction of Acetylenes with CO and Alcohol (Carbonylation)

Introduction

In this reaction, the elements of formic acid or formate ester, H— and —$CO_2H(R)$, add across a triple bond. This is not a vinylation by strict definition, but is so closely related that it is discussed in connection with vinylation.

1. NICKEL CATALYSTS

Reppe discovered the reaction of acetylenes with alcohols and CO over nickel catalysts and developed commercial methods for preparing acrylate esters.[1] Bird[2] reviewed the literature through 1960 and listed the acetylenes which have been carbonylated. The most important reaction is with acetylene itself, to form acrylate esters. The acrylate synthesis can be accomplished three ways:

(1) *Stoichiometric*: Acetylene reacts with nickel carbonyl (and ROH) around 40°. All of the CO is furnished by $Ni(CO)_4$, and HCl is usually present.

(2) *Low-pressure catalytic*: By proper choice of conditions, most of the CO can be furnished by CO gas.

(3) *High-pressure catalytic*: Nickel catalyst is used, and nickel carbonyl probably forms *in situ*. The reaction is usually carried out at 150° and 30 atmospheres pressure.

1.1. Stoichiometric CO Reactions

This is not an interesting commercial system, but it has been used to carbonylate a large number of substituted acetylenes. Yields vary from 1–50%.[2] Carbonylation of substituted acetylenes almost always involves *cis* addition of the elements of formic acid to form Markownikoff products. Diphenylacetylenes with *meta* or *para* substituents usually carboxylate in accordance with the Markownikoff rule, but in *ortho*-substituted diphenylacetylenes the direction of addition is determined by steric factors.[2a] As the size of the *ortho*-substituent on one of the rings increases, the amount of addition of $-CO_2R'$ to the β-acetylenic carbon increases:

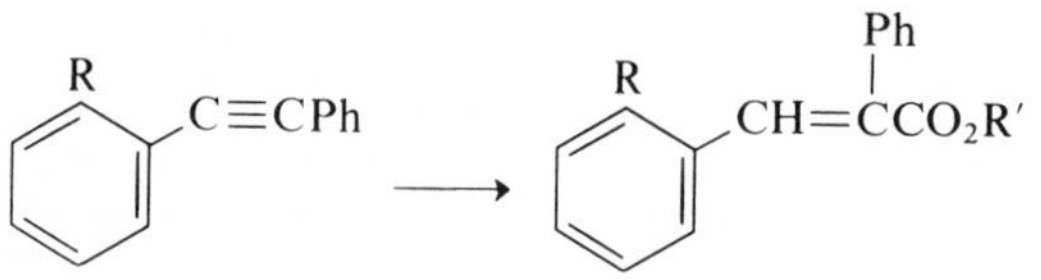

The electronic and steric effects determine which direction the reaction will take, according to the mechanism on page 309:[2a]

Propargylic chlorides react with nickel carbonyl to form allenic acids.[3] Propargylic alcohols also react in the presence of HCl to give allenic acids (which sometimes lactonize by intramolecular addition of the carboxyl group across the allenic bond). In this carbonylation, $^-C\equiv O^+$ adds during concerted displacement of Cl^- from propargyl chloride, instead of the addition first of H^+ as in the usual Reppe carbonylation.

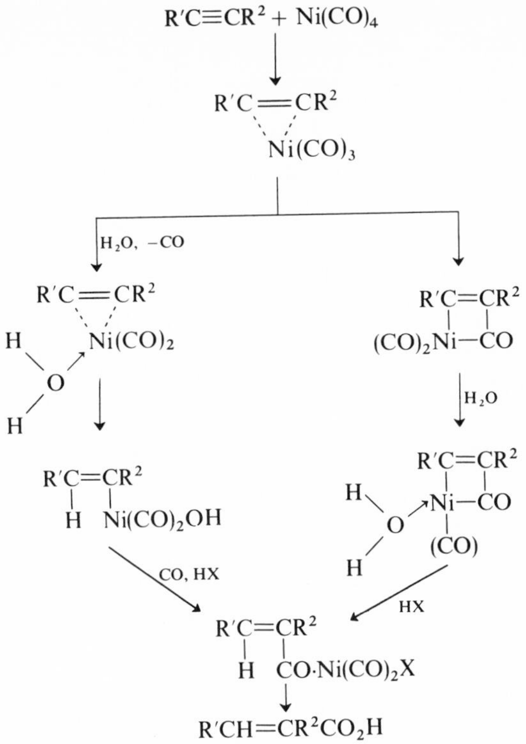

$$R'C\equiv CR^2 + Ni(CO)_4$$

$$\downarrow$$

$$\underset{\overset{|}{Ni(CO)_3}}{R'C=CR^2}$$

$$\underset{\underset{H}{\overset{|}{O}}\nearrow}{\overset{H}{\underset{Ni(CO)_2}{R'C=CR^2}}} \qquad \underset{(CO)_2Ni-CO}{R'C=CR^2}$$

H₂O, –CO (left branch)

H₂O (right branch)

$$\underset{\overset{|}{H}\ \ \overset{|}{Ni(CO)_2OH}}{R'C=CR^2} \qquad \underset{\underset{H}{\overset{H}{O}}\nearrow Ni-CO}{\overset{R'C=CR^2}{}} \atop (CO)$$

CO, HX   HX

$$\underset{\overset{|}{H}\ \ \overset{|}{CO\cdot Ni(CO)_2X}}{R'C=CR^2}$$

$$\downarrow$$

$$R'CH=CR^2CO_2H$$

Allylic halides react with acetylenes and Ni(CO)₄ to give lactones. Esters form if alcohols are present:

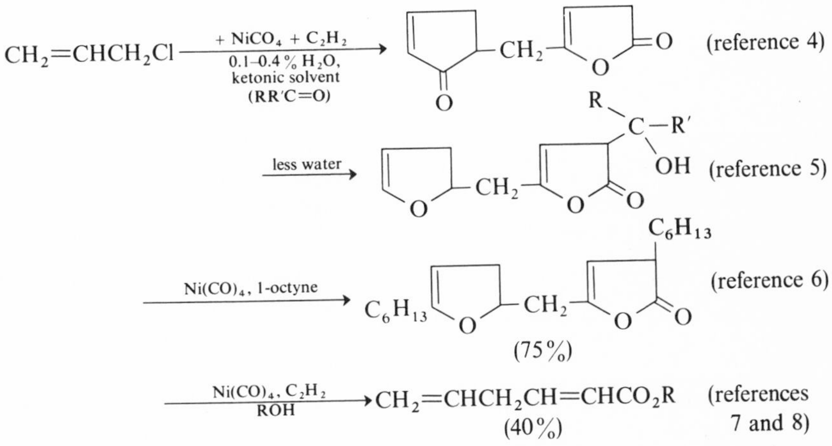

$$CH_2=CHCH_2Cl \xrightarrow[\substack{0.1-0.4\%\ H_2O, \\ \text{ketonic solvent} \\ (RR'C=O)}]{+ NiCO_4 + C_2H_2}$$

(reference 4)

less water → (reference 5)

Ni(CO)₄, 1-octyne → (75%) (reference 6)

$$\xrightarrow[\text{ROH}]{Ni(CO)_4,\ C_2H_2} CH_2=CHCH_2CH=CHCO_2R$$
(40%)

(references 7 and 8)

Iodobenzene reacts with acetylene, CO and alcohol in the presence of $Ni(CO)_4$ at 130° and 30 atmospheres to give

$$\underset{\displaystyle PhCCH_2CH_2CO_2R,}{\overset{\displaystyle O \atop \displaystyle \|}{}}$$

and no

$$\underset{\displaystyle PhCCH=CHCO_2R.[9]}{\overset{\displaystyle O \atop \displaystyle \|}{}}$$

The reaction can be operated under pressure using nickel salts as catalysts. Other "active" halogen compounds such as acyl halides can be used.

Compounds which contain two allylic substituents can react at one or both substituents. cis-2-Butene-1,4-diol reacts with acetylene and CO in methanol in the presence of HCl and nickel carbonyl to form methyl 7-hydroxy-2,5-heptadienoate (1) and dimethyl 2,5,8-decatriene-1,10-dioate (2):[9a]

$$HOCH_2CH=CHCH_2OH + HCl \xrightarrow[HC\equiv CH, CH_3OH]{Ni(CO)_4, CO}$$

$$HOCH_2CH=CHCH_2CH=CHCOOCH_3$$

$$(1) \quad \Big\downarrow \begin{smallmatrix} Ni(CO)_4, CO \\ HC\equiv CH, CH_3OH \end{smallmatrix}$$

$$H_3COOCCH=CHCH_2CH=CHCH_2CH=CHCOOCH_3$$

$$(2)$$

Yields of (1) and (2) can be varied by adding different inorganic halides to the reaction. Lithium chloride, for example, allows 1% yield of (1) and 52% yield of (2). Sodium bromide causes 14% yield of (1) and 47% yield of (2). Butadiene monoepoxide reacts under similar conditions to give up to 47% yields of (1). No (2) is found.

$$\underset{O}{\overset{\displaystyle CH_2CHCH=CH_2}{\diagdown\diagup}} \xrightarrow{HCl} HOCH_2CH(Cl)CH=CH_2$$

$$\Big\downarrow \begin{smallmatrix} Ni(CO)_4, HC\equiv CH, CO \\ CH_3OH, CaCO_3 \end{smallmatrix}$$

$$HOCH_2CH=CHCH_2CH=CHCOOCH_3$$

$$(1)$$

The stoichiometry of the butadiene monoepoxide reaction, assuming 75% reaction of the epoxide, shows that carbonylation is semicatalytic (1 CO from $Ni(CO)_4$, 3 from CO gas):

$$16\underset{O}{\overset{\displaystyle CH_2CHCH=CH_2}{\diagdown\diagup}} + Ni(CO)_4 + 2HOAc + 16HC\equiv CH + 12CO + 16CH_3OH \xrightarrow{LiCl}$$

$$Ni(OAc)_2 + 2[H] + 16HOCH_2CH=CHCH_2CH=CHCOOCH_3$$

Chiusoli and Cassar, who have reported extensive work with the reaction of allylic compounds published a review in 1967.[10]

1.2. Low-pressure Catalytic Carbonylation

This version of the carbonylation reaction is practiced commercially. Its success depends on careful control of reaction conditions. Erreich[11] and co-workers studied the mechanism of the reaction in an integral reactor, which was better than a differential reactor because carbon monoxide partial pressure has a very large effect on the stability of the catalytic reaction. The most important variable in the system was stirring rate. If excess CO is present in the liquid homogeneous system, the catalyst is deactivated. The reaction is characterized by an induction period, just as the stoichiometric reaction is, during which the active catalyst is probably forming. The rate then increases and finally settles to a steady rate. The reaction is relatively insensitive to temperature. In reaction at the steady state with 75% of the CO coming from carbon monoxide, gradually varying the temperature from 2–77° has no effect. A solution which contains considerable acrylate product is the best solvent for a stable catalytic reaction. Iron carbonyl and cobalt carbonyl are not catalysts or promoters under these conditions. Oxidizing agents such as nitrobenzene are poisons, as is carbon tetrachloride. The maximum catalytic level is around 85%, i.e., only 15% of the CO is furnished by nickel carbonyl.

A mechanism to account for all the observations has been proposed:[11]

Formation of acrylate ester:

$$Ni(CO)_4 \rightleftharpoons Ni(CO)_3 + CO$$
$$\quad (1) \qquad\qquad (2)$$

$$(2) + H_2C=CH-CO_2Et \rightleftharpoons$$

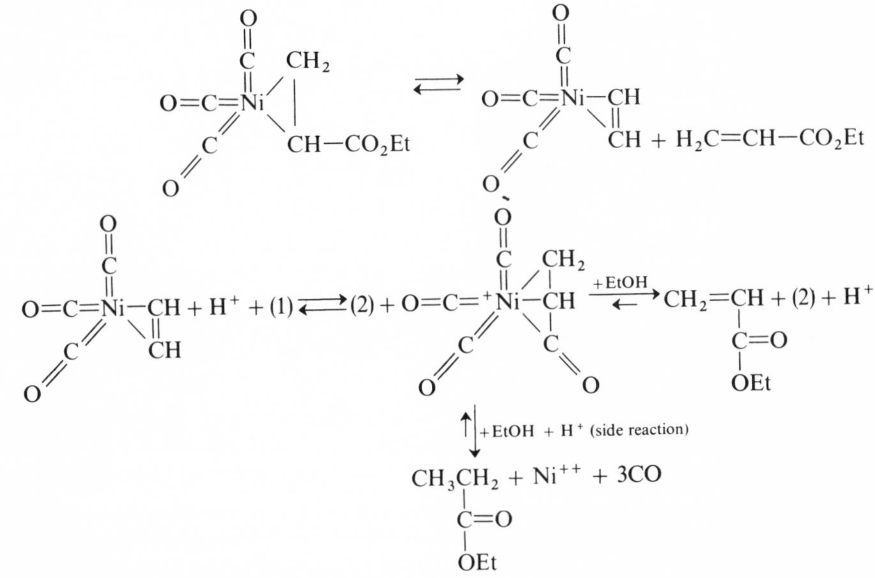

Other side reactions:

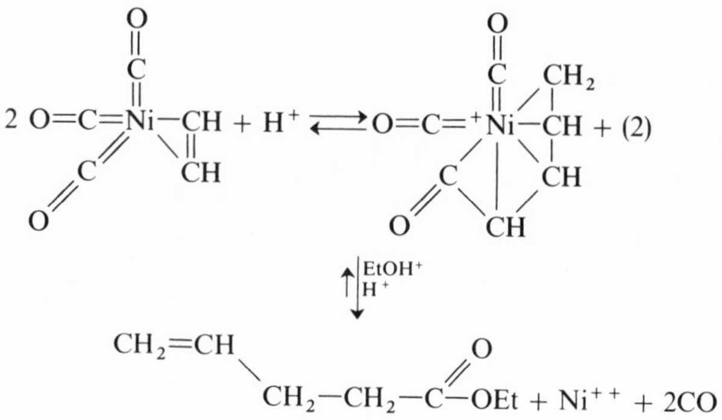

$$2 \ O=C=Ni-CH \ + \ H^+ \ \rightleftharpoons \ O=C=^+Ni-CH \ + \ (2)$$

$$CH_2=CH \diagdown CH_2-CH_2-C \overset{O}{=} OEt \ + \ Ni^{++} \ + \ 2CO$$

$$Ni(CO)_3 \ \rightleftharpoons Ni^\circ \ + \ 3CO$$
$$(2)$$
$$\uparrow +2H^+$$
$$Ni^{++} \ + \ H_2$$

$$xNi(CO)_3 \ + \ yC_2H_2 \longrightarrow red \ polymer$$

$$Ni \ + \ zC_2H_2 \longrightarrow red \ polymer$$

The cyclopropenone formed by addition of CO across a triple bond is proposed as a reaction intermediate in the carbonylation of acetylenes.[12] Diphenylpropenone does not react with HCl in benzene, but it does react with nickel carbonyl in benzene to form tetraphenylcyclopentadienone. Thus, the cyclopropenone is probably not intermediate to the formation of carbonylation products.[13]

Mixtures of acetylenes carbonylate independently. For example, acetylene and propyne carbonylate to form a mixture of acrylate and methacrylate esters.[14]

Mueller[15] obtained anomalous product from the reaction of diphenylacetylene with CO in presence of nickel carbonyl and HCl. The product was 2,3,4,5-tetraphenylcyclopenta-2-en-1-one, 37% yield.

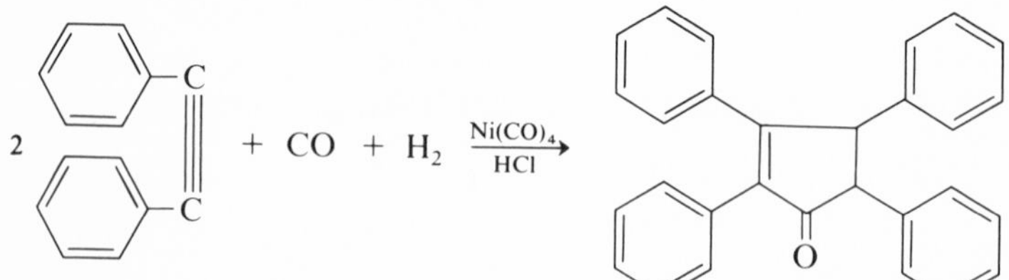

$$2 \ \text{(diphenylacetylene)} \ + \ CO \ + \ H_2 \ \xrightarrow[\text{HCl}]{Ni(CO)_4} \ \text{(tetraphenylcyclopentenone)}$$

(37%)

Hydrogen is always formed in carbonylations in the presence of nickel carbonyl and HCl, and it is likely that the hydrogen reduced the tetraphenylcyclopentadienone formed as the primary product to give the product actually isolated (tetraphenylcyclopentenone). The hydrogen is formed in the carbonylation by reaction between metallic nickel and acid:

$$RC \equiv CR + 4R'OH + Ni(CO)_4 + 2HCl \longrightarrow 4RHC = CRCOOR' + NiCl_2 + H_2$$

Such reductions can also occur in other carbonylations.

In alkaline medium, diphenylacetylene reacts with nickel carbonyl to give 25% yield of α-phenyl-*trans*-cinnamic acid and 67% yield of 1,2,3,4-tetraphenylbutadiene, formed by dimerization within the complex. The source of CO in this reaction is $Ni_3(CO)_8^{-2}$.[16]

1.3. High-pressure Catalytic Carbonylation

Bhattacharyya[17] determined the optimum conditions for reaction of acetylene with CO and methanol in the liquid phase. The calculated equilibrium constant K_p decreases as the temperature increases, but it depends less on pressure as the pressure is increased. The results of a study of the variables are:

(1) *Catalysts*: Halides of nickel, cobalt, and iron deposited on silica gel, and the naphthenates of these metals, are active. Nickel is better than iron, which is a little better than cobalt. Iodide is better than bromide or chloride. Maximum yield is at an $Ni:SiO_2$ ratio of 1.

(2) *Temperature*: Each catalyst has its own optimum. For $Ni-SiO_2$, 170° is best, and is the lowest optimum found.

(3) *Pressure*: Optimum pressure at optimum temperature depends on the catalyst. For the best catalyst, the optimum pressure is 520 psig.

(4) *Time*: The best time for $Ni-SiO_2$ is 4 hours.

(5) *Amount of catalyst*: 5 grams of $Ni-SiO_2$ per 0.1 mole of acetylene is best.

(6) *Charge gas*: An acetylene:CO ratio of 1 is optimum.

A little water decreases the yield. The catalyst loses 80% of its activity in one cycle. In the best reaction, conversion of methanol to acrylate is 12.8%, and conversion of acetylene is 47.3%, along with 3.2% conversion to acrylic acid. Liquid by-products are acetaldehyde and hydrocarbons. Gaseous by-products are carbon dioxide, ethylene, ethane, methane and hydrogen.

Other catalysts have been claimed: $NiBr_2$ + a phosphorus, sulfur or selenium compound;[18] nickel complex catalysts containing mercuric bromide and butyl bromide;[19] and nickel sulfonates.[20]

2. OTHER CATALYSTS

2.1. Dicobalt Octacarbonyl

Dicobalt octacarbonyl does not catalyze the reaction to form acrylates. Products depend on conditions:

$RC\equiv CH + CO + H_2 \xrightarrow{CO_2(CO)_8} RCH_2CH_2CH_2OH$ (same product as $RCH=CH_2$ but lower yield)[21]

$C_2H_2 + CO + ROH \longrightarrow$ acrylates, fumarates, succinates[22]

$C_2H_2 + CO + H_2 \longrightarrow$ complex mix

$C_2H_2 + 2CO + H_2 + EtOH \longrightarrow 60\%$ $\begin{cases} \text{ethyl propionate, propionaldehyde, ethyl} \\ \text{acrylate, diethylsuccinate, ethyl } \gamma,\gamma\text{-} \\ \text{diethoxybutyrate}^{22} \end{cases}$

$C_2H_2 + 2CO + H_2 + i\text{-PrOH} \longrightarrow$ very low yield,[22] some $O=\overset{\overset{\displaystyle H}{|}}{C}CH_2CH_2CO_2-i\text{-Pr}$

$C_2H_2 + CO \xrightarrow[\substack{\text{acetic} \\ \text{anhydride}}]{\text{acetone}}$ [23]

2.2. Palladium Chloride

2.2.1. CARBONYLATION OF ACETYLENIC HYDROCARBONS

Tsuji[24] studied the palladium chloride-catalyzed reaction of olefins with CO. He also carried out a reaction with acetylene, using $PdCl_2 \cdot 2$ benzonitrile as catalyst. Muconyl chloride was the major product, along with fumaryl chloride and maleyl chloride:

$$C_2H_2 + CO + PdCl_2 \cdot 2PhCN \xrightarrow[100°, 100\,kg]{\text{benzene}} \begin{matrix} CH=CHCOCl \\ | \\ CH=CHCOCl \end{matrix}$$
$$\text{muconyl chloride}$$

Palladium chloride and diphenylacetylene react easily to form cyclobutadiene ring complexes, which have been carbonylated to form 2,3,4,5-tetraphenyl-2-cyclopentenone and a little tetraphenylcyclone. Direct carbonylation in the presence of nickel carbonyl gives the same tetraphenylcyclopentenone,[15] and ethyl 2,3-diphenylacrylate can be made in benzene-ethanol solvent.

Tsuji[25] later extended his carbonylation study. In ethanol, acetylene and carbon monoxide react in the presence of $PdCl_2$ to form acrylate, maleate and fumarate esters. Diphenylacetylene gave some unexpected results. Hydrogen chloride and metallic palladium are essential for catalytic carbonylation of olefins. Carbonylation of diphenylacetylene in ethanol, HCl present, at $100\,kg/cm^2$ CO pressure gives lactone (1) in high yield. The concentration of HCl is critical. At lower concentration of HCl, the yield of dicarbonylation products decreases drastically, and a little ethyl 2,3-diphenylacrylate forms.

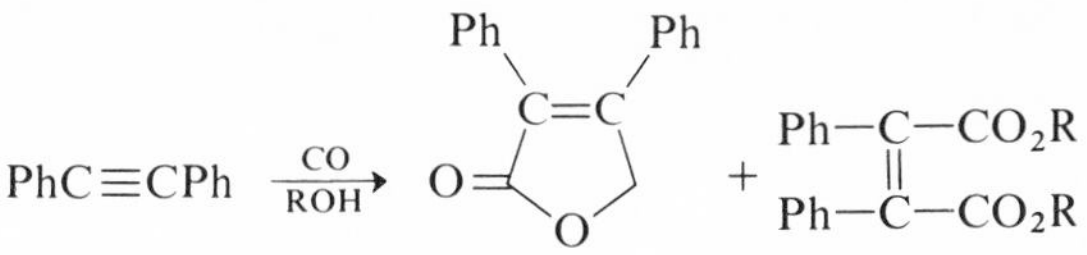

$$\text{PhC}\equiv\text{CPh} \xrightarrow[\text{ROH}]{\text{CO}} \quad (1) \quad + \quad \begin{array}{c} \text{Ph}-\text{C}-\text{CO}_2\text{R} \\ \| \\ \text{Ph}-\text{C}-\text{CO}_2\text{R} \end{array}$$

(α,β-diphenyl-γ-crotonolactone)

Lactone (1) is probably formed by *cis* addition of 2 moles of CO to the triple bond. The product is presumably hydrogenolyzed to lactone like Sternberg's[26] dicobalt carbonyl adduct:

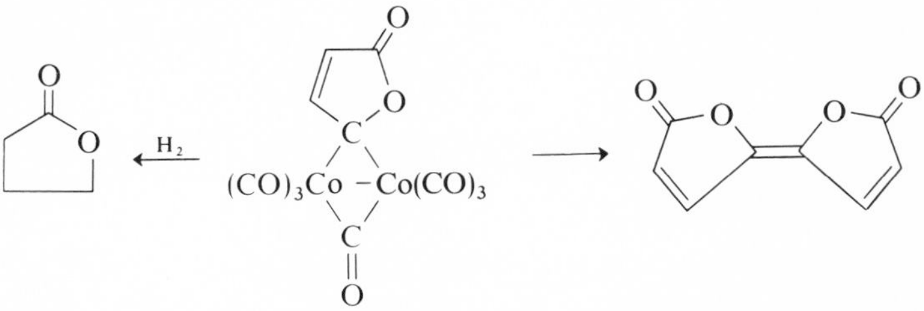

Palladium metal and HCl are the true catalysts. Thus, Pd—C and HCl give lactone. In benzene saturated with HCl, no lactone forms; the product is hexaphenylbenzene.

Diphenylacetylene with an equivalent of PdCl$_2$ in ethanol plus HCl was heated for an hour at 50°, and then CO was added. No lactone formed. The products were tetraphenylfuran and desoxybenzoin. The furan formed from the tetraphenylcyclobutadiene complex, and the desoxybenzoin from addition of ethanol across the triple bond. This shows that CO must be present before complex formation if lactone is to form.

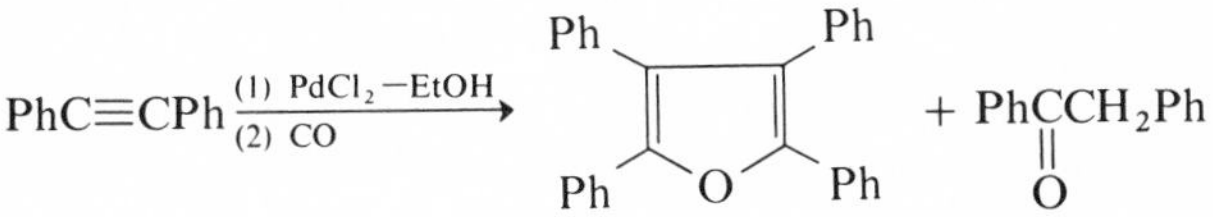

$$\text{PhC}\equiv\text{CPh} \xrightarrow[\text{(2) CO}]{\text{(1) PdCl}_2 - \text{EtOH}} \quad + \quad \text{PhCCH}_2\text{Ph}$$

2.2.2. CARBONYLATION OF ACETYLENIC ACIDS

Jones and co-workers[27] found that acetylenic acids are very difficult to carbonylate in the presence of nickel catalysts and concluded that the carboxyl group deactivates the triple bond. However, acetylenic acid esters are carbonylated readily in the presence of Pd-HCl.[28] See equations on p. 316.

The reaction goes rapidly at room temperature: the ester group increases the reactivity of the triple bond in the Pd-HCl system. The carbonylation and dimerization reactions [leading to (4)] are along the same reaction pathway.

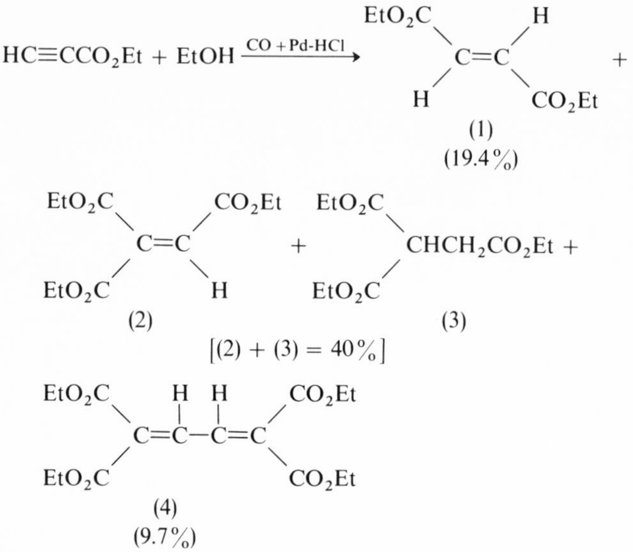

$$HC \equiv CCO_2Et + EtOH \xrightarrow{CO + Pd\text{-}HCl}$$

EtO$_2$C, H
   C=C
H, CO$_2$Et
(1)
(19.4%)

EtO$_2$C, CO$_2$Et   EtO$_2$C,
   C=C   +   CHCH$_2$CO$_2$Et +
EtO$_2$C, H   EtO$_2$C
(2)   (3)

[(2) + (3) = 40%]

EtO$_2$C, H H, CO$_2$Et
   C=C–C=C
EtO$_2$C, CO$_2$Et
(4)
(9.7%)

Diethyl acetylenedicarboxylate also reacts readily at room temperature, to give the expected products. Some of the product is saturated ester (tetraethyl ethanetetracarboxylate), and saturated product increases as the concentration of HCl increases.

References

1. Copenhaver, J. W., and Bigelow, M. H., "Acetylene and Carbon Monoxide Chemistry," New York, Reinhold Publishing Corp., 1949.
2. Bird, C. W., *Chem. Rev.*, **62**, 283, 289 (1962).
2a. Bird, C. W., and Briggs, E. M., *J. Chem. Soc. (C)*, 1265 (1967).
3. Jones, E. R. H., Whitham, G. H., and Whiting, M. C., *J. Chem. Soc.*, 4628 (1957).
4. Cassar, L., and Chiusoli, G. P., *Chim. Ind. (Milan)*, **48**, 323 (1966).
5. Cassar, L., and Chiusoli, G. P., *Tetrahedron Letters*, 3295 (1965).
6. Chiusoli, G. P., and Bottaccio, G., *Chim. Ind. (Milan)*, **47**, 165 (1965).
7. Chiusoli, G., Bottaccio, G., and Cameroni, A., *Chim. Ind. (Milan)*, **44**, 131 (1962).
8. Chiusoli, G., and Merzoni, S., *Chem. Ind. (Milan)*, **45**, 6 (1963).
9. Chiusoli, G. P., Merzoni, S., and Mondelli, G., *Tetrahedron Letters*, 2777 (1964).
9a. Mettalia, J. B., Jr., and Specht, E. H., *J. Org. Chem.*, **32**, 3941 (1967).
10. Chiusoli, G. P., and Cassar, L., *Angew. Chem. Intern. Ed.*, **6**, 124 (1967).
11. Erreich, J. E., Nickerson, R. G., and Ziegler, C. E., *Ind. Eng. Chem. Process Design Develop.*, **4**, 77 (1965).
12. Reppe, W., *Ann. Chem.*, **582**, 1 (1953).
13. Bird, C. W., and Hudec, J., *Chem. Ind. (London)*, 570 (1959).

14. Sakakibara, Y., *Bull. Chem. Soc. Japan*, **37**, 1601 (1964); *Chem. Abstr.*, **62**, 5187 (1965).
15. Mueller, G. P., and MacArtor, F. L., *J. Am. Chem. Soc.*, **76**, 4621 (1954).
16. Sternberg, H. W., Markby, R., and Wender, I., *J. Am. Chem. Soc.*, **82**, 3638 (1960).
17. Bhattacharyya, S. K., and Sen, A. K., *J. Appl. Chem. (London)*, **13**, 498 (1963).
18. Dunn, J. T., and Brodhag, A. E., Jr. (to Union Carbide Corp.), British Patent 885,048 (July 13, 1959); *Chem. Abstr.*, **57**, 4550 (1962).
19. Dunn, J. T., (to Union Carbide Corp.), U.S. Patent 3,013,067 (Dec. 12,1961); *Chem. Abstr.*, **56**, 9973 (1962).
20. Anderson, R. D., and Smolin, E. M. (to American Cyanamid Co.), U.S. Patent 3,025,319 (Mar. 13, 1962); *Chem. Abstr.*, **57**, 11027 (1962).
21. Greenfield, H., Wotiz, J. H., and Wender, I., *J. Org. Chem.*, **22**, 542 (1957).
22. Crowe, B. F., *Chem. Ind. (London)*, 1000, 1506 (1960).
23. Albanesi, G., and Tovagliere, M., *Chim. Ind. (Milan)*, **41**, 189 (1959); *Chem. Abstr.*, **53**, 19872 (1959).
24. Tsuji, J., Morikawa, M., and Iwanamoto, N., *J. Am. Chem. Soc.*, **86**, 2095 (1964).
25. Tsuji, J., and Noge, T., *J. Am. Chem. Soc.*, **88**, 1289 (1966).
26. Sternberg, H. W., *et al.*, *J. Am. Chem. Soc.*, **78**, 3621 (1956).
27. Jones, E. R. H., Shen, T. Y., and Whiting, M. C., *J. Chem. Soc.*, 48 (1951).
28. Tsuji, J., and Nogi, T., *J. Org. Chem.*, **31**, 2641 (1966).

PART FOUR

Vinylation of HCN and Nitrogen Compounds

1. VINYLATION OF HCN—ACRYLONITRILE SYNTHESIS

This commercially important liquid phase vinylation is discussed in detail by Miller.[1] Virtually no work on mechanisms has been published. Catalyst solutions are nearly identical to the "Nieuwland" solutions used for vinylacetylene. The solution is "activated" by acetylene, and forms acrylonitrile when acetylene and HCN are added.[2] For example, when acetylene and HCN are added to a solution of cuprous chloride and sodium chloride, acrylonitrile forms and continues to form for 70 minutes after addition is stopped.[3] Various complexing agents have been used in the solution, including cyanide ion and chloride ion. Increasing the concentration of Cu_2Cl_2 relative to complex-forming agents increases the space-time yield of acrylonitrile from 40 up to 60 g/h/liter.[4] At 70–110°, and 1–4 atmospheres pressure, the yields are above 90%.

2. VINYLATION OF NITROGEN COMPOUNDS

2.1. Aliphatic Amines (For a recent review, see reference 5)

Reppe[6] believed that primary aliphatic amines vinylate in the presence of zinc or cadmium acetates to form N-vinylamines. Kruse[7] proved that the products are ethylidene imines $(RN=CHCH_3)$. According to Reppe,[6] secondary amines react with acetylene in the presence of bases or cadmium or zinc salts to give unidentified resins, but cuprous chloride catalyzes the formation of N,N-dialkyl-1-methyl-2-propynylamines,

$$\underset{MeCHC\equiv CH}{\overset{\overset{\textstyle NR_2}{|}}{}}$$

Kruse[8] used propyne, which has only one acetylenic hydrogen, and was able to react it with secondary amines in the presence of zinc or cadmium salts:

$$R_2NH + \text{propyne} \xrightarrow[17\,h]{120°} \left[\underset{MeC=CH_2}{\overset{\overset{\textstyle NR_2}{|}}{}} \right] \xrightarrow{\text{propyne}} \underset{Me_2C-C\equiv CMe}{\overset{\overset{\textstyle NR_2}{|}}{}}$$
$$\text{(70\% yield)}$$

The second molecule of propyne adds across the intermediate vinyl group.

2.2. Cyclic Amines

Carbazole vinylates in the presence of KOH, or ZnO, or 3KOH:1ZnO at 175–190°.[9] Vinylation is fastest when the alkali is suspended in a saturated hydrocarbon. Vinyl carbazole has been studied fairly extensively as a monomer.[5]

Indole is difficult to vinylate. Pyrrole and acetylene over KOH on MgO or CaO at 215° give 50% yield of N-vinylpyrrole.[10]

Sauer[11] reported a new reaction. Secondary amines react with CS_2 and acetylene in tetrahydrofuran at 130° and 15 atmospheres. The products are vinyl N,N-dialkyldithiocarbamates, $R_2N\overset{\overset{\textstyle S}{\|}}{C}SCH=CH_2$. When R is ethyl, butyl or propyl, the yields are around 50–60%.

2-Vinylimidazoles have been made by elimination reactions[12] and so have N-vinyltetrazoles.[13]

2.3. Cyclic Lactams

Elimination methods have been frequently used.[5] Direct vinylation in the presence of KOH with acetylene under pressure has been used for pyrrolidone, piperidone and caprolactam.[14] N-Vinylpyrrolidone is a commercial material, and its polymers and copolymers are widely used.

2.4. Sulfonamides

Cairns[15] prepared a series of new vinylsulfonamides. Sulfonamides react with acetylene in water solutions of KOH at 7–16 atmospheres and 180° to give N-vinylsulfonamides (Table 3-10). Benzene can also be used as diluent. Mercuric chloride or cuprous chloride do not catalyze the vinylation.

TABLE 3-10[15]

$RSO_2NR'CH{=}CH_2$

R	R'	% Yield[a]	b.p. (°C, @ mm Hg)
Ph	Me	74	120, @ 2
Cyclohexyl	Me	62	115–118, @ 3
Bu	Me	45	120–123, @ 2
p-Tolyl	Et	85	133–139, @ 1.5

* 16 hr reaction.

N-Vinylsulfonamides isomerize when exposed to electrons from an electron generator:[16]

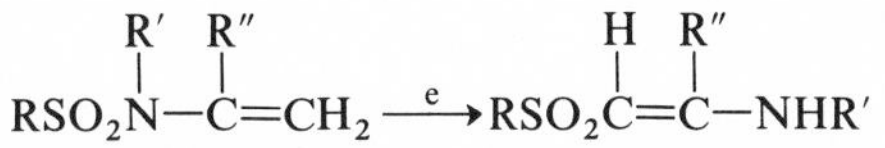

$$\underset{RSO_2N-C=CH_2}{\overset{R'\quad R''}{|\quad\ \ |}} \xrightarrow{\ e\ } \underset{RSO_2C=C-NHR'}{\overset{H\quad R''}{|\quad\ \ |}}$$

2.5. Urethanes and Oxazolidones

Vinyl isocyanate has been converted to N-vinylurethanes and ureas, which polymerize in the presence of radical catalysts.[17] Peppel[18] prepared N-vinyl-urethanes and oxazolidones by vinyl interchange with vinyl isobutyl ether, using mercuric acetate and a trace of sulfuric acid. Some interchange products are:

3-Vinyl-5-methyl-2-oxazolidone, 86% yield, b.p. 105–108 at 2.5 mm
3-Vinyl-2-oxazolidone, b.p. 76.5–78 at 0.5 mm
Ethyl N-methyl-N-vinylurethane, b.p. 35–36 at 4.5 mm

References

1. Miller, S. A., "Acetylene. Its Properties, Manufacture and Uses," Vol. 2, p. 156, New York, Academic Press, 1966.
2. Masui, M., *Kogyo Kagaku Zasshi*, **63**, 1955 (1960); *Chem. Abstr.*, **58**, 963 (1963).
3. Masui, M., Sakurai, E., and Kobayashi, S., *Kobunshi Kagaku* **18**, 487 (1961); *Chem. Abstr.*, **55**, 27041 (1961).

4. Sennewald, K., Legutke, G., and Ohorodnik, A. (to Knapsack A.-G.), German Patent 1,211,159 (Feb. 24, 1966); *Chem. Abstr.*, **64**, 17433 (1966).
5. Miller, S. A., "Acetylene. Its Properties, Manufacture and Uses," Vol. 2, p. 325, New York, Academic Press, 1966.
6. Reppe, W., *et al., Ann. Chem.*, **596**, 1 (1955).
7. Kruse, C. W., and Kleinschmidt, R. F., *J. Am. Chem. Soc.*, **83**, 213 (1961).
8. *Ibid.*, 216 (1961).
9. Otsuka, S., and Murahashi, S., *Kogyo Kagaku Zasshi*, **59**, 511 (1956); *Chem. Abstr.*, **52**, 3818 (1958).
10. Sims, V. A. (to Air Redn. Co.), U.S. Patent 3,047,583 (July 31, 1962); *Chem. Abstr.*, **58**, 509 (1963).
11. Sauer, J. C., *J. Org. Chem.*, **24**, 1592 (1959).
12. Lawson, J. K., Jr., *J. Am. Chem. Soc.*, **75**, 3398 (1953).
13. Finnegan, W. G., and Henry, R. A., *J. Org. Chem.*, **24**, 1565 (1959).
14. Shostakovskii, M. F., Didel'kovskaya, F. P., and Zelenskaya, M. G., *Izv. Akad. Nauk SSSR Otd. Khim. Nauk*, 1406, 1457 (1957); *Chem. Abstr.*, **52**, 7270 (1958).
15. Cairns, T. L., and Sauer, J. C., *J. Org. Chem.*, **20**, 627 (1955).
16. Stacey, F. W., Sauer, J. C., and McKusick, B. C., *J. Am. Chem. Soc.*, **81**, 987 (1959).
17. Weizel, G., and Greber, G., *Makromol. Chem.*, **31**, 230 (1959).
18. Peppel, W. J., and Watkins, J. D. (to Jefferson Chem. Co.), U.S. Patent 3,019,231 (Jan. 30, 1962); *Chem. Abstr.*, **56**, 12748 (1962).

PART FIVE

Vinylation at "Activated" C—H Bonds

Introduction

C—H bonds in some molecules are acidic enough to add across acetylene. Ethynyl hydrogen is an example: one molecule of acetylene adds across another molecule to form vinylacetylene.

1. VINYLATION OF β-DICARBONYL COMPOUNDS

C—H groups activated by two adjacent carbonyl groups can be vinylated in the presence of zinc salts and cadmium salts at 150–200° and 25 atmospheres pressure.[1] Bases or mercury salts do not catalyze this vinylation. Alkyl-substituted malonic esters give products $CH_2{=}CH{-}\overset{\displaystyle R}{\underset{\displaystyle |}{C}}(CO_2Et)_2$ in 73–89%

yields. Cyclic β-diketones form products such as

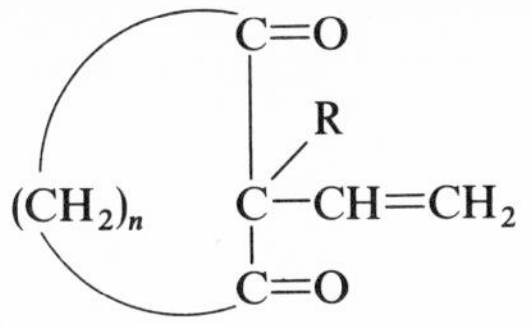

(n = 3 or 4, R = OMe, 32% yield)

and

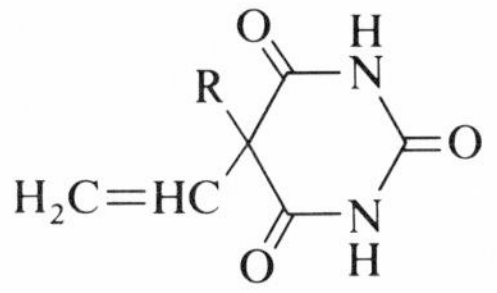

(R = n-butyl, etc.)

Allylmalonic esters give remarkable products, accounted for by this series of reactions:

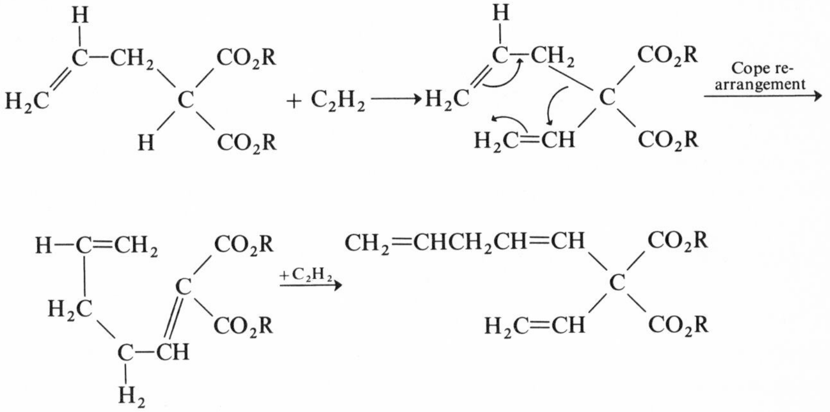

Vinylation of malonic esters is regarded as an electrophilic substitution. Monosubstituted malonic or cyanoacetic esters give best results. Decarboxylation of vinylmalonic acids gives α-substituted crotonic acids:

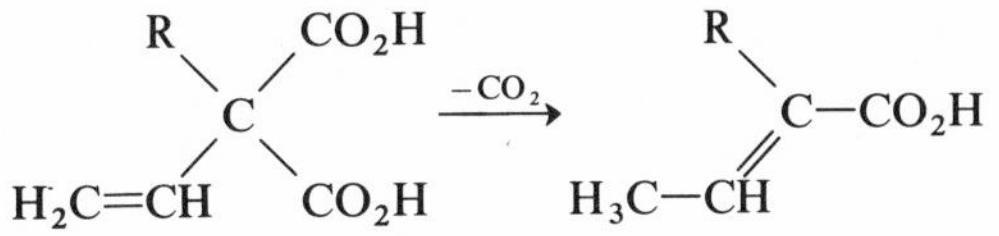

Some vinyl diethylmalonates prepared at 180° and 20 atmospheres pressure in the presence of 3–4 weight % zinc stearate are listed in Table 3–11.

TABLE 3-11
Vinyl Diethylmalonates[1]

$$H_2C=CHC(CO_2Et)_2 \overset{R}{|}$$

R	% Yield	b.p. (°C, @ mm Hg)
Amyl	69	128–129
2-Amyl	80	130–131, @ 10
Butyl	73	132–134, @ 20
Isoamyl	79	128–129, @ 10
2-Hexyl	67	137–138, @ 11
2-Methylamyl	55	133–134, @ 9
4-Penten-1-yl	89	151–153, @ 20
Cyclooctyl	80	133, @ 6

2. VINYLATION OF AROMATIC HYDROCARBONS

Fluorene is vinylated at 210° and 26 atmospheres pressure in the presence of sodium hydroxide, water and toluene. The yield of 9-vinylfluorene is 79% after 7 hours reaction.[2] Vinylacenaphthene (65–79% yield) and 9-vinylanthracene (80% yield) are formed under similar conditions.[3] In a reaction similar to vinylation, tetrahydronaphthalene and acetylene react in the presence of di-*t*-butyl peroxide to form vinyltetrahydronaphthalene.[4] Cyclohexane gives 27% yield of vinylcyclohexane. Decalin vinylates under similar conditions.[5] Dioxane gives 17% yield of vinyldioxane, and tetrahydrofuran gives 2-vinyltetrahydrofuran.

Acid-catalyzed reactions between aromatic hydrocarbons and acetylene give 1,1-diarylethanes, which can be cracked to vinylaromatics. Franzen[6] reviewed these reactions in 1964. Phenols require mercuric ion catalyst for reasonably good reaction.

Phenylacetylene reacts with 2 moles of phenol in the presence of BF_3 etherate at 55° to form 61% yield of 1-phenyl-1,1-bis(4-hydroxyphenyl)ethane.[7] The catalyst can be reused several times. The activity drops a little after the first reaction, but it remains practically constant thereafter.

The condensation product of acetylene with 2 moles of phenol, 1,1-(4,4'-dihydroxy)diphenylethane, cracks to vinylphenol and phenol over silica-alumina catalyst.[8] At 550°, the yield is 43.3% of *p*-vinylphenol.

Phenol reacts with acetylene in the presence of $HgO-H_2SO_4$ to form

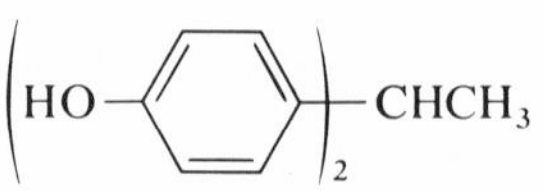

$$\left(HO-\!\!\left\langle\!\!\!\bigcirc\!\!\!\right\rangle\!\!-\right)_2 CHCH_3$$

in 35% yield.[9]

3. VINYLATION AT ALLYLIC HYDROGENS

3.1. Allylic Addition of Olefins to Activated Acetylenes

Although addition of allylic hydrogen atoms of olefins to negatively substituted olefins is well known, in 1962 the addition of an allylic hydrogen to benzyne was the only known allylic addition to an acetylene.[10]

In 1962, Sauer[11] reported the first allylic addition to a negatively substituted acetylene. When he heated isobutylene and hexafluoro-2-butyne at 145° and autogenous pressure, he obtained 80 % yield of 1,1,1-trifluoro-3-trifluoromethyl-5-methyl-2,5-hexadiene:

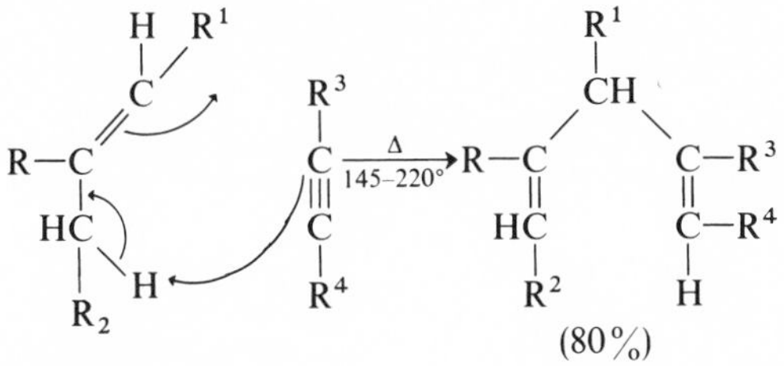

$R = CH_3$
R^1 and $R^2 = H$
R^3 and $R^4 = CF_3$

At 145% only a trace of diadduct formed, but at 240° the yield of diadduct was 14%:

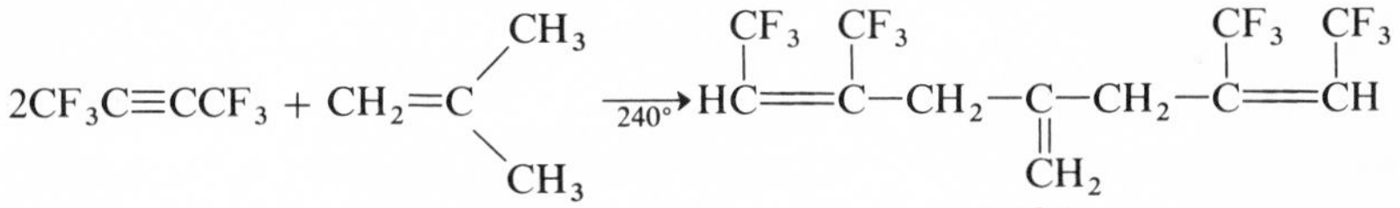

Other reactions are listed in Table 3-12.

TABLE 3-12
Allylic Addition of Olefins to Activated Acetylenes[11]

Activated Acetylene	Olefin	Product	% Yield
Hexafluorobutyne-2	2-Butene	1,1,1-Trifluoro-3-trifluoromethyl-4-methyl-2,5-hexaduene	32
Trifluoropropyne	Isobutylene	1,1,1-Trifluoro-5-methyl-2,5-hexadiene	15 (including isomers)
Dimethyl acetylene-dicarboxylate	Isobutylene	Methyl 3-methoxycarbonyl-5-methyl-2,5-hexadienoate	41
Methyl propiolate	Isobutylene	Methyl 5-methyl-2,5-hexadienoate	Low

3.2. Thermal Addition of Olefins to Acetylene to Form 1,4-Dienes

Cywinski[12] reacted olefins with acetylene in a flow system at 2500 psig and 350°. The effluent gas contained less than 10% 1,4-diolefin. The most likely mechanism is addition of allylic hydrogen to acetylene, followed by allylic rearrangement of the double bond:

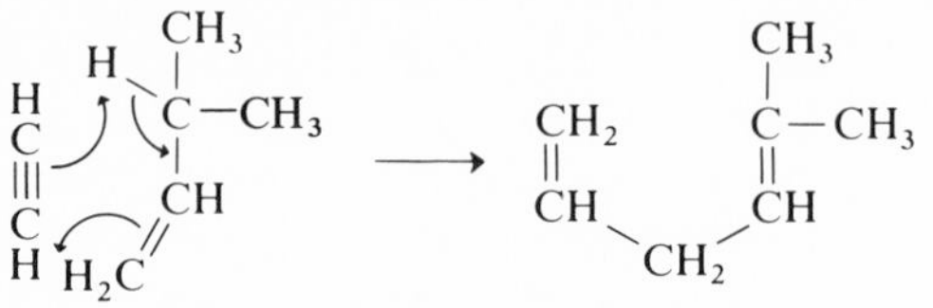

When more than one allylic hydrogen is available, as is frequently the case, the amount of each entering the reaction can be determined by analysis of the isomers produced (Table 3-13).

TABLE 3-13
Per Cent of Allylic Hydrogen Reacting with Acetylene to Produce 1,4-Diolefins[12]

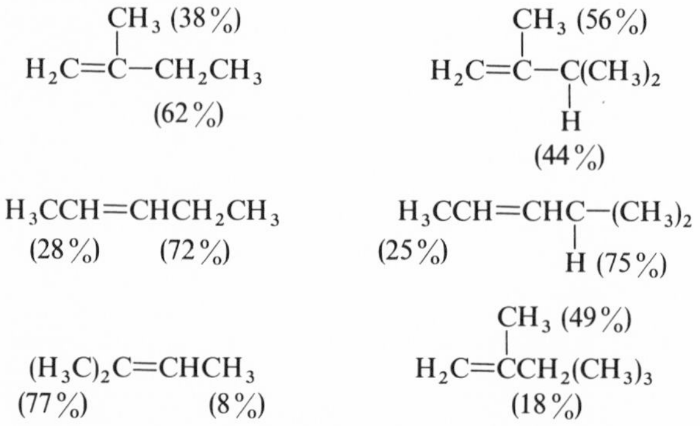

$$\underset{(62\%)}{H_2C=\underset{\underset{(38\%)}{\overset{|}{CH_3}}}{C}-CH_2CH_3}$$

$$\underset{(44\%)}{H_2C=\underset{\underset{H}{\overset{|}{C}(CH_3)_2}}{C}-\underset{\underset{(56\%)}{\overset{|}{CH_3}}}{\overset{}{}}}$$

$$\underset{(28\%)\quad(72\%)}{H_3CCH=CHCH_2CH_3}$$

$$\underset{(25\%)}{H_3CCH=CH}\underset{H\,(75\%)}{C-(CH_3)_2}$$

$$\underset{(77\%)\qquad(8\%)}{(H_3C)_2C=CHCH_3}$$

$$\underset{(18\%)}{H_2C=\underset{\underset{(49\%)}{\overset{|}{CH_3}}}{C}CH_2(CH_3)_3}$$

3.3. Catalytic Addition of Olefins to Acetylene

Mixtures of acetylene and propylene over various catalysts at 250–430° form traces of isoprenoid hydrocarbons.[13] Alumina, and alumina as carrier for copper, cuprous chloride, cupric chloride, nickel oxide and silver are slightly active. The reaction produces much hydrogen, methane, ethane, ethylene, propylene and 2-methyl-1-butene.

Acetylene does not react with propane to form isoprene.[14] Molybdena-silica catalysts, or molybdena-alumina, treated with ferric oxide, cobalt oxide, titania, calcium oxide or tin oxide are inactive. KOH is inactive, as is zinc oxide-titanium oxide. Above 210°, reaction occurs, but the product does not contain any isoprene or piperylene.

4. SELF-VINYLATION OF ACETYLENE TO FORM VINYLACETYLENE

Addition of one molecule of acetylene across another to form vinylacetylene can be regarded as a vinylation. This is one of the most important commercial vinylations. As in the case of vapor phase vinylation, and the acrylonitrile synthesis, the commercial importance is not reflected by mechanism studies in the literature. Most of the recently published mechanism work is by Russian authors, who studied by various means the changes which occur in the catalyst solution during vinylacetylene formation.

Early studies of the vinylacetylene reaction are covered by Nieuwland's book,[15] and the addition of HCl to vinylacetylene to form chloroprene is also described. Chloroprene is the monomer for neoprene rubber.

The "Nieuwland" catalyst solution is generally cuprous chloride, ammonium chloride and HCl, although many variations have been investigated. The Russian workers have concluded that the reaction is ionic in the activation and dimerization stages.[16,17] When acetylene is added to catalyst solution, the pH decreases to a constant value. Heating to remove acetylene returns the pH to its original value.[18] The pH change is thus due to complex formation. The nature of the complex is unknown, but its lability indicates that it is a coordination complex, probably involving the acetylenic bond.

Many patents have appeared to claim improvements in rates, stability, yields, etc. (e.g., reference 19), but the actual mode of operation of vinylacetylene production units has not been revealed.

5. CATALYTIC DIMERIZATION OF 1-ACETYLENES

In the "Straus" coupling of terminal acetylenes using cuprous oxide in glacial acetic acid, branched enynes never form:[20]

$$RC\equiv CH \xrightarrow[HOAc]{Cu_2O} RC\equiv CCH=CHR$$
$$(5–65\% \text{ yields})$$

In the reaction catalyzed by $3Et_2Zn:1Cr(O-t-Bu)_4$ at 30°, products are exclusively branched enynes:[21]

$$PrC\equiv CH \longrightarrow PrC\equiv C-\overset{\overset{\displaystyle Pr}{|}}{C}=CH_2$$
$$(100\%)$$

1-Hexyne and t-butylacetylene give similar results.

References

1. Seefelder, M., *Ann. Chem.*, **652**, 107 (1962).
2. Karpukhin, P. P., and Levchenko, A. I., *Ukr. Khim. Zh.*, **24**, 544 (1958); *Chem. Abstr.*, **53**, 11325 (1959).
3. Karpukhin, P. P., Levchenko, A. I., and Dudko, E. V., *Zh. Prikl. Khim.*, **34**, 1117 (1961); *Chem. Abstr.*, **55**, 22259 (1961).
4. Shuikin, N. I., Lebedev, B. L., and Nikol'ski, V. G., *Izv. Akad. Nauk SSSR, Ser. Khim.*, 351 (1965); *Chem. Abstr.*, **62**, 14589 (1965).
5. *Idem.*, *Dokl. Akad. Nauk SSSR*, **158**, 692 (1964); *Chem. Abstr.*, **62**, 3998 (1965).
6. Franzen, V., in (Olah, G. A., editor) "Friedel-Crafts and Related Reactions," Vol. 2, p. 413, New York, Interscience Publishers, 1964.
7. Vaiser, V. L., *et al.*, *Tr., Mosk. Inst. Neftekhim. i Gaz. Prom.*, **51**, 25 (1964); *Chem. Abstr.*, **62**, 14545 (1965).
8. Vaiser, V. L., *et al.*, *Proc. Acad. Sci. USSR Chem. Sect. (English Transl.)*, **132**, 505 (1960).
9. Furukawa, J., *et al.*, *Kogyo Kagaku Zasshi*, **60**, 803 (1957); *Chem. Abstr.*, **53**, 10121 (1959).
10. Arnett, E. M., *J. Org. Chem.*, **25**, 324 (1960).
11. Sauer, J. C., and Sausen, G. N., *J. Org. Chem.*, **27**, 2730 (1962).
12. Cywinski, N. F., *J. Org. Chem.*, **30**, 361 (1965).
13. Katagiri, T., *Bull. Inst. Chem. Res. Kyoto Univ.*, **41**, 159 (1963); *Chem. Abstr.*, **60**, 7905 (1964).
14. Nazarov, I. N., Kravchenko, N. A., and Klabunovskii, E. I., *Izv. Akad. Nauk SSSR Otd. Khim. Nauk*, 2171 (1959); *Chem. Abstr.*, **54**, 10822 (1960).
15. Nieuwland, J. A., and Vogt, R. R., "The Chemistry of Acetylene," New York, Reinhold Publishing Co., 1945.
16. Vartanyan, S. A., and Pirenyan, S. K., *Dokl. Akad. Nauk Armyan. SSR*, **23**, 23 (1956); *Chem. Abstr.*, **51**, 2522 (1957).
17. Khazhakyan, L. V., *Izv. Akad. Nauk Armyan. SSR Ser Khim. Nauk*, **10**, 77 (1957); *Chem. Abstr.*, **52**, 6907 (1958).
18. Vartanyan, S. A., Pirenyan, S. K., and Musakhanyan, G. A., *Dokl. Akad. Nauk Armyan. SSR*, **27**, 81 (1958); *Chem. Abstr.*, **53**, 6055 (1959).
19. Apotheker, D., (to E. I. du Pont de Nemours & Co.), U.S. Patent 2,934,575 (Apr. 26, 1960); *Chem. Abstr.*, **54**, 15992 (1960).
20. Garwood, R. F., Oskay, E., and Weedon, B. C. L., *Chem. Ind. (London)*, 1684 (1962).
21. Hagihara, N., *et al.*, *Bull. Chem. Soc. Japan*, **34**, 892 (1961); *Chem. Abstr.*, **56**, 4598 (1962).

Chapter Four

CYCLIZATION AND OLIGOMERIZATION OF ACETYLENES. METAL COMPOUND-ACETYLENIC COMPOUND COMPLEXES

1. THERMAL AND PHOTOCHEMICAL CYCLOOLIGOMERIZATION REACTIONS

1.1. Thermal

In 1866 Berthelot[1] reported thermal trimerization of acetylene to benzene. Nieuwland[2] discussed this reaction in detail in his book. Substituted acetylenes cyclize thermally with more or less ease, depending on the structure of the acetylene. Some of the cyclization products are substituted benzenes, but four-carbon and eight-carbon rings can also form. The benzene ring may be formed by reaction of acetylene with diacetylene (formed thermally) to form benzene, which reacts further.[3]

Hexafluoro-2-butyne gives both benzene and eight-carbon rings: at 280°, hexakis(trifluoromethyl)benzene and octakis(trifluoromethyl)bicyclo[4.2.0]-octa-2,4,7-triene form.[4] Dichloroacetylene trimerizes to hexachlorobenzene.[5] Fluoroacetylene gives 1,2,4-trifluorobenzene.[6] *t*-Butylfluoroacetylene spontaneously trimerizes at 0° in the liquid phase to give the unusual rearranged product 1,2,3-tri-t-butyl-4,5,6-trifluorobenzene.[7] Phenyldiethylaminoacetylene and two molecules of dimethyl acetylenedicarboxylate react directly via a cyclobutene ring intermediate to form 1-phenyl-2-diethylamino-3,4,5,6-tetrakis-(carboxymethyl)benzene.

Thermal cyclodimerization and addition reactions frequently lead to cyclobutane or cyclobutene rings. Alkene plus alkene gives cyclobutane rings, and alkene plus alkyne gives cyclobutene rings. Alkynes and allenes give saturated or unsaturated rings. Roberts and Sharts[8] reviewed this reaction recently.

Hexafluoro-2-butyne held at 320° for 31 hours gives dimer plus 7% tetramer:[9]

dimer:

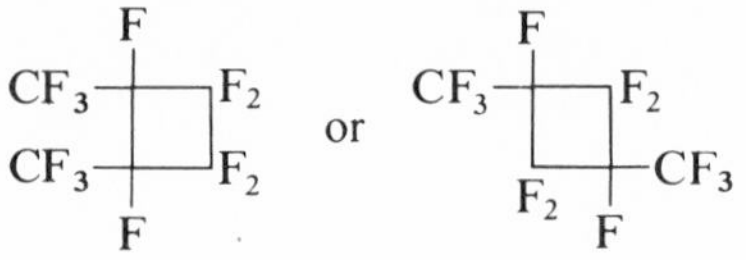

tetramer:

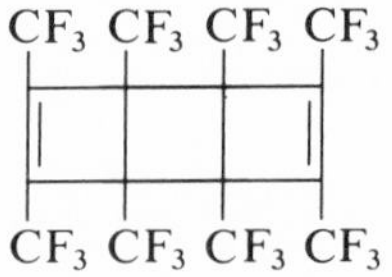

Nonconjugated enynes and diynes also rearrange and cyclize thermally. 6-Octen-1-yne forms 1-methylene-2-vinylcyclopentane.[10] 1,5-Hexadiyne cyclo-rearranges cleanly to 3,4-dimethylenecyclobutene in the gas phase:[11]

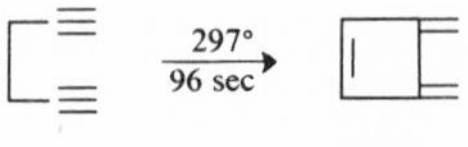

(80%)

Attempts to do liquid phase rearrangements resulted in explosions. 1,5-Heptadiyne cyclizes more slowly to form 1-methyl-3,4-dimethylenecyclobutene:

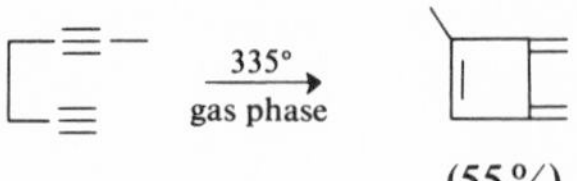

(55%)

2,6-Octadiyne is even slower. 3,4-Dimethyl-1,5-hexadiyne gives a 3,4-bis-(ethylidene)cyclobutene whose steric configuration is doubtful. The mechanism is also uncertain. A Cope rearrangement via the conjugated bis(allene) is one possibility.

When dimethyl acetylenedicarboxylate is heated at 120° for several hours, an unusual trimer forms, perhaps via 1,3-dipolar addition of dimethyl acetylenedicarboxylate to itself:[10a]

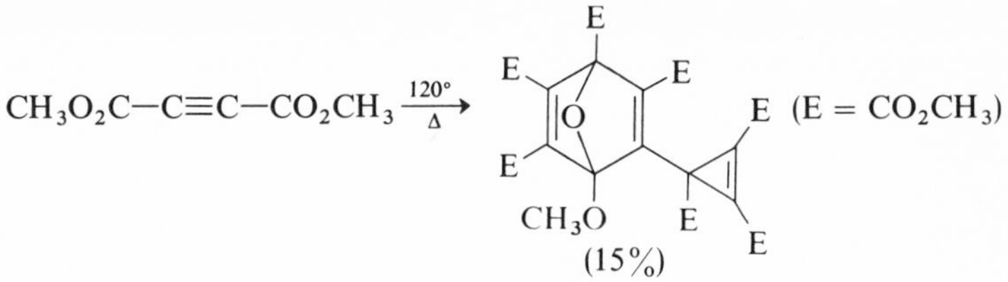

$$CH_3O_2C-C\equiv C-CO_2CH_3 \xrightarrow[\Delta]{120°} \quad (E = CO_2CH_3)$$

(15%)

1.2. Photochemical

o-Styrylstilbene photodimerizes to a molecule which has two cyclobutane rings. *o*-Di(phenylethynyl)benzene photodimerizes to form a different ring system, a bis(azulene), called "Verdene" because of its deep blue-green color:[12]

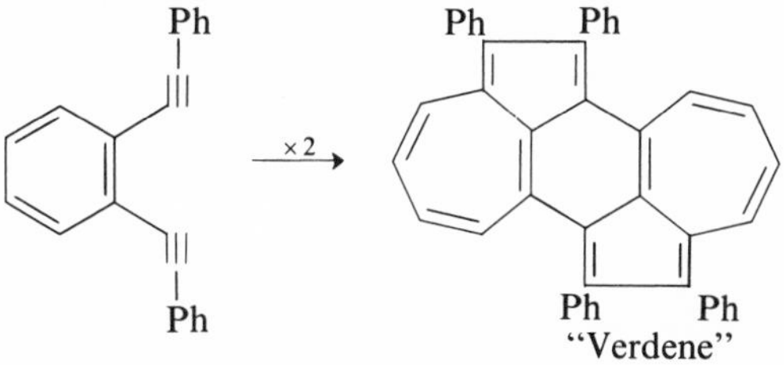

"Verdene"

Three molecules of acetylene and one molecule of disubstituted acetylene co-react to form cyclooctatetraene derivatives when irradiated by ultraviolet light:[13]

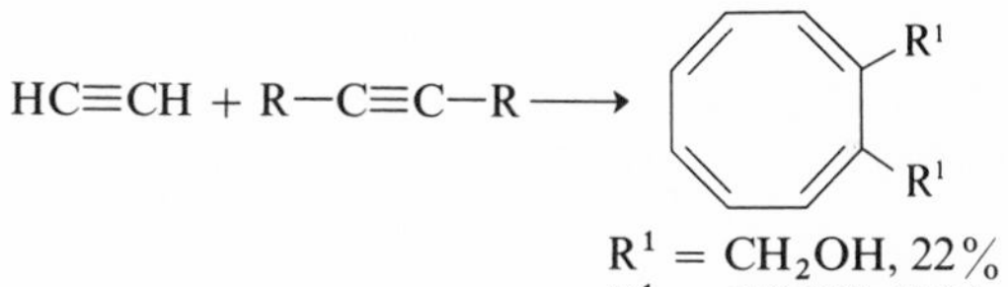

$$HC\equiv CH + R-C\equiv C-R \longrightarrow$$

$R^1 = CH_2OH, 22\%$
$R^1 = COOH, 10\%$

When 2,2'-di(phenylethynyl)biphenyl is heated at 120° in isooctane for 20 days, or is irradiated by an ultraviolet lamp for 2 minutes (in "Pyrex" at 25° in ethanol), intramolecular dimerization-cyclization occurs. The product is identified as 9-phenyldibenz[a,c]anthracene:[13a]

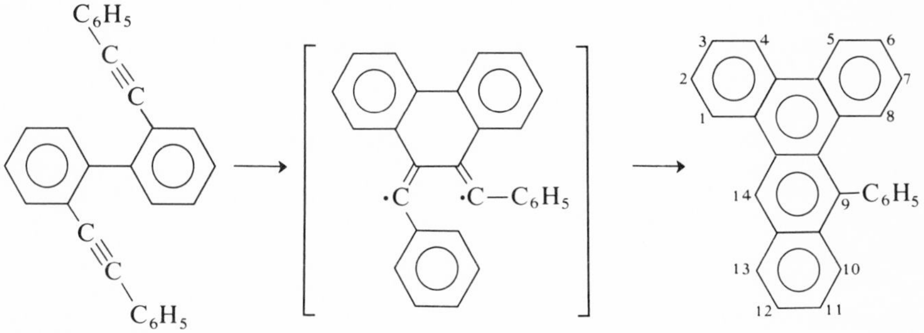

1,8-Diiodonaphthalene reacts with cuprous phenylacetylide in boiling pyridine to form two products:[14a]

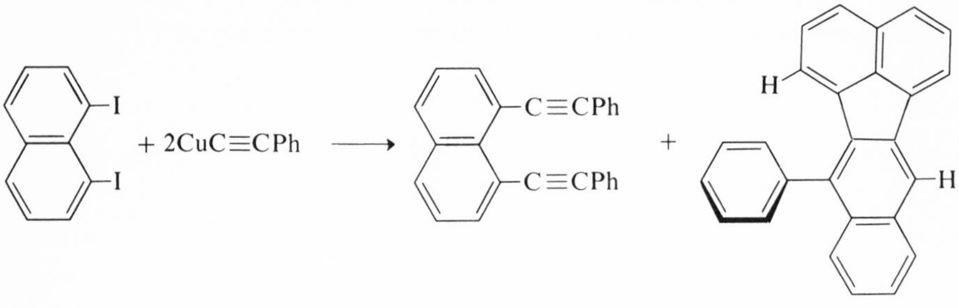

| | (1) | (2) |
| | (1) | (2) |

| Reaction Time (hr) | % Yield | |
	(1)	(2)
5	62	19.5
10	40	35
25	18	64

The increased yield of 1-phenyl-11,12-benzofluoranthrene is the result of a thermal intramolecular cyclization, similar to the one described above.

2. METAL-ACETYLENE COMPLEXES—CATALYZED CYCLIZATION AND OLIGOMERIZATION REACTIONS

2.1. Background

Metal-acetylene complexes are involved in most catalytic addition reactions of acetylenes. These include: cyclization to four-, five-, six-, seven- and eight-membered ring compounds, oligomerization, polymerization, and many other addition reactions. The keys to most future commercial acetylene reactions are undoubtedly among the numerous possible metal complexes which can catalyze reactions at low temperature and pressure. Complexes are discussed in connection with cyclization, oligomerization and polymerization (Chapter 5) because complexes have been studied more for these reactions than for any others.

Several recent reviews are available: In 1966, Reikhsfel'd and Makovetskii[14] published a discussion on mechanism and catalysis of cyclotrimerization of acetylenes, and Kritskaya[15] wrote a review of σ-bonded complexes of acetylenes with Group V, VI, VII and VIII metals. In 1960, Coates and Glocking[16] reviewed the chemistry of transition metal alkyls and aryls, and included some acetylene complexes.

Acetylene complexes can be classified as:

(1) Nast products, such as $K_6[Ni_2(C\equiv CPh)_8]$, in which the metal-acetylene bonds are normal σ bonds.[17]

(2) Complexes in which the triple bond is retained, as in $(PtCl_2 \cdot t\text{-}BuC\equiv C\text{-}t\text{-}Bu)_2$.[18]

(3) Complexes in which the acetylenic compound is a chelating ligand, as in $(Ph_3P)_2Pt(PhC\equiv CPh)$,[19] and

$$X-Cu\begin{array}{c} \diagup DPPA \diagdown \\ -DPPA- \\ \diagdown DPPA \diagup \end{array}Cu-X \qquad (DPPA = Ph_2PC\equiv CPPh_2)^{[19a]}$$

(4) Complexes in which the acetylene forms cyclic structures, such as cyclobutadiene, cyclopentadienone, benzene and cyclooctatetraene rings.

(5) Complexes in which the acetylenic compound forms linear structures with conjugated double bonds, which lead to linear oligomers or high polymers.

Metal-acetylenic compound complexes are discussed in the above order. Only the complexes which have shown catalytic application are emphasized, even though the others are interesting, and some may yet be found to be catalysts for certain reactions.

2.2. Nast Products

Nast has reported many interesting complex acetylenes, usually made by reacting potassium or sodium acetylide in ammonia with an inorganic complex of a transition metal. Since Nast products have not yet been reported as catalysts or as intermediates in catalytic reactions, only a few of them are listed here as examples:

Compound	Properties	Reference
$Ni(C\equiv CR)_2 \cdot 4NH_3$	Pyrophoric	20
$K_6[Ni_2(C\equiv CR)_8]$	Explosive	17
$Na_2[Mn(C\equiv CR)_4]$	Pyrophoric	21
$K_3[Mn(C\equiv CH)_6]$	Highly explosive	21
$K_3[Cr(C\equiv CH)_6]$	Explosive	22
$Ba_3[Cu(C\equiv CPh)_3]_2$	Not explosive	23
$K_3[Cr(CO)_3(C\equiv CR)_3]$	Not pyrophoric or explosive	24
$K_3[Mo(CO)_3(C\equiv CPh)_3]$	Pyrophoric	25
$K_3[W(CO)_3(C\equiv CPh)_3]$	Pyrophoric	25

2.3. Complexes in Which the Triple Bond Is Unchanged

Only a few examples of these complexes are included, because their catalytic significance has not been proved in very many cases. It seems likely that some of these complexes form in reactions involving transition metal compounds, and perhaps in reactions using copper, silver and mercury salts as catalysts (e.g., hydration, vinylation, Cadiot-Chodkiewicz coupling, etc.).

Complex	Reference

$$RC{\equiv}CR' + aq.\ Ag^+ \overset{Ag^+}{\rightleftharpoons} RC{\dot{\equiv}}CR'$$

26

t-Bu
|
C
|||···PtCl$_3$
C
|
t-Bu

18

$(Me_2\overset{\overset{\displaystyle OH}{|}}{C}C{\equiv}CCH_2CH_2NMe_2)_2$

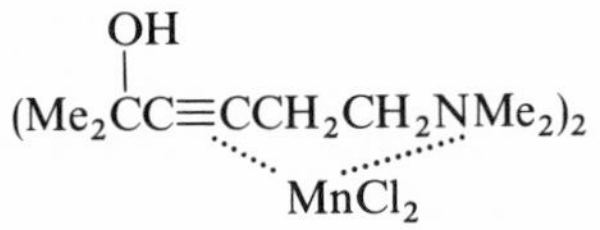

27

MnCl$_2$

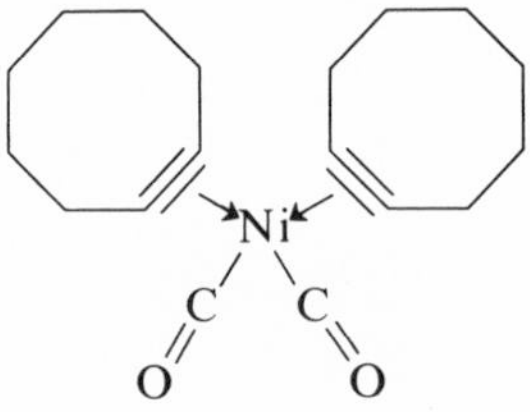

27a

2.4. Complexes in Which the Acetylenic Compound Is a Chelating Ligand

2.4.1. PLATINUM COMPLEXES

Chatt[19] in 1957 reported an extensive new series of acetylenic (ac) complexes of platinum: Pt(PPh$_3$)$_2$ac. The complexes form when an alcoholic suspension of cis-(PPh$_3$)$_2$PtCl$_2$ is reduced in the presence of an acetylene. They are much more stable to air and moisture than the olefin-platinum(II) complexes. One acetylenic compound can displace another from a complex in solution at room temperature:

$$Pt(PPh_3)_2ac + ac' \rightleftharpoons Pt(PPh_3)_2ac' + ac$$

Here, ac and ac' are different acetylenes. The rate of the exchange varies only with the concentration of the complex.[28] The more electron attracting the acetylene, the more stable is its complex.

Possible structures are:

$$\begin{array}{ccc}
\text{Ph}_3\text{P} & \overset{\displaystyle R}{\underset{\displaystyle R'}{\overset{\displaystyle C}{\underset{\displaystyle C}{\text{Pt} \cdots \; \big|\big|\big|}}}} & \text{and} \\
\text{Ph}_3\text{P} & &
\end{array} \qquad \begin{array}{c}
\text{Ph}_3\text{P} \\
\\
\text{Ph}_3\text{P}
\end{array} \; \overset{\displaystyle R}{\underset{\displaystyle R'}{\text{Pt}}}$$

                   (1)   (2)

Structure (1) is more likely, but each of the structures can exist and share some of the character of the other. Chatt[16,29] also prepared complexes Na(acPtCl$_3$), ac$_2$Pt$_2$Cl$_4$, and *trans*-(ac·piperidine·PtCl$_2$) from di-*t*-butylacetylene, *t*-butyl-isopropylacetylene, *t*-butylethylacetylene, and other acetylenes.

2.4.2. NICKEL COMPLEXES

Some nickel complexes are known in which an acetylene is a chelating ligand: (PR$_3$)$_2$Ni(C≡CR')$_2$ is more stable than the cobalt complexes.[30,31] (Cyclopentadiene)$_2$Ni$_2$(C≡CR') is pictured as:[32]

$$\text{(CPD)Ni} \overset{\displaystyle R \diagdown \quad \diagup R'}{\underset{\displaystyle \diagup \qquad \diagdown}{\overset{\displaystyle C-C}{\boxtimes}}} \text{Ni(CPD)}$$

The cyclopentadienylnickel-acetylene complex probably has the same structure.[33]

2.4.3. IRON COMPLEXES

Two ferrocenyl complexes have been reported:[34]

Fc—C≡CCO$_2$H   and   Fc—C≡C—Fc   (Fc = ferrocenyl)

2.4.4. METAL CARBONYL COMPLEXES

2.4.4.1. *Iron.* Simple complexes such as RC≡CR'Fe$_3$(CO)$_9$ are probably intermediates in the formation of more complicated complexes. The simple complexes are trinuclear organoiron complexes in which the acetylenic compound, for example diphenylacetylene, has replaced from Fe$_3$(CO)$_{12}$ two of the bridging carbonyls and one of the carbonyls of the *cis*-Fe(CO)$_4^-$ fragment.[35] The alkyne has four π electrons available for bonding, but has replaced three carbonyl groups with six bonding σ-type electrons, so the product complex is electron deficient and has two three-center bonds.

A hydrocarbon solution of 1,6-dichloro-2,4-hexadiyne and triiron dodecacarbonyl heated in the presence of zinc gives the hexapentaenyliron carbonyl complex.[36]

2.4.4.2. *Niobium.* Cyclopentadienylniobium tetracarbonyl and diphenyl-acetylene under ultraviolet light react to liberate carbon monoxide and form 41 % yield of a complex, cyclopentadiene(diphenylacetylene)Nb(CO)$_2$, which is unstable in solution in air.[37]

2.4.4.3. *Tungsten.* Tate[38] discovered a unique group of tungsten complexes: tris(hexyne)tungsten(0) monocarbonyl, tris(diphenylacetylene)tungsten(0) monocarbonyl, and tris(phenylmethylacetylene)tungsten(0) monocarbonyl. The ligands are arranged symmetrically around the tungsten and are doubly π complexed through the C≡C. These complexes form when nonterminal acetylenes react with acetonitriletungsten carbonyls. Similar reactions with corresponding chromium and molybdenum complexes are different—the acetylenic hydrocarbon cyclizes to hexasubstituted benzene or tetrasubstituted cyclopentadienone. Similar complexes of hexafluoro-2-butyne, $(CF_3C_2)_2W$-(acetonitrile), have been made. The complex $(PhC≡CPh)_3WCO$ forms when tungsten hexacarbonyl and diphenylacetylene are irradiated with ultraviolet light until 3 moles of carbon monoxide evolve.[39]

3. COMPLEXES HAVING CYCLIC STRUCTURES

3.1. Cyclobutadiene Rings

Transition metal complexes of cyclobutadiene have been postulated as inter-mediates to further products in some reactions of acetylenes.[40] The proof of such structures is difficult and has been reported in only a few cases.

3.1.1. TETRAPHENYLCYCLOBUTADIENE-PdCl$_2$

Diphenylacetylene in benzene cyclotrimerizes to hexaphenylbenzene in the presence of bis(benzonitrile)palladium chloride catalyst.[41] This is one of the easiest syntheses of hexaphenylbenzene. Bis(benzonitrile)palladium chloride and diphenylacetylene also form a little bis(cyclobutadiene) complex.[42] Bis(p-chlorophenylacetylene) forms 43 % yield of complex, which reacts with tributylphosphine to give octa-p-chlorophenylcyclooctatetraene.[42a]

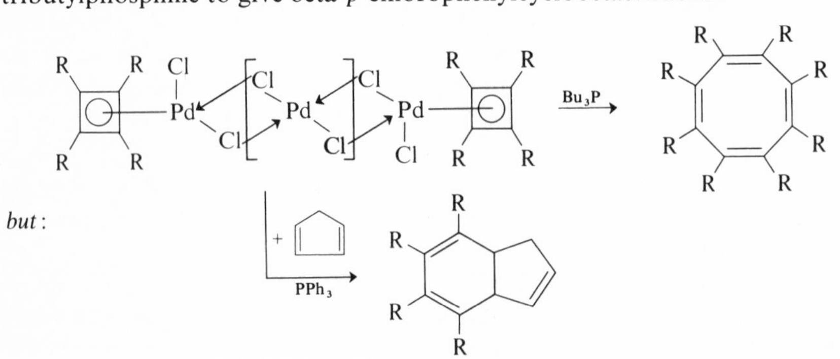

but :

(R = *p*-ClPh)

Diphenylacetylene reacts with palladium chloride in alcoholic HCl to form a bis(butadienyl)palladium chloride complex:[41]

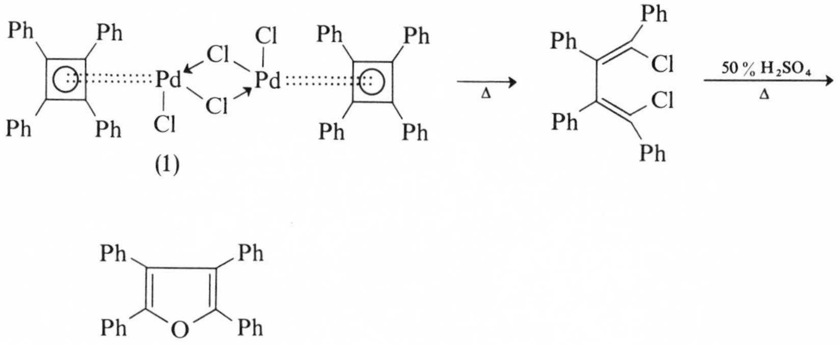

(1)

tetraphenylfuran

Complex(1) does not react with more diphenylacetylene to give hexaphenylbenzene, so it is not an intermediate in cyclotrimerization. The actual cyclotrimerization intermediate is probably a precursor to complex (1), pictured as complex (2):

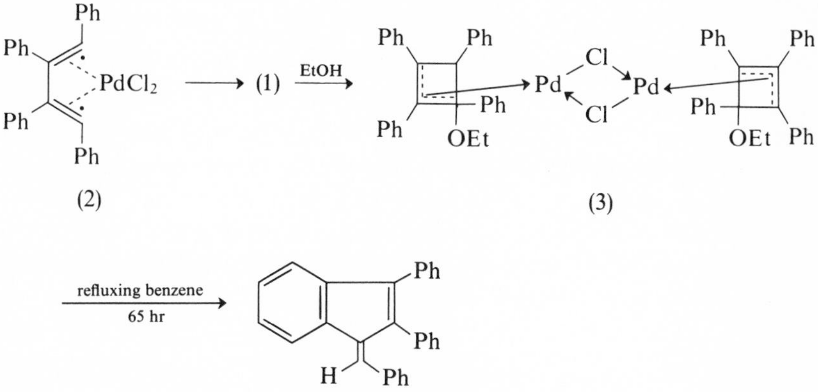

(2)   (3)

2,3,8-triphenylbenzofulvene

Complex (3) is the same as Malatesta's[43] "complex B."

Two moles of 1,2-bis(phenylethynyl)benzene form a similar complex with one mole of $PdCl_2$. The complex decomposes in hot dimethylformamide to give 6–8% yield of 5-phenylindeno[2.1-a]indene. Platinum chloride reacts with 1,2-bis(phenylethynyl)benzene to form a different kind of complex which decomposes to form the new indene in 85% yield.[43a]

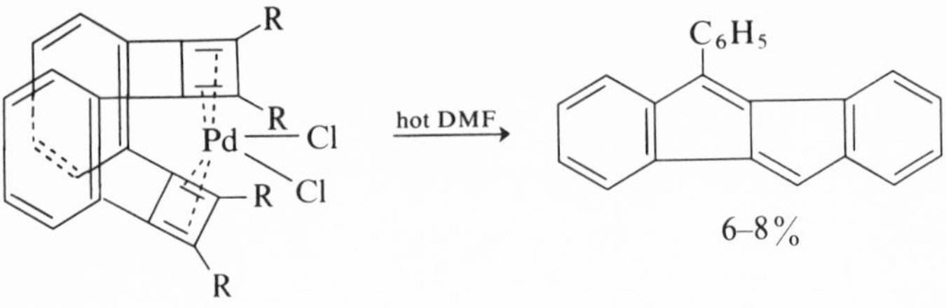

6–8%

R = C_6H_5

3.1.2. CYCLOBUTADIENES AS INTERMEDIATES IN THE HEPTATRIENENITRILE SYNTHESIS

Two molecules of acetylene react with a molecule of acrylonitrile or acrylate ester in the presence of complex catalysts such as $(Ph_3P)Cu[Ni(CN)_4]$ to form hepta-2,4,6-triene-1-nitrile or ester in 70–90% yield.[44] Schrauzer[45] reported that bis(acrylonitrile)nickel catalyzes this reaction at 15° and 13.5 atmospheres acetylene pressure. In tetrahydrofuran at 80° and 20 atmospheres, the products are resin, cyclooctatetraene and higher-boiling liquids. Schrauzer believes that the catalyst is $Ni(an)_2(C_2H_2)_2$, where an = acrylonitrile. Longuet-Higgins[40] suggested that cyclobutadiene complexes may be intermediates in various acetylene reactions, such as cyclotetramerization to cyclooctatetraene.

Sauer and Cairns[46] prepared ^{14}C-labeled 2,4,6-heptatrienenitrile by reacting labeled acrylonitrile ($\overset{*}{C}H_2{=}\overset{*}{C}HCN$) with acetylene; they then removed the terminal methylene group by ozonolysis. Most of the ^{14}C was nonterminal. This is strong evidence against the cyclobutadiene complex as an intermediate. Bieber[47] felt that this experiment was not conclusive. In another reaction, Sauer and Cairns[46] reacted acetylene with 2-butyne. The cyclobutadiene mechanism should give some p-xylene. They found no p-xylene, but did identify o-xylene.

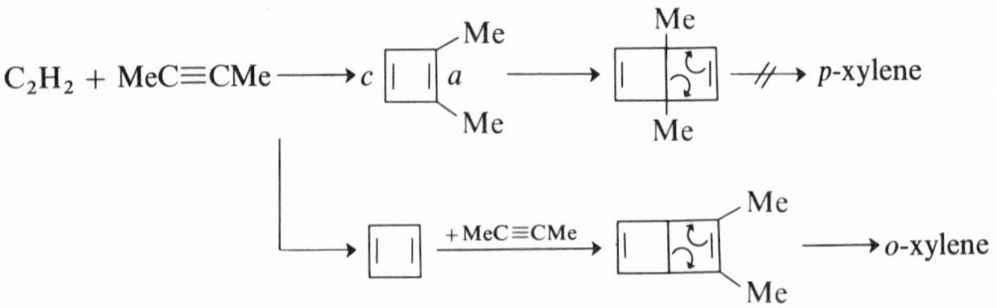

Bieber[47] also commented on this experiment. Failure to detect p-xylene was not crucial to the mechanism. Approach of acetylene at a, to give the intermediate for *p*-xylene, is least favored sterically, and the other approaches can give only *o*-xylene. He suggested using a large ratio of 2-butyne to acetylene and looking for durene as a product. This would form by attack by 2-butyne at *c*, the most sterically favorable approach.

3.1.3. TETRAPHENYLCYCLOBUTADIENE DIMER

Tsuji[48] reported a surprising cyclization reaction. When he heated 1 mole of phenylmagnesium bromide and 2 moles of diphenylacetylene for a short time, he obtained 10% yield of a mixture of products. Hexaphenylbenzene constituted two-thirds of the product, and the rest was tetraphenylbutadiene dimer, identified as:

Diphenylmagnesium gave the same products in lower yield, but anhydrous magnesium bromide did not catalyze the reaction under these conditions.

Drefahl[49] treated benzene solutions of diphenylacetylene at 60°–70° with $Et_3Al\text{-}TiCl_4$ in ratios from 1:1 to 3:1. The major product was hexaphenyl-benzene, maximum yield at 2:1 ratio. At higher ratios, the yield decreased drastically. At 1:1 ratio, he also obtained 4% yield of tetramer, formulated as Tsuji's product.

3.1.4. CYCLIZATION BY METAL CARBONYLS

Huebel[50] used $Mo(CO)_6$ or diglyme·$Mo(CO)_3$ to cyclize diphenylacetylene. The tetraphenylcyclobutadiene ring is ligand in the complex. Quite surprisingly, one of the products is pentaphenylcyclopentadiene, apparently from $2\frac{1}{2}$ moles of diphenylacetylene! Some cleavage apparently occurred. Another product is bis(pentaphenylcyclopentadienyl)molybdenum.

Hexafluoro-2-butyne reacts with cyclopentadienylnickel carbonyl and cyclopentadienyliron carbonyls to give complexes containing either the cyclobutadiene ring or the cyclopentadienone ring.[51]

3.1.5. CYCLOBUTADIENE COMPLEXES OF COBALT

Arylacetylenes, $ArC\equiv CR$ ($R = Si(Me)_3$, $SnPh_3$, $COCH_3$, CHO and CF_3) react with π-cyclopentadienyl-π-1,5-cyclooctadienecobalt(I) or with cobaltacene to form cyclobutadiene complexes.[51a]

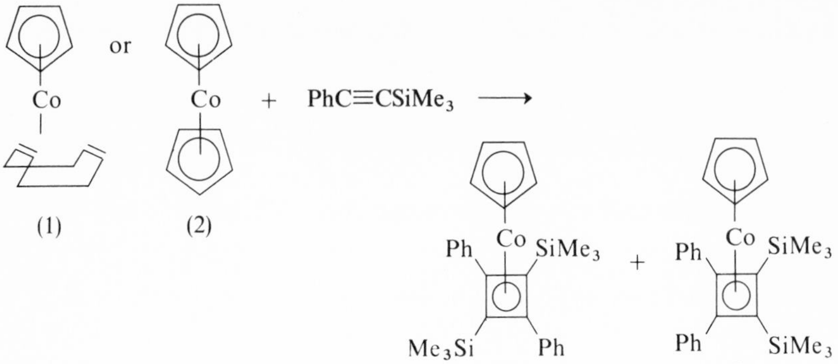

$[(1)$ gives 59% (3) + (4) in refluxing
xylene; (2) gives 38% yield in
refluxing diethylbenzene]

(1) reacts with $PhC{\equiv}CSnPh_3$ to form

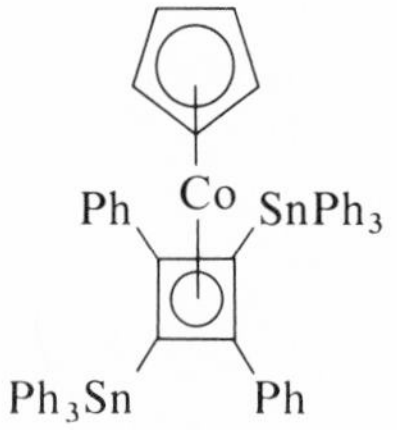

in 53% yield, but (2) does not react. Compounds (3) and (5) react with HCl
to give

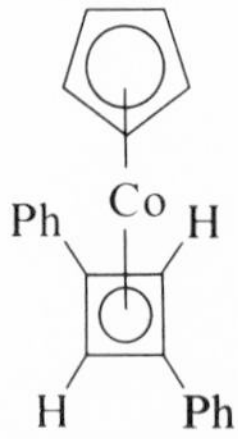

one of the first examples of a cyclobutadiene complex which contains unsubstituted positions in the cyclobutadiene ring.

4. COMPLEXES HAVING FIVE- AND SIX-MEMBERED RINGS

Most of the complexes of acetylenes and transition metal compounds have
five- or six-membered rings. Some complexes spontaneously decompose to
form lactone, cyclopentadienone, quinone or benzene derivatives. Some of the
same complexes also form linear oligomers.

4.1. Metal Carbonyl-Acetylenic Compound Complexes

Acetylenes react with metal carbonyls to give three kinds of complexes:

(1) The acetylene is bonded to a single metal atom, without ring formation.

(2) Acetylenic compound is a dinuclear bridging ligand.

(3) The acetylene reacts to form complexes which have homocyclic rings (such as cyclopentadiene, benzene, cyclopentadienone, quinone and tropone) or heterocyclic rings (such as lactones). Excess acetylenic compound may displace the final cyclic product, but frequently treatment with a reducing agent or a halogen (usually bromine) is necessary to liberate the cyclic product.

4.1.1. BONDING

Zerovalent metals in hexacarbonyl derivatives of alkenes and alkynes result from removal of the negative charge from the metal by back-donation of charges from a filled metal atom d orbital to vacant antibonding π orbitals of the ligands.[52] Bonding is thus more extensive than from σ bonding alone and causes the infrared carbonyl stretching frequencies to be slightly lower than in the parent carbonyl. Acetylenic ligands are stronger charge donors than ethylenic ligands. The carbonyl stretching frequencies of many complexes agree with this explanation.[52]

4.1.2. COBALT CARBONYLS

The relative rates of reaction of acetylenes with dicobalt octacarbonyl correlate with steric factors, but not with electronic effects.[53] Highly hindered acetylenes react more slowly than less hindered ones.

In one cobalt complex, $H_3Co_3(CO)_9C_2H_2$, an acetylene molecule is probably associated with three cobalt atoms, instead of the usual one or two. A definite structure cannot be assigned to this complex.[54]

Allenes with at least one hydrogen on allenic carbon form complexes with dicobalt octacarbonyl.[55] If a large excess of allene is added, an allene polymer is obtained.

4.1.3. DINUCLEAR BRIDGING COMPLEXES

Sternberg[56] prepared Reppe's[57] iron hydrocarbonyl-acetylene complex, and assigned this structure:

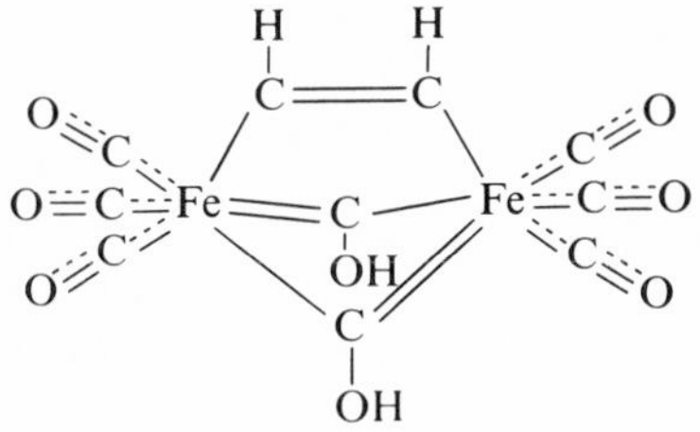

On the basis of the nonequivalence of the hydroxyl groups in esterification reactions, Clarkson[58] proposed another structure:

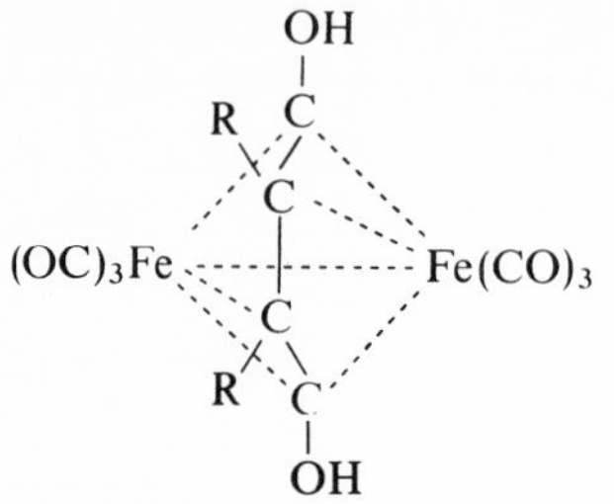

— = π electron framework

--- = π and d bonding which give aromatic character

Nickel carbonyl heated with hexafluoro-2-butyne forms a complex in 31% yield:[59] Four nickel atoms and three acetylenic bridging ligands are present in the complex.

4.1.4. NON-RING COMPLEXES

When $Fe(CO)_3(PhC{\equiv}CPh)_2$ is treated with $LiAlH_4$ in tetrahydrofuran and then hydrolyzed, 92% yield of 1,2,3,4-tetraphenyl-1,3-butadiene is obtained[60] (compare Section 3.1.4). SO_2Cl_2 reacts with $Fe_2(CO)_6(PhC{\equiv}CPh)_2$ to form cis-dibenzoylstilbene.[61]

4.1.5. COMPLEXES HAVING LACTONE RINGS

Sternberg[62] made a new complex by reacting dicobalt octacarbonyl with acetylene and CO, or by reacting the dicobalt hexacarbonyl-acetylene complex with CO. The lactone structure is supported by the reaction of dicobalt octacarbonyl, acetylene and CO in methanol to give a dilactone:[63]

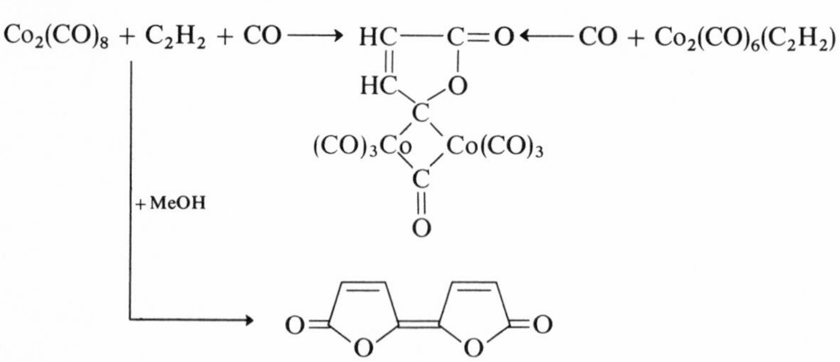

dilactone

Acetylene can react with dicobalt octacarbonyl, CO and methanol to form dimethyl succinate.[64] Sternberg showed that methyl acrylate is not intermediate and felt his new complex might be the intermediate to dimethyl succinate.

Some unique complexes form when acylcobalt tetracarbonyls react with substituted acetylenes.[65] The complexes have a lactone structure:

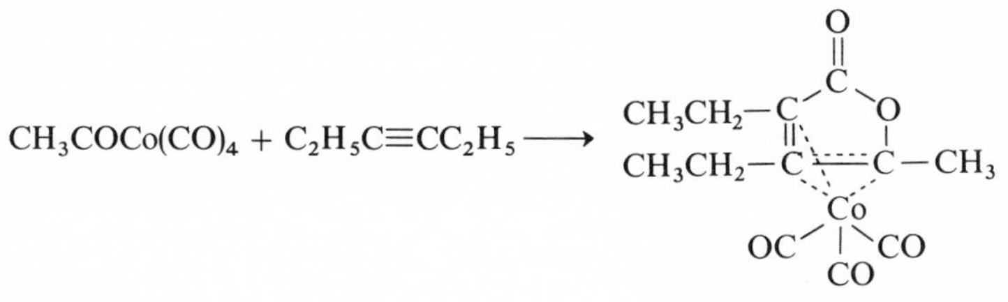

$$CH_3COCo(CO)_4 + C_2H_5C{\equiv}CC_2H_5 \longrightarrow$$

(74% of the product)

Hindered amines displace the ring structure from the complex and liberate free unsaturated lactone:

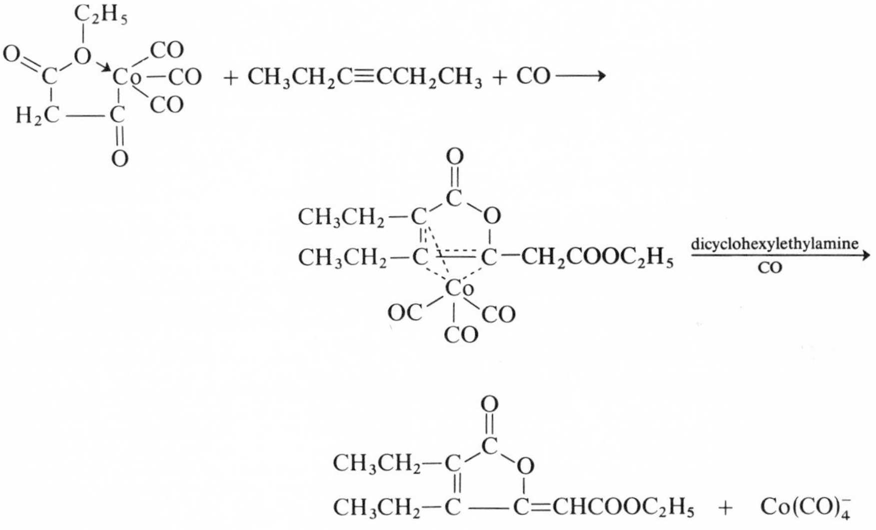

$$+ \; CH_3CH_2C{\equiv}CCH_2CH_3 + CO \longrightarrow$$

4.1.6. COMPLEXES HAVING CYCLOPENTADIENONE RINGS

Photochemical reaction of diphenylacetylene with iron pentacarbonyl gives tetraphenylcyclopentadienone as the main product and tetraphenylbutadiene as a minor product. Schrauzer[66] reported this reaction in 1959 and noted that the intermediate complex [(1) in the following equations] is the first known complex in which a cyclopentadienone bonded via π electrons to a transition metal. Products form under mild conditions: diphenylacetylene and iron pentacarbonyl in refluxing benzene were irradiated with ultraviolet light for

20 hours. The transformations leading to the observed products are outlined:

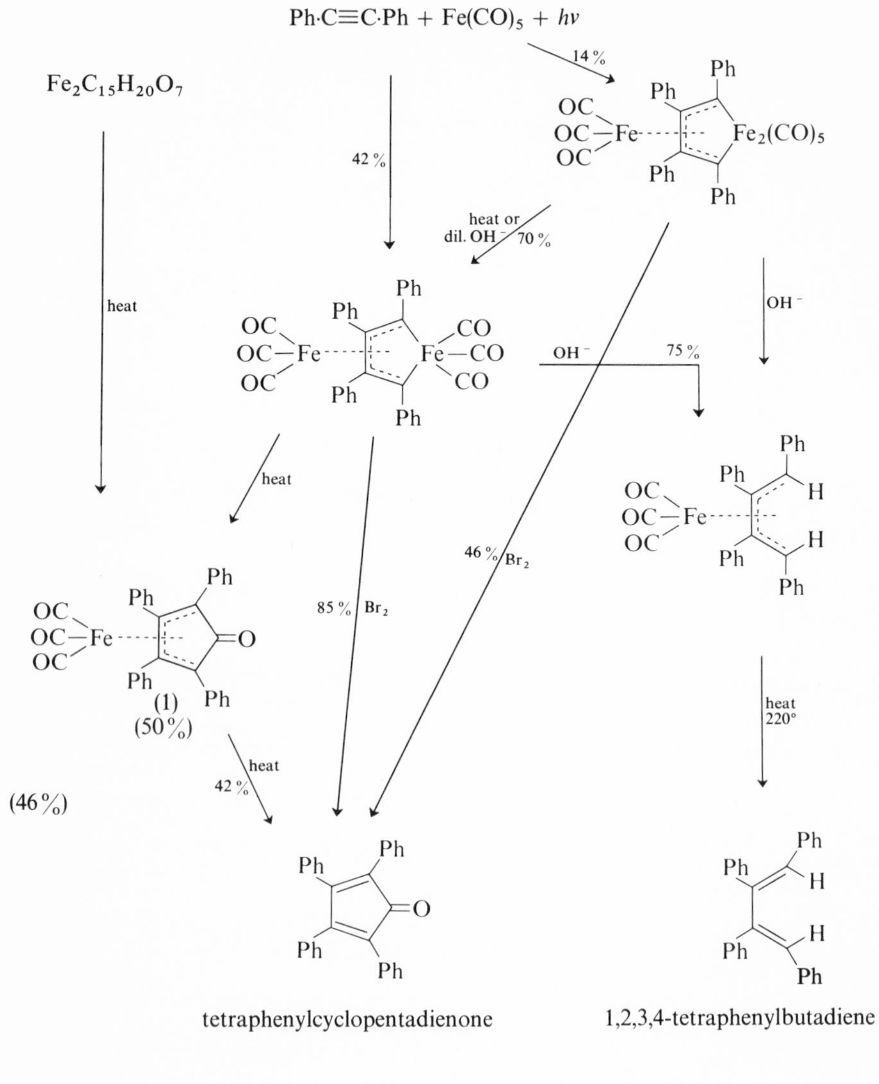

$$Ph \cdot C \equiv C \cdot Ph + Fe(CO)_5 + h\nu$$

tetraphenylcyclopentadienone   1,2,3,4-tetraphenylbutadiene

Iron pentacarbonyl reacts with 2-butyne, 2-pentyne and 3-hexyne in sunlight to give quinone complexes.[67] Phenylacetylene and diphenylacetylene form cyclopentadienone complexes.[66] Chromium hexacarbonyl, molybdenum hexacarbonyl, dimanganese decacarbonyl, and nickel tetracarbonyl do not form complexes with 2-butyne. Cyclopentadienylcobalt dicarbonyl and 2-butyne give this complex:

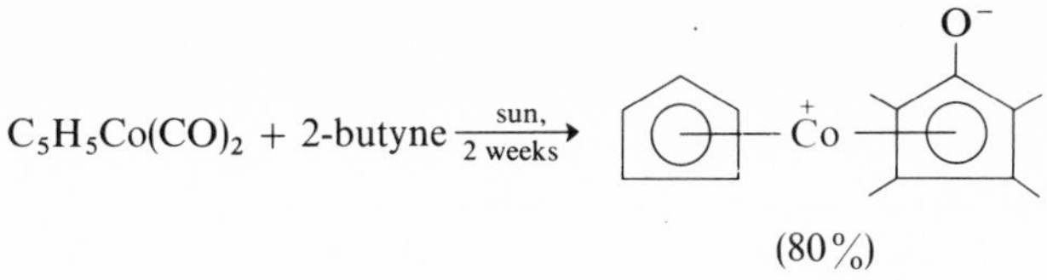

$$C_5H_5Co(CO)_2 + \text{2-butyne} \xrightarrow[\text{2 weeks}]{\text{sun,}}$$

(80%)

Manganese carbonyl anion and 2-butyne give duroquinone:

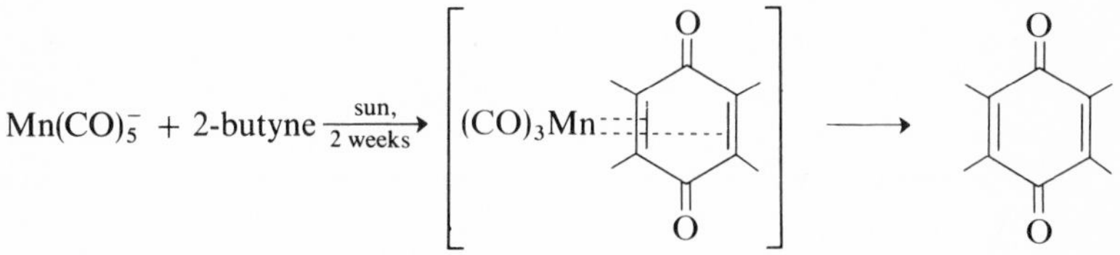

$$Mn(CO)_5^- + \text{2-butyne} \xrightarrow[\text{2 weeks}]{\text{sun,}} \left[(CO)_3Mn \cdots \right] \longrightarrow$$

Dahl and his co-workers[68] reported four new cyclopentadienone complexes from methyl phenylpropiolate and iron carbonyl. The structure and reactions of the most interesting of the new complexes are:

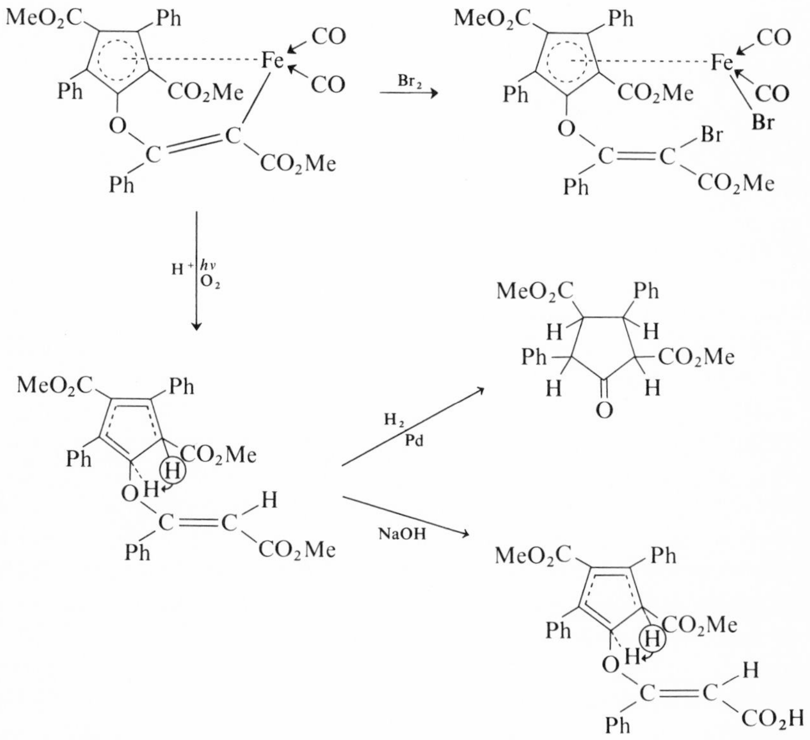

4.2. Nickel Carbonyl-Phosphine Complexes

4.2.1. BACKGROUND

Reppe[69] discovered that nickel carbonyl-triphenylphosphine complexes catalyze the trimerization of acetylenes to substituted benzenes. Acetylene in acetonitrile gave a product which was 88% benzene and 12% styrene. Propargyl alcohol in refluxing benzene gave quantitative yields a mixture of 1,2,4- and 1,3,5-trimethylolbenzene, but butynediol did not trimerize.

The literature reports since Reppe's paper in 1949 are sometimes contradictory and confusing. For example, Kleinschmidt[70] claimed that Reppe's catalyst trimerized butynediol. Rose[71] found that some reactions gave only 1,2,4-trisubstituted benzenes, while others gave only the 1,3,5-derivatives. Rose could not trimerize 3-methyl-1-butyn-3-ol, but McKeever could.[72] Harris[73] trimerized hexafluoro-2-butyne. Sauer[46] made o-xylene from 2-butyne and acetylene, but no p-xylene. Cairns[74] reported linear polymerization of acetylene with nickel carbonyl-phosphine catalysts. He also reacted acetylene and acrylonitrile to form 2,4,6-heptatrienenitrile.

Since trimerizations can be very sensitive to poisons,[75] some of the discrepancies may be due to poisoning effects.

4.2.2. SCOPE OF THE OLIGOMERIZATION-CYCLOOLIGOMERIZATION REACTIONS CATALYZED BY NICKEL CARBONYL-PHOSPHINE COMPLEXES

Meriwether and his co-workers[76] studied the scope of the cyclization-oligomerization reaction, and the products from it, to try to resolve some of the confusion. Their reports are the most comprehensive published so far, and their results must be discussed in some detail.

Meriwether used a simple procedure: the acetylenic compound and catalyst, $Ni(CO)_2(Ph_3P)_2$, were dissolved in benzene, cyclohexane or ethanol, about 250 moles of solvent per mole of acetylenic, and the solution was refluxed for 1–24 hours. For gaseous or low-boiling acetylenes, the reactor was a stainless steel autoclave.

Most monosubstituted acetylenes react. Disubstituted acetylenes are usually inert, but butynediol does react to form hexamethylolbenzene in 78% yield. Initially, the solutions are pale yellow, but suddenly turn dark red or brown as the reaction starts. "This was accompanied by a violent exotherm which sometimes carried the reaction mixture out of the reflux condenser."[76] The acetylene usually reacts completely, but maximum product isolation is generally around 35% because much tar forms. Catalyst is consumed, and CO is evolved. Small amounts of carbonyl products are frequently detected. Product trimers are 1,2,4- and 1,3,5-trisubstituted benzenes. The linear products are "Nieuwland-type condensation"[2] enynes. Meriwether is the first to report linear homooligomers from substituted acetylenes and nickel carbonyl-phosphine catalysts. Monoalkylacetylenes give the greatest variety of linear products, with oligomers up to heptamers definitely identified, and chains of perhaps 20 monomer units indicated.

With 1-alkynes, the amount of aromatic trimer decreases as the size of the alkyl group increases, but total conversion is constant. The proportion of 1,2,4- and 1,3,5-trisubstituted benzene was 50–50 to 65–35, the unsymmetrical isomer usually being favored, but no real trend emerged. Aromatization apparently has more severe steric requirements than linear oligomerization does.

A study of 35 acetylenic compounds allowed these conclusions:[76]

(1) The linear oligomers of monosubstituted acetylenes are a new class of products.

(2) The reaction is general for monosubstituted acetylenes. The qualitative order of reactivity is: esters, ethers, ketones, and vinylacetylenes > arylacetylenes > acetylenic alcohols > higher alkynes > lower alkynes, acetylene ≫ acetylenic acids, halides, and some other derivatives.

(3) Some monosubstituted acetylenes are substantially inert: amides, nitriles, halides and some highly hindered acetylenes. Also, all disubstituted acetylenes are inert except butynediol and 2-butyn-1-ol.

(4) Monomers which are similar form co-oligomers.

(5) The more active acetylenes usually give aromatic products, and the less active generally give a large proportion of linear oligomer.

4.2.3. STRUCTURE OF LINEAR OLIGOMERS

Meriwether[77] used GLC and infrared analysis to identify and quantitate the 1,2,4- and 1,3,5-trialkylbenzenes. Infrared and ultraviolet spectra show that the nonaromatic products are linear, substituted vinylacetylenes, butadienylacetylenes, etc. Infrared indicates: internal acetylenic bonds, internal *trans*-olefin, no internal *cis*-olefin, and some terminal olefin. Dimers and higher oligomers are distinguished by the position of the *trans*-olefin band (956 ± 4 versus $970 \pm 5\,cm^{-1}$). The shift is probably caused by conjugation of the terminal olefin with a triple bond in the dimers and with a double bond in the higher oligomers.

Dimers can have only three structural and geometrical isomers:

(3) $RC{\equiv}C$ ... $\xrightarrow{3H_2} R(CH_2)_2\overset{\displaystyle CH_3}{CH}{-}R$ (5)

Structure: $RC{\equiv}C$ and H on $C{=}C$, R and H — $\xrightarrow{3H_2} R(CH_2)_2\underset{}{CH}{-}R$ with CH_3 on the CH, (5)

GLC of dimers from 1-butyne, 1-pentyne and 1-heptyne gives only two components each. Infrared indicates that these are mixtures of (1) and (3).

Linear trimers are more complex, since there are four possible structural isomers, each of which can have at least two geometrical isomers. Trimers from these same 1-alkynes do not separate cleanly from aromatics in GLC, but hydrogenated trimers separate well. Complete hydrogenation of the four possible structural isomers should give only three saturated hydrocarbons:

(6) $RC{\equiv}CC{=}CHCH{=}CHR$
     with R below

(7) $RC{\equiv}CCH{=}CCH{=}CHR$
     with R below

$\begin{Bmatrix} cis\text{-}cis \\ cis\text{-}trans \\ trans\text{-}cis \\ trans\text{-}trans \end{Bmatrix} \xrightarrow{4H_2} R(CH_2)_2\underset{R}{CH}(CH_2)_3R$ (10)

(8) $RC{\equiv}CC{=}CHC{=}CH_2$ (R, R below)   ($cis\text{-}trans$) $\xrightarrow{4H_2} R(CH_2)_2\underset{R}{CH}CH_2\underset{R}{CH}CH_3$ (11)

(9) $RC{\equiv}CCH{=}C{-}C{=}CH_2$ (R R below)   ($cis\text{-}trans$) $\xrightarrow{4H_2} R(CH_2)_3\underset{R}{CH}\underset{R}{CH}CH_3$ (12)

Butyne trimer hydrogenates to a single component, and 1-pentyne and 1-heptyne trimers give only two nonaromatic peaks after hydrogenation. The hydrogenated trimer of 1-pentyne is represented by (10) and (11). Structure (9) is probably not present in the trimers.

The terminal olefinic unit is probably incorporated by a cis addition to the triple bond. The central monomer unit probably enters intact by cis addition of the catalyst-polymer chain to its triple bond.

Higher oligomers can be represented by these structures:

$RC{\equiv}C\left[\begin{array}{c} C{=}C \\ R \ H \end{array}\; \begin{array}{c} R \ H \\ C{=}C \end{array}\right]_{n/2} C{=}CHR$ and $RC{\equiv}C\left[\begin{array}{c} C{=}C \\ R \ H \end{array}\; \begin{array}{c} R \ H \\ C{=}C \end{array}\right]_{n/2} C{=}CH_2$ (with H, R)

4.2.4. KINETICS AND "POLYMERIZATION" MECHANISMS

Meriwether and co-workers[78] studied the effects of solvent, temperature, gas sweeping, acetylenic:catalyst ratio and catalyst composition. They also determined the kinetics, effect of pretreatment of catalyst and deuterium isotope effects. Two representative acetylenes were used: ethyl propiolate, which reacts very rapidly to form only aromatics, and 1-heptyne, which reacts more slowly to form mostly linear dimers and trimers. The catalyst was $Ni(CO)_2(PPh_3)_2$, except in experiments to determine the effect of catalyst structure.

Sixteen different solvents were used for the ethyl propiolate cyclization. No direct relation between rate and solvent polarity is apparent, so the reaction is not a simple anionic or cationic process. For 1-heptyne, conversions are greater than 95% in cyclohexane, benzene, acetonitrile or methanol. Reactions are negligible at 25° and at 75° in unstirred solutions. If solvent cyclohexane is refluxed or stirred at 75°, 1-heptyne reacts very rapidly. Sweeping with CO inhibits the reaction, presumably by preventing loss of CO from the initial complex to form the active catalyst. Oxygen inhibits the reaction, but nitrogen and carbon dioxide do not change the rate. Changing the ratio of acetylenic compound to catalyst from 1000:1 to 10,000:1 decreases the rate significantly.

With different catalyst compositions, the length of the induction period and the ultimate activity are unrelated. For the dicarbonyl-diphosphine catalysts $Ni(CO)_2(PR_3)_2$, the activity varies with R: $C_2H_4CN > Ph > H > OEt > n$-alkyl > OPh > Cl. This suggests that the phosphine ligands are not lost from nickel in the formation of active catalyst. The phosphine ligands are essential: nickel carbonyl alone is inactive. Free phosphines alone are not catalysts either. Nickel tricarbonyl-monophosphine gives slightly better conversions than nickel dicarbonyl-diphosphine. Monocarbonyl-triphosphine and nickel-tetra-kisphosphine complexes are inactive for 1-heptyne reactions. Some tetrakis-phosphine-nickel complexes are active for the ethyl propiolate reaction, however.[79] Ethyl propiolate is so active it gives nearly quantitative aromatiza-tion with every catalyst tried. With phenylacetylene, linear trimerization is favored by the cyanoethylphosphine complexes, and aromatization is favored by the triphenylphosphine complex.

During the first 50 minutes of the reaction of 1-heptyne, using $Ni(CO)_2(PPh_3)_2$ as catalyst, in refluxing cyclohexane, the 1-heptyne is largely consumed. The amount of trimer increases and then levels off, and the amount of dimer increases very slowly and then also levels off. At high 1-heptyne:catalyst ratios, the rate is independent of 1-heptyne concentration up to 75–80% conversion. At lower ratios, the rate is proportional to 1-heptyne concentration for more than 60% of the reaction. Significantly, the addition of fresh monomer to spent catalysts causes resumption of activity, and marked decline in activity is noted only after several such cycles.

The observed kinetics and product distributions eliminate a stepwise process in which trimer forms by reaction of monomer with dimer, or with complexed dimer in equilibrium with free dimer.

With deuterated 1-heptyne, after the induction period rapid deuterium exchange occurs, and deuterium substitution decreases the oligomerization rate. This indicates that hydrogen atom transfer is involved in the rate-determining step. Since only the *active* catalyst causes deuterium exchange, this supports the conclusion that the active catalyst is different from the original complex.

Ethyl propiolate aromatizes so fast that kinetic data are impossible. Phenylacetylene reacts at a moderate rate and gives isolated product which is 30% aromatic and 70% linear trimer. The relative rates of formation of aromatic and linear products are nearly constant, and this may mean that a common active catalyst is involved. Deuterium exchange and the polymerization studies indicate that cyclization and linear oligomerization have a common step which involves a hydrogen transfer.

Catalysts can be pretreated with inactive acetylenes, and then used to oligomerize and cyclize reactive acetylenes with no induction period. The inactive acetylene is unchanged. Pretreatment must be done at the reflux temperature of the solvent. The linear 1-pentyne trimer, a disubstituted acetylene, is an active pretreating agent. This explains why spent linear polymerization reaction mixtures can polymerize additional fresh monomers. However, the aromatic products do not stabilize the active catalyst for aromatization of freshly added monomer.

The active forms of the catalysts are so unstable that they cannot be isolated. Their ultraviolet spectra are similar to spectra of a series of bisphosphine-platinum-acetylene complexes made by Chatt,[19] so they may have a similar structure:

$$
\begin{array}{c}
R_3P \diagdown \quad \vdots \quad \diagup CR' \\
\qquad Ni \diagup \!\!\! \diagdown \; \| \\
R_3P \diagup \quad \vdots \quad \diagdown CR'
\end{array}
$$

The mechanism must account for these facts about the catalyst: It is a nickel-phosphine-acetylenic material, which has its original phosphine ligands but has lost all of the CO ligands, and thus has two of the original four coordination sites ready for bonding with the acetylenes. It is in reversible equilibrium with the monomer and the original nickel carbonyl-phosphine complex. It has a high "turnover" number, and retains its original structure and activity after each reaction sequence. It catalyzes the exchange of acetylenic hydrogens between different monosubstituted acetylenes. The catalyst may form this way:

$$
\begin{array}{ccc}
\begin{array}{c}
R_3P \diagdown \quad \diagup CO \\
\quad Ni \\
R_3P \diagup \quad \diagdown CO
\end{array}
+ R'C\!\equiv\!CH \rightleftharpoons 2CO +
\begin{array}{c}
R_3P \diagdown \quad \diagup CR' \\
\quad Ni \quad \| \\
R_3P \diagup \quad \diagdown CH
\end{array}
& \rightleftharpoons &
\begin{array}{c}
R_3P \diagdown \quad \diagup C\!\!\nearrow^{CR'} \\
\quad Ni \\
R_3P \diagup \quad \diagdown H
\end{array}
\\
& (1) & (2)
\end{array}
$$

(1) and (2) are planar, and have a vacant p_z orbital for π bonding with another acetylene molecule. The catalyst (2) then reacts with another acetylenic compound to form product:

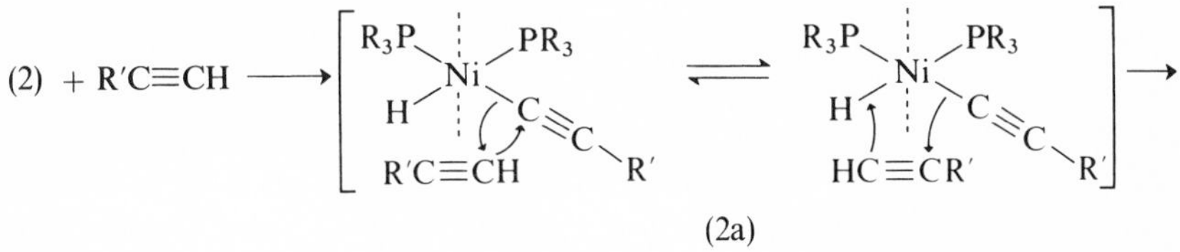

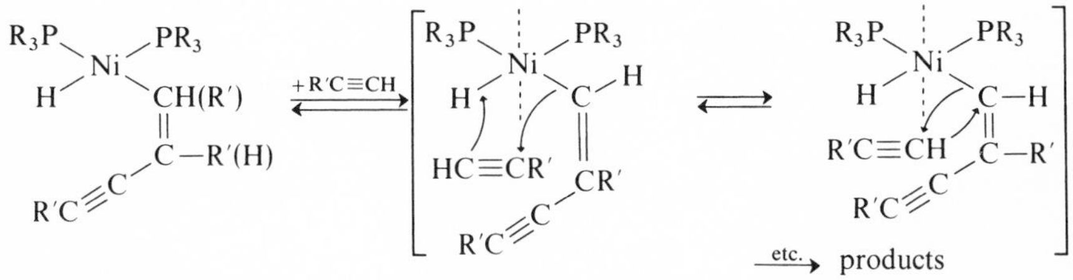

(2a)

etc.→ products

The isolated products are pictured as forming by insertion reactions which are similar to those proposed for olefin polymerization on Ziegler catalysts.

A mechanism for the aromatization is incorporated in this sequence by assuming that there is a favored conformation of the intermediate formed next in the above scheme, in which the carbon atoms in positions 1 and 6 of the butadienylacetylene group can get within bonding distance. Concerted hydrogen transfer and ring closure give an aromatic product:

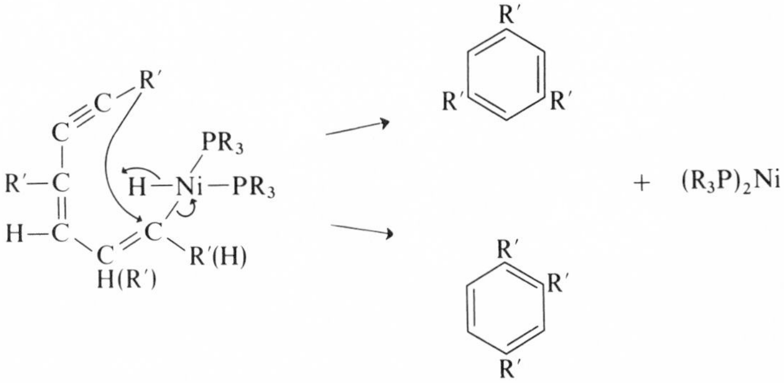

$+$ $(R_3P)_2Ni$

The predominance of 1,2,4-trisubstituted benzenes over 1,3,5-compounds is explained by the stability and relative amounts of the two possible structures of the complex.

The formation of hexasubstituted benzenes from some disubstituted acetylenes noted by Meriwether, and more generally by Schrauzer[80] in reactions catalyzed by bis(acrylonitrile)nickel-phosphine, might best be explained by this similar mechanism:

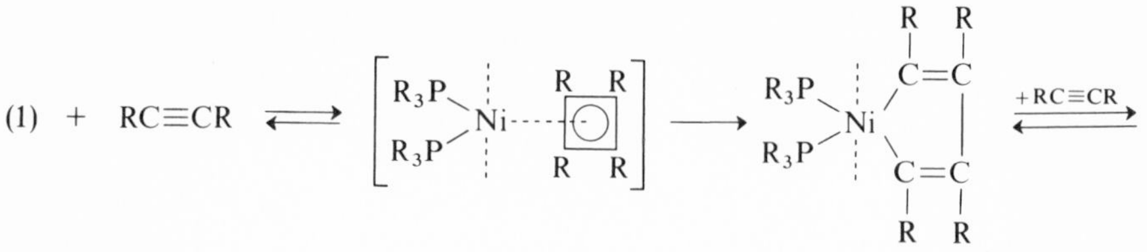

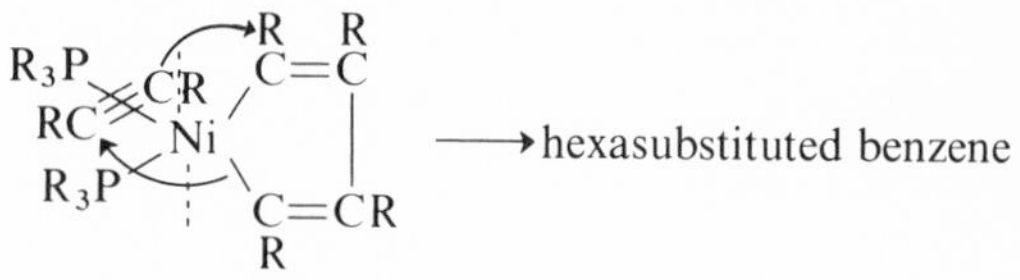

The formation of condensed aromatics from the unconjugated diacetylenes may go similarly through a complex such as (3):

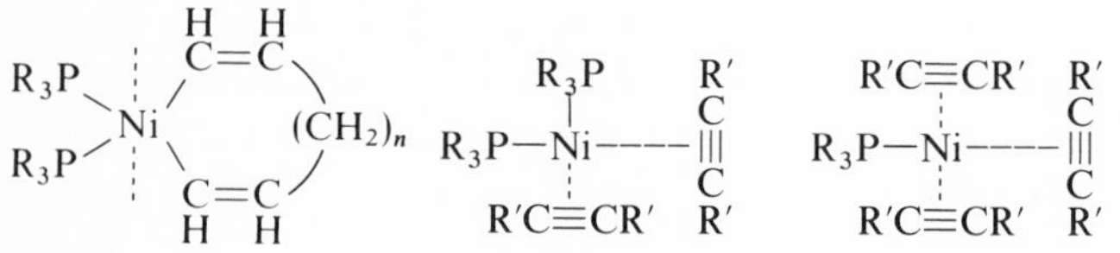

Catalysts which have only π-bond connections between the metal and the acetylene are unlikely.[4,5] They do not allow prediction of the high steric specificity usually observed, there is no way of showing activation of the third acetylenic monomer, and R_3PNi probably would not survive more than one cyclization series.

4.3. Bis(acrylonitrile)nickel Catalysts

Norbornadiene reacts with metal carbonyls much like acetylenes do. For example, nickel carbonyl and norbornadiene under Reppe acrylate ester conditions give bicycloheptenecarboxylic acid ester. Schrauzer[81] found that bis(acrylonitrile)nickel catalyzes the dimerization of norbornadiene at 50–120°. The addition of triphenylphosphine stabilizes the complex intermediate and increases the yield of dimer. Bis(acrylonitrile)nickel converts diphenylacetylene into hexaphenylbenzene and 2,3,4,5-tetraphenylbenzonitrile.[80] Thus, the norbornadiene and diphenylacetylenes are roughly parallel.

Catalyst activity is due to some specific property of nickel complexes. Both Ni(0) in bis(acrylonitrile)nickel and Ni(II) in bis(triphenylphosphine)dicyano-nickel are active. The latter is very easy to make, stable in air and, therefore, the catalyst of choice.

Schrauzer[82] noted that the only previous noncatalytic addition of an acetylene to norbornadiene was with highly activated hexafluoro-2-butyne.[83] This reaction gave 78% yield of adduct after 6 hours at 150°. In contrast, the reaction of norbornadiene with dimethyl acetylenedicarboxylate goes at 120°. Diphenylacetylene is slower, even at 160°. The products from these acetylenes are:

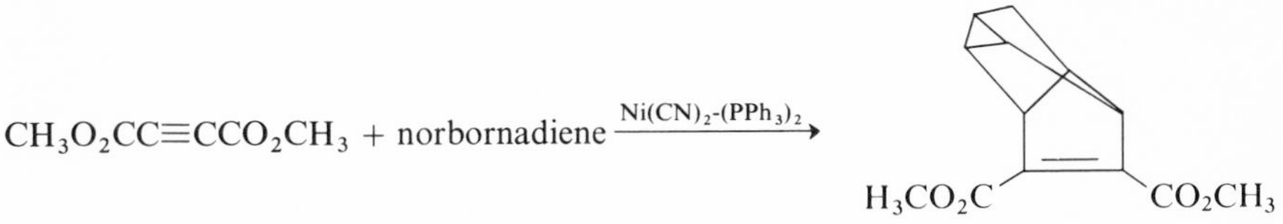

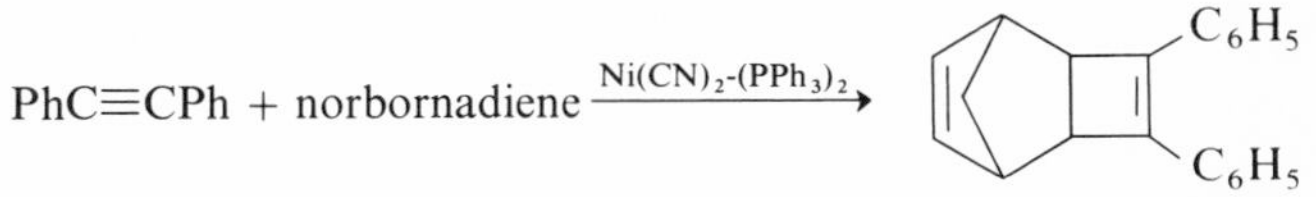

$$CH_3O_2CC{\equiv}CCO_2CH_3 + \text{norbornadiene} \xrightarrow{\text{Ni(CN)}_2\text{-(PPh}_3)_2}$$

$$PhC{\equiv}CPh + \text{norbornadiene} \xrightarrow{\text{Ni(CN)}_2\text{-(PPh}_3)_2}$$

Acetylene does not react thermally with norbornadiene. In the presence of the nickel catalysts, two molecules of acetylene react with one molecule of norbornadiene:

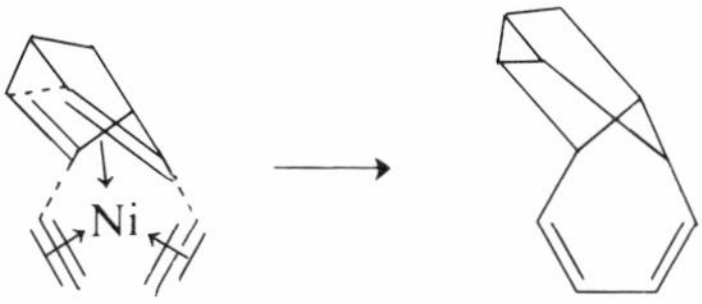

Phenylacetylene trimerizes to 1,2,4-triphenylbenzene under these conditions and also gives some linear polymer. Propargyl alcohol, propargylic acid ester, 1-hexyne and 1-heptyne do not give definite norbornadiene adducts. 2-Butyne, butynediol and butynediol diacetate are recovered unchanged.

4.4. Metal Carbonyl Cyclotrimerization Catalysts

In 1960, Hübel and Hoogzand[84] reported extensive experiments using metal carbonyls to cyclize alkynes and disubstituted acetylenes to benzenes. In the first experiments, they heated neat diphenylacetylene with $Fe_3(CO)_{12}$. At 260–280° a vigorous reaction occurred, forming hexaphenylbenzene in 75% yield. The activity of other metal carbonyls and their complexes varied (Table 4-1). Other acetylenes also trimerized to benzenes (Table 4-2).

The relative rates of trimerization of some acetylenes are: 3-hexyne > diphenylacetylene = di(p-chlorophenyl)acetylene > dimethyl acetylenedicarboxylate. If the reaction time is extended too long, the reactive acetylenes polymerize to resins. Co-trimerization of a slow and a fast acetylene gives a

TABLE 4-1
Cyclotrimerization of Diphenylacetylene
by Metal Carbonyls[84]

Metal Carbonyl	Temperature (°C)	% Yield of Hexaphenylbenzene
$Fe_3(CO)_{12}$	260–280	75
$Fe_2(CO)_9$	250	25
$Fe(CO)_5$	270	20
$Fe(CO)_4Hg$	285	50
$Fe_2(CO)_6(C_6H_5C_2C_6H_5)_2$	270	60
$Fe_2(CO)_7(C_6H_5C_2C_6H_5)_2$	250–270	15
$Fe_3(CO)_8(C_6H_5C_2C_6H_5)_2$	260–280	50
$(Tetracyclon)Fe(CO)_3$	280	1
$[C_5H_5Fe(CO)_2]_2$	270	6
$Co_2(CO)_8$	280	60
$[Co(CO)_4]_2Hg$	270	70
$Co_2(CO)_6C_6H_5C_2C_6H_5$	150	70
$Ni(CO)_4$	260	5
$Mn_2(CO)_{10}$	270	55
$Mo(CO)_6$	270	50
$W(CO)_6$	270	15

mixture of products, and two acetylenes of equal reactivity give the statistically predicted trimers.

The most striking feature of these catalysts is that they trimerize acetylenes in a completely ordered way. For example, methylphenylacetylene gives only one product, 1,2,4-trimethyl-3,5,6-triphenylbenzene. Meriwether's[76] catalysts give mixtures of trisubstituted benzenes. With various cobalt carbonyl catalysts, the ordered substituted benzenes given in Table 4-3 were obtained.

Another example of ordered cyclotrimerization induced by dicobalt octa-carbonyl is the formation of 1,2,4-triferrocenylbenzene from ferrocenylacetylene. Refluxing a dioxane solution for 3 hours gives 64% yield. Diferrocenylacetylene

TABLE 4-2
Cyclotrimerization of Other Acetylenes by Metal Carbonyls[84]

Acetylene	Catalyst	Temperature (°C)	% Yield
$ClC_6H_4C{\equiv}CC_6H_4Cl$	$[Co(CO)_4]_2Hg$	270	45
$ClC_6H_4C{\equiv}CC_6H_4Cl$	$Fe_3(CO)_{12}$	320	40
$ClC_6H_4C{\equiv}CC_6H_4Cl$	$Mo(CO)_6$	270	40
$ClC_6H_4C{\equiv}CC_6H_4Cl$	$W(CO)_6$	270	25
$CH_3O_2CC{\equiv}CCO_2CH_3$	$[Co(CO)_4]_2Hg$	200	25
$C_6H_5C{\equiv}CH$	$[Co(CO)_4]_2Hg$	143	8

TABLE 4-3
Trimerization of Acetylenes by Cobalt Carbonyls in Solution[84]

Acetylene	Catalyst	Trimer	% Yield
$C_6H_5C\equiv CH$	$[Co(CO)_4]_2Hg$	1,2,4-Triphenyl-benzene	70
$BrC_6H_4C\equiv CH$	$[Co(CO)_4]_2Hg$	1,2,4-Tris(p-bromophenyl)-benzene[a]	65
$(CH_3)_3SiC\equiv CH$	$Co_2(CO)_6(CH_3)_3SiC_2H$	1,2,4-Tris(tri-methylsilyl)-benzene[a]	55
$HOC_2H_4C\equiv CH$	$Co_2(CO)_6HOC_2H_4C_2H$	1,2,4-Tris(ethylol)-benzene[a]	14
$CH_3OCH_2C\equiv CH$	$Co_2(CO)_6CH_3OCH_2C_2H$	1,2,4-Tris(methoxy-methyl)benzene	17
$C_3H_7C\equiv CH$	$[Co(CO)_4]_2Hg$	1,2,4-Tri-n-propyl-benzene	11
$C_6H_5C\equiv CC_6H_5$	$[Co(CO)_4]_2Hg$	Hexaphenylbenzene	90
$ClC_6H_4C\equiv CC_6H_4Cl$	$[Co(CO)_4]_2Hg$	Hexakis(p-chloro-phenyl)benzene[a]	95
$CH_3O_2CC\equiv CCO_2CH_3$	$[Co(CO)_4]_2Hg$	Mellitic acid hexa-methyl ester	80
$C_2H_5C\equiv CC_2H_5$	$[Co(CO)_4]_2Hg$	Hexaethylbenzene	75
$CH_3C\equiv CC_6H_5$	$[Co(CO)_4]_2Hg$	1,2,4-Trimethyl-3,5,6-triphenylbenzene[a]	90
$C_6H_5C\equiv CCO_2H$	$Co_2(CO)_6C_6H_5C_2CO_2H$	1,2,4-Triphenyl-3,5,6-tricarboxybenzene[a]	11
$C_6H_5C\equiv CCO_2CH_3$	$[Co(CO)_4]_2Hg$	1,2,4-Triphenyl-3,5,6-tricarbo-methoxybenzene[a]	55
$C_6H_5C\equiv CCl$	$[Co(CO)_4]_2Hg$	1,2,4-Triphenyl-3,5,6-trichlorobenzene[a]	14

[a] New compounds.

gives a green complex. The complex reacts with another molecule of diferrocenylacetylene to form tetraferrocenylcyclopentadienone (81 % yield).[84a]

Kruerke and Hübel[85] reported a detailed study of complexes and cyclization reactions of other cobalt carbonyl-acetylenic complexes in 1961. Some of the complexes gave unsymmetrical hexasubstituted benzenes as sole products:

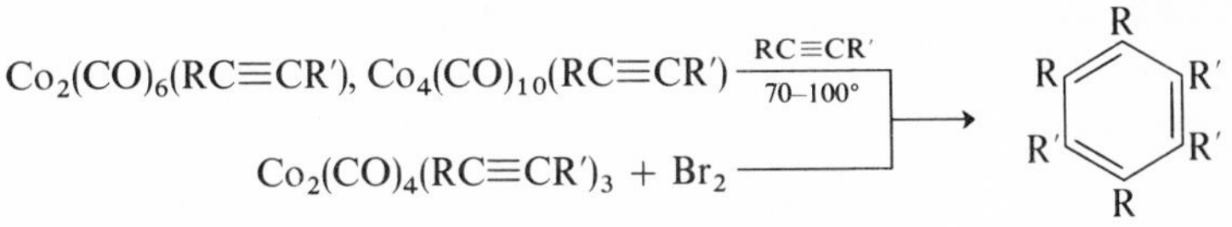

With most catalysts, silylacetylenes give both benzene and cyclopentadienone products.

$Co(CO)_3NO$ is an excellent catalyst for cyclizing acetylenes to unsymmetrical products. 3-Methyl-1-butyn-3-ol in n-hexane at 30–120° gives 90% conversion to the 1,2,4-trisubstituted benzene. $Ni(CO)_3(PPh_3)$ gives only 63% conversion.[86]

Other Ni(0)-complexed catalysts are also effective for cyclotrimerizing acetylenic alcohols[86] and are much better than the cobalt carbonyl complexes of Hübel and Hoogzand:[84] $Ni(CO)_2(PPh_3)_2$ and $Ni(CO)_3(PPh_3)$. Combinations of sodium borohydride and cobalt salt or nickel salt (2:1 ratio) are also excellent catalysts if a NO-forming substance (a nitrite) or NO itself is added. The reaction with methylbutynol and with propargyl alcohol goes easily at 80° in 6 hours in 95% ethanol or acetonitrile. Air decreases the yield.

Ni(II) catalysts are quite active: Nickel(II) acetylacetonate:$PPh_3(1:1)$ causes a violent reaction which converts 91% of the propargyl alcohol to a mixture containing 45% tars, 25% linear crystalline oligomer (4–6 units of propargyl alcohol), 27% 1,3,5-trimethylolbenzene, and 3% of the 1,2,4-isomer.

Nickelous salts are effective catalysts for cyclotrimerizing cyclooctyne in boiling tetrahydrofuran.[27a] No complexing agent is needed.

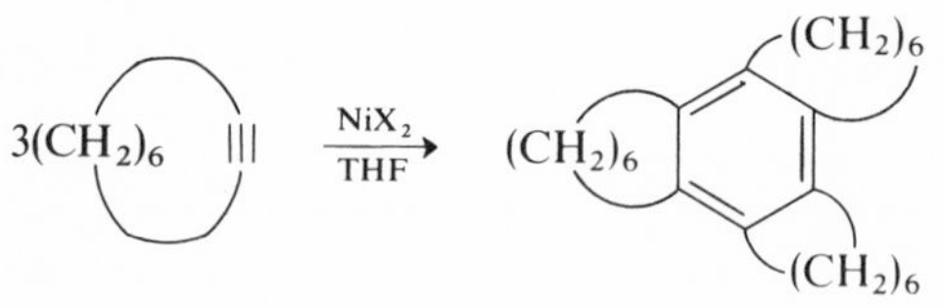

X	% Yield
CN	71
Br	96
I	83

If a little water is added to the $NiBr_2$ or NiI_2 system, a nickel complex forms as a low yield by-product:

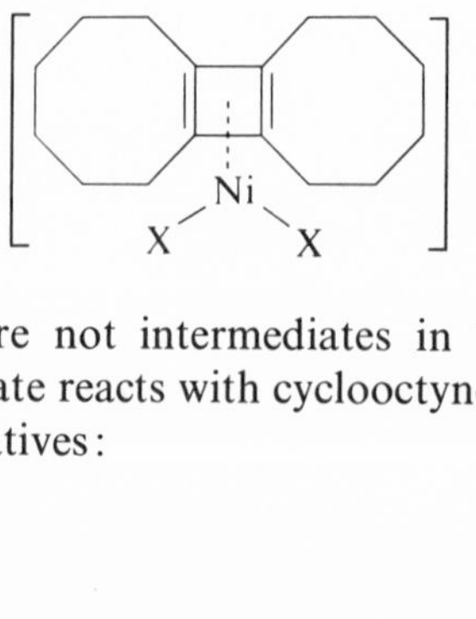

$(X = I, Br)$.

These complexes are not intermediates in the cyclotrimerization reaction. Nickel acetylacetonate reacts with cyclooctyne in boiling THF to form benzo-dicyclooctene derivatives:

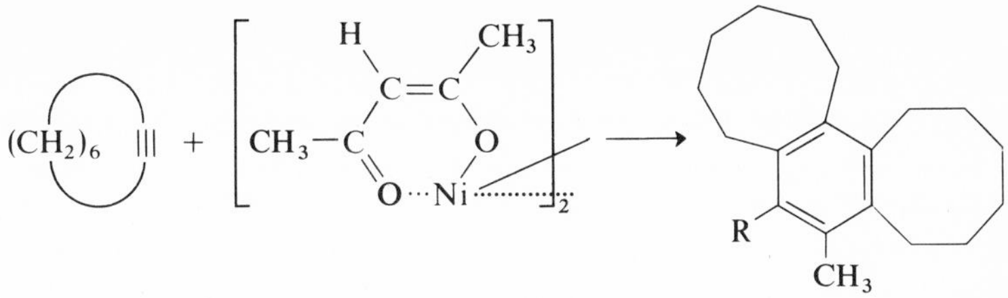

R	% Yield
H	25
$\overset{\overset{\text{O}}{\|}}{\text{CCH}_3}$	40

Cyclotrimer also forms (4% yield).

4.5. Bis(arene) Complexes of Chromium, Manganese, Nickel and Cobalt

In 1960, Zeiss[87] reviewed his work with bis(arene) complexes as catalysts for cyclizing acetylenes. The bis(arene) complexes are easily made and display some remarkable catalytic effects.

Phenylmagnesium bromide reacts with $CrCl_3$ in ether or tetrahydrofuran to give the triphenylchromium trietherate[88,89] which reacts with acetylenes. The acetylenes initially displace ether or tetrahydrofuran via the acetylenic π electrons. 2-Butyne reacts to form hexamethylbenzene, 1,2,3,4-tetramethyl-naphthalene and the complexes of these with chromium. Cobalt and manganese also give complexes with the hexamethylbenzene formed from bis(arene) complexes of these metals and 2-butyne. Acetylene or monosubstituted acetylenes react with triarylchromium to form polymers, and decompose the chromium complex. Disubstituted acetylenes react rapidly to form their complexes:

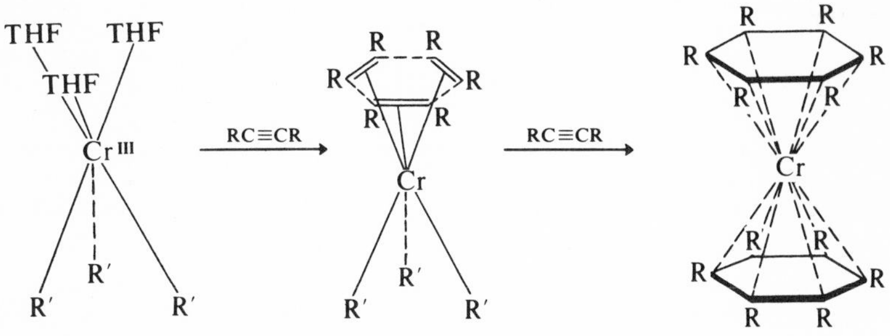

Cyclotrimerization of 2-butyne-1,1,1-d$_3$ indicates that a cyclobutadiene-Cr complex is not an intermediate.[89a]

Addition of excess nonterminal alkyne gives aromatic hydrocarbons as primary products.[90] With 2-butyne, as the molar excess of alkyne increases, the yield of hexamethylbenzene increases to 55%, and the yield of 1,2,3,4-tetramethylnaphthalene increases to 40%. Formation of the naphthalene ring is intriguing. The chromium in the complex is able to dehydrogenate an *ortho* position of one of the phenyl groups, and the favorable positions of the complexed 2-butyne and the dehydrogenated phenyl allow condensation to form the naphthalene ring.

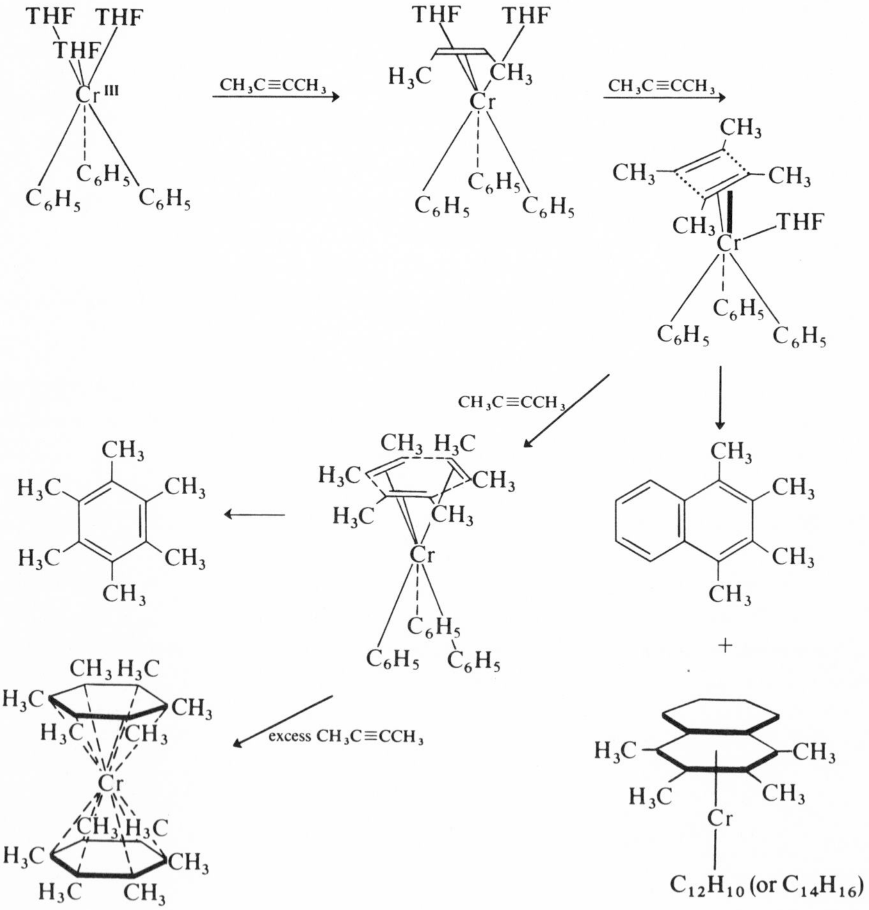

The dehydrogenating-cyclizing effect is general. Naphthylchromium complexes give anthracenes and phenanthrenes in the same way:

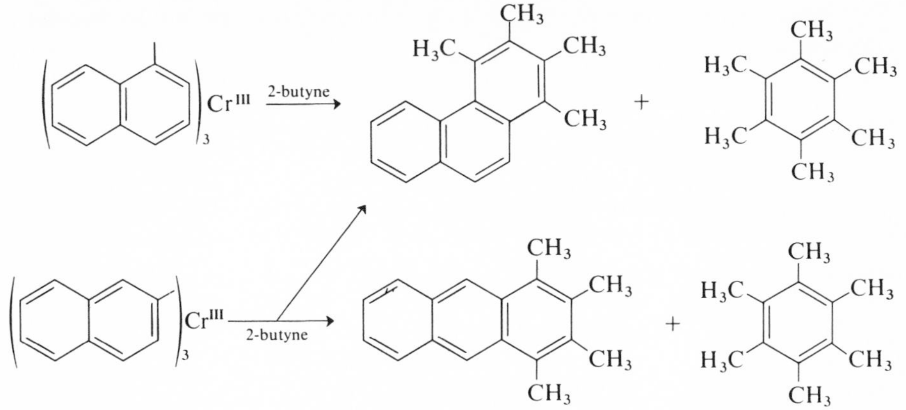

The chromium atom is a very strong hydrogen acceptor. Triethylchromium with diphenylacetylene gives hexaphenylbenzene and 1,2,3,4-tetraphenylbenzene. The latter is from two diphenylacetylene molecules and one ethyl group from the catalyst. Here the chromium not only removes hydrogen from the CH_3— of the ethyl group, but also dehydrogenates the resulting product to give the aromatic ring. Trimethylchromium gives 1,2,3,4-tetraphenylcyclopentadiene. This dehydrogenating effect is unique to the chromium complexes and promises to be a powerful new reaction system for other syntheses.

The complexed acetylene molecule can also undergo addition reactions. Trimesitylchromium and 2-butyne give the normal cyclization product hexamethylbenzene and also 2-mesityl-2-butene.[91] Dimethyl acetylenedicarboxylate has delocalized π bonds and does not form cyclic products. Instead, two phenyl groups from triphenylchromium add to the triple bond of complexed dimethyl acetylenedicarboxylate to give cis-dimethyl diphenylmaleate.

Dimesitylchromium cyclizes 2-butyne to hexamethylbenzene and its π complex in 70–80% yield.[92] Dimesityliron is similar, and gives 20–30% yields. Dimesitylnickel, on the other hand, gives a little benzene and a large amount of amorphous nickel-containing polymer. In the same way, addition of tolan to dimesitylnickel in tetrahydrofuran gives only a little hexaphenylbenzene plus much polymer. Reverse addition gives up to 85% yield of hexaphenylbenzene. Dimesitylcobalt can trimerize at least 40 moles of 2-butyne to hexamethylbenzene. The amount of bis(hexamethylbenzene)cobalt remains constant, indicating that this complex is probably part of the cyclization sequence.[93]

4.6. Cyclization Catalyzed by "Active" MnO_2

"Active" MnO_2 is commonly used to oxidize secondary ethynylcarbinols to ketones. If the reaction is done in refluxing benzene or in chloroform at room temperature, the resulting ketones can cyclize. This unusual cyclization reaction is illustrated by the oxidation of ethynyl-2-furylcarbinol:[93a]

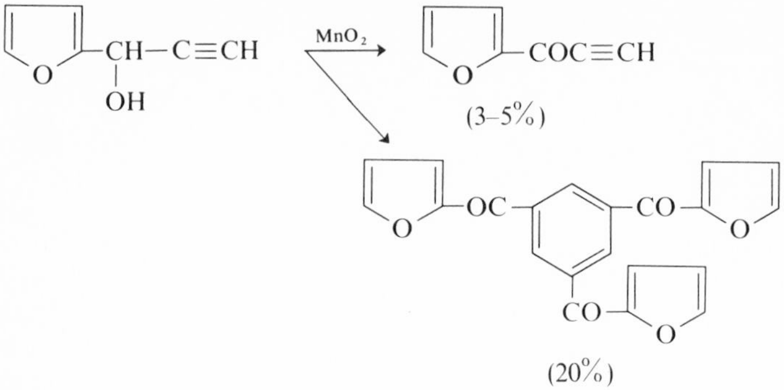

furyl–CH(OH)–C≡CH $\xrightarrow{MnO_2}$ furyl–COC≡CH

(3–5%)

furyl–OC–C₆H₃(–CO–furyl)–CO–furyl

(20%)

5. COMPLEXES HAVING EIGHT-MEMBERED RINGS

5.1. Nickel Complex Catalysts

Reed[75] in 1954 reported that $Ni(CO)_2(PPh_3)_2$ co-cyclizes acetylene and butadiene to form *cis,cis*-cycloocta-1,5-diene in 30–40% yields. The reaction is very sensitive to poisons. Starting with a clean, sand-blasted autoclave, yields from successive runs decrease. This is a common occurrence in Reppe cyclo-octatetraene synthesis. Since water and formaldehyde inhibit the cyclo-octatetraene reaction, Reed used a little calcium carbide in his mixtures to keep them anhydrous. Alumina, silica gel, ethylene oxide and PCl_5 inhibit the reaction. Oxygen is an inhibitor. Since calcium carbide contains inorganic impurities which give arsine, stibine, sulfides, etc., on reaction with water, Reed may have inadvertently added potent poisons by using calcium carbide.

Nickel carbonyl, $Ni(PCl_3)_4$ and $(PF_3)_2Ni(CO)_2$ are inactive. Phosphite ligands give more active catalysts than phosphines, and butadiene dimerizes in the absence of acetylene to form the cyclooctadiene. $[(PhO)_3P]_2Ni(CO)_2$ also gives some butadiene trimer, 1,5,9-cyclododecatriene. Pretreatment of the catalyst with acetylene causes it to be active for making divinylcyclohexenes. From his study of catalysts, Reed concluded that the catalyst must have the form:

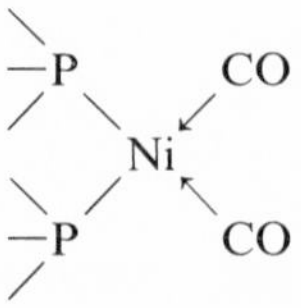

$$
\begin{array}{c}
P \\
\quad \backslash \;\; CO \\
\qquad Ni \\
\quad / \;\; CO \\
P
\end{array}
$$

Formation of eight-membered rings from acetylene is unique to nickelous cation complexes in this study (but see reference 86). The acetylene probably associated with the catalyst via an electron pair from the triple bond. Reed proposed this mechanism:

$$
\begin{array}{ccc}
\underset{\text{HC}\equiv\text{CH}}{\overset{R}{\overset{\downarrow}{\text{Ni}^{++}}}} \longrightarrow
\underset{\underset{\overset{+}{\text{CH}}}{\overset{\text{HC}^-}{\diagdown}}}{\overset{R}{\overset{\downarrow}{\text{Ni}^{++}}}}
\xrightarrow{\text{C}_2\text{H}_2}
\underset{\underset{\text{CH}-\text{CH}}{\overset{^-\text{CH}}{\diagdown}}}{\overset{R}{\overset{\downarrow}{\text{Ni}^{++}}}}
& \xrightarrow{\text{C}_2\text{H}_2} \text{benzene} \\
\end{array}
$$

$$\overset{+}{\text{CH}}\diagup\diagup \xrightarrow[\text{2 steps}]{2\text{C}_2\text{H}_2} \text{cyclooctatetraene}$$

Formation of the large rings from butadiene is possible because the absence of conjugated double bonds in the complex minimizes ring contraction by isomerization. Reed found no open-chain products from any of his reactions, so he assumed that the only way the carbon chain can detach from the catalyst is by reaction at the other end of the chain. If the cyclization does not occur, and it probably becomes more difficult as chain length increases, the complex could give long open-chain molecules.

Ni(0) complexes can cyclize some acetylenes to eight-membered rings. Tetrakis(phosphorus trihalide)nickel(0) catalysts cyclize propiolic acid esters to cyclooctatetraenes,[79] but are inactive for cyclizing acetylene itself. NMR studies with ^{31}P complexes indicate this equilibrium exchange:

$$\text{Ni(PX}_3)_4 \rightleftharpoons [\text{Ni(PX}_3)_n] + (4-n)\text{PX}_3$$

The forward step forms the ligand-deficient Ni(0) compound which is probably the actual catalyst. When $X = F$, Cl or Br, and when the catalyst is $\text{Ni(PhPCl}_2)_4$, ethyl propiolate gives 1,2,4,6- and 1,3,5,7-tetracarbethoxycyclooctatetraene [(1) and (2), respectively]. The reactions are homogeneous in cyclohexane or benzene, and go at room temperature under nitrogen. With $\text{Ni(PCl}_3)_4$ catalyst and a monomer:catalyst ratio of 1000, the conversion is 73% in only 3 minutes. The yield of (1) is 28%, and the yield of (2) is 1%; 71% of the mixture is 1,2,4- and 1,3,5-tricarbethoxybenzene. $\text{Ni(PF}_3)_4$ gives 53% conversion of ethyl propiolate to a mixture of 30% (1) and 70% benzenetricarboxylic acid esters. $\text{Ni(PBr}_3)_4$ and $\text{Ni(PhPCl}_2)_4$ cause only 1% conversion. Addition of 5 moles of PCl_3 to $\text{Ni(PCl}_3)_4$ decreases the yield, but product distribution stays the same. This supports the dissociation pictured in the equation.

Propiolate esters, allene and butadiene are the only monomers which form trimers and tetramers with these catalysts. Propiolic esters can co-tetramerize, but will not induce an inactive acetylene to enter the tetramerization. The unusual reactivity of propiolate esters in this system cannot be explained.[79]

The nickel-phosphorus-halogen complexes are unique. Phosphorus halides form stronger π bonds with transition metals than phosphites do, and this could increase catalytic activity by decreasing the Ni—P σ bond strength, or by stabilizing the ligand-deficient nickel complex to favor conjugation of acetylene with the nickel catalyst. The results of this study are summarized:[79]

(1) Monomers which give tetramers and trimers with $Ni(PCl_3)_4$	Methyl propiolate, ethyl propiolate, allene, butadiene
(2) Inactive monomers	$RC{\equiv}CH$: R = H, amyl, phenyl, CO_2H, CO_2K, CH_2Cl, CH_2OH, CH_2OCOCH_3, $CONH_2$, CN
	$RC{\equiv}CR$: R = CO_2Et, ethyl, 1,6-heptadiyne
(3) Catalysts which form both tetramers and trimers: order of reactivity with ethyl propiolate	$Ni(PCl_3)_4 > Ni(PF_3)_4 > Ni(PBr_3)_4 > Ni(PhPCl_2)_4$
(4) Catalysts which form trimers only	$Ni[P(OEt)_3]_4$, $Ni(PhCN)_4$
(5) Inactive catalysts	$Ni(C_5H_6)_2$, $Zn[OP(Ph)_3]_4(ClO_4)_2$

The tetracarboxyesters of cyclooctatetraene react with 3 moles of hydrogen at ambient conditions over Pd—C catalyst to give 100% yields of the cyclo-octene tetraester. Hydrogen uptake stops after 1 hour. After another 72 hours, the cyclooctane derivative is formed. The tetracarboxy acids titrate and show four inflections at successively lower pK_a values. 1,3,5,7-Tetracarboxy-cyclooctatetraene gives only one inflection on titration (all carboxyls are equivalent).

Chini and co-workers[86] found several Ni(0) and Ni(II) catalysts for cyclizing acetylenic alcohols such as 3-methyl-1-butyn-3-ol to tetrasubstituted cyclo-octatetraenes. Nickel carbonyl at 60–80° cyclizes methylbutynol easily; 1 mole of $Ni(CO)_4$ makes about 200 moles of the cyclooctatetraene. Bis(cyclooctadiene)-nickel(0) is even more active. The major product is the 1,2,4,6-cyclooctatetraene derivative.

Sodium borohydride-nickelous salts, without nitrite or NO, but preferably in the presence of triphenylphosphine, are active. Linear oligomers form. These are amorphous hygroscopic solids containing 4–6 alcohol units.

The Ni(II) systems which are effective[86] are $Ni(CN)_2$ or nickelous acetyl-acetonate. A mixture of $Ni(OEt)_2$ and NaCl, formed by anhydrous $NiCl_2$ and NaOEt in ethanol, is an especially active catalyst. The catalysts effective for cyclizing acetylenic alcohol (methylbutynol) to benzene derivatives and to cyclooctatetraene derivatives are summarized:[86]

Catalyst	Main Product
$Co(NO)(CO)_3$, or $2NaBH_4 + Co^{++} + NO$, or $Co(acac)_2 + NO$	1,2,4-Aromatic isomer
$Ni(CO)_3PPh_3$, or $2NaBH_4 + Ni^{++} + PPh_3$, or $Ni(acac)_2 + PPh_3$	1,3,5-Aromatic isomer + linear trimer
$Ni(CO)_4$, or $2NaBH_4 + Ni^{++}$, or $Ni(acac)_2$	Mixture of cyclooctatetraene isomers

5.2. Condensation-Cyclization to Ten-Carbon Rings

Mono- or disubstituted acetylenes cyclo-condense with two molecules of butadiene in the presence of Ni(0):tri(2-biphenylyl) phosphite (1:1) at 40°. Products are 4- or 5-mono- or 4,4-disubstituted cis-1, cis-4,-trans-7-cyclo-decatrienes.[94] 2-Butyne gives 95% yield. Conversion is 75%, and 80% of the reacted butadiene forms cyclooctadiene as by-product:

5.3. Bis(acrylonitrile)nickel Catalyst

5.3.1. CYCLOTRIMERIZATION AND CYCLOTETRAMERIZATION

Schrauzer[80] in 1961 reviewed briefly the status of the reaction of acetylenes over transition metal complexes, particularly the metal carbonyls. The observed reactions can sometimes be explained equally well by radical and ionic mechanisms. The reactions do not depend very much on the polarity of the reaction medium. Some reactions go as well without solvent as with, so a polar reaction appears unlikely or unobvious. Some evidence argues against the radical mechanism: inhibitors for radical reactions have no effect on the reactions, and the reaction of diphenylacetylene with $Fe_3(CO)_{12}$ in ethyl acrylate, vinyl acetate and other vinyl monomers does not cause any vinyl polymerization. The reaction product, hexaphenylbenzene, is formed via the known iron carbonyl-tolan complex. Thus, the solvent has little influence on this particular reaction. Irradiation also has no effect. These facts indicate a multicenter reaction mechanism involving a complex of the metal carbonyl and the acetylenic compound. The primary adduct with the reactive metal carbonyl fragment depends on the reaction conditions, the metal carbonyl and the acetylene. Most of the acetylenic compound adducts are unstable and decompose rapidly to their end products. The cyclobutadiene intermediate widely speculated on is only one possibility. Complexes can have five-, six- and seven-atom ring systems.

The reaction of acetylenes through metal π complexes must show a dependence between the reaction product and the composition of the intermediate π complex. Bis(acrylonitrile)nickel(0) complex is ideal for studying the possibilities. This complex is very reactive, its ligands are attached only to the nickel atom, and it has two coordination sites available. For example, with triphenylphosphine (TPP) it forms a diamagnetic complex $Ni(AN)_2(TPP)_2$, in which nickel has the sp^3 configuration. From this, the expected acetylenic compound complex should be $Ni(AN)_2(ac)_2$. The nickel is not charged, and if the adduct has the expected sp^3 configuration, the molecule must be tetrahedral. The product from the reaction of one molecule of acrylonitrile and two molecules of acetylenic compound is a benzonitrile, undoubtedly formed by this sequence:

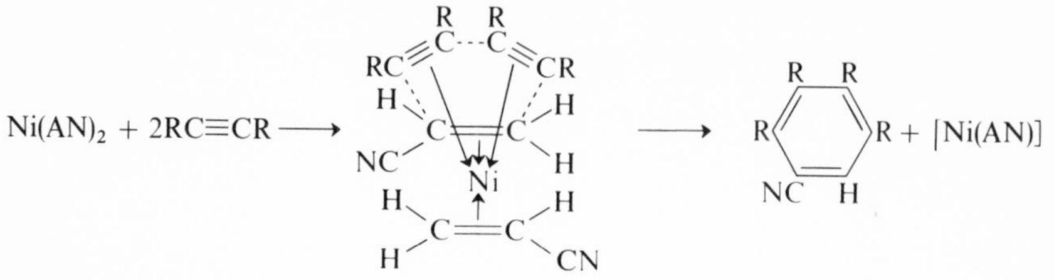

Three molecules of the acetylene can react to give benzenes:

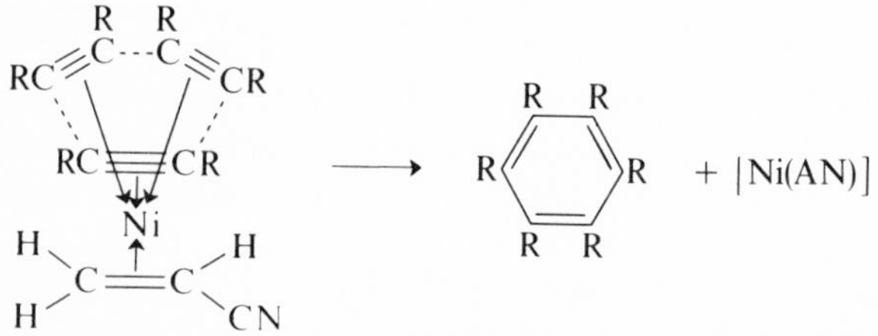

Tolan with a stoichiometric amount of $Ni(AN)_2$ gives tetraphenylbenzonitrile and hexaphenylbenzene in good yield. In the same way, tolan and bis(acrolein)-nickel [$Ni(AC)_2$] react to form tetraphenylbenzaldehyde and hexaphenyl-benzene. At higher temperature, the product is mostly hexaphenylbenzene. Tolan is also trimerized by bis(fumaronitrile)nickel and bis(cinnamonitrile)-nickel, and 2-butyne forms hexamethylbenzene.

Phenylacetylene reacts easily at room temperature with $Ni(AN)_2$ and $Ni(AC)_2$, and very rapidly at 70°. Some trimer, probably 1,2,4-triphenylbenzene, forms but the main product is solid polymer. Propargyl alcohol, phenylpro-pargylaldehyde, propiolic ester and acetylenedicarboxylic acid ester react even more vigorously, sometimes explosively, to form polymers.

The complex $Ni(AN)_2(TPP)_2$ reacts with acetylene and excess acrylonitrile to give heptatrienenitrile, so the complex may be the intermediate in Cairns' synthesis also.[74] Schrauzer's experiments are pictured:

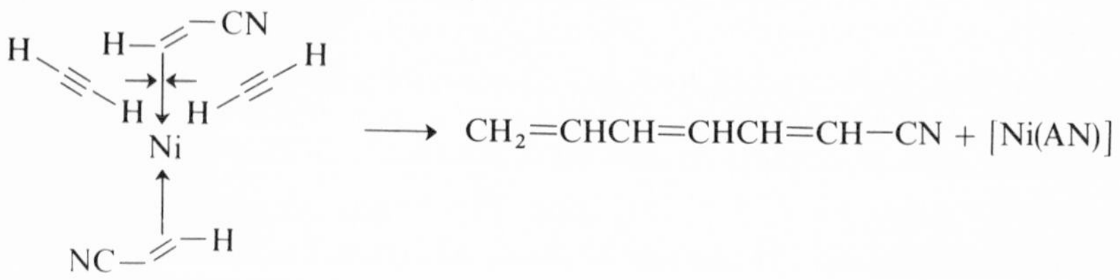

$$\longrightarrow \quad CH_2=CHCH=CHCH=CH-CN + \lceil Ni(AN) \rceil$$

If the triphenylphosphine is omitted, activity to form heptatrienenitrile is slight.

The acrylonitrile-nickel complex can also catalyze the cyclooctatetraene reaction. In tetrahydrofuran suspension with 15–25 atmospheres of acetylene at 70–85°, the main product is benzene, but some cyclooctatetraene forms. The reaction is essentially stoichiometric. Apparently the cyclooctatetraene reaction goes stepwise to form the $Ni(AN)_2(ac)_2$ complex first, followed by further displacement of AN by acetylene to form $Ni(ac)_4$:

Addition of a nickel complexer should inhibit the cyclooctatetraene reaction if it does require $Ni(ac)_4$ complex formation. Addition of TPP increases the activity, but the only product is benzene. No tetramerization occurs. Thus, the required $Ni(ac)_4$ cannot form in the presence of TPP.

5.3.2. MECHANISM OF THE CYCLOOCTATETRAENE SYNTHESIS

Reppe[95] discovered the nickel-catalyzed cyclization of acetylene to cyclooctatetraene. Schröder's book in 1965 on cyclooctatetraene includes most of the data known about it.[96] Conditions are much milder than conditions used for trimerization-cyclization of acetylene to benzene, and Ni(II) complex catalysts are usually used. Reppe thought the eight-membered ring formed within labile acetylene-nickel complexes, but only recently have detailed studies of mechanisms been published. This hinges on better understanding of the chemistry of nickel complexes.

Schrauzer summarized earlier work, and discussed mechanisms and structures of intermediates in detail.[45,80,97] For cyclooctatetraene, the best temperature is 85–95°, in the presence of a nickel complex such as acetylacetonate, cyanide or salicylaldehydate, at 15–25 atmospheres acetylene pressure. Generally, the yield of cyclooctatetraene is 70%, with up to 15% benzene, some resin and powdery polymer similar to cuprene (usually called "niprene"). Small amounts of styrene, 1-phenyl-1,3-butadiene, vinylcyclooctatetraene and azulene also frequently form. Catalysts have short active lives, mostly because niprene fouls them. Acetylene is the best monomer; 1 mole of a mono- or disubstituted acetylene can react with 3 moles of acetylene, to give a variety of 1,2-disubstituted cyclooctatetraenes.

The only nickel complexes which are active catalysts are those which have octahedral configuration or can assume this configuration easily. Schrauzer tabulated the characteristics of catalysts and non-catalysts:[97]

	Active Weak Ligand Field	Active Intermediate Range	Inactive Strong Ligand Field
Configuration	Octahedral	Octahedral and Planar	Planar
Magnetic behavior	Paramagnetic	Anomalous	Diamagnetic
Ligand exchange	Rapid	Moderately rapid to rapid	Very slow
Catalytic activity	Active	Active	Inactive
Examples	Ni-salicylaldehyde	Ni-bis-(N-alkylsalicyl- aldimine)	Ni-phthalocyanine
	Ni-acetylacetonate	Ni(CN)$_2$	Ni-dimethylglyoxime

Ligand exchange must be fairly fast, otherwise acetylene cannot displace ligands and form the necessary complexes. Ligand exchange is assumed to be the rate-determining step. Polar solvents participate in the ligand exchange of the chelates, so reactions are faster in tetrahydrofuran than in benzene. Dioxane is also a good solvent, but stronger ligand solvents like pyridine or benzonitrile form strong complexes with the nickel and prevent coordination with the acetylene molecules. Four coordination sites are made available by cleaving the coordinate linkages, and four acetylenes can occupy these sites to form a nearly octahedral complex. The acetylenes can assume the steric configuration necessary to form cyclooctatetraene:

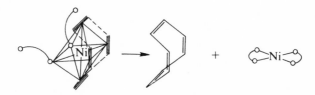

The magnetic moments of bis(N-alkylsalicylaldiminato)nickel(II) chelates alternate parallel with the base strengths of the alkylamines, and the yield of cyclooctatetraene also varies with effective ligand field strength. N-Methyl is the most strongly paramagnetic chelate of the series and the best cyclo-octatetraene catalyst.

Inhibition of cyclooctatetraene catalysts by adding strong ligands in controlled amounts gives catalysts still quite active for benzene formation. For example, adding 1 mole of triphenylphosphine to the nickel catalyst destroys its activity for cyclooctatetraene, but not for benzene. Addition of difunctional electron donors which can block two *cis* positions in the active complex, such as a,a'-dipyridyl and 1,10-phenanthroline, also destroy catalyst activity for benzene:

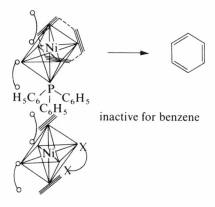

inactive for benzene

Thus, acetylene and the Ni(II) ion form relatively weak σ-bonds, with only small π-bond contributions, since cyclooctatetraene formation is inhibited by trimethylamine and by pyridine, both of which are weak π-bonding ligands but are σ donors.

If the ligands around Ni(II) are too bulky, or if chelates are cyclic, catalysts are inactive. Thus,

is inactive, but when $n = 7$ or 8, complexes become slightly active.

Formation of by-products mentioned before is now easily understood. Benzene can be formed by slightly "switched-over" catalyst. Styrene and the other products can be formed by mixed cyclooligomerization of acetylene with vinylacetylene and higher linear oligomers. (See structures on page 366.)

Cyclobutadiene has long been postulated as an intermediate in cyclo-octatetraene synthesis. For example, Longuet-Higgins[40] proposed that two cyclobutadiene molecules on the surface of the catalyst dimerize to form cyclooctatetraene. Schrauzer feels that his data argue strongly against this explanation. Strongly π-bonding cyclobutadiene should form complexes with Ni(II) which are diamagnetic and kinetically more stable than original catalyst molecules. The rate of cyclooctatetraene formation would be controlled by consecutive reactions of the cyclobutadiene complex, which Schrauzer could not reconcile with the data.

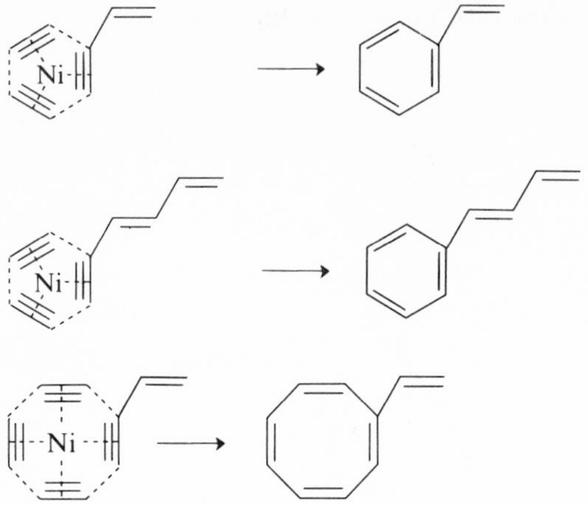

Schrauzer made cyclooctatetraene, heptatrienenitrile and benzene over a single catalyst: bis(acrylonitrile)nickel(0).

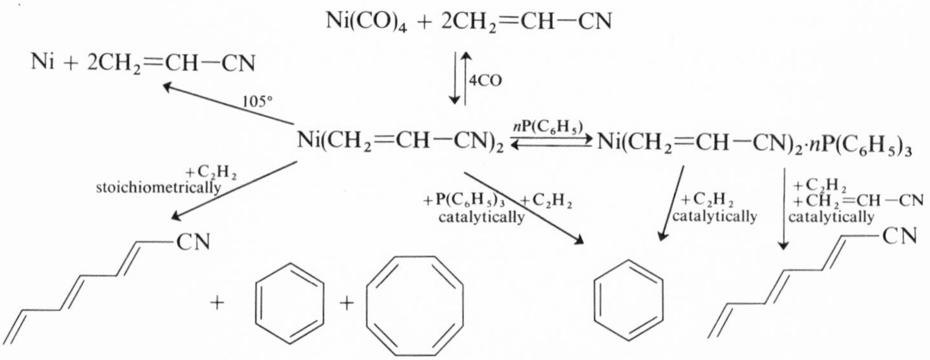

$$Ni(CO)_4 + 2CH_2=CH-CN$$

$$Ni + 2CH_2=CH-CN$$

$$105°$$

$$\downarrow 4CO$$

$$Ni(CH_2=CH-CN)_2 \xrightleftharpoons{nP(C_6H_5)} Ni(CH_2=CH-CN)_2 \cdot nP(C_6H_5)_3$$

stoichiometrically $+C_2H_2$

$+P(C_6H_5)_3 +C_2H_2$ catalytically

$+C_2H_2$ catalytically

$+C_2H_2$ $+CH_2=CH-CN$ catalytically

The active catalyst is zerovalent Ni, which rapidly becomes inactive because it is difficult for acetylene to resolvate it. Electron donors like triphenylphosphine stabilize atomic nickel, and catalyze the formation of heptatrienenitrile and benzene, but inhibit or completely suppress cyclooctatetraene formation. The polarity of the solvent does not affect reactions with bis(acrylonitrile)nickel; therefore, the active species is the uncharged nickel atom.

5.3.3. MECHANISM OF FORMATION OF BY-PRODUCT CUPRENE

The mechanism of formation of cuprene (or "niprene") in the Reppe[98] cyclo-octatetraene synthesis has not been studied much. Fredericks[99] prepared cuprene by the Reppe method (acetylene in tetrahydrofuran at 90° for 72 hours in the presence of anhydrous nickel cyanide catalyst). The cuprene was a dark brown powder, which showed practically no crystallinity by x-ray analysis. He also prepared crystalline polyacetylene, using nickel dihalide-bis(tertiary phosphine) catalysts at room temperature. Crystalline polyacetylene does not change on long heating *in vacuo* at 200°, but when heated in air it changes rapidly. Within 2 hours at 205°, polyacetylene goes completely to cuprene. The cuprene prepared this way has nearly the same x-ray pattern as Reppe cuprene.

The origin of cuprene in the cyclooctatetraene synthesis can be explained in the following way: polyacetylene is first formed by the action of the nickel catalyst, and is then converted to cuprene by the action of residual oxygen in the system.

6. OLIGOMERIZATION AND CYCLIZATION BY LEWIS ACIDS, METAL ALKYLS, AND COMBINATIONS OF METAL ALKYLS AND LEWIS ACIDS (ZIEGLER CATALYSTS)

6.1. Silica-Alumina Catalysts

Clark[100] passed acetylenes diluted with isobutane over acidic chromia-silica-alumina catalysts similar to those used to polymerize ethylene. Acetylene at 40° gives benzene, 90% conversion. Propyne gives 1,2,4-trimethylbenzene and a little 1,3,5-trimethylbenzene. 1-Pentyne and 1-hexyne also give trialkylbenzenes. Nonterminal alkynes are relatively unreactive. Silica, silica gel and carbon are active aromatization catalysts only at 250–400°.[101]

Silica-alumina alone at room temperature is not a catalyst for the acetylene-to-benzene trimerization. When the silica-alumina is exposed to diborane, two hydrogen molecules form per diborane reacted.[102] This indicates a reaction which forms surface boron hydride groups ($M-OBH_2$). When diborane-treated silica-alumina is exposed to acetylene, it becomes purple, the temperature rises rapidly, and benzene forms. Dideuteroacetylene gives hexadeuterobenzene. Mixtures of acetylene and dideuteroacetylene give various deuterobenzenes. The benzenes contain both odd and even numbers of D atoms, showing exchange between acetylene and dideuteroacetylene.

6.2. Metal Halide Lewis Acids

Stannic chloride in benzene complexes 2 moles of phenylacetylene or 1-pentyne.[103] The phenylacetylene complex has a terminal vinyl group, and phenylacetylene is not regenerated by addition of water. Stannic chloride apparently activates one molecule of phenylacetylene so it reacts with the second molecule by hydrogen migration to form a complex which may be represented as $SnCl_4 \cdot PhC{\equiv}C\overset{\overset{\displaystyle Ph}{|}}{C}{=}CH_2$. The complex from 1-pentyne has no terminal vinyl groups, and water regenerates 1-pentyne. Olefins and stannic chloride usually form carbonium ions.

A mixture of aromatic sulfenic acid and aluminum chloride dimerizes phenylacetylene at 0°. Aluminum bromide and aromatic sulfenyl chlorides are not catalysts. The dimer is 1,2,3-triphenylazulene, 25% yield. Without the aluminum chloride, sulfenyl chlorides add across the triple bond. Aluminum bromide plus a little V^{++} or Ni^{++} gives 41% yield of 1,2,3-triphenylazulene. Very pure aluminum bromide, refined by zone melting, is necessary.[104] Other metal ions decrease the yields. Formation of the fused ring azulene requires a dehydrogenation and enlargement of one of the benzene rings of diphenyl-acetylene. This is similar to the dehydrogenation obtained with bis(arene)-chromium complexes (see Section 4.5).

Acetylene over $NbCl_5$ at 140-200° gives 38% yield of benzene.[105] Propyne at 70° gives 51% yield of mesitylene and pseudocumene, ratio 1:2. 1-Butyne gives 70% yield of 1,3,5- and 1,2,4-triethylbenzene, ratio 1:2.6. 2-Butyne gives hexamethylbenzene, 32% yield. Diacetylene-acetylene mixture at 0° over $TaCl_5$ and $NbCl_5$ gives diphenyl, terphenyl and higher polyphenylenes, but no condensed aromatic systems.

Aluminum chloride catalyzes the cyclotetramerization of 2-butyne to form syn-octamethyltricyclo[4.2.0.0^{2,5}]octadiene:[105a]

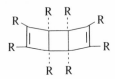

$(R = CH_3)$. Ferric chloride, a similar Lewis acid, does not cyclotetramerize diethyl acetylenedicarboxylate, however. Reaction for 45 hours gives 30% yield of diethyl dichlorofumarate, an example of stereospecific chlorination.[105b]

Berkoff and co-workers[107] tried unsuccessfully to prepare hexamethyl-"Dewar" benzene. 3,4-Dichlorotetramethylcyclobutene reacts with zinc in the presence of 2-butyne to form hexamethylbenzene. Diphenylacetylene and potassium metal give hexaphenylbenzene and other products.

Schafer[107] reported the first successful preparation of hexamethyl-"Dewar" benzene. He simply treated 2-butyne in benzene with 5 weight % of dry $AlCl_3$ below 35°. At higher temperature, ordinary hexamethylbenzene forms. In one reaction, the conversion of 2-butyne was 16%, and the yield of hexamethyl-"Dewar" benzene, hexamethylbicyclo[2.2.0]hexa-2,5-diene, was 20%:

$$CH_3-C\equiv C-CH_3 \xrightarrow[\substack{\text{benzene} \\ < 35°}]{AlCl_3} \left[\begin{array}{c} H_3C \quad\quad CH_3 \\ \square\cdots AlCl_3 \\ H_3C \quad\quad CH_3 \end{array} \right]$$

$$\downarrow + \text{2-butyne}$$

(1,2,3,4-tetramethyl structures)

The product can be stored for months in a refrigerator. It boils at 43–45° at 15 mm of Hg.

At $-20°$ in the presence of BF_3·etherate and a trace of water, 2 moles of 2-butyne and 1 mole of chlorine react to give 30% yield of 1,2,3,4-tetramethyl-3,4-dichlorocyclobutene.[108] It is likely that cyclodimerization precedes chlorination.

6.3. Metal Alkyls

Triethylaluminum is a catalyst for oligomerization of diphenylacetylene.[109] After 6 hours at 120°, the products are 28% 1,2,3,4-tetraphenyl-1,3-butadiene, 24% 1,2-diphenyl-1-butene (ethyl group from the catalyst), and some 1,2,3,4-tetraphenyl-1,3-hexadiene. $TiCl_4$-$AlEt_3$ cyclizes acetylene to trimethylbenzenes. The ethyl groups from the catalyst must participate since cleavage of acetylene to give the odd number of carbons in the product is highly unlikely.[110]

6.4. Metal Alkyl-Lewis Acid Catalysts (Ziegler Catalysts)

In 1960, Ziegler[111] filled an entire issue of *Annalen der Chemie* with the chemistry of alkylaluminum compounds. The reaction of acetylenes to give olefins and aromatics was covered.

In 1959, Natta[112,113] reviewed the polymerization of α-olefins over Ziegler catalysts and described the polymerization of acetylenes to form long chains of conjugated double bonds. Thus, acetylene and terminal acetylenic compounds polymerize to high polymers in the presence of $Al(i\text{-}Bu)_3$-$TiCl_4$ (a Ziegler catalyst). Trimethylsilylacetylene cyclotrimerizes to 1,3,5-trimethylsilylbenzene in the presence of $AlMe_3$-$TiCl_4$. $Al(i\text{-}Bu)_3$-$TiCl_4$ can trimerize other acetylenes if the $Al:TiCl_4$ ratio is held between 1.5 and about 2.8.[114] Diphenylacetylene cyclizes to hexaphenylbenzene, 60–80% yield. At other ratios of catalyst ingredients, conversion drops rapidly. Neither the catalyst ingredients alone nor TiCl$_3$ can cyclize diphenylacetylene. Any symmetrical dialkyl- or diaryl-acetylene will cyclotrimerize if the proper catalyst ratio is used. 2-Butyne gives 100% yield of hexamethylbenzene at Al:Ti ratio of 1.8. The crude product is very pure. 3-Hexyne gives 100% yield of hexaethylbenzene, 17 moles per mole of catalyst. Organochromium compounds in reagent quantities give similar results. This indicates the similarity of the two systems and may be a clue to the structure of the Al:Ti complex catalysts. Al(i-Bu)$_3$-TiCl$_4$ in a ratio of 3 catalyzes the cyclization of 1-acetylenes to 1,3,5-trisubstituted benzenes in 90% yield.[115] This is a little higher than the range which Franzus found to be best.[114]

Lutz[116] noted that the literature up to 1961 suggested that 1-acetylenes are too reactive to cyclotrimerize and are more likely to form high polymers. He found that he could trimerize 1-acetylenes over Ziegler catalysts (e.g., AlEt$_3$-TiCl$_4$) by using high dilution of the 1-acetylene, small catalyst particle size and good stirring. In heptane solvent, acetylene gives 49% yield of benzene. Methylacetylene gives 40% yield of mesitylene and 21% yield of pseudocumene. Other acetylenes give comparable yields. Lutz proposes that the reaction is surface catalyzed: the acetylene is chemisorbed to form a polarized intermediate which interacts with its polarized neighbors to form either trimer or polymer:

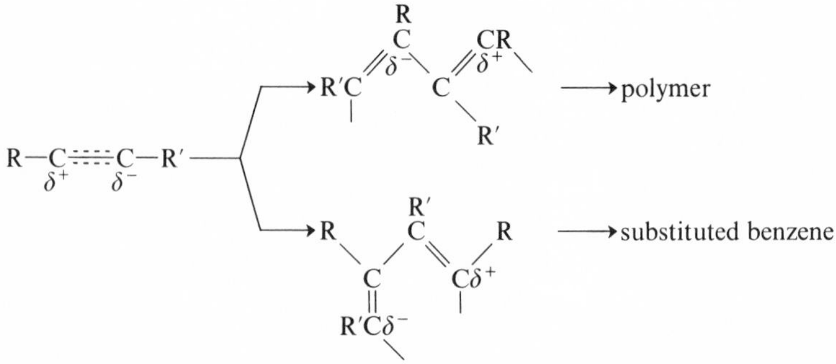

The easy trimerization may be due to the two π orbitals of the triple bond. One of the orbitals forms a bond with the catalyst surface, while the other is available to react with other activated chemisorbed molecules. Strong Lewis acid catalysts give different results.

6.4.1. CYCLOTRIMERIZATION OF NONCONJUGATED POLYYNES BY ZIEGLER CATALYSTS

Hubert and Dale[117] found that Ziegler catalysts cyclotrimerize the acetylenic groups in terminal diacetylenes to form trimers of two types: (1) dumbbell shaped and (2) cage trimers.

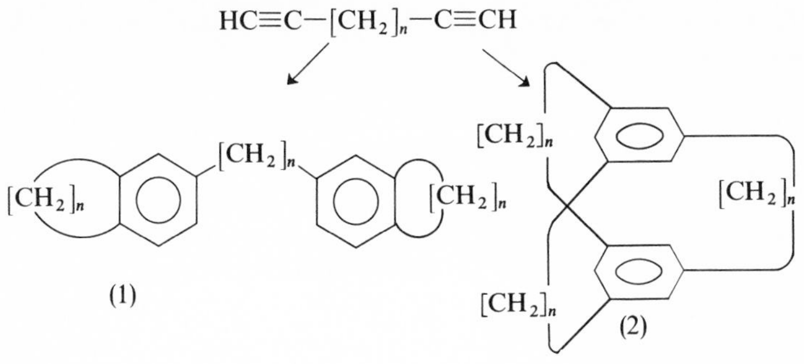

If $n < 5$, isomer (1) is the only product; if $n = 5$ or 6, both trimers form; if $n = 7$, only the cage trimer (2) forms. The yield of cage trimer is less than 1% in all cases, and yields are not improved by high-dilution procedures.

Hubert[118] later developed a better synthesis for (2), the cage trimer (yields up to 50%):

$$3HC{\equiv}C(CH_2)_nCl \xrightarrow[\text{5 min, low temperature}]{TiCl_4\text{-}Al{-}i\text{-}Bu_3} 1,3,5\text{- and } 1,2,4\text{-}C_6H_3[(CH_2)_nCl]_3 \xrightarrow{NaC{\equiv}CH{-}NH_3}$$

$$30\text{–}60\%\ 1,3,5\text{- and } 1,2,4\text{-}C_6H_3[(CH_2)_nC{\equiv}CH]_3 \xrightarrow[\text{high dilution}]{\text{Ziegler catalyst}}$$

cage trimer (2), 1,3,5- and 1,2,4-isomers

If the diyne has an aromatic ring, the cage trimer is a pentabenzenoid compound. Cyclotrimerization of *p*-bis(3-butynyl)benzene gives such a cage compound. The compound is able to clathrate dioxane into the cavity, but does not clathrate *n*-hexane. Hubert carried this novel approach a step further. He prepared hexakis(4-pentynyl)benzene and made the cage trimer. In this molecule, the trimerization is intramolecular:

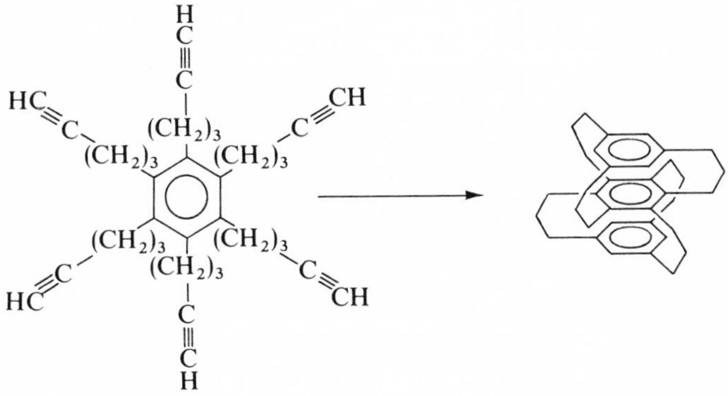

The same catalysts cause an internal trimerization reaction with non-conjugated triynes:

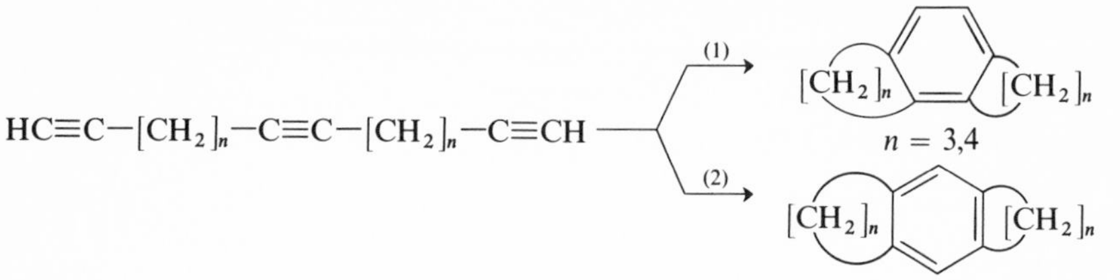

$$HC\equiv C-[CH_2]_n-C\equiv C-[CH_2]_n-C\equiv CH$$

(1) → $[CH_2]_n$ ⬡═⬡ $[CH_2]_n$　$n = 3,4$

(2) → $[CH_2]_n$ ⬡⬡ $[CH_2]_n$

$n = 4$(octahydroanthracene)

If $n = 3$, 1,2,3,6,7,8-hexahydro-*as*-indacene forms in good yield according to the first equation. If $n = 4$, both reactions occur to give about equal amounts of the two products. The octahydroanthracene may involve a Dewar-like intermediate on the surface of the catalyst:[118a]

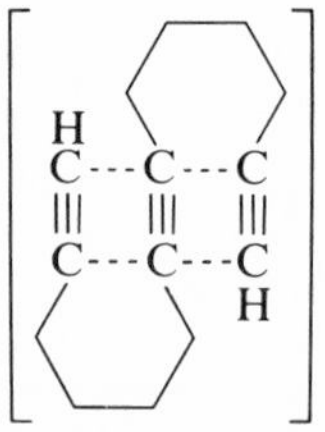

$$
\begin{bmatrix}
& \begin{array}{c} H \\ C---C---C \\ ||| \quad ||| \quad ||| \\ C---C---C \\ \qquad\qquad H \end{array} &
\end{bmatrix}
$$

Ziegler catalysts polymerize olefins and cyclize acetylenes, so vinylacetylenes might be expected to give complex products.[119] However, the vinyl group does not react at all. Vinylacetylene at $-10°$ gives 74% yield of a 9:1 mixture of 1,2,4- and 1,3,5-trivinylbenzene. Divinylacetylene and acetylene co-trimerize to give *o*-divinylbenzene and dimethyl-*o*-divinylbenzene isomers. With a large excess of acetylene, *o*-divinylbenzene forms in 30% yield (on divinylacetylene). With Meriwether's catalysts,[78] the yields are lower and the products are more difficult to purify.

6.4.2. ZIEGLER CATALYSTS VERSUS LUTTINGER'S CATALYST

Luttinger's catalyst [$(Ph_3P)_2NiCl_2$-$NaBH_4$,[120] see Chapter 5] cyclotrimerizes acetylenes in addition to polymerizing them. Donda and Moretti[121] compared the activity of Ziegler and Luttinger catalysts under the same conditions. $TiCl_4$-$AlEt_3$ is more active than Luttinger's catalyst: the yields of aromatics are 70–80% versus 50%. Luttinger's catalyst gives a greater percentage of unsymmetrical trisubstituted benzenes. Luttinger's system offers the advantage of stability toward oxygen and moisture.

6.4.3. MECHANISM OF THE ZIEGLER-CATALYZED CYCLOTRIMERIZATION OF ACETYLENES

Ikeda and Tamaki[122] used $TiCl_4$-$AlEt_3$ to cyclotrimerize acetylene. The catalyst and benzene do not react to form ethylbenzene, but ethylbenzene is a product from the acetylene cyclotrimerization. Benzene yield is 72%; ethylbenzene yield 0.22%. Thus, the bond between the acetylene and the alkyl group of the catalyst (ethyl) must form before cyclization occurs. Tracer studies show that ethyl groups labeled with D and ^{14}C enter the ethylbenzene as ethyl groups. This sequence pictures the reaction:

(1)　$[Cat]-R + HC\equiv CH \longrightarrow [Cat]-CH=CH-R$

　　$[Cat]-CH=CH-R + HC\equiv CH \longrightarrow [Cat]-CH=CH-CH=CH-R$

　　$[Cat]-CH=CH-CH=CH-R + HC\equiv CH \longrightarrow [Cat]-H + C_6H_5-R$

(2)　$[Cat]-H + HC\equiv CH \longrightarrow [Cat]-CH=CH_2$

　　$[Cat]-CH=CH_2 + HC\equiv CH \longrightarrow [Cat]-CH=CH-CH=CH_2$

　　$[Cat]-CH=CH-CH=CH_2 + HC\equiv CH \longrightarrow [Cat]-H + C_6H_6$

The second sequence leading to the major product, benzene, assumes that the catalyst AlR_3 is changed into a catalyst with at least one Al—H bond. If this is true, catalysts like Et_2AlH-$TiCl_4$ should give substantially more benzene and less ethylbenzene. The use of such catalysts has not been reported.

6.5. Grignard Reagents as Cyclotrimerization Catalysts

When diphenylacetylene and 0.5 mole of phenylmagnesium bromide are refluxed in xylene for 4 hours, hexaphenylbenzene and octaphenylcubane are formed.[123] Other acetylenes also give substituted benzenes. Many organometallics are inactive: dimethylmagnesium, divinylmagnesium, ethynylmagnesium bromide, etc.

References

1. Berthelot, M., *Compt. Rend.*, **62**, 905 (1866).
2. Nieuwland, J. A., and Vogt, R. R., "The Chemistry of Acetylene," New York, Reinhold Publishing Corp., 1945.
3. Fujiki, S., *Nippon Kagaku Zasshi*, **87**, 189 (1966); *Chem. Abstr.*, **65**, 16215 (1966).
4. Harris, J. F., Jr. (to E. I. du Pont de Nemours & Co.), U.S. Patent 2,923,746 (Feb. 2, 1960); *Chem. Abstr.*, **54**, 9799 (1960).
5. Riemschneider, R., and Brendel, K., *Ann. Chem.*, **640**, 5 (1961).
6. Middleton, W. J., and Sharkey, W. H., *J. Am. Chem. Soc.*, **81**, 803 (1959).
7. Viehe, H. G., *et al.*, *Angew. Chem.*, **76**, 888 (1964).
8. Roberts, J. D., and Sharts, C. M., *Org. Reactions*, **12**, 1 (1962).

9. Brown, H. C., *J. Org. Chem.*, **22**, 1256 (1957).

10. Huntsman, W. D., and Hall, R. P., *J. Org. Chem.*, **27**, 1988 (1962).

10a. Le Goff, E., and La Count, R. B., *Tetrahedron Letters*, 2333 (1967).

11. Huntsman, W. D., and Wristers, H. J., *J. Am. Chem. Soc.*, **89**, 342 (1967).

12. Müller, E., Sauerbier, M., and Heiss, J., *Tetrahedron Letters*, 2473 (1966).

13. Cope, A. C., and Meili, J. E., *J. Am. Chem. Soc.*, **89**, 1883 (1967).

13a. White, E. H., and Sieber, A. F., *Tetrahedron Letters*, 2713 (1967).

14. Reikhsfel'd, V. O., and Makovetskii, K. L., *Russ. Chem. Rev. (English Transl.)*, **35**, 510 (1966).

14a. Ipaktschi, J., and Staab, H. A., *Tetrahedron Letters*, 4403 (1967).

15. Kritskaya, I. I., *Russ. Chem. Rev. (English Transl.)*, **35**, 167 (1966).

16. Coates, G. E., and Glocking, F., in "Organometallic Chemistry," p. 458, New York, Reinhold Publishing Corp., 1960.

17. Nast, R., and Kasperl, H., *Chem. Ber.*, **92**, 2135 (1959).

18. Chatt, J., Guy, R. G., and Duncanson, L. A., *J. Chem. Soc.*, 827 (1961).

19. Chatt, J., Rowe, G. A., and Williams, A. A., *Proc. Chem. Soc.*, 208 (1957).

19a. Carty, A. J., and Efraty, A., *Can. J. Chem.*, **46**, 1598 (1968).

20. Nast, R., Vester, K., and Griesshammer, H., *Chem. Ber.*, **90**, 2678 (1957).

21. Nast, R., and Griesshammer, H., *Chem. Ber.*, **90**, 1315 (1957).

22. Nast, R., and Sirtl, E., *Chem. Ber.*, **88**, 1723 (1955).

23. Nast, R., *et al.*, *Chem. Ber.*, **96**, 3302 (1963).

24. Nast, R., and Koehl, H., *Z. Anorg. Allgem. Chem.*, **319**, 135 (1963).

25. Nast, R., and Koehl, H., *Chem. Ber.*, **97**, 207 (1963).

26. Helmkamp, G. K., Carter, F. L., and Lucas, H. J., *J. Am. Chem. Soc.*, **79**, 1306 (1957).

27. Avakyan, S. N., and Voskerchyen, S. V., *Armyansk. Khim. Zh.*, **19**, 19 (1966); *Chem. Abstr.*, **65**, 3308 (1966).

27a. Wittig, G., and Fritze, P., *Ann. Chem.*, **712**, 79 (1968).

28. Allen, A. D., and Cook, C. P., *Can. J. Chem.*, **42**, 1063 (1964).

29. Chatt, J., Duncanson, L. A., and Guy, R. G., *Chem. Ind. (London)*, 430 (1959).

30. Chatt, J., and Shaw, B. L., *Chem. Ind. (London)*, 675 (1959).

31. Chatt, J., and Shaw, B. L., *J. Chem. Soc.*, 1718 (1960).

32. Tilney-Bassett, J. F., and Mills, O. S., *J. Am. Chem. Soc.*, **81**, 4757 (1959).

33. Dubeck, M., *J. Am. Chem. Soc.*, **82**, 502 (1960).

34. Schloegl, K., and Steyrer, W., *Monatsh. Chem.*, **96**, 1520 (1965); *Chem. Abstr.*, **65**, 5132 (1966).

35. Blount, J. F., *et al.*, *J. Am. Chem. Soc.*, **88**, 292 (1966).

36. Nakamura, A., *Bull. Chem. Soc. Japan*, **38**, 1868 (1965); *Chem. Abstr.*, **64**, 5132 (1966).

37. Nesmeyanov, A. N., *et al.*, *Izv. Akad. Nauk SSSR Ser. Khim.*, 774 (1966); *Chem. Abstr.*, **65**, 8229 (1966).

38. Tate, D. P., *et al.*, *J. Am. Chem. Soc.*, **86**, 3261 (1964).

39. Strohmeier, W., and Hobe, D. V., *Z. Naturforsch.*, **19b**, 959 (1964); *Chem. Abstr.*, **62**, 1681 (1965).

40. Longuet-Higgins, H. C., and Orgel, L. E., *J. Chem. Soc.*, 1969 (1956).

41. Blomquist, A. T., and Maitlis, P. M., *J. Am. Chem. Soc.*, **84**, 2329 (1962).

42. Maitlis, P. M., *et al.*, *Can. J. Chem.*, **43**, 470 (1965).

42a. Cookson, R. C., and Jones, D. W., *J. Chem. Soc.*, 1881 (1965).

43. Malatesta, L., *et al.*, *Angew. Chem.*, **72**, 34 (1960).
43a. Müller, E., *et al.*, *Ann. Chem.*, **713**, 40 (1968).
44. BASF, A.-G., German Patent 1,005,954 (April 11, 1957).
45. Schrauzer, G. N., *J. Am. Chem. Soc.*, **81**, 5310 (1959).
46. Sauer, J. C., and Cairns, T. L., *J. Am. Chem. Soc.*, **79**, 2659 (1957).
47. Bieber, Th. I., *Chem. Ind. (London)*, 1126 (1957).
48. Tsuji, J., and Susuki, T., *Tetrahedron Letters*, 3024 (1965).
49. Drefahl, G., Hoerhold, H. H., and Bretschneider, H., *J. Prakt. Chem.*, **25**, 113 (1964); *Chem. Abstr.*, **62**, 463 (1965).
50. Heubel, W., and Merenyi, R., *J. Organometall. Chem.*, **2**, 213 (1964).
51. Boston, J. L., Sharp, D. W. A., and Wilkinson, G. A., *J. Chem. Soc.*, 3488 (1962).
51a. Helling, J. F., Rennison, S. C., and Merijan, A., *J. Am. Chem. Soc.*, **89**, 7140 (1967).
52. Stolz, I. W., Dobson, G. R., and Sheline, R. K., *Inorg. Chem.*, **2**, 1264 (1963).
53. Tirpak, M. R., Hollingsworth, C. A., and Wotiz, J. H., *J. Org. Chem.*, **25**, 687 (1960).
54. Markby, R., *et al.*, *J. Am. Chem. Soc.*, **80**, 6529 (1958).
55. Greenfield, H., Wender, I., and Wotiz, J. H., *J. Org. Chem.*, **21**, 875 (1956).
56. Sternberg, H. W., *et al.*, *J. Am. Chem. Soc.*, **78**, 3621 (1956).
57. Reppe, W., and Vetter, H., *Ann. Chem.*, **582**, 133 (1953).
58. Clarkson, R., *et al.*, *J. Am. Chem. Soc.*, **78**, 6206 (1956).
59. King, R. B., *et al.*, *Inorg. Chem.*, **5**, 684 (1966).
60. Huebel, K. W., and Braye, E. H. (to Union Carbide Corp.), U.S. Patent 3,097,153 (July 9, 1963); *Chem Abstr.*, **60**, 461 (1964).
61. Huebel, R. W., and Braye, E. H., U.S. Patent 3,096,265 (July 2, 1963); *Chem. Abstr.*, **60**, 2848 (1964).
62. Sternberg, H. W., *et al.*, *J. Am. Chem. Soc.*, **81**, 2339 (1959).
63. Albanesi, G., and Tovagliere, M., *Chim. Ind. (Milan)*, **41**, 189 (1959); *Chem. Abstr.*, **53**, 19872 (1959).
64. Pino, P., and Miglierina, A., *J. Am. Chem. Soc.*, **74**, 5551 (1952).
65. Heck, R. F., *J. Am. Chem. Soc.*, **86**, 2819 (1964).
66. Schrauzer, G. N., *J. Am. Chem. Soc.*, **81**, 5307 (1959).
67. Markby, R., Sternberg, H. W., and Wender, I., *Chem. Ind. (London)*, 1381 (1959).
68. Dahl, L. F., *et al.*, *J. Am. Chem. Soc.*, **88**, 446 (1966).
69. Reppe, W., and Schweckendieck, W. J., *Ann. Chem.*, **560**, 104 (1948).
70. Kleinschmidt, R. F. (to General Aniline & Film Corp.), U.S. Patent 2,542,417 (Feb. 20, 1951); *Chem. Abstr.*, **45**, 7594 (1951).
71. Rose, J. D., and Statham, F. S., *J. Chem. Soc.*, 69 (1950).
72. McKeever, C. H., and VanHook, J. O. (to Rohm and Haas Co.), U.S. Patent 2,542,551 (Feb. 20, 1951).
73. Harris, J. F., Jr., Harder, R. J., and Sausen, G. N., *J. Org. Chem.*, **25**, 633 (1960).
74. Cairns, T. L., *et al.*, *J. Am. Chem. Soc.*, **74**, 5636 (1952).
75. Reed, H. W. B., *J. Chem. Soc.*, 1931 (1954).
76. Meriwether, L. S., *et al.*, *J. Org. Chem.*, **26**, 5155 (1961).
77. Meriwether, L. S., Colthup, E. C., and Kennerly, G. W., *J. Org. Chem.*, **26**, 5163 (1961).
78. Meriwether, L. S., *et al.*, *J. Org. Chem.*, **27**, 3930 (1962).
79. Leto, J. R., and Leto, M. F., *J. Am. Chem. Soc.*, **83**, 2944 (1961).
80. Schrauzer, G. N., *Chem. Ber.*, **94**, 1403 (1961).

81. Schrauzer, G. N., and Eichler, S., *Chem. Ber.*, **95**, 2764 (1962).

82. Schrauzer, G. N., and Glockner, P., *Chem. Ber.*, **97**, 2451 (1964).

83. Krespan, C. G., McKusick, B. C., and Cairns, T. L., *J. Am. Chem. Soc.*, **83**, 3428 (1961).

84. Huebel, W., and Hoogzand, C., *Chem. Ber.*, **93**, 103 (1960).

84a. Rosenblum, M., Braun, N., and King, B., *Tetrahedron Letters*, 4421 (1967).

85. Kruerke, U., and Huebel, W., *Chem. Ber.*, **94**, 2829 (1961).

86. Chini, P., Santambrogio, A., and Palladino, N., *J. Chem. Soc.* (*C*), 830, 836 (1967).

87. Zeiss, H., in "Organometallic Chemistry," p. 406, New York, Reinhold Publishing Corp., 1960.

88. Zeiss, H., and Herwig, W., *J. Am. Chem. Soc.*, **80**, 2913 (1958).

89. Herwig, W., Metlesics, W., and Zeiss, H. H., *J. Am. Chem. Soc.*, **81**, 6203 (1959).

89a. Whitesides, G. M., and Ehmann, W. J., *J. Am. Chem. Soc.*, **90**, 804 (1968).

90. Strohmeier, W., and Barbeau, Cl., *Z. Naturforsch.*, **19b**, 262 (1964).

91. Metlesics, W., and Zeiss, H. H., *J. Am. Chem. Soc.*, **81**, 4117 (1959).

92. Tsutsui, M., and Zeiss, H., *J. Am. Chem. Soc.*, **82**, 6255 (1960).

93. *Ibid.*, **83**, 825 (1961).

93a. Sasaoki, T., and Suzuki, Y., *Tetrahedron Letters*, 3137 (1967).

94. Heimbach, P., *Angew. Chem. Intern. Ed.*, **5**, 961 (1966).

95. Reppe, W., *et al.*, *Ann. Chem.*, **560**, 1 (1948).

96. Schroder, G., "Cyclooctatetraen," Weinheim/Bergstr., Germany, Verlag Chemie, 1965.

97. Schrauzer, G. N., *Angew. Chem., Intern. Ed.*, **3**, 185 (1964).

98. Reppe, W., P. B. Report 40837-T, Office of Technical Services, Dept. of Commerce, Washington, D.C.

99. Fredericks, R. J., Lynch, D. G., and Daniels, W. E., *J. Polymer Sci. Part B*, **2**, 803 (1964).

100. Clark, A., *et al.*, *World Petrol Congr. Proc.*, *5th, New York*, **4**, 267 (1959).

101. Massimilla, L., Saracini, L., and Mastrovita, E., *Chem. Ind.* (*Milan*), **44**, 341 (1962); *Chem. Abstr.*, **57**, 11060 (1962).

102. Shapiro, I., and Weiss, H. G., *J. Am. Chem. Soc.*, **79**, 3294 (1957).

103. Evans, A. G., James, E. A., and Phillips, B. D., *J. Chem. Soc.*, 1016 (1965).

104. Meijer, H. J. DeL., Pauzenga, U., and Jellinek, F., *Rec. Trav. Chim.*, **85**, 634 (1966).

105. Ciba, Ltd., British Patent 973,934 (Nov. 4, 1964); *Chem. Abstr.*, **62**, 11734 (1965).

105a. Rosenberg, H. M., and Eimutis, E. C., *Can. J. Chem.*, **45**, 2263 (1967).

105b. Fox, B. L., and Rosenberg, H. M., *J. Org. Chem.*, **33**, 1292 (1968).

106. Berkoff, C. E., *et al.*, *J. Chem. Soc.*, 194 (1965).

107. Schafer, W., *Angew. Chem., Intern. Ed.*, **5**, 669 (1966).

108. Criegee, R., and Moschel, A., *Chem. Ber.*, **92**, 2181 (1959); *Org. Syn.*, **46**, 34 (1966).

109. Nesmeyanov, A. N., Borisov, A. E., and Savel'eva, I. S., *Izv. Akad. Nauk SSSR Otd. Khim. Nauk*, 1034 (1959); *Chem. Abstr.*, **54**, 1366 (1960).

110. Smith, W. S. (I.C.I.), British Patent 802,510 (Oct. 8, 1958); *Chem. Abstr.*, **53**, 8070 (1959).

111. Ziegler, K., *et al.*, *Ann. Chem.*, **629**, 1 (1960).

112. Natta, G., *et al.*, *Gazz. Chim. Ital.*, **89**, 465 (1959); *Chem. Abstr.*, **54**, 11967 (1960).

113. Natta, G., and Pasquon, I., *Advan. Catalysis*, **11**, 1 (1959).

114. Franzus, B., Canterino, P. J., and Wickliffe, R. A., *J. Am. Chem. Soc.*, **81**, 1514 (1959).

115. Reikhsfel'd, V. O., Makovetskii, K. L., and Erokhina, L. L., *Zh. Obshch. Khim.*, **32**, 653 (1962); *Chem. Abstr.*, **58**, 459 (1963).
116. Lutz, E. F., *J. Am. Chem. Soc.*, **83**, 2551 (1961).
117. Hubert, A. J., and Dale, J., *J. Chem. Soc.*, 3160 (1965).
118. Hubert, A. J., *J. Chem. Soc.* (*C*), 6, 11, 13 (1967).
118a. Hubert, A. J., *J. Chem. Soc.* (*C*), 1984 (1967).
119. Hoover, F. W., Webster, O. W., and Handy, C. T., *J. Org. Chem.*, **26**, 2234 (1961).
120. Luttinger, L. B., *Chem. Ind.* (*London*), 1135 (1960); *J. Org. Chem.*, **27**, 1591 (1962).
121. Donda, A. F., and Moretti, G., *J. Org. Chem.*, **31**, 985 (1966).
122. Ikeda, S., and Tamaki, A., *J. Polymer Sci. B, Polymer Letters*, **4**, 605 (1966).
123. Throndson, H. P., and Zeiss, H., *J. Organometall. Chem.*, **1**, 301 (1964).

Chapter Five

POLYMERS FROM ACETYLENIC COMPOUNDS

Introduction

Acetylenic monomers have been studied much less than olefinic monomers. One obvious reason is that many acetylenic monomers are relatively hard to make, expensive and not commercially available. Acetylene itself is the only very cheap acetylenic monomer. While the polyenes made by polymerizing acetylene are interesting, apparently none of them is in commercial use. Certainly not enough work has been published to allow the conclusion that acetylenic monomers have little potential. Industrial polymer chemists, who do most of the extensive work necessary to develop a commercial polymer, are understandably reluctant to devote time to polymers from expensive and unavailable monomers. It seems obvious that new chemistry and new systems must be devised by acetylene chemists if they are to make economical acetylenic monomers available to the polymer chemists.

Acetylenic monomers are polymerized by Ziegler-Natta catalysts, by complexed nickel catalysts, and by anionic, free radical and cationic catalysts, among others. Acetylenes can polymerize thermally and under the influence of irradiation. Many of the catalysts which oligomerize and cyclotrimerize acetylenes (see Chapter 4) are also active polymerization catalysts.

Addition across the acetylenic bond gives saturated or olefinic polymers. 1,3-Dipolar addition can form polymers containing heterocyclic rings. Oxidative coupling has been used as a polymer-forming reaction to give conjugated polyynes. Acetylenic acids and alcohols polymerize to form acetylenic polyesters and polyurethanes. The Mannich reaction of diacetylenes and diamines has not yet been reported as a polymerization system, although diacetylenes give high yields in the usual Mannich reaction.[1] Thus, the triple bond allows a variety of polymerization reactions not possible with other monomers.

The conjugated olefinic polymers formed by homo- and co-polymerization of acetylenes are interesting because they are organic semiconductors. The polymer from hexafluoro-2-butyne is remarkable: it does not melt at 800°. Many of the addition polymers show good thermal stability. The condensation polymers containing triple bonds were made mainly to investigate the effect of rigid rod-like segments on polymer properties.

PART ONE

Addition Polymerization to Form Polymers Containing
Conjugated Olefinic Bonds

1. GENERAL PROPERTIES OF POLYMERS

Acetylene polymers[2] and allene polymers[3] were reviewed recently. In 1964, Pen'kovskii reviewed the synthesis and properties of polymers with conjugated double bonds.[4] Generally, the polymers from acetylene and various catalysts contain 12–15 units from acetylene, are black, and are insoluble. Benzene is a high-yield by-product in many polymerizations. Liquid acetylene at low temperature is polymerized by radiation to form brown *cis* polymer, and solid acetylene is polymerized to red insoluble *trans* polymer.[5] Propargyl alcohol

$$CH_2OH$$

gives dark red *trans* polymer, $-(CH=C-)_n$. Phenylacetylene is polymerized by fast electrons or gamma rays to *trans* polymers with molecular weights up to 1700. The rate increases linearly with the integral dose of irradiation between wide limits and is not temperature sensitive. The polymers are assumed to be head-to-tail.

Phenylacetylene forms polymers with a mean molecular weight of 5000 with Ziegler catalysts, and polymers with a mean molecular weight of 7000 over copper oxide in a flow system at 300°. *p*-Diethynylbenzene polymerizes thermally to a three-dimensional polymer containing conjugated unsaturation. β-Chlorovinyl methyl ketone polymerizes rapidly in the presence of ferric

$$COCH_3$$

chloride to form $-(CH=CH-)_n$, losing HCl in the polymerization.

The most characteristic feature of polymers with many conjugated double bonds is a single narrow EPR signal with a g factor of 2.003, corresponding to a free electron. Most of these polymers are semiconductors. Generally, the polyenes:

(1) Are stable to air up to 200–250°.

(2) Are easily oxidized by benzoyl peroxide to form acids.

(3) Add chlorine to give white, insoluble products containing about 50% Cl.

(4) Do not react with maleic anhydride.

(5) Hydrogenate slowly in the presence of Raney nickel.

(6) Are free radicals, stable because of their molecular size and because the unpaired electrons are delocalized.

(7) Can be compressed to give positive temperature coefficients of electrical conductance. When polyacetylene is compressed at 1000–4000 atmospheres, the electrical resistivity decreases from 10–15 megohm/cm down to 4.8 megohm/cm.[6]

2. ORGANIC SEMICONDUCTORS

Organic semiconductors are interesting because they offer such broad ranges of electronic properties and because they have the highest known piezoconductive coefficients. Semiconductors have conduction intermediate between metals and insulators (10^3–10^{-10} mho/cm), they supply their own carriers, have positive temperature coefficients of conductivity, are highly sensitive to impurities or to morphology, and exhibit photoconduction and photovoltaism.

Ziegler catalysts polymerize alkynes at 20–70° to semiconducting polymers[7] (Table 5-1). Polypropyne is an orange, brittle solid; polybutyne-1 and polypentyne-1 are deep blue-black, soft and rubbery; polyphenylacetylene is a deep orange powder.

TABLE 5-1
Properties of Polymers from Various Acetylenic Hydrocarbons[7]

Polymer	Intrinsic Viscosity	% Soluble in Toluene	Resistivity, 25° (ohm/cm)	ΔE eV
Polypropyne	0.07	100	1.5×10^{11}	0.645
Polybutyne-1	0.16	99		
Polypentyne-1	0.21	93	$3 \times 10^9 (10–15°)$	
Polyphenylacetylene	0.16	96	4.8×10^{10}	0.432

So-called ekaconjugation probably exists in some of the polymers.[7] Ekaconjugation means that the molecular systems have considerable excited electronic states at room temperature; thus it is a measure of the mobility of electrons. In conjugated systems, the electronic orbitals overlap so that continuous orbitals exist along long segments of the molecule. This increases conductance.

Beck[8] prepared polyphenylene semiconductors by aluminum chloride-catalyzed dehydrogenation-polymerization of benzene, naphthalene, anthracene, etc. He prepared polyvinylene by polymerizing acetylene in solution with active cobalt or Ziegler catalysts. The conductivity of powdered samples under pressure does not change above 300 atmospheres. The conductivity increases as the temperature of polymerization increases and as the oxygen content of the polymer decreases. The sign of the thermopotential is positive with respect to the cold pole, indicating p-conduction (defect electron). The polyacetylenes autooxidize 1000 times as fast as the polyphenylenes.

3. ANIONIC POLYMERIZATION

3.1. Acetylenic Hydrocarbons

Phenylacetylene in toluene at 30° polymerizes in the presence of triethylboron and an electron donor such as oxygen. Maximum conversion is at oxygen:triethylboron ratio of 0.2.[9] Tetrahydrofuran, methanol and trimethylpyridine are also effective electron donors; THF is best. With THF, conversion is independent of the oxygen:triethylboron ratio. Maximum conversion with THF-Et_3B is 19%. With oxygen as donor, the polymer is an amorphous orange solid, melting around 150°. The molecular weight is around 1500. Infrared shows conjugated double bonds and aromatic rings.

deVries[10] prepared polymers by an unusual reaction. When he hydrolyzed the product from 2,4-hexadiyne (nonterminal acetylene groups) and sodium amide in ammonia, black insoluble polymer formed spontaneously. He prepared products in several ways and found the resistivity to be 10^6–10^{17} ohm/cm. One of the products contained 50 double bonds per molecule.

3.2. Other Acetylenic Compounds

Benes[11] in 1962 reported one of the first attempts to polymerize acetylenic compounds other than hydrocarbons. Dicyanoacetylene polymerizes easily in the presence of an anionic catalyst: triethylphosphine, butylmagnesium bromide, potassium benzophenone, sodium naphthalene, butyllithium or sodium cyanide (in dimethylformamide). Butyllithium at −73° gives 90% yield of black powder, soluble in most common solvents. ESR shows 10^{18} unpaired electrons per gram. Infrared shows conjugated double bonds and nitrile groups. The ebullioscopic molecular weight is 550, but the actual weight is higher (some low molecular weight catalyst fragments in the sample). Several other anionic catalysts can form black polymers, different from "black Orlon," made by pyrolyzing polyacrylonitrile.[12] The product from butyllithium is:

$$Bu-\underset{\underset{CN}{|}}{\overset{\overset{CN}{|}}{C}}{=}C{\left[\underset{\underset{CN}{|}}{\overset{\overset{CN}{|}}{C}}{=}C\right]}_n\overset{\overset{CN}{|}}{C}{=}\underset{\underset{CN}{|}}{C^-}\quad Li^+$$

MacNulty[13] reviewed the formation of "black Orlon" from "Orlon," and then described the polymerization of cyanoacetylene, propiolamide and methyl propiolate. Methyl propiolate polymerizes spontaneously with bases such as triethylamine to give 50–60% yield of polymer. The polymer is soluble in polar solvents. It sinters at 220° and melts to a firm tar between 254 and 280°. Cyanoacetylene gives a polymer which melts above 400°. It is insoluble in most organic solvents and is stable to strong acids except concentrated nitric acid or aqua regia. Propiolamide polymerizes to a solid which partly sinters at 310°. A little liquid distills out, leaving a solid which does not melt at 360°. MacNulty could not polymerize propiolamide and cyanoacetylene with Meriwether's catalyst[14] [(Ph$_3$P)$_2$Ni(CO)$_2$].

Elemental analyses and the end group analyses indicate that the polymers have chains with four to ten conjugated double bonds. The polymer from cyanoacetylene is most heat resistant. It decomposes around 600° and has electron spin concentrations nearly the same as "black Orlon."

1,6-Dichloro-2,4-hexadiyne is polymerized by a solution or slurry of sodium cyanide or cuprous cyanide.[15] The polymers are insoluble, are stable to sulfuric acid at 100°, and do not decompose at 600°. The polymers contain cyano and imino groups, and are probably aromatic. Propargyl bromide polymerizes, but propargyl alcohol does not.

4. FREE RADICAL POLYMERIZATION

4.1. Thermal

Okamoto[16] reviewed the literature on acetylene polymers in 1961 and reported some thermal polymerizations of phenylacetylene. Polymers form at 130–195°, either neat or in solvents. Nearly quantitative yields are obtained after 18–65 hours. Crude polymers are deep red, and are yellow or orange powders after reprecipitation from solvent. No cyclization occurs, even in the neat polymerization. Phenylacetylene polymers are *trans* $+\overset{\overset{Ph}{|}}{C}{=}CH)_n$, $n = 5$–10 (molecular weight = 500–1000). The terminal group is solvent molecule, except when polymer is made in chlorobenzene or neat. Room temperature resistivity is

10^{16} ohm/cm, and the apparent energy gap is 0.963 eV. Thus the polymer is a dielectric, although 1000 times as good a conductor as polystyrene.

When α,β-dihalo compounds are heated to 200–300° in the presence of bases (CaO or tertiary amines), polymers form. Acetylenes are presumed to be the intermediates, since the polymers contain conjugated double bonds.[16a] For example, dibromopropionitrile gives a new polymer, $+C=CH+_n$ at 200°
$\quad\quad\quad\quad\quad\quad\quad\quad\quad\quad\quad\quad\quad\quad\quad\quad\quad\quad$ |
$\quad\quad\quad\quad\quad\quad\quad\quad\quad\quad\quad\quad\quad\quad\quad\quad\quad\quad$ CN

At 300°, cyclization occurs to form a polymer similar to "black Orlon":

Copolymers can also be formed:

$$\begin{array}{c} \text{Br} \quad \text{Br} \\ | \quad\quad | \\ n\text{CH}-\text{CH}_2 \\ | \\ \text{CN} \end{array} + \begin{array}{c} \text{Br} \quad \text{Br} \\ | \quad\quad | \\ n\text{CH}-\text{CH}_2 \\ | \\ \text{C}_6\text{H}_5 \end{array} \longrightarrow \begin{array}{c} \\ +\text{C}=\text{CH}-\text{C}=\text{CH}+_n \\ | \quad\quad\quad | \\ \text{CN} \quad\quad \text{C}_6\text{H}_5 \end{array}$$

If the dihalide, or an aldehyde or a ketone, are heated in the presence of calcium carbide, the acetylene produced by the elimination reaction between the calcium carbide and the other compound enters the reaction and copolymers form:

$$n\text{CaC}_2 + \begin{array}{c} \text{Br} \quad \text{Br} \\ | \quad\quad | \\ n\text{CH}-\text{CH}_2 \\ | \\ \text{R} \end{array} \xrightarrow[-n\text{CaBr}_2]{200°} \left[\text{HC}\equiv\text{CH} + \begin{array}{c} \text{C}\equiv\text{CH} \\ | \\ \text{R} \end{array}\right]_n \longrightarrow \begin{array}{c} \\ +\text{CH}=\text{CH}-\text{C}=\text{CH}+_n \\ | \\ \text{R} \end{array}$$

A wide variety of polymers and copolymers have been made by these methods.[16a] Molecular weights are usually fairly low, indicating that 10–30 monomer units have entered the polymer chain.

4.2. Radiation

Ivanov[5] reviewed radiation polymerization of monomers, including alkynes, in 1966. At low concentrations, acetylene is a retarder and chain transfer agent for the γ-ray induced polymerization of ethylene, but at higher concentrations, acetylene-ethylene copolymer forms.[17] γ-Rays polymerize phenylacetylene to a red polymer with a G value of 11, and 1-butyne to a polymer with $G = 186$.[18] Acetylenes irradiated with 1.5-MeV electrons at -196 to 80° give polymers (Table 5-2).[19] The polymer from phenylacetylene contains some chains terminated by cyclic trimers (triphenylbenzenes). The 1-hexyne polymer has a double bond in the side chain.

TABLE 5-2
Polymerization of Acetylenes by Electron Irradiation[19]

Acetylene	Moles Polymerized per 100 eV	Average M.W.	Degree of Polymerization
Phenylacetylene	9	1100	11
1-Hexyne	7	900	11
Cyclohexylacetylene	5	1400	13

A solution of 80% bromoacetylene-20% dibromoethylene in vinyl acetate copolymerizes in a sealed ampule in sunlight. The brown amorphous polymer probably is: $+CH(OAc)-CH_2CH=CBr+_n$.[20]

γ Irradiation of liquid propiolic acid gives a polymer with a molecular weight of 1320 (degree of polymerization = 19). The polymerization is independent of dose rate and temperature.[20a] Solid state polymerization allows a higher degree of polymerization (up to 30). All of the carboxyl groups are on the same side of the polymer backbone. At 220°, the carboxyl groups undergo anhydrization. The polymer from liquid phase reaction is less ordered; dehydration gives acyclic intermolecular anhydrides.

4.3. Catalytic

Radical polymerization of acetylene in the presence of vinyl chloride gives crystalline polyacetylene, along with PVC or a copolymer.[58] The PVC dehydrohalogenates at 180° over solid NaOH to give similar crystalline polyacetylene. These polyacetylenes are semiconductors.

5. ZIEGLER-NATTA POLYMERIZATION
5.1. Acetylene

In 1958, Natta[21] described homopolymerization of acetylene by Ziegler catalysts. In heptane at -20 to $+80°$, Et_3Al-$Ti(OPr)_4$ gives 90–95% yield of dark crystalline polymer, completely insoluble. At Al:Ti ratio of 2.5, 98.5% of the acetylene is polymerized, but 1 gram of catalyst produces only 2 grams of polymer. The polymer suspended in carbon tetrachloride adds chlorine to form a white amorphous solid containing 64% chlorine. Similar Ziegler catalysts homopolymerize other acetylenic hydrocarbons.[22] The polymers can be used to reinforce natural or synthetic rubber when mixed in before vulcanization.

Ethylene and ethane evolve when triethylaluminum and vanadyl bis(acetylacetonate) react to form Ziegler catalyst.[23] Magnetic moments indicate that the vanadium is reduced from $+4$ to $+2$. The maximum rate of acetylene polymerization occurs at an Al:V ratio of 4. Polymerization is first order on acetylene, and the activation energy is nearly zero between 20 and 40°. Infrared shows that the polymer is a conjugated *trans*-polyene. The polymer oxidizes easily in air at room temperature. After 10 days, 9% of the calculated oxygen is absorbed. This polymer can also be chlorinated.[21] The degree of polymerization is 40–50.

Nicolescu[24] studied the effect of various ratios of Al:Ti in Et_3Al-$Ti(OBu)_4$ on the rate of polymerization of acetylene. Rate, composition and electrical conductivity of catalysts are related. The value of the activation energy of the electrical conductivity correlates with the energy required to break the complex ion aggregate of catalyst, and thus correlates with catalyst activity. In another study[25] the yield of polyacetylene increased from 0.5 to 5.6% in 4 hours as the Al:Ti ratio was increased from 3.6–15.5.

$Al(i\text{-}Bu)_3$-pyridine with dimethylglyoxime-ferrous complex catalyzes the room temperature polymerization of acetylene in heptene or benzene.[26] The rate of polymerization increases gradually as the ratio of $Al(i\text{-}Bu)_3$ increases. The most active catalysts form when the ingredients are mixed in the presence of acetylene. Oxygen decreases catalyst activity, and ethylene increases it.

The low molecular weight polyacetylenes obtained with Ziegler catalysts contain up to 12 conjugated double bonds. The group at one end is always R from AlR_3; R inserts itself between the growing chain and the end attached to the catalyst. The other end group is vinyl.[27]

In 1961, Watson[28] reported one of the first systematic studies of Ziegler catalysts for polymerizing acetylene and 1-alkynes. He determined the following optimum ratios of organometallic to $TiCl_4$: $Al(i\text{-}Bu)_3$:$TiCl_4$ = 2.0–2.5 (11.4 mmoles $TiCl_4$ + 28.6 mmoles $Al(i\text{-}Bu)_3$ polymerize 7.6 grams of acetylene); $BuLi$:$TiCl_4$ = 4.0; Et_2Zn:$TiCl_4$ = 2.5–3.0. As the reaction temperature increases from 21–62°, more acetylene polymerizes and the percentage of soluble polymer increases. The apparent activation energy of acetylene polymerization is 7.0 kcal/mole, compared to 10 for ethylene and 12–14 for propylene with the same catalysts. The ratio of soluble to insoluble polymer is a linear function of catalyst concentration and increases with increased catalyst concentration. Thus, more catalyst gives shorter and/or less cross-linked polymers.

The most active catalysts are made by saturating the metal alkyl solution with acetylene before adding $TiCl_4$. Propyne and 1-butyne give polymers of lighter color and lower molecular weight. The polymers from $TiCl_4$-$Al(i\text{-}Bu)_3$ have both *cis*- and *trans*-ethylenic groups, while polymers from $TiCl_4$-$BuLi$ are mostly *trans*. EPR measurements of polymers give values of 0.94–14.6×10^{18} spins per gram. Resistance is about 10^4 ohm-cm at 25°. Temperature dependence is typical of semiconductors.

Generally the yield of polyacetylene per gram of Ziegler catalyst is low. With $Al(i\text{-}Bu)_3$-$TiCl_4$ catalyst at Al:Ti = 6, the amount of polymer per gram of $TiCl_4$ increases from 2–10 grams as the reaction temperature decreases from

+60 to −60°. The rate also increases. Dioxane increases catalyst productivity. At an Al:Ti ratio 3.5, adding 0.05–0.1 mole of dioxane per mole of $TiCl_4$ increases productivity from 3.4–7.2 grams of polymer per gram of catalyst.[29]

5.2. Acetylenic Hydrocarbons

Noguchi[30] in 1963 reviewed his earlier work with triethylaluminum-metal acetylacetonate catalysts. These are better for polymerizing phenylacetylene than standard Ziegler catalysts. In toluene at 30°, a combination of 3 $AlEt_3$ and 1 VO(salicylaldehyde)$_2$ gives the polymer fraction (2) (Table 5-3) in 58.3% yield (number average molecular weight = 7700). Noguchi also described a new system: the metal alkyl-metal dimethylglyoximates. Using these catalysts, he made polymers and fractionated them according to solubility. Fraction (1) was insoluble in benzene, (2) was soluble in benzene and insoluble in methanol, and (3) was soluble in methanol (Table 5-3).

For dimethylglyoximate complexes at 30° the best catalyst is $3AlEt_3 : 1Fe(di-$ methylglyoximate)$_2$, 99% total yield; 40% yield of fraction (2), number average molecular weight 8000. The catalyst system $AlEt_3 : Fe(dmg)_2 : 2pyridine$ gives the results shown in Table 5-3 in 1 hour at 30°. $Fe(dmg)_2$ is a better co-catalyst than $Fe(dmg)_2 : 2pyridine$, which is better than $Fe(dmg)_2 : 2NH_3$. The chelates of iron, cobalt and nickel have the same orders of co-catalyst activity, regardless of the ligand. Donor molecules coordinated to the metal dimethylglyoximate

TABLE 5-3
Polymerization of Phenylacetylene by
AlEt$_3$: Fe(dmg)$_2$:2Pyridine[30]

AlEt$_3$: Fe(dmg)$_2$:2Py (mole ratio)	% Yield Fraction[a]			M_n Fraction (2) $\times 10^{-3}$
	1	2	3	
2.10	0	0	0.5	
4.20	0.2	3.0	2.3	4.0
5.25	2.4	5.2	4.8	3.7
6.30	4.1	15.8	6.4	8.0
8.39	32.2	20.8	14.0	4.5
9.44	38.4	37.2	14.1	7.8
10.49	22.0	48.5	20.9	5.1
11.54	4.4	34.3	15.5	8.9
12.59	1.4	16.5	12.4	6.5
15.73	Trace	6.1	11.3	3.8

[a] Fraction (1): insoluble in benzene.
Fraction (2): soluble in benzene, insoluble in methanol.
Fraction (3): soluble in methanol.

sharply decrease the co-catalytic activity. Thus, coordination of phenylacetylene or any monosubstituted acetylene with the transition metal of the catalyst is a very important factor.

Phenylacetylene and p-diethynylbenzene polymerize and copolymerize in the presence of R_3Al and $TiCl_3$ or $TiCl_4$.[31] The yield, molecular weight and viscosity of the polymer increase as temperature increases in the range between 70 and 400°. Diphenyldiacetylene, a nonterminal diyne, polymerizes in the presence of $Al(i\text{-}Bu)_3\text{-}TiCl_4$ to form a polymer which melts at 270°.[32] When the polymer is heated to 400° under nitrogen, the melting point increases to 480°. The polymer is photoelectric and semiconducting.

9-Vinylanthracene is polymerized 1,6 through the central ring of the anthracene to form poly-9,10-dimethylenanthracene. 9-Ethynylanthracene, however, is polymerized 1,2 by either Ziegler or acid catalysts to give

$$\text{-}\hspace{-2pt}+\hspace{-2pt}C\hspace{-2pt}=\hspace{-2pt}C\hspace{-2pt}+\hspace{-2pt}\text{-}$$
$$|$$
$$\text{anthracene}$$

The highest molecular weight (around 2840) is obtained by cationic polymerization in the presence of $TiCl_4$ at $-70°$.[32a]

Ethoxyacetylene polymerizes in the presence of Ziegler catalysts to give 5% dark, waxy solid, molecular weight 500–600, and 5% resin, molecular weight 1600.[33]

5.3. Copolymerization of Acetylenes with Olefins

Copolymers of acetylenes with olefins are valuable because incorporation of a little acetylene into the polyolefin gives enough conjugated double bonds to make the polymer reactive to acid anhydrides,[34,35] acidic chlorine compounds[34] and tetracyanoethylene.[35] The modified polyolefins accept dyes more easily than unmodified polyolefins.

With $VOCl_3\text{-}i\text{-}Bu_3Al$ catalyst, polymer from a charge of 930 moles of ethylene per mole of acetylene has a glass transition temperature of $-71°$, while the polymer from 17.7 moles of ethylene per mole of acetylene has a glass transition temperature of $-30°$. Increasing the amount of acetylene in the copolymer causes darker color.[35]

Phenylacetylene copolymerizes with an equal mixture of butene-1 and butene-2 to form an unsaturated polymer.[36] After 20 hours at 60° with $TiCl_3$-$AlEt_3$, 21% of the solid polymer is copolymer, 54% is polyphenylacetylene and 23% is polybutene. Styrene and acetylene give a solid polymer. The copolymer constitutes 20% of total polymer, and 32% of the copolymer is soluble in hot benzene. It contains 36% (weight) of acetylene. Copolymers of phenylacetylene and 1-pentene, and of propylene and phenylacetylene, have also been made.[37,38]

5.4. Nonconjugated Terminal Diynes

Nonconjugated diolefins polymerize to linear polymers, probably by alternating intramolecular-intermolecular propagation, since the polymers have no cross-linking and no cyclic recurring units. Some dienes, such as 1,6-heptadiene, biallyl and bimethallyl, give polymers containing cyclic recurring units.[39] Polybiallyl has methylenecyclopentane recurring units. Polymers of higher α,ω-diolefins have more noncyclic cross-linked chains.[40]

Stille[41] polymerized α,ω-diacetylenes to make a polymer containing alternating double bonds and single bonds along the backbone, and pendant cyclic units. He chose 1,6-heptadiyne as the first monomer, because he expected it to polymerize most easily by an internal head-to-tail propagation to form a six-membered carbocyclic recurring unit. $Al(i\text{-}Bu)_3\text{-}TiCl_4$ was the best catalyst, giving 80–90% conversion. Soluble $Al(i\text{-}Bu)_3\text{-}Ti(OEt)_4$ gave a large amount of cross-linked polymer, in addition to soluble polymer. Polymers were dark red or black, indicating much conjugation. The number average molecular weight was 13,500. On standing in solution, the dark color faded to pale yellow, probably because oxidation interrupted the conjugated sequences.

Poly-1,6-heptadiyne dehydrogenated at 350° over Pd catalyst to give a soluble polymer, with infrared consistent with m-xylene recurring units. The polymer added chlorine to form white solid containing 91% of the Cl calculated for addition of Cl_2 to each double bond. Oxidative decomposition of the ozonide gave 85% yield of glutaric acid. All evidence indicates that the polymer is

$$HC\equiv C(CH_2)_3C\equiv CH \longrightarrow$$

Poly-1,6-heptadiyne had a resistivity of 10^{10}–10^{13} ohm-cm. This high resistance is probably caused by instability of the polymer to many reagents and to air. A few reactions at some points in the chain break the conjugation.

1,7-Octadiyne and 1,8-nonadiyne gave 84 and 92% conversion, respectively, to polymer. Polymers were not highly colored and showed no absorption in the visible region, as poly-1,6-heptadiyne did. Extraction with boiling xylene dissolved 49% of the poly-1,7-octadiyne, and the soluble fraction showed only the highly conjugated olefinic bond absorption at 1605 cm^{-1}. Only 22% of the poly-1,8-nonadiyne was soluble, and it showed absorption for conjugated olefinic bonds and for acetylenic bonds.

Since active hydrogens in an olefinic monomer destroy Ziegler catalysts, it is unusual that 1-acetylenes polymerize instead of destroying the catalyst.[41] Stille proposed an intramolecular-intermolecular mechanism for the propagation step:

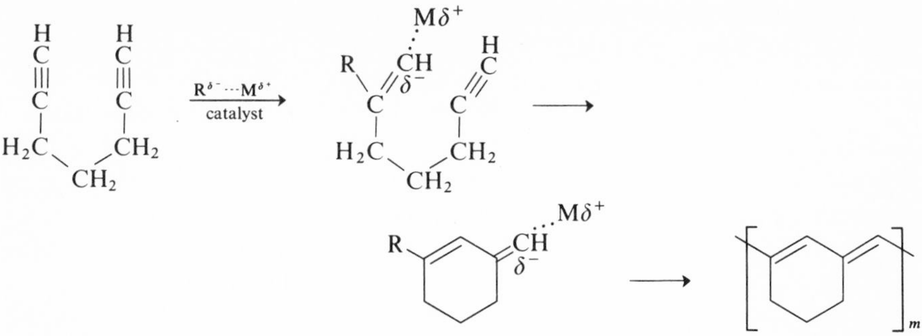

1,7-Octadiyne and 1,8-nonadiyne can form only seven- and eight-membered rings by this head-to-tail propagation. Polymerization of these monomers occurs via the independently acting triple bonds, to form cross-linked polymers.

6. POLYMERIZATIONS CATALYZED BY TRANSITION METAL COMPLEXES

6.1. Nickel and Cobalt Complexes

6.1.1. MONOACETYLENES

In 1960, Luttinger[42] reported a new system for polymerizing acetylenes: freshly prepared mixtures of hydridic reducing agent and a salt or complex of a Group VIII metal. This is the simplest polymerization system so far reported and is not sensitive to moisture or air. In one experiment, Luttinger saturated 100 cc of ethanol with acetylene in an open vessel at 25°, and added 0.026 mole of sodium borohydride and 0.0024 mole of cobaltous nitrate. He then added acetylene to maintain saturation. Smooth but slow polymerization occurred: only 1 gram of polymer formed in 3–5 hours. The polymer is like the one Natta obtained with Ziegler catalysts. Small amounts of oily, low molecular weight polyenes also formed. Water, acetonitrile and diglyme are satisfactory solvents, but the reaction is influenced by the solubility of the acetylene. The catalyst is much more efficient with substituted acetylenes. One mole of reduced $(Bu_3P)_2$-$NiCl_2$ can polymerize several thousand moles of phenylacetylene in a few minutes. 1-Heptyne gives mostly linear oligomer. Under some conditions phenylacetylene gives mostly triphenylbenzene. Disubstituted acetylenes do not react.

The catalyst composition can vary widely. All Group VIII metals are catalysts, and alkali borohydrides, lithium aluminumhydride and diborane are good reducing agents. The catalysts also polymerize ethylene, butadiene and allene. Cuprene is a byproduct from the acetylene polymerization.

The polymers from reduced nickel salts contain nickel and boron, up to 7 weight %, Ni:B ratio = 2. This corresponds to Ni_2B, a product known to form when some nickel salts are reduced by $NaBH_4$. Nickel and boron can be removed by washing with HCl.

The polymer from acetylene does not melt. It decomposes very slowly, sometimes as low as 130°, but not completely even at 300°. Continuous extraction with hot acetone slowly swells and decomposes the polymer to soluble fractions. During the extraction, the polymer becomes lighter in color, and the infrared band for conjugated unsaturation disappears. Amorphous segments in the linear polyacetylene chain may be the points of some oxidative attack. The stability of the polymer depends on close packing of chains, and perhaps also on some resonance stabilization and interchain forces. As chains become solvated and peel away from the main polymer body, they become susceptible to oxidation. The soluble fragments are polyenes with 3–12 conjugated double bonds. The polyenes oxidize rapidly even in solution. Natta[43] obtained similar polyenes as by-products.

1,6-Heptadiyne gives mostly dimer. 1,7-Octadiyne forms an insoluble rubbery polymer. Propargyl chloride and propiolic acid do not react. Propargyl alcohol forms trisubstituted benzenes, reacting with nearly explosive violence. Allene gives mostly trimers (trimethylenecyclohexanes) and tetramers (tetramethylenecyclooctanes).

Green[44] simultaneously reported a similar system: phosphine complexes of nickel or cobalt salts, $(R_3P)_2MX_2$, reduced with $NaBH_4$, in tetrahydrofuran. The catalyst $(Pr_3P)_2NiCl_2$ shows well-defined proton resonance absorption in the region characteristic of hydrogen bound directly to transition metals. This indicates that the active catalyst is a species with metal-hydrogen bonds.

Daniels[45] discovered that unreduced nickel halide-tertiary phosphine complexes are catalysts for polymerizing acetylene, phenylacetylene, 1-hexyne and propargyl alcohol. Unreduced cobalt chloride complexes are inactive. The polymerizing activity of unreduced catalysts is apparently specific to nickel halide-phosphine complexes. Nickel chloride is slightly active. Nickel bromide and iodide are much more active. Tributylphosphine and triphenylphosphine are good complexing agents. Acetylene polymerizes to the black polymer obtained with Ziegler catalysts and with Luttingers catalyst.

Dangerous exotherms sometimes occur in tetrahydrofuran. Ethanol is a better solvent, and polymers prepared in ethanol are more ordered. Propargyl alcohol gives trimethylolbenzenes and insoluble polymer, probably polypropynol. 1-Hexyne gives small amounts of linear dimers, trimers and tetramers. Vinyl monomers do not homopolymerize or copolymerize with acetylene.

Daniels' best result with acetylene was with $NiBr_2 \cdot (Ph_3P)_2$ in 10% tetrahydrofuran-90% ethanol at 17°. One gram of this catalyst gives 12.7 grams of polymer. One gram of catalyst gives 1 gram of polymer from phenylacetylene, and 6 grams of polymer from propynol. (Luttinger's catalyst is much more efficient for phenylacetylene polymerization.) The reaction slows down after 24 hours. Filtered solutions after polymer removal do not contain any active catalyst. Polymer always contains nickel and halogen. Nickel can be removed by washing with HCl. The halogen is apparently in the polymer chain. In Daniels' mechanism, acetylene and catalyst form a π complex, which leads to polymer by a series of ligand insertion and monomer complexing steps:

$$
H-C{\equiv}C-H + R_3P{\rightarrow}\underset{\underset{(1)}{\overset{|}{Br}}}{\overset{\overset{R_3P}{\overset{\downarrow}{|}}}{Ni}}-Br \longrightarrow R_3P{\rightarrow}\underset{\overset{|}{Br}}{\overset{\overset{H-C{\equiv}C-H}{\overset{\downarrow}{|}}}{Ni}}-Br \quad or \quad \underset{\underset{(4)}{R_3P}}{\overset{R_3P}{\diagdown}}\overset{\overset{HC{\equiv}CH}{\overset{\downarrow}{\diagup\;\diagdown}}}{\underset{\diagup\;\diagdown}{Ni}}\overset{Br}{\underset{Br}{}} + R_3P
$$

$$
(1) \longrightarrow \left[R_3P{\rightarrow}\underset{\overset{|}{Br}}{Ni}-\overset{\overset{H}{|}}{C}{=}\overset{\overset{H}{|}}{C}-Br \right]
$$
$$(2)$$

$$
(2) + H-C{\equiv}C-H \longrightarrow R_3P{\rightarrow}\underset{\overset{|}{Br}}{\overset{\overset{HC{\equiv}C-H}{\overset{\downarrow}{|}}}{Ni}}-CH{=}CHBr
$$
$$(3)$$

$$
(3) \longrightarrow \left[R_3P{-}\underset{\overset{\downarrow}{Br}}{Ni}{-}CH{=}CH{-}CH{=}CHBr \right]
$$

This accounts for the halogen in the polymer, here shown as a terminal group. Termination occurs at least partly because the polymer becomes insoluble.

6.1.2. NONCONJUGATED TERMINAL DIYNES

Colthup and Meriwether[46] used their nickel carbonyl-phosphine complex catalysts (Chapter 4) to polymerize terminal nonconjugated diacetylenes. $(Ph_3P)_2Ni(CO)_2$ was the catalyst in most experiments, with a monomer to catalyst ratio of 100:1 to 300:1. Refluxing cyclohexane was generally the solvent. Reactions using dicarbonyl bis(or tris)(2-cyanoethyl)phosphine-nickel catalysts give the same results in refluxing acetonitrile.

All of the diacetylenes studied give high molecular weight polymers [$HC{\equiv}C(CH_2)_nC{\equiv}CH$, $n = 2$–5]. 1,5-Hexadiyne, $n = 2$, forms a yellow insoluble cross-linked cork-like polymer which melts above 300°. The polymer probably has short alkylene chains between the groups of conjugated double and triple bonds, and this causes the insolubility and rigidity:

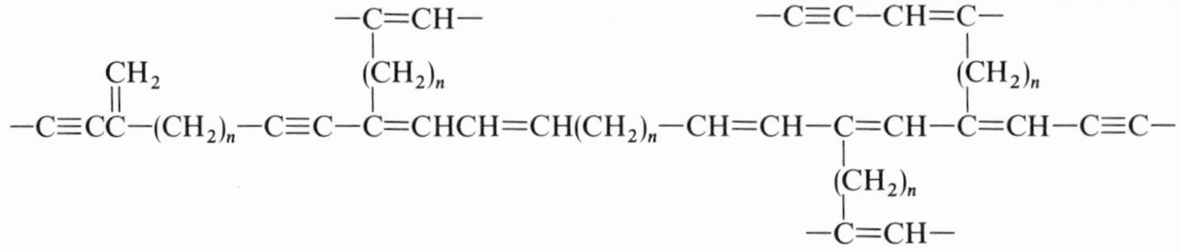

$$-C \equiv C\overset{\overset{\displaystyle CH_2}{\|}}{C}-(CH_2)_n-C\equiv C-\overset{\overset{\displaystyle -C=CH-}{|}}{\underset{\underset{\displaystyle -CH=CH-}{|}}{C}}=CHCH=CH(CH_2)_n-CH=CH-\overset{\overset{\displaystyle -C \equiv C-CH=C-}{|}}{\underset{\underset{\displaystyle -C=CH-}{|}}{C}}=CH-\overset{\overset{\displaystyle (CH_2)_n}{|}}{C}=CH-C\equiv C-$$

1,6-Heptadiyne ($n = 3$) gives two types of product: brown undistillable tar, 75–80% yield, and aromatic trimer (20–25% yield) identified as 1,3-bis(5-indanyl)propane. The trimer forms by a new cyclization reaction in which two triple bonds in one molecule react with one triple bond of another:

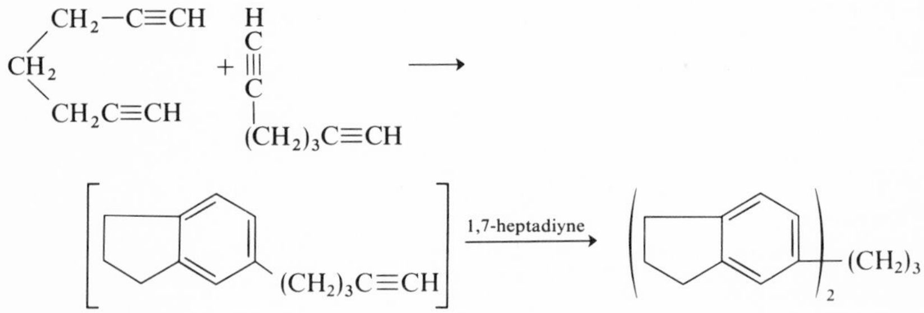

The tarry brown polymer shows infrared absorption for both 1,2,4-trisubstituted aromatic groups and linear conjugated groups. The most likely structural characteristics are:

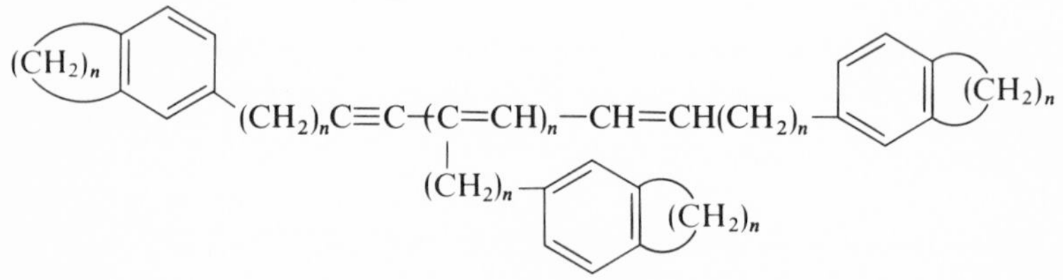

($n = 3$ for heptadiyne)

The product probably has low molecular weight, formed from less than 20 monomer units. It has high aromatic content, is soluble in benzene, and resinifies in air, probably by oxidative cross-linking at the double bonds.

The diynes can copolymerize with an equimolar amount of l-heptyne. 1,5-Hexadiyne and 1-heptyne give a dark brown, undistillable liquid copolymer which has both aromatic and conjugated linear groups. Mass spectra show some dimers and trimers of 1-heptyne, but most of the reactants, including all of the diyne, form higher polymer.

The 1,6-heptadiyne–1–heptyne copolymer is more stable, and can be fractionated to give mostly 1:1 adduct, along with 1-heptyne dimers and trimers, and 1,6-heptadiyne trimer. The 1:1 product is 5-amylindane. 1-Pentyne and 1,6-heptadiyne copolymerize to form 5-propylindane in lower yield.

1,7-Octadiyne copolymerizes with 1-heptyne to give a brown viscous oil, which hardens to a rubbery high polymer. Infrared shows 1,2,4- and 1,3,5-trisubstituted benzene rings and linear conjugated groups. Permanganate oxidation gives a 4:1 mixture of trimellitic and trimesic acids. A little yellow oil can be extracted from the rubbery high polymer. It is a 1:1 adduct, probably 6-amyltetralin.

When the ratio of mono- to diacetylene is 9:1 to 20:1, some high polymers form. The presence of the diacetylene molecule does not change the number of monoacetylene molecules which react linearly before polymerization stops.

The conclusions from this study[46] of nonconjugated diynes are:

(1) The terminal unconjugated diacetylenes studied can polymerize in two different ways: by linear condensation and by intramolecular aromatization.

(2) Each acetylenic group can react independently of the other to form linear acetylene polymers. The number of carbons between the acetylenic groups is unimportant.

(3) Cyclization decreases in the order 1,6-heptadiyne > 1,7-octadiyne > 1,5-hexadiyne, 1,8-nonadiyne.

(4) During the cyclization reaction, both ends of the diacetylene are coordinated with the nickel catalyst. Molecular models support this idea, since 1,6-heptadiyne, and to a lesser extent 1,7-octadiyne, are the only diynes which can have the two acetylene groups in a conformation in which they can both coordinate by π bonds to the nickel and have the two penultimate carbons close enough for bond formation.

(5) Two π-bonded complex structures are possible. In (1) the diacetylene is a bidentate chelating ligand, and in (2) the two acetylene groups coordinate as a π-cyclobutadiene ring:

$$
\begin{array}{cc}
(H_2C)_n \overset{\displaystyle C\equiv CH}{\underset{\displaystyle C\equiv CH}{\diagup\,\,Ni\,\diagdown}} & \quad \overset{\boxed{\bigcirc}(CH_2)_m}{\underset{\underset{\diagup\,\diagdown}{Ni}}{|}} \\[2ex]
(1) & (2)
\end{array}
$$

(6) In the most likely cyclization mechanism another acetylene group attacks the intermediate complex (1) or (2).

6.1.3. HEXAFLUORO-2-BUTYNE

Hexafluoro-2-butyne reacts with iron pentacarbonyl to give stable tricarbonyl tetrakis(trifluoromethyl)cyclopentadienone-iron complex.[47] Other transition metal carbonyls give similar complexes. Bis(acrylonitrile)nickel gives mostly trimer, with little polymer. $(Ph_3P)_2Ni(CO)_2$, $(Ph_3P)_2NiCl_2$-$NaBH_4$, triphenyl-chromium and diethylnickel are polymerization catalysts, and give mostly white solid polymers. The polymers are almost completely insoluble in organic solvents, do not melt at 850°(!), and are inert to acids and bases. Triphenyl-chromium gives 10% yield, and bis(acrylonitrile)nickel gives 20% yield of polymer and 60% yield of hexakis (trifluoromethyl)benzene.

6.2. Rhodium Catalysts

Hydrated rhodium chloride in benzene-ethanol polymerizes phenylacetylene.[48] The rate is 3.2×10^{-7} l/mole-sec, and the activation energy is 15 kcal/mole. The mechanism is probably similar to the emulsion polymerization of butadiene by $RhCl_3$.

$RhCl(PPh_3)_3$ polymerizes hexafluoro-2-butyne in benzene to a white solid polymer which does not melt at 800°.[49]

6.3. Palladium Catalysts

Palladium chloride in acetic acid polymerizes acetylene.[50] At 120° under slight pressure, 1 gram of $PdCl_2$ gives 1.1 grams of *trans*-polyacetylene and a little benzene. $(PhCN)_2PdCl_2$ gives 21.6 grams of polymer per gram of catalyst. Benzene is a poor solvent for this polymerization. Propyne in benzene, however, with $PdCl_2$ catalyst, gives 28% yield of polymer after reaction at 160°. Phenylacetylene gives 1,2,4-triphenylbenzene, 1,3,5-triphenylhexa-2,5-dien-1-yne, and 1.1 gram of polymer per gram of $PdCl_2$ (molecular weight of polymer = 966). The mechanism is probably similar to the mechanism in the Ziegler system, in which polymer forms by insertion of monomer molecules.

6.4. Copper Catalysts

Although copper is not a transition metal, it gives polymers similar to the ones obtained with nickel, cobalt, etc. Korshak[51] used an addition polymerization technique to convert acetylene to a black polymer which contained 98% carbon. He added acetylene to an ammoniacal solution of $CuCl_2$ and oxidized the solution with potassium ferricyanide at reflux. The polymer, probably linear polyvinylene containing a little residual copper, had 10^{19} unpaired electrons per gram. At 2300°, or on prolonged boiling in HCl, the polymer could be converted to graphite. The original polymer is an *n*-type semiconductor, and after heating to 2000°, it is a *p*-type semiconductor.

7. POLYMERIZATION OF PHENYLACETYLENE BY DIFFERENT METHODS

Phenylacetylene has been polymerized in numerous systems. The properties of the polymers obtained by the different methods are compared:

Polymerization System	Catalyst	Conditions	Polymer Properties and Remarks	Reference
Thermal		100–160°	Faster at higher temperature	52
		130–170°	m.p. 50–120°, M.W. 400	53
Cationic	BF$_3$·ether, or CF$_3$CO$_2$H, or SnCl$_2$		Low M.W., proton addition complex	54 55
	Cl$_3$CCO$_2$H	Cold	Yellow M.W. 560–1050	56
	BF$_3$		M.W. 2000	52
Radical	Benzoyl peroxide	60–160°	Faster at higher temperature	52
Ziegler-Natta	Ti-dicyclopentadiene	Toluene	Linear polymer	57
	Et$_3$Al-TiCl$_4$	Benzene, 50°	Al:Ti > 5, partly insoluble in benzene, M.W. 300–1000	53
Ni catalysts	NiX$_2$-NaBH$_4$	Acetonitrile or THF	Mostly triphenylbenzene	42
	NiX$_2$-PPh$_3$	EtOH-THF	Black, 1 g/g of catalyst	45
Rhodium catalyst	RhCl$_3$	Benzene-EtOH		48
Pd catalyst	PdCl$_2$·(PhCN)$_2$	HOAc, 160°	Low yield, M.W. 966	50

References

1. Kotlyarevskii, I. L., and Andrievskaya, E. K., *Izv. Akad. Nauk SSSR Ser. Khim.*, 546 (1966); *Chem. Abstr.*, **65**, 8797 (1966).
2. Smolin, E. M., and Hoffenberg, D. S., *Encycl. Polymer Sci. Technol.*, **1**, 46 (1964).
3. Baker, W. P., Jr., *Encycl. Polymer Sci. Technol.*, **1**, 746 (1964).
4. Pen'kovskii, V. V., *Russ. Chem. Rev.*, (*English Transl.*), **33**, 532 (1964).
5. Ivanov, V. S., *Russ. Chem. Rev.* (*English Transl.*), **35**, 49 (1966).
6. Montecatini, Italian Patent 566,623 (Sept. 23, 1957); *Chem. Abstr.*, **53**, 13661 (1959).
7. Pohl, H. A., and Chartoff, R. P., *J. Polymer Sci.* (*A*), **2**, 2787 (1964).
8. Beck, F., *Ber. Bunsenges. Physik. Chem.*, **68**, 558 (1964); *Chem. Abstr.*, **62**, 13261 (1965).
9. Isfendiyaroglu, A., *et al.*, *Bull. Soc. Chim. France*, 3155 (1965).
10. deVries, G., and vanBeek, L. K. H., *Rec. Trav. Chim.*, **84**, 184 (1965).
11. Benes, M., Peska, J., and Wichterle, O., *Chem. Ind.* (*London*), 562 (1962).
12. Benes, M., Peska, J., and Wichterle, O., *J. Polymer Sci.* (*C*), 1377 (1964).
13. MacNulty, B. J., *Polymer*, **7**, 275 (1966).
14. Meriwether, L. S., *et al.*, *J. Org. Chem.*, **26**, 5155 (1961).
15. Lee, S. M., *J. Appl. Polymer. Sci.*, **9**, 1431 (1965).
16. Okamoto, Y., *et al.*, *Chem. Ind.* (*London*), 2004 (1961).
16a. Paushkin, Ya. M., *et al.*, *J. Polymer Sci.* (*A-1*), **5**, 1203 (1967).
17. Munari, S., Castello, G., and Russo, S., *Chim. Ind.* (*Milan*), **47**, 26 (1965); *Chem. Abstr.*, **62**, 16388 (1965).
18. Utley, L. W., Jr., U.S. Atomic Energy Comm., **TID-21389**, 73 pp. (1964); *Chem. Abstr.*, **62**, 14833 (1965).
19. Backalov, I. M., *et al.*, *Tr. Z-go Vses. Soveshch. po Radiats. Khim. Akad. Nauk SSSR Otd. Khim. Nauk Moscow*, 455 (1960); *Chem. Abstr.*, **58**, 4652 (1963).
20. Oppenheim, W., and Shorr, L. M., *Israel J. Chem.*, 2, 121 (1964); *Chem. Abstr.*, **61**, 14791 (1964).
20a. Davidov, B. E., Krentsel, B. A., and Kchutareva, G. V., *J. Polymer Sci.* (*C*), 1365 (1967).
21. Natta, G., Mazzanti, G., and Corradini, P., *Atti accad. nazl. Lincei Rend. Classe sci. fis. mat. e nat.*, **25**, 3 (1958); *Chem. Abstr.*, **53**, 13985 (1959).
22. Monetcatini, British Patent 826,674 (Jan. 20, 1960).
23. Nasirov, F. M., Krentsel, B. A., and Davydov, B. E., *Izv. Akad. Nauk SSSR Ser. Khim.*, 1009 (1965); *Chem. Abstr.*, **63**, 10071 (1965).
24. Nicolescu, I. V., and Angelescu, Em., *J. Polymer Sci.* (*A*), **3**, 1227 (1965).
25. Higashimura, K., and Oiwa, M., *J. Chem. Soc. Japan*, **69**, 109 (1966).
26. Matkovskii, P. E., and Zavorohkin, N. D., *Izv. Akad. Nauk Kaz. SSR Ser. Khim. Nauk.*, **15**, 70 (1965); *Chem. Abstr.*, **63**, 10066 (1965).
27. Lombardi, E., and Giuffre, L., *Atti accad. nazl. Lincei Rend. Classe sci. fis. mat. e nat.*, **25**, 701 (1958); *Chem. Abstr.*, **53**, 15948 (1959).
28. Watson, W. H., Jr., McMordie, W. C., Jr., and Lands, L. G., *J. Polymer Sci.*, **55**, 137 (1961).
29. Noskova, N. F., *et al.*, *Vysokomolekul. Soedin.*, **8**, 1524 (1966); *Chem. Abstr.*, **66**, 1110 (1967).
30. Noguchi, H., and Kambara, S., *J. Polymer Sci.* (*B*), **1**, 553 (1963).

31. Berlin, A. A., *et al.*, *Vysokomolekul. Soedin.*, **1**, 1817 (1959); *Chem. Abstr.*, **54**, 25948 (1960).
32. Korn-Girard, A., and Teyssie, P. (to Inst. Français du Petrole, des Carburants et Lubrificants), French Patent 1,402,817 (June 18, 1965); *Chem. Abstr.*, **65**, 3789 (1966).
32a. Mickel, R. H., *J. Polymer Sci. (A-1)*, **5**, 920 (1967).
33. Murahashi, S., and Ketsuki, H. (to Mitsubishi Chem. Ind.), Japanese Patent 3713 ('66) (March 3, 1966); *Chem. Abstr.*, **65**, 828 (1966).
34. Stedefeder, J., and Daeuble, M. (to BASF), Belgian Patent 655,390 (May 6, 1965); *Chem. Abstr.*, **64**, 19686 (1966).
35. Matkovskii, P. E., *et al.*, *Vysokomolekul. Soedin.*, **8**, 1712 (1966); *Chem. Abstr.*, **66**, 1110 (1967).
36. Montecatini, British Patent 834,820 (May 11, 1960).
37. Natta, G., *et al.*, (to Montecatini), Italian Patent 536,899 (Dec. 12, 1955); *Chem. Abstr.*, **53**, 1837 (1959).
38. Natta, G., *et al.*, (to Montecatini), Italian Patent 530,753 (July 15, 1955); *Chem. Abstr.*, **52**, 15128 (1958).
39. Marvel, C. S., and Stille, J. K., *J. Am. Chem. Soc.*, **80**, 1740 (1958).
40. Marvel, C. S., and Garrison, W. E., Jr., *J. Am. Chem. Soc.*, **81**, 4737 (1959).
41. Stille, J. K., and Frey, D. A., *J. Am. Chem. Soc.*, **83**, 1697 (1861).
42. Luttinger, L. B., *Chem. Ind. (London)*, 1135 (1960); *J. Org. Chem.*, **27**, 1591 (1962).
43. Natta, G., *et al.*, *Gazz. Chim. Ital.*, **89**, 465 (1959); *Chem. Abstr.*, **54**, 11967 (1960).
44. Green, M. L. H., Nehme, M., and Wilkinson, G., *Chem. Ind. (London)*, 1136 (1960).
45. Daniels, W. E., *J. Org. Chem.*, **29**, 2936 (1964).
46. Meriwether, L. S., *J. Org. Chem.*, **26**, 5169 (1961).
47. Blount, J. F., *et al.*, *J. Am. Chem. Soc.*, **88**, 292 (1966).
48. Teyssie, P., *et al.*, *Compt. Rend.*, **261**, 997 (1965).
49. Mays, M. J., and Wilkinson, G., *J. Chem. Soc.*, 6629 (1965).
50. Odaira, Y., Hara, M., and Tsutsumi, S., *Technol. Rept. Osaka Univ.*, **16**, 325 (1965); *Chem. Abstr.*, **65**, 10670 (1966).
51. Korshak, V. V., *et al.*, *Dokl. Akad. Nauk SSSR*, **136**, 1342 (1961); *Chem. Abstr.*, **55**, 18554 (1961).
52. Liu, Yu. C., *et al.*, *Ko Fen Tzu T'ung Hsum*, **6**, 71 (1964); *Chem. Abstr.*, **63**, 11708 (1965).
53. Higashiwa, K., *et al.*, *Kogyo Kagaku Zasshi*, **66**, 374, 379 (1963); *Chem. Abstr.*, **63**, 5753 (1965).
54. Lee, B. E., and North, A. M., *Makromol. Chem.* **79**, 135 (1964).
55. Bawn, C. E. H., Lee, B. H., and North, A. M., *J. Polymer Sci. (B)*, **2**, 263 (1964).
56. Evans, A. G., and Phillips, B. D., *J. Polymer Sci.*, **3**, 77 (1965).
57. Yokokawa, K., and Azuma, K., *Bull. Chem. Soc. Japan.*, **38**, 859 (1965); *Chem. Abstr.*, **63**, 5753 (1965).
58. Amagi, Y., and Murayama, N. (to Kureha Chem. Ind.), Japanese Patent 10, 590 ('65) (May 28, 1965); *Chem. Abstr.*, **63**, 16499 (1965).

PART TWO

Addition Across the Triple Bond as a Polymer-forming
Reaction

1. SATURATED POLYMERS

Silicon, boron and tin hydrides add across the triple bond to form vinyl- and alkylsilanes, boranes and stannanes (see Chapter 2). Difunctional hydrides react in the same way to form polymers.

Acetylene reacts with tetramethyldisiloxane in isopropanol (chloroplatinic acid catalyst) to form polymeric siloxanes which are heavy oils:[1]

$$C_2H_2 + Me_2\overset{H}{\underset{|}{Si}}-O-\overset{H}{\underset{|}{Si}}Me_2 \xrightarrow[\text{4 hr}]{120°} \left[\overset{Me}{\underset{Me}{\overset{|}{\underset{|}{Si}}}}-O-\overset{Me}{\underset{Me}{\overset{|}{\underset{|}{Si}}}}-CH_2CH_2\right]_n$$

Similar polymerization reactions occur with substituted acetylenes.[2] $MePhSiH_2$ and $Ph_2Si(C{\equiv}CH)_2$ give a dark brown brittle polymer, molecular weight 5800. With *p*-diethynylbenzene, the polymer is dark brown and hard, and has an elastic deformation range of 400–600°.

Polymers are formed by reaction of a boron-hydrogen compound with an aliphatic hydrocarbon which has at least one acetylenic bond, two olefinic bonds, or one acetylenic and one olefinic bond. The polymers are not pyrophoric like boron trialkyls, and they slowly oxidize in air to form liquids.[3] Sodium borohydride·BF_2·Et_2O and acetylene in tetrahydrofuran at 40° give a rubbery, waxy solid containing 11.5% boron. The presumed intermediate is trivinylborane, which slowly polymerizes.

Acetylene reacts with diborane in the gas phase to form polymers of uncertain structure. Diacetylene also forms polymers which decompose violently at 80° to give yellow product plus a black solid, perhaps carbon.[4]

2. UNSATURATED POLYMERS

2.1. Diynes Plus Organometal Dihydrides

Diacetylenes react faster than diolefins with organotin dihydrides.[5,6] For preparing polystannanes, diyne and organotin dihydride are refluxed in hexane for 2–4 hours, the solvent is evaporated, and the residue is heated at

100° for 2–5 hours. Polymers are rubbery or plastic masses. The properties of some polystannanes are given in Table 5-4.[5,6] Many polystannanes are insoluble because of cross-linking. Disproportionation of dihydride to trihydride and monohydride would cause cross-linking and chain termination.

Solid germanium and lead polymers are made by similar reactions. Diphenylgermanium dihydride reacts slowly with 1,5-hexadiyne to form a polymer, molecular weight 10,000.

TABLE 5-4
Polymers from Organotin Dihydrides and Diynes

$$\left[\begin{array}{c} R \\ | \\ Sn-CH{=}CH-R'-CH{=}CH \\ | \\ R \end{array}\right]_n$$

R	R′	Nature of Polymer	$\overline{M}_w \times 10^{-3}$
Me	1,4-Tetramethylene	Soft, rubbery	Slightly cross-linked
Et	1,4-Tetramethylene	Rubberlike	Slightly cross-linked
Pr	1,4-Tetramethylene	Soft, slightly elastic	50
Bu	1,4-Tetramethylene	Soft, plastic	50
Ph	1,4-Tetramethylene	Rubber-like	75
Bu	1,5-Pentamethylene	Viscous oil	45
Ph	1,5-Pentamethylene	Soft, plastic	100
Ph	1,4-Phenylene	Rubber above 160°	65

Chloroplatinic acid catalyzes the emulsion copolymerization of methylphenylsilane and 1,6-heptadiyne.[7] Polymers are elastic gums which cross-link on storage at room temperature. The type of emulsifier has considerable effect on the molecular weight of the polymer. Anionic soaps give highest molecular weights, but not necessarily with other catalysts or monomer combinations. Hydrogenation of the unsaturated rubbers gives nonrubbery solids.

The polymers are probably similar to the polymers from organotin dihydrides and diynes:[5]

$$\left[\begin{array}{c} Ph \\ | \\ Si-CH{=}CH-(CH_2)_3CH{=}CH \\ | \\ Ph \end{array}\right]_n$$

2.2. Addition of Polyols to Diacetylenic Esters

Butler[8] claimed an unusual variation of vinylation of hydroxyl groups as a polymer-forming reaction:

$$HC\equiv C-\overset{\overset{\displaystyle O}{\|}}{C}-ORO-\overset{\overset{\displaystyle O}{\|}}{C}-C\equiv CH + HOR'OH \xrightarrow{\text{base}}$$

$$\left[OR'OCH=CH\overset{\overset{\displaystyle O}{\|}}{C}OROC\overset{\overset{\displaystyle O}{\|}}{}-CH=CH\right]_n$$

The dipropiolate esters of various diols react with glycols, such as ethylene glycol, diethylene glycol, hydroquinone and bisphenol A, to give polymers. The resins are liquids or solids, and can be molded, cast and spun. They can copolymerize with vinyl monomers and can be cured by radical catalysts.

2.3. Polymerization of Propiolamide

The $-NH_2$ group of one molecule of propiolamide adds across the acetylenic bond of another molecule to form a dark brown polymer. After two weeks at 135° in the presence of NaCN catalyst, propiolamide in dimethylformamide

gives 24% yield of polymer. Units in the chain are $\left[CH=CH\overset{\overset{\displaystyle O}{\|}}{C}NH\right]_n$. Prolonged hydrolysis with hot dilute HCl gives acetaldehyde.[9]

2.4. Polymers from Diels-Alder Reactions of Diacetylenes

The Diels-Alder reaction between acetylenes and tetraphenylcyclopentadienone is practically quantitative, and is irreversible because CO is eliminated and a stable benzene ring is formed (see Chapter 2). Both ethynyl groups of 1,4-diethynylaromatics react with substituted cyclopentadienones.[10]

Stille and co-workers[11] adapted this reaction to form polymers containing predetermined numbers of aromatic segments. The polymerization reaction is represented as:

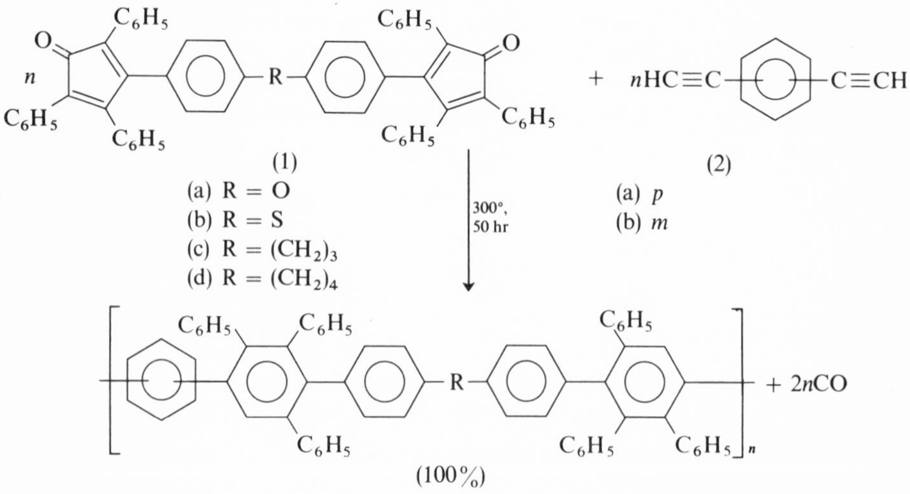

(1)

(a) R = O
(b) R = S
(c) R = (CH$_2$)$_3$
(d) R = (CH$_2$)$_4$

300°,
50 hr

(2)

(a) p
(b) m

$+ 2nCO$

(100%)

Properties of the polymers are summarized in Table 5-5. The polymers are light tan solids. The low viscosities are probably due to a little chain terminating monotetracyclone mixed with the ditetracyclone used. The polymers have excellent thermal stability, even in air.

TABLE 5-5
Diels-Alder Polymers

Bistetracyclone, R	Diethynyl-benzene	$[\eta]_{sp}\,(dl/g)^a$	TGA Break (°C), in Air
O	m-	0.31	550
S	p-	0.40	550^b
S	m-	0.26	550
$(CH_2)_3$	p-	0.17	490
$(CH_2)_3$	m-	0.16	490
$(CH_2)_4$	p-	0.36	490
$(CH_2)_4$	m-	0.14	470

^{a} In toluene at 25°.
^{b} Nitrogen.

2.5. Polymers from 1,3-Dipolar Additions

1,3-Dipolar addition reactions of acetylenes are often quantitative, and the 1,3-dipole compounds are relatively easy to make (see Chapter 2). Thus, the reaction can be applied to difunctional reagents to make polymers. Frazza[12] reacted p-phenylene dicyanate (a nitrile oxide) with p-diethynylbenzene to form a polymer which loses only 10% of its weight when heated at 428° in air.

Stille and Bedford[13] used the sydnone dipole and the nitrilimine dipole to prepare polymers. The sydnone reaction is:

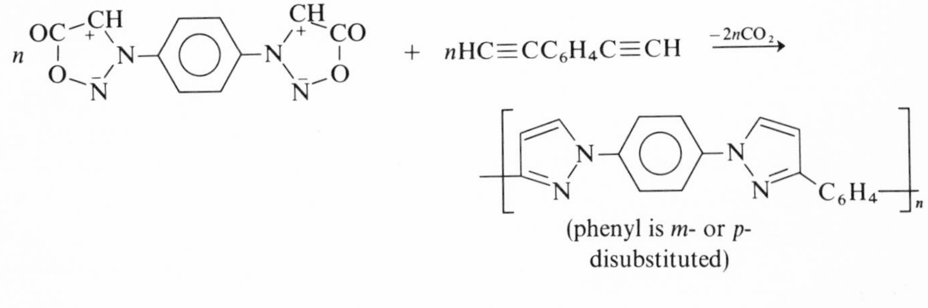

(phenyl is m- or p-disubstituted)

The reaction is easily followed by measuring the CO_2 evolved. Polymers are soluble in polar solvents such as dimethylformamide, and have inherent viscosities of 0.4. Thermogravimetric analysis showed peaks at 420° in air and 460° in nitrogen.

The nitrilamine reaction gives polypyrazoles which have thermal stabilities similar to the sydnones. The polymer with *m*-phenylene linkages is more soluble in organic solvents than the *p*-substituted compound:

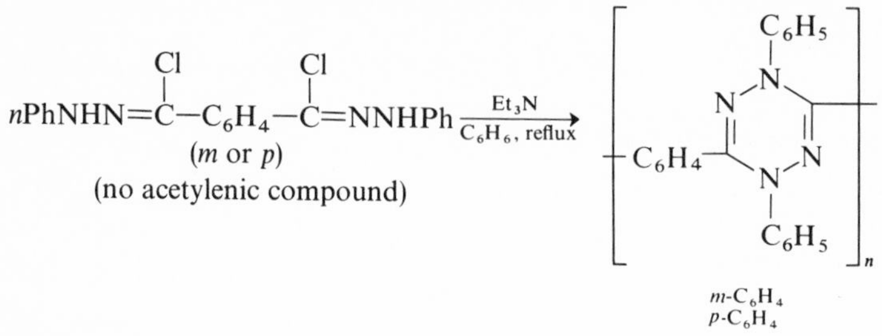

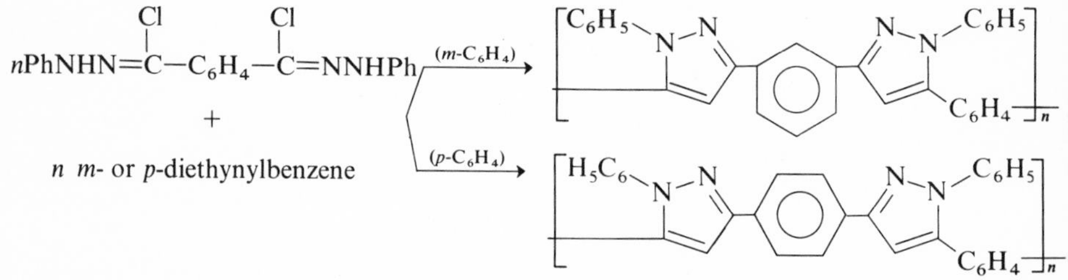

Johnson and co-workers[14] reported the first polymerization involving 1,3-dipolar addition of an azido group to an acetylenic bond. They polymerized 4-azido-1-butyne at 70°. This is also the first report of 1,3-dipolar addition polymerization in which the dipolarophile and the 1,3-dipole are in the same molecule. In the bulk polymerizations at 70°, the monomer becomes viscous and ultimately sets to a hard white glass. Polymer prepared by refluxing the monomer in benzene precipitates as a hard yellow solid. The polymers contain both 1,4- and 1,5-disubstituted triazole rings:

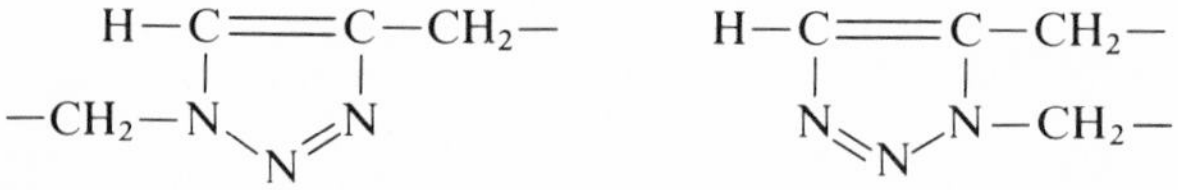

The polymers are thermally stable, with no exothermic decomposition below 320°. They are soluble in dimethyl sulfoxide. Later, this group of workers prepared polymers from 3-azido-1-propyne (a very sensitive explosive), 2-propynyl azidoacetate and *p*-azidophenylacetylene.[14a]

References

1. Polyakova, A. M., *et al., Izv. Akad. Nauk SSSR Otd. Khim. Nauk,* 2257 (1959); *Chem. Abstr.,* 54, 10915 (1960).
2. Korshak, V. V., Sladkov, A. M., and Luneva, L. K., *Izv. Akad. Nauk SSSR Otd. Khim. Nauk,* 2251 (1962); *Chem. Abstr.,* 58, 12685 (1963).

3. Schubert, F., and Nuetzel, K., (to Farbenfabriken Bayer A.-G.), German Patent 1,070,384 (Dec. 3, 1959).
4. Lindner, H. L., and Onak, T., *J. Am. Chem. Soc.*, **88**, 1886 (1966).
5. Morgan, P. W., "Condensation Polymers," p. 429, New York, Interscience Publishers, 1965.
6. Noltes, J. G., and van der Kerk, G. J. M., *Rec. Trav. Chim.*, **80**, 623 (1961).
7. Freeman, R. R., and Spearman, E. I., AD 609,002, Available CFSTI, 19pp. (1964); *Chem. Abstr.*, **63**, 1969 (1965).
8. Butler, J. M., Miller, L. A., and Wesp, G. L., U.S. Patent 3,201,370 (Aug. 17, 1965); *Chem. Abstr.*, **63**, 15009 (1965).
9. Allison, J. P., and Michel, R. E., *Chem. Commun.*, 762 (1966).
10. Ried, W., and Freitag, P. D., *Naturwissenschaften* **53**, 306 (1966).
11. Stille, J. K., *et al.*, *J. Polymer Sci.* (*B-4*), 791 (1966): *J. Polymer Sci.* (*A-1*), **5**, 2721 (1967).
12. Frazza, E. J. (to American Cyanamid Co.), U.S. Patent 3,213,068 (Oct. 19, 1965); *Chem. Abstr.*, **64**, 2189 (1966).
13. Stille, J. K., and Bedford, M. A., *J. Polymer Sci.* (*B-3*), 329, 333 (1966).
14. Johnson, K. E., *et al.*, *J. Polymer Sci.* (*B-4*), 977 (1966).
14a. Baldwin, M. G., *et al.*, *J. Polymer Sci.* (*B-5*), **5**, 803 (1967).

PART THREE

Polymers Containing Acetylenic Bonds

1. OXIDATIVE COUPLING AS A POLYMER-FORMING REACTION

1.1. Hydrocarbons

Oxidative coupling is frequently a high-yield reaction and thus can be used as a polymerization system. Conjugated diyne groups form rigid, rod-like segments in polymer molecules.

Glaser coupling of *p*-diethynylbenzene gives orange-red oligomers, insoluble in most solvents: these oligomers usually explode at 120°.[1] In Hay's cupric acetate-amine system, *m*-diethynylbenzene couples to form quantitative yield of pale yellow polymer.[2] Terminal ethynyl groups determined by infrared indicate a molecular weight of 7000. The polymer is stable at room temperature, but on heating rapidly *in vacuo* it abruptly decomposes at 180° to form hydrogen, a little methane and carbon. The polymer explodes when ignited at room temperature. *p*-Diethynylbenzene couples to form a bright yellow polymer, insoluble in all solvents: it decomposes rapidly at 100°.

1,1-Bis(p-ethynylphenyl)ethane couples in pyridine-CuCl-O_2 to form a polymer which is not melted at 500°.[3] Tris(p-ethynylphenyl)methane also gives a polymer infusible at 500°. Bis(p-butenynyl)benzene gives a polymer with 9×10^{18} unpaired electrons per gram. p-Bis(1-butynyl)benzene couples to form a product which melts at 250°.

1.2. Diethynyl Aromatic Ethers and Esters

p-Ethynyl aromatic ethers oxidatively couple to form polymers:[4]

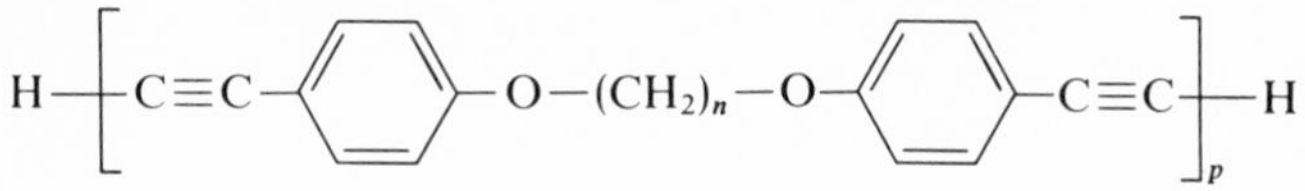

n	Cu(OAc)$_2$	Cu$_2$Cl$_2$-O$_2$
1	10	32
2	10	10
3	34	21

The degree of polymerization varies with the coupling system. The polymers are pale yellow powders, insoluble in common solvents. Above 200° in air they turn orange and then black. In the infrared, they absorb at 3300 cm^{-1} for free ethynyl groups and therefore are linear. The degree of crystallinity is $n = 1,3 > 2$, and the degree of polymerization is $3 > 1 > 2$. The higher the degree of polymerization, the higher is the crystallinity.

Polymers containing conjugated diacetylene groups and ester groups can also be made by a Glaser cross-coupling.[5] Thus, a solution of 1 gram of p-diethynylbenzene and 1 gram of dipropargyl isophthalate in 150 ml of pyridine and 50 ml of methanol containing 0.2 gram of cuprous chloride was refluxed for 7 hours while air bubbled through. The product was a yellow powdery polymer:

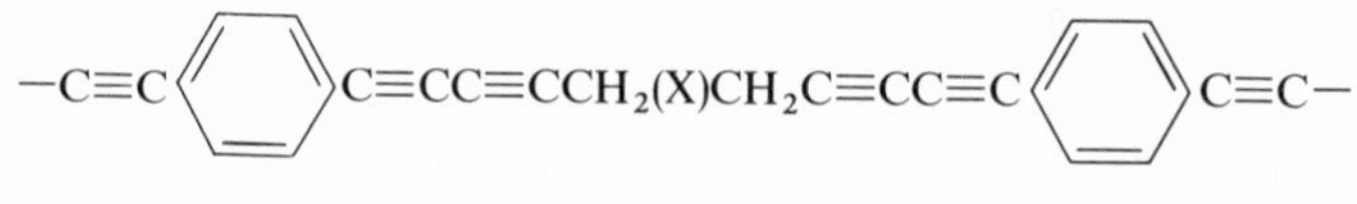

where X is

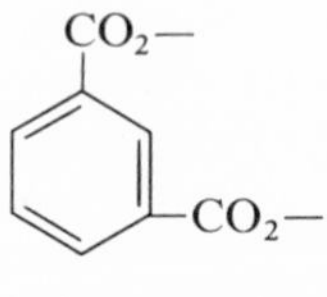

The unstable compound 4,4′-di(1-butadiynyl)diphenyl ether couples in the presence of cuprous chloride-pyridine and oxygen to give a black-violet insoluble polymer which explodes on heating. EPR shows that the polymer has 8×10^{17} unpaired electrons per gram.[6]

2. POLYMERS FROM METALLOACETYLENES

Polydimethyleneacetylene was prepared by adding 1-chloro-4-bromo-2-butyne to a large excess of magnesium.[7] The Grignard reagent forms rapidly and couples with itself on standing to form an insoluble polymer mixture. The untreated crude product has a Grignard end group or some adsorbed Grignard reagent. The insoluble precipitate begins to form when about one-third of the dihalide is added, and the final product is the same even if no more halide is added. The polymer is a powdery, infusible, insoluble solid. It is stable indefinitely in nitrogen, but rapidly absorbs oxygen from the air.

About 10% of the polymer is extractable with hot xylene. The soluble fraction does not change below 400° and darkens a little at 550°. The molecular weight of the soluble polymer is 2500. The unusual properties of the soluble polymer can be explained by assuming that the triple bonds produce a rigid, symmetrical structure in which each monomer unit is completely linear. The total structure would look like a series of sticks joined together at the ends. The chains will be very close together, and can form strong crystallites. The reaction giving this polymer is:

$$ClCH_2C{\equiv}CCH_2Br + Mg \longrightarrow [ClCH_2C{\equiv}CCH_2MgBr] \longrightarrow polymer \xrightarrow[\text{xylene}]{\text{boiling}}$$

10% soluble polymer: $ClCH_2C{\equiv}CCH_2(CH_2C{\equiv}CCH_2)_xCH_2C{\equiv}CCH_3$

3. METAL SUBSTITUTION REACTIONS TO FORM ACETYLENIC POLYMERS

Dilithio-p-diethynylbenzene refluxed in ether with mercuric chloride gives a polymer containing mercury between the triple bonds.[8] Silicon and arsenic compounds can react like $HgCl_2$. The mercury polymer at 200° decomposes to form a conducting carbon product. The polymers are:

$$HC{\equiv}C\left\langle\bigcirc\right\rangle C{\equiv}C{-}\left[M{-}C{\equiv}C\left\langle\bigcirc\right\rangle C{\equiv}C\right]_n{-}H$$

where M = Hg, Ph_2Si, PhAs.

4. POLYMERIZATION OF SUBSTITUTED VINYLACETYLENES

4.1. Vinylethynylcarbinols

Vinylethynylcarbinols polymerize either spontaneously or with catalysts to form soluble linear polymers. A cyclization-polymerization mechanism is reasonable for the radical-catalyzed polymerization.[9] Addition of initiator radical gives an allenic radical, which adds to the double bond of a second molecule of the vinylethynylcarbinol:

$$CH_2=CHC\equiv C-\underset{\underset{R}{|}}{\overset{\overset{OH}{|}}{C}}-R' \xrightarrow{Z\cdot} ZCH_2CH=C=\underset{R'}{\overset{R}{C}}\cdot \longrightarrow ZCH_2CH=C=C$$

Z· = Initiator

1-Vinylethynylcyclopentanol polymerizes to a powdery solid which softens between 150 and 200°.[10] Isopropenylethynylcarbinols polymerize more slowly. Polymers have low average molecular weight,[11] but careful fractionation gives fractions with molecular weights of 10^4–10^6. The radicals may add to 2 moles of vinylethynylcarbinol in the 1,4- and 1,2-positions to form unsaturated diradicals, stabilized by intramolecular cyclization and isomerization, and these initiate polymerization at both ends of the molecule.[12]

In bulk polymerizations with benzoyl peroxide or azobisisobutyronitrile catalyst, tertiary vinylethynylcarbinols polymerize faster than secondary, which are faster than primary.[13] In the tertiary carbinols, the rate increases as the size of the alkyl group increases, but rate decreases with increasing size of the alkyl group in secondary carbinols. Products are white poly(keto alcohols), soluble in polar solvents, but insoluble in water and in hydrocarbons. The ethers and esters polymerize in the same order as the alcohols, but the rate decreases as the size of the alkyl groups increases. These polymers are rubbery, or powdery, and are soluble in benzene. The molecular weight and softening point are lower than the corresponding carbinol polymers.

Matsoyan[14] polymerized various aliphatic vinylethynylcarbinols, and copolymerized them with styrene, acrylonitrile, butadiene and other vinyl monomers. He used benzoyl peroxide at 70°, and determined relative reactivity ratios. As the amount of carbinol in the monomer mix increases, molecular weight and softening point of the copolymers increase. Polymers are linear, and are soluble in alcohols, acetone and acetic acid. They contain cyclic structures rather than cross-links.

4.2. Glycidyl Ethers of Vinylethynylcarbinols

Glycidyl ethers of vinylethynylcarbinols are prepared by reaction with epichlorohydrin in the usual way.[15] Yields are usually 35–65%. The glycidyl ethers polymerize easily in the air and in the presence of radical initiators to form linear solid rubbery or powdery polymers. The polymers contain the epoxide group.

4.3. Other Vinylethynyl Monomers

Davis and Hunter[16] studied polymerization of some monomers derived from vinylacetylene. p—[1—(3'-Buten-1'-ynyl)cyclohexyl]phenol (1) does not polymerize on prolonged heating with benzoyl peroxide or on ultraviolet irradiation. The acetate (1a) slowly polymerizes in the presence of benzoyl peroxide to form a polymer with an average of 12 monomer units. The methyl ether (1b) gives a polymer with 6–30 monomer units. Ether (1b) copolymerizes with styrene to form a glassy clear resin, whose properties indicate a copolymer.

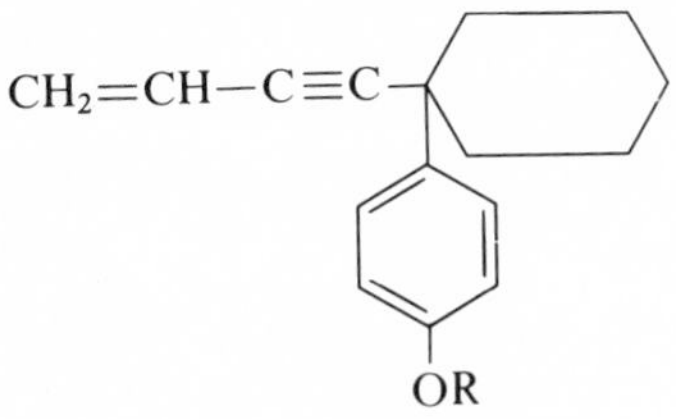

$$CH_2{=}CH{-}C{\equiv}C{-}$$

OR

(1) R = H
(1a) R = OAc
(1b) R = Me

4-Penten-2-yn-1-yl acetate and 4-penten-2-yn-1-yl benzoate polymerize in the presence of benzoyl peroxide to form soft gels which are insoluble in common solvents. These monomers do not copolymerize with styrene.

Dipent-4-en-2-yn-1-yl adipate (from the carbinol and adipyl chloride) polymerizes in the presence of diazoaminobenzene at 65–95° to form a tough insoluble polymer which does not melt at 360°.[17] The polymer is probably cross-linked by 1,2-addition of ethylenic bonds to form structures like those from divinylbenzene and diallyl phthalate. The terephthalate ester polymerizes in the crystalline state on standing at room temperature.

The adipic ester of hex-5-en-3-yn-2-ol $[CH_2{=}CHC{\equiv}C\underset{H}{\overset{Me}{C}}{-}O\overset{O}{C}(CH_2)_2\,{+}_2$,

which might be more thermally stable because of the methyl groups adjacent to the acetylenic bonds, gives brown infusible resins which are softer than the polymer from pentenynyl adipate.

Reaction of vinylacetylene with glycerol in the presence of $HgO\text{-}BF_3\cdot Et_2O$ and trichloroacetic acid gives a polymer with molecular weight about 1300.[18] The polymer contains dioxolane units:

$$\text{vinylacetylene + glycerol} \longrightarrow \quad \underset{\underset{\overset{|}{CH_2}}{\overset{|}{O_{\diagdown}\diagup CHCH_2OH}}}{CH_2{=}CH{-}\overset{\overset{\displaystyle Me}{|}}{C}{-}O} \quad \longrightarrow$$

$$HO\underset{\underset{H_2C \underline{\quad\quad} O}{\overset{|}{\diagup}}}{\underset{\diagdown}{\left[CH_2{-}\overset{\overset{O}{\diagdown}}{CH}\overset{\overset{\displaystyle Me}{|}}{C}{-}CH_2O\right.}}\underset{\quad}{\left.\vphantom{X}\right]_n}\overset{\overset{\displaystyle OH}{|}}{CHCH_2OH}$$

5. CONDENSATION POLYMERS CONTAINING ACETYLENIC BONDS

5.1. Butynediol Polyester and Polyurethane

Marvel and Johnson[19] prepared saturated and acetylenic polyesters and polyurethanes of about the same molecular weight and compared their low-temperature behavior. Using butanediol and butynediol, they prepared the sebacate polyesters and the polyurethane from hexamethylene diisocyanate. Direct esterification was best for polybutynediol sebacate, since ester interchange gave insoluble polymer, probably by secondary reactions at the triple bond. They also prepared polyurethane copolymers using mixtures of butanediol and butynediol. X-ray patterns of the polymers indicated the high degree of regularity expected. Intrinsic viscosities, capillary melting points and second-order transition temperatures are listed in Table 5-6.

TABLE 5-6
Properties of Polyesters and Polyurethanes

Polymer	Intrinsic Viscosity (dl/g)	Capillary m.p. (°C)	Second-order Transition Temperature (°C)
Polysebacate, butanediol	0.98	64–65	−75
Polysebacate, butanediol	0.26	64–64.5	−57
Polysebacate, butynediol	0.19	51–53	−27
Polysebacate, butynediol	0.20	51–53	−26
Polyurethane, butanediol	0.25	176–177	−58
Polyurethane, butynediol	0.23	149–150	−45

Capillary melting points of the acetylenic polymers are lower, but second-order transition temperatures are higher. The triple bond in these polymers makes them less useful at low temperatures. Polyurethanes with acetylenic bonds have lower calculated heats of fusion and entropies of fusion than the saturated polyurethanes.

5.2. Polyesters from Acetylenedicarboxylic Acid

Acetylenedicarboxylic acid and the usual glycols form polyesters which are hard plastics but have no fiber-forming properties.[20] One difficulty encountered in preparing polyesters from unsaturated acids is the great tendency to cross-link. Some of the properties of polyesters from acetylenedicarboxylic acid are:[20]

(1) They show no tendency to crystallize.

(2) They do not polarize light.

(3) They have dense folded or compact coiled chains.

(4) Hydrogenation to succinic polyester does not change the molecular weight.

(5) The triple bonds show no evidence of association in solutions.

(6) Higher molecular weight polyesters are hydrogenated more slowly than lower molecular weight ones because the chains are more intertwined, so hydrogen cannot easily reach the inner parts of the chains. Low molecular weight polyesters show little change in viscosity when hydrogenated, but higher ones increase slightly in viscosity when hydrogenated.

5.3. Polyesters from Dibasic Acids, Acetylenedicarboxylic Acid, Glycols and Butynediol

Feit[21] prepared three series of polyesters:

(1) Butynediol and ordinary dicarboxylic acids.

(2) Acetylenedicarboxylic acid and ordinary glycols.

(3) Acetylenedicarboxylic acid and butynediol.

(1) *Butynediol polyesters*: A two-stage direct esterification gave butynediol polyesters. After heating glycol and acid for 2–8 hours at 180–190°, and removing water as it formed, the reaction was completed at 1–10 mm for 10–20 hours. Linear soluble polyesters formed even under the relatively drastic conditions. Linear soluble maleate-butynediol polyesters required milder conditions to prevent cross-linking. Properties of the butynediol polyesters are listed in Table 5-7.

(2) *Acetylenedicarboxylic acid polyesters*: These polyesters formed only under the mild conditions used for the maleate-butynediol polyester. Acid and glycol polymerized at 80° during 9–48 hours under nitrogen, followed by 4–12 hours at 80–140° under vacuum. Polyesters of diethylene glycol, and

TABLE 5-7
Polyesters from Dibasic Acids and Butynediol[21]

$$H\text{---}\!\!\!\left[O\text{---}CH_2\text{---}C\equiv C\text{---}CH_2\text{---}O\text{---}\underset{\underset{O}{\|}}{C}\text{---}R\text{---}\underset{\underset{O}{\|}}{C}\right]_{\!n}\!\!\!\text{---}OH$$

Polyester R	Solubility	Softening Temperature (°C)	$\overline{M}_n$	n
$-CH_2-CH_2-$	Soluble: chloroform, DMF, DMSO Insoluble: acetone, nitrobenzene, THF	40	3330	19.8
$-CH_2-(CH_2)_2-CH_2-$	Soluble: acetone, chloroform, DMF, THF Insoluble: ethanol, benzene	40–44	1970	10.0
$-CH_2(CH_2)_6-CH_2-$[a]	Soluble: chloroform, nitroethane, DMF Insoluble: ethanol, benzene	54	2240	8.8
(cyclohexane ring structure)	Soluble: acetone, chloroform, DMF Insoluble: methanol	Viscous product	4820	22.3
$-CH=CH-$	Soluble: acetone, DMF Insoluble: chloroform, toluene	Viscous product	381	2.3

[a] Polysebacate ester prepared according to reference 19.

tetra- and hexamethylene glycols were yellow, spongy, rubber-like insoluble solids. 1,3-Cyclohexanedimethanol and 2-butene-1,4-diol gave linear, powdery polyesters, which were soluble or insoluble depending on conditions. Cyclohexanedimethanol linear polyesters showed little tendency to cross-link and remained linear after long storage. Most of the other polyesters from acetylenedicarboxylic acid were cross-linked and insoluble, in contrast to the linear polyesters from butynediol (Table 5-8).

(3) *Polyesters from acetylenedicarboxylic acid and butynediol*: These polyesters were even more difficult to prepare. The reaction was quite sensitive, and frequent spontaneous highly exothermic decompositions gave carbon. A two-stage reaction was used to control the exotherm. First, equimolar diol and diacid in dioxane, acetone or tetrahydrofuran were refluxed for 24 hours. Second, solvent was removed, and the residue was heated for 20-24 hours at 80° and 690 mm of Hg, followed by 4 hours at 5–70 mm of Hg.

The resulting "polymers" had a degree of polymerization of only 2–3. They were soluble in many solvents and had relatively high softening points (150–175°). The high softening points are probably caused by the extreme rigidity of the triple bonds.

TABLE 5-8
Polyesters from Acetylenedicarboxylic Acid

$$H{+}O{-}R{-}O{-}\underset{O}{C}{-}C{\equiv}C{-}\underset{O}{C}{+}_n OH$$

Polyester R	Solubility	Softening Temperature (°C)	$\overline{M}_n$	n
$-CH_2-CH_2-O-CH_2-CH_2-$	Insoluble: m-cresol, dichloroacetic acid, DMF, trifluoroacetic acid	—	1,367	7.4
$-CH_2-(CH_2)_2-CH_2-$	Insoluble: m-cresol, dichloroacetic acid	—	19,050	113.4
$-CH_2-(CH_2)_4-CH_2-$	Insoluble: m-cresol, dichloroacetic acid, chloroform, acetone, DMF	—	7,767	39.6
	Soluble: acetone, DMF, DMSO			
$-CH_2-CH=CH-CH_2-$	Insoluble: m-cresol, dichloroacetic acid, trifluoroacetic acid	—	1,070	6.4
$-CH_2-\!\!\bigcirc\!\!-CH_2-$	Soluble: chloroform, dioxane, THF	80–120	4,324	19.5
	Insoluble: acetone,-nitroethane, DMF			

One explanation of the strong exotherm is that 2-butyne-1,4-diol undergoes the strongly exothermic Meyer-Schuster rearrangement in the presence of strong acids. The product is hydroxymethyl vinyl ketone. Acetylenedicarboxylic acid, $pKa = 1.74$, is apparently strong enough to cause the Meyer-Schuster rearrangement, as evidenced by the fact that small quantities of hydroxymethyl vinyl ketone distilled out during the exothermic decompositions. None of the ketone was incorporated in the polymer, however.

5.4. Linear Polyesters from 4,4′-Tolandicarboxylic Acid

Laakso[22] made polyesters from 4,4′-tolandicarboxylic acid. The polyesters melted above 200–300°, and had exceptionally high tensile strength and elasticity. They showed excellent resistance to swelling by water and to attack by organic solvents. The starting diethyl 4,4′-tolandicarboxylate was made from 1,2-diphenylethane in 5 steps (overall yield only 6%).

5.5. Polyacetylenic Polyurethanes

Slezak[23] made three new polyacetylenic α,ω-diols by reacting the diGrignard reagent of 1,7,13,19-eicosatetrayne with formaldehyde, ethylene oxide and trimethylene oxide. He used the diols to prepare a series of polyurethanes by bulk reactions at 80–110° (Table 5-9). Most of the polyurethanes were hard and tough. Generally, the acetylenic bonds increased the softening point. Marvel[19] had noted the reverse tendency with butynediol-hexamethylene diisocyanate polyurethanes. The polyurethanes from the polyacetylenic glycols have around 6.5% higher density, and about 20% higher heat of combustion, in calories per milliliter, than polyurethanes from polypropylene glycol. The polymers were suggested as binders for solid rocket propellants.

5.6. Organotin Polyester from Acetylenedicarboxylic Acid

Frankel[24] prepared organotin polyesters by interfacial condensation of dialkyltin dihalides with aqueous solutions of salts of acetylenedicarboxylic acid. The polymers are:

$$\left[-O\overset{O}{\overset{\|}{C}}-C\equiv C-\overset{O}{\overset{\|}{C}}-O-\overset{Bu}{\underset{Bu}{\overset{|}{\underset{|}{Sn}}}}- \right]_n$$

Dipotassium acetylenedicarboxylate and dibutyltin dichloride at 0° for 1 hour gave 81% yield of polyester. The polymer did not melt at 300°. The fumarate polyester was similar. Polymers were insoluble in most solvents. Films on glass were not wetted by water. The tin may confer antifungal properties on the polymers.

TABLE 5-9
Polyurethanes from Polyacetylenic Glycols[23]

Diol	Diisocyanate	$\eta_{inh}{}^a$ (dl/g)	Softening Temperature (°C)
2,8,14,20-Docosa-tetrayne-1,22-diol	Hexamethylene	0.37	250–283
	1,16-Hexadecane	0.26	280–300
	4-Methyl-1,3-phenylene	0.65	274–287
	Bis(4-isocyanatophenyl)-methane	0.45	133–190
	5,11-Hexadecadiyne-1,16	—	<50
3,9,15,21-Tetracosa-tetrayne-1,24-diol	Hexamethylene	0.62	150–280
	1,16-Hexadecane	0.23^b	335–345
	4-Methyl-1,3-phenylene	0.31	235–245
	Bis(4-isocyanatophenyl)-methane	0.70	274–277
	5,11-Hexadecadiyne-1,16	0.17	70–75
4,10,16,22-Hexacosa-tetrayne-1,26-diol	Hexamethylene	0.38	126–138
	4-Methyl-1,3-phenylene	0.41	226–238
	Bis(4-isocyanatophenyl)-methane	0.41	136–150
	5,11-Hexadecadiyne-1,16	0.19	70–80
1,24-Tetracosanediol	1,16-Hexadecane	0.62^b	160–360

^{a} In DMF at 25°, except as noted.
^{b} In m-cresol.

6. POLYMERS FROM ACRYLATES OR STYRENES WHICH CONTAIN A TRIPLE BOND

Acrylate and methacrylate esters of acetylenic alcohols which do not contain a terminal ethynyl group were synthesized and polymerized recently.[25] Radical initiators and cationic initiators cause polymerization through both the vinyl and the acetylenic groups, and result in insoluble cross-linked polymers. Anionic initiators selectively attack the vinyl group, however. The resulting polymers contain pendant acetylenic side chains:

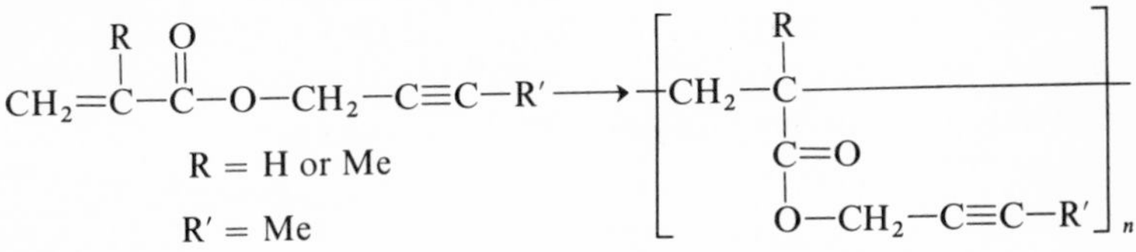

$$CH_2{=}\overset{R}{\underset{}{C}}{-}\overset{O}{\underset{}{C}}{-}O{-}CH_2{-}C{\equiv}C{-}R' \longrightarrow \left[CH_2{-}\overset{R}{\underset{\overset{|}{\underset{\overset{C=O}{\underset{|}{O-CH_2-C{\equiv}C-R'}}}}}{C}} \right]_n$$

R = H or Me
R' = Me

Polymerization for 1–1½ hours in tetrahydrofuran or toluene at −20 to −78°, in the presence of the radical-anion initiators sodium naphthalene or sodium benzalaniline, gives 60–99% yield of polymer. Molecular weight spread is 15,000–175,000 for 2-butynyl methacrylate, and 13,500–140,000 for 2-butynyl acrylate. The polymers do not cross-link in air at room temperature as poly(allyl acrylate) does.

The linear polymers add 1 mole of bromine per triple bond to form products which are self-extinguishing. Ignited in a flame, they stop burning when withdrawn from the flame. The polymers can be cross-linked by heating at 100° or by benzoyl peroxide at 60–80°. Decaborane adds to the triple bonds in the polymers to form products which remain solid at 300°, even if only about 37% of the acetylenic groups react.

$$
\left[\begin{array}{c} R \\ | \\ \text{-CH}_2\text{-C-} \\ | \\ \text{C}=\text{O} \\ | \\ \text{O} \\ | \\ \text{CH}_2 \\ | \\ \text{C} \\ ||| \\ \text{C} \\ | \\ \text{CH}_3 \end{array}\right]_n \xrightarrow[\text{CH}_3\text{CN}]{\text{B}_{10}\text{H}_{14}} 2\text{H}_2 + \left[\begin{array}{c} R \\ | \\ \text{-CH}_2\text{-C-} \\ | \\ \text{C}=\text{O} \\ | \\ \text{O} \\ | \\ \text{CH}_2 \\ | \\ \text{C} \\ | \text{B}_{10}\text{H}_{10} \\ \text{C} \\ | \\ \text{CH}_3 \end{array}\right]_n
$$

Acetonitrile is the catalyst. Decaboronated polymers mixed with ammonium perchlorate burn with an intense greenish-white flame, and thus may have utility as propellants.

Acrylate and methacrylate esters of acetylenic alcohols which do contain a terminal ethynyl group are polymerized by the same anionic initiators,[26] despite the possibility of reaction of the initiator with the ethynyl hydrogen. Propargyl acrylate and methacrylate, and 1- and 2-acetoxy-2-butyne, do not polymerize at all under comparable conditions. n-Butyllithium and lithium naphthalene also catalyze the polymerization of the acetylenic acrylates with the terminal ethynyl groups. The lithium organics are more polar than the radical-anion initiators, and probably react with the monomers to form the lithium acetylide which is the active initiator. Post-reactions of the polymers are similar to the brominations, cross-linking reactions and decaboronations of the nonterminal acetylenic acrylate polymers.

The use of anionic initiators to polymerize vinyl groups selectively in monomers containing acetylenic groups is successful also with styrenes which contain p-acetylenic groups.[27] These monomers can be represented as:

$$\text{CH}_2\text{=CR}$$

$$R'$$

where R = H or Me, and R′ = —CH$_2$C≡CMe, —CO$_2$CH$_2$C≡CMe or —CH$_2$OCH$_2$C≡CMe. The vinyl group is sufficiently activated by the benzene ring to undergo anionic polymerization selectively. The presence of acetylenic groups in the polymers is established by the same post-reactions used for the acrylates and methacrylates.

References

1. Kotlyarevskii, I. L., Fisher, L. B., and Domnina, E. S., *Izv. Akad. Nauk SSR Otd. Khim. Nauk*, 1905 (1961); *Chem. Abstr.*, **56**, 7234 (1962).
2. Hay, A. S., *J. Org. Chem.*, **25**, 1275 (1960); (to General Electric Co.), U.S. Patent 3,300,456 (Jan. 24, 1967).
3. Kotlyarevskii, I. L., *et al.*, *Izv. Akad. Nauk SSSR Ser Khim.*, 902 (1966); *Chem. Abstr.*, **65**, 15251 (1966).
4. Tani, H., Murayama, K., and Toda, F., *Chem. Ind.* (*London*), 1980 (1962).
5. Korshak, V. V., *et al.*, *Izv. Akad. Nauk SSSR Ser Khim.*, 1852 (1963); *Chem. Abstr.*, **60**, 5650 (1964).
6. Shvartsverg, M. S., Kotlyarevskii, I. L., and Andrievski, V. N., *Izv. Akad. Nauk SSSR Otd. Khim. Nauk*, 575 (1963); *Chem. Abstr.*, **59**, 5272 (1963).
7. Bailey, W. J., and Fujiwara, E. J., *J. Org. Chem.*, **24**, 545 (1959).
8. General Electric Co., Netherlands Appl. 287,353 (Feb. 25, 1965); *Chem. Abstr.*, **64**, 6848 (1966).
9. Matsoyan, S. G., *Russ. Chem. Rev.*, (*English Transl.*), **35**, 35 (1966).
10. Meier, J., *et al.*, *Compt. Rend.*, **245**, 1634 (1957).
11. Matsoyan, S. G., *et al.*, *Izv. Akad. Nauk Arm SSR Khim. Nauk*, **17**, 676 (1964); *Chem. Abstr.*, **63**, 7114 (1965).
12. Matsoyan, S. G., *et al.*, *Izv. Akad. Nauk Arm SSR Khim. Nauk*, **15**, 405 (1962); *Chem. Abstr.*, **58**, 6930 (1963).
13. Matsoyan, S. G., *et al.*, *Izv. Akad. Nauk Arm SSR Khim. Nauk*, **17**, 319, 329 (1964); *Chem. Abstr.*, **61**, 13430 (1964).
14. Matsoyan, S. G., *et al.*, *Izv. Akad. Nauk Arm SSR Khim. Nauk*, **17**, 522 (1964); *Chem. Abstr.*, **62**, 14832 (1965).
15. Matsoyan, S. G., *et al.*, *Izv. Akad. Nauk Arm SSR Khim. Nauk*, **17**, 703 (1964); *Chem. Abstr.*, **63**, 4230 (1965).
16. Davis, A. C., and Hunter, R. F., *J. Appl. Chem.* (*London*), **9**, 137 (1959).
17. *Ibid.*, **9**, 364 (1959).
18. *Ibid.*, **9**, 660 (1959).
19. Marvel, C. S., and Johnson, J. H., *J. Am. Chem. Soc.*, **72**, 1674 (1950).
20. Batzer, H., and Weissenberger, G., *Makromol. Chem.*, **12**, 1 (1954).
21. Feit, B.-A., *et al.*, *J. Appl. Polymer Sci.*, **9**, 2379 (1965).
22. Laakso, T. M., and Reynolds, D. D. (to Eastman Kodak), U.S. Patent 2,856,384 (Oct. 14, 1958).
23. Slezak, F. B., *et al.*, *J. Org. Chem.*, **26**, 3137 (1961).
24. Frankel, M., *et al.*, *J. Appl. Polymer Sci.*, **9**, 3383 (1965).
25. D'Alelio, G. F., and Evers, R. C., *J. Polymer Sci.* (*A-1*), **5**, 813 (1967).
26. *Ibid.*, **5**, 999 (1967).
27. D'Alelio, G. F., and Hoffend, T. R., *J. Polymer Sci.* (*A-1*), **5**, 1245 (1967).

INDEX